Innovative Economic Symposium (IES2020)

Stable Development in Unstable World

SHS Web of Conferences Volume 91 (2021)

Samara, Russia
13 November 2020

Editors:

Jakub Horák
Zuzana Rowland
Petr Šuleř

ISBN: 978-1-7138-2413-8

Printed from e-media with permission by:

Curran Associates, Inc.
57 Morehouse Lane
Red Hook, NY 12571

Some format issues inherent in the e-media version may also appear in this print version.

No changes have been made to the content of these proceedings. There may be changes to pagination and minor adjustments for aesthetics.

Printed with permission by Curran Associates, Inc. (2021)

For additional information, please contact EDP Sciences – Web of Conferences at the address below.

EDP Sciences – Web of Conferences
17, Avenue du Hoggar
Parc d'Activité de Courtabœuf
BP 112
F-91944 Les Ulis Cedex A
France

Phone: +33 (0) 1 69 18 75 75
Fax: +33 (0) 1 69 28 84 91

contact-edps@webofconferences.org

Additional copies of this publication are available from:

Curran Associates, Inc.
57 Morehouse Lane
Red Hook, NY 12571 USA
Phone: 845-758-0400
Fax: 845-758-2633
Email: curran@proceedings.com
Web: www.proceedings.com

TABLE OF CONTENTS

TAKEOVER AS A FORM OF CORPORATE CONFLICT AND THE PROCESS OF TRANSFER OF STOCK OWNERSHIP'S RIGHTS 1
Belayeva I. Yu., Danilova O. V., Uskov K. V.

THE RELATIONSHIP BETWEEN CAPITAL STRUCTURE AND FINANCIAL PERFORMANCE OF THE COMPANY 12
Tretiakova V. V., Shalneva M. S., Lvov A. S.

SOURCES OF INVESTMENT SUPPORT OF PRODUCTION AND BUSINESS ACTIVITIES IN THE RUSSIAN ECONOMY 21
Streltsov A. V., Yakovlev G. I., Kosyakova I. V.

PROJECT-BASED LEARNING APPROACH TO MARKETING COMPETENCIES DEVELOPMENT 30
Dušek Radim

FUZZY ESTIMATE OF THE DEVELOPMENT OF PASSENGER CARS PRODUCTION IN THE CZECH REPUBLIC 37
Hašková Simona

PREDICTION OF FINANCIAL HEALTH OF BUSINESS ENTITIES OF SELECTED SECTOR USING BALANCE ANALYSIS II. BY RUDOLF DOUCHA AND VERIFICATION OF ITS PREDICTIVE ABILITY THROUGH ROC 46
Karpac Dusan, Bartosova Viera

JONES\' MODEL AND ITS MODIFICATIONS IN THE CONDITIONS OF THE SLOVAK REPUBLIC 52
Strakova Lenka, Svabova Lucia

AN EMPIRICAL VIEW ON THE DETERMINANTS OF SUSTAINABLE ECONOMIC DEVELOPMENT: EVIDENCE FROM EU(28) MEMBER STATES 65
Sofrankova Beata, Kiselakova Dana, Onuferova Erika

VALUE ADDED TAX AND ITS INFLUENCE ON PRICING AND PRICE DECISION MAKING OF COMPANIES – A CASE STUDY 78
Kubjatkova Anna, Krizanova Anna, Jurickova Veronika

THE IMPORTANCE OF PROBABILISTIC ANALYSIS IN THE APPLICATION OF THE MARKETING STRATEGY OF THE SELECTED BRAND 85
Vrtana David, Krizanova Anna

ON THE RELATIONSHIP BETWEEN BANK LENDING INDICATORS AND GENERAL ECONOMIC INDICATORS 95
Korobov Yury, Bogomolov Sergey, Ilyina Larisa, Plotnikova Marina

A BIBLIOMETRIC ANALYSIS OF BRANDING TRENDS: DOES IRRATIONALITY REALLY MATTER? 104
Majerova Jana, Gajanova Lubica, Nadanyiova Margareta

IMPACT OF DIVIDEND POLICY ON THE MARKET VALUE OF THE COMPANY 113
Eryomin Ivan, Likhacheva Olga, Chernikova Lyudmila

BENEFIT'S PORTFOLIO IN A COMPANY IN CONNECTION TO THE GENERATION OF MANAGERIAL APPLICANTS 126
Janèíková Kateøina, Milichovský František

NORMALITY AND ITS VERIFICATION AS A BASIC PREREQUISITE FOR THE APPLICATION OF VAR 136
Chutka Jan, Vagner Ladislav

PROMISING AREAS FOR DEVELOPING HR MANAGERS\' COMPETENCIES IN THE DIGITAL AGE 144
Kapitanov V. A., Osipova O. S., Chikileva L. S.

RESOURCE POTENTIAL AS A KEY FACTOR OF UKRAINIAN MINING ENTERPRISES COMPETITIVE POSITION 154
Pochtovyuk Andriy, Latyshev Kostyantyn, Zaika Kateryna

SUSTAINABLE GROWTH OF AIRLINE COMPANIES: CHALLENGES AND PROSPECTS ASSESSMENT 165
Slepneva T. A., Svishcheva N. V.

SCORING METHOD OF ASSESSMENT: RELEVANCE AND PERSPECTIVE 174
Gusev Andrey, Invdeeva Alina, Motuzenko Polina

DECENT WORK DURING THE PANDEMIC: INDICATION AND PROFILING MATTERS 186
Simonova M. V., Sankova L. V., Mirzabalaeva F. I.

INNOVATION AND ITS IMPORTANCE IN THE CONTEXT OF A COMPANY'S MARKETING STRATEGY 197
Olexova Veronika, Gogolova Martina

IMPACT OF CREDIT MARKET DEVELOPMENT ON PROJECT FINANCING 206
Poradova Monika, Janoskova Katarina

COMPARISON OF SELECTED SUBJECTS OF THE NON-PROFIT SECTOR IN THE CONDITIONS OF THE SLOVAK REPUBLIC 217
Rosnerova Zuzana, Hraskova Dagmar

THE USAGE OF THE CSR CONCEPT AND ITS COMMUNICATION IN SLOVAK COMPANIES IN THE PROCESS OF GLOBALIZATION 228
Nadanyiova Margareta, Majerova Jana, Gajanova Lubica

THE AGILE CONTENT MARKETING ROADMAP BASED ON THE B2B BUYER'S JOURNEY – THE CASE STUDY OF THE SLOVAK REPUBLIC 238
Gajanova Lubica

FACTORS OF ENTREPRENEURIAL SUCCESS: A QUALITATIVE MULTIPLE CASE STUDY OF NEW SMALL BUSINESSES 247
Urban Jan

NON-PROFIT ORGANIZATION'S ENDOWMENT AS A SOURCE OF FINANCING TO IMPROVE ITS SUSTAINABILITY 256
Rebetak Filip, Bartosova Viera

IMPACTS OF COVID-19 ON THE REAL ESTATE MARKET IN THE CZECH REPUBLIC 264
Hromada Eduard

LONG MEMORY IN STOCK RETURNS: EVIDENCE FROM THE EASTERN EUROPEAN MARKETS 271
Dias Rui, Heliodoro Paula, Alexandre Paulo, Santos Hortense, Farinha Ana

CURRENT NEEDS OF BUSINESS OWNERS AND MANAGERS, BUSINESS SCHOOLS' VIEW 281
Mantulenko V. V., Aleshkova D. V.

PROSPECTS FOR THE DEVELOPMENT OF THE ENTREPRENEURIAL POTENTIAL OF THE REGIONS OF THE RUSSIAN FEDERATION 291
Ashmarina S. I., Murzagalina G. M.

SUSTAINABILITY OF PASSENGER TRANSPORT AS A PRECONDITION FOR FULFILLING THE 4.0 COMPANY CONCEPT – ANALYSIS OF TRENDS IN THE CZECH REPUBLIC 307
Hinke Jana, Komorousová Veronika

IOT - TOOL OF SOCIETY 4.0 - LITERARY REVIEW 318
Èerná Marie, Vallišová Lucie

METHODOLOGY FOR CALCULATING THE REPRODUCTION VALUE OF BUILDINGS 328
Hromada Eduard, Schneiderová Heralová Renáta, Dlask Petr, Krulický Tomáš

STABLE DEVELOPMENT OF TRAVEL BUSINESS IN THE ODESSA REGION: ESTIMATE USING MULTIVARIATE STATISTICAL ANALYSIS 340
Matskul Valerii, Kovalyov Anatoliy, Chernyshev Valentyn, Okara Diana

BIG FIVE IN UNIVERSITY STUDENTS IN RELATION TO PSYCHOLOGICAL AND ENTREPRENEURIAL COMPETENCES 354
Kmecová Iveta, Kajanová Alena

THE STATE'S RESPONSIBILITY TOWARDS ENTREPRENEURS IN THE CZECH REPUBLIC FOR THE MEASURES ADOPTED IN RELATION TO THE EPIDEMIC OF CORONAVIRUS 364
Bílek Stanislav, Caha Zdenìk, Velková Vendula

SUBJECTIVE EVALUATION OF SOCIO-ECONOMIC SITUATION OF RESIDENTS OF THE SOUTH BOHEMIAN REGION AND FACTORS IMPACTING IT 375
Kajanova Alena, Urban David

BUSINESS VALUATION IN TIMES OF CRISIS 383
Krulicky Tomas, Horak Jakub, Skulcova Kristyna

DETERMINING BUSINESS VALUE ACCORDING TO FCFE 392
Rowland Zuzana, Stanek Tomas

A BASIC TECHNICAL ANALYSIS OF SHARES ON THE EXAMPLE OF A SPECIFIC COMPANY 401
Horak Jakub, Sulek Jiri

THE APPLICABILITY OF FCFF METHOD EVALUATING AN ENTERPRISE OF REAL ESTATE SEGMENT 412
Vrbka Jaromír, Vitková Pavla

APPLICATION OF THE CLASSICAL PAYOUT RATIO MODEL AND DETERMINATION OF THE LIMITATIONS OF THE LINTNER MODEL DEMONSTRATED ON THE EXAMPLE OF KOMERÈNÍ BANKA, PLC 421
Machova Veronika, Geierova Tereza

ANALYSIS OF GOVERNMENT BONDS AND PREDICTION OF THEIR DEVELOPMENT AFTER THE PANDEMIC CAUSED BY COVID-19........ 430
Suler Petr, Polan Vaclav

METHODOLOGY OF CALCULATING RISK PREMIUMS IN THE ENVIRONMENT OF THE CZECH REPUBLIC AND ITS COMPARISON WITH DAMODARAN........ 446
Kucera Jiri, Maskova Lenka

AN ANALYSIS OF THE TIME SERIES OF THE PEPSICO, INC. (PEP) SHARE PRICE AND A PREDICTION OF ITS DEVELOPMENT........ 458
Kalinová Eva, Tlustý Michal

CREATING EFFICIENCY MODEL OF HIGHER EDUCATION INSTITUTIONS 469
Janikova Jana

CONFLICT BETWEEN THE ACCOUNTING ACT AND CZECH ACCOUNTING STANDARDS IN LAND........ 480
Kabourkova Kristina

Author Index

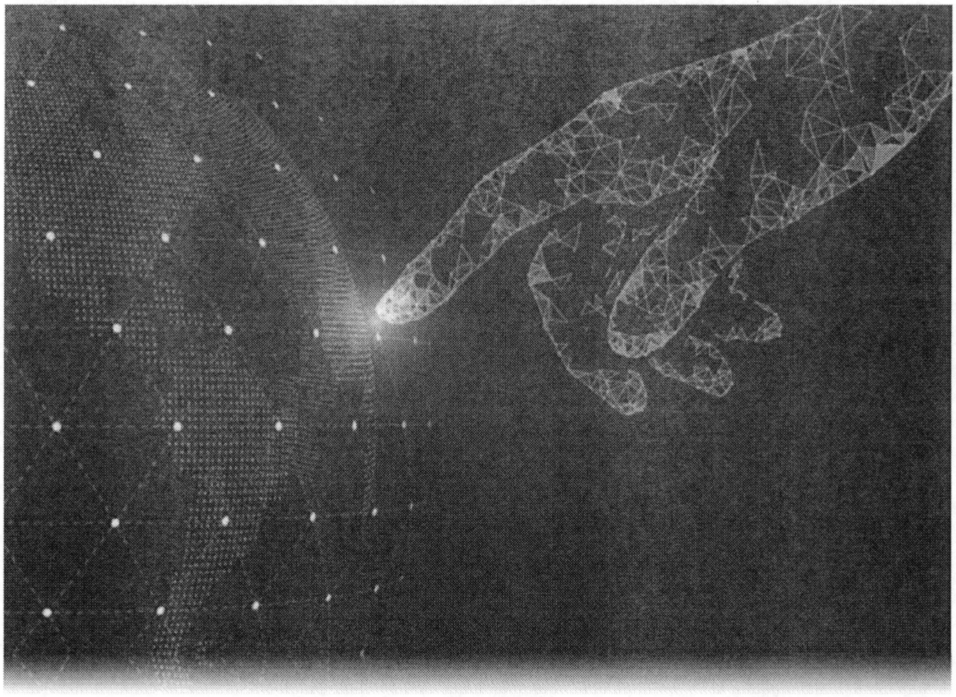

Innovative Economic Symposium 2020

Stable Development in Unstable World

Innovative Economic Symposium 2020

Stable Development in Unstable World

http://ies.vstecb.cz/

Editors: Jakub Horák, Zuzana Rowland, Petr Šuleř

About the conference IES 2020

Date of the conference: 2020-11-13

Motto

"Coming together is a beginning; keeping together is progress; working together is success." Henry Ford

Target

The IES 2020 conference brings together scientists and practitioners in economy, international relations, finance and marketing from Asia and Europe in order to collect innovative ideas and advance knowledge of globalization and deglobalization trends in the current situation of the world economy.

Thematic Areas

Stable Development in Unstable World

- Where is the world moving to and where is the economy in it?
- Institutionalization of innovations
- Network architecture of economic relations
- Competences for the future
- Smart change management
- Financial management of the company, investment decision making

Globalization and Fragmentation Forces of the Current World Economy

- Monetary and fiscal policy development as a factor of economic modernization
- Role of international trade in the economy globalization
- Impact of globalization and economic fragmentation on the enterprise's internal environment
- Financial conditions for entrepreneurship under the economic modernization
- Impact of scientific and technological progress on globalization and fragmentation of the economy.

About organizers

Institute of Technology and Business in České Budějovice (VŠTE)

The primary focus of VŠTE activities is the implementation of full-time and part-time studies for professional bachelor's and master's degree programmes. The professional study programmes are supported by cooperation with regional entrepreneurs. VŠTE´s other priorities include applied research, the implementation of innovative technologies and knowledge transfer into areas that are current and crucial with regards to the regions needs. More information about VŠTE at www.vstecb.cz.

North China University of Technology (NCUT)

North China University of Technology (NCUT), located on the western side of Beijing, is a multi-disciplinary university that combines the natural sciences and engineering with liberal arts, economics, management and law. NCUT targets the development of qualified, high-level, application-oriented professionals with solid basic theoretical knowledge and strong practical skills and creative abilities. A quality education is the university's lifeline. Teaching is a primary focus at NCUT and the center of its activities, while the university also emphasizes the importance of research and the contribution research can make both to the educational environment and to address local and national needs. The under graduate program offers a full range of undergraduate degrees; there are also numerous graduate degree programs, as well as continuing education programs. The academic education is complemented with moral, intellectual, physical and aesthetic training in order to build well-rounded students prepared to meet the challenges of modern society. More information about NCUT at www.ncut.ciss.org.cn.

Kookmin University

Kookmin University, the first private university after Korea's independence, was established by Mr. Shin, Ik Hee and other main members of the Shanghai interim government of Korea. Deeply rooted in the four founding principles, humanism, nationalism, culturalism, and industrialism that Mr. Kim, Sung Gon had emphasized, Kookmin University has grown into a fixture of the academia of Korea with 24,000 students enrolled. Now universities of a knowledge based society have entered into a competition that is more severe than ever before. KMU is trying to solidify its educational foundation to maintain its role as a leading player by employing operative goals. The three operative goals are attracting outside funding, establishing the standards of reasonable distribution of resources, and building an efficient structural system. More information about Kookmin University at https://english.kookmin.ac.kr/.

Samara State University of Economics

The Samara State University of Economics was founded in 1931. Today SSEU is the leading university in the Volga region. The Ministry of Education and Science of the Russian Federation acknowledged SSEU to be an effective university. All educational programs have licenses and accreditation certificates and correspond to the international standards.A practical-oriented approach in teaching is implemented in Samara State University of Economics. The University has basic departments at regional enterprises, that allow the students to be involved into the practical activity during the educational process. The university is constantly developing. Foreign teachers, international internships and double degree programs are involved in the educational process. The graduates of SSEU get the European supplement to their diplomas, that confirms knowledge and skills, the level of specialist preparation and is recognized all over the world. The students and teachers of SSEU are provided with free electronic access to educational resources of leading universities in Russia and other countries, that allows them to get the latest information and scientific trends very fast. The cutting-edge educational technologies, including distance learning technology and e-learning are widely used in SSEU. More information about Samara State University of Economics at http://www.sseu.ru/en.

Henan University

Developing for nearly 110 years, HENU is a comprehensive university with 12 academic fields of study covering Humanities, History, Philosophy, Economy, Management, Law, Science, Engineering, Medicine, Agriculture, Education and Arts. Henan University has 37 schools and departments, 43 First-Level Disciplines for Master's Degree Granting and 20 First-Level Disciplines for Doctoral Degree Granting. HENU has 6 national key scientific research platforms, 2 national key social science research platforms and 4 key scientific research platforms of the Ministry of Education. Actively participating in internationalized education, HENU continuously broadens the channels of international exchange and cooperation, and has established partnership with more than 170 universities from 36 countries and regions around the world. More information about Samara State University of Economics at http://iao.henu.edu.cn/yw/Home.htm.

Conference report

The notion of an unstable world mentioned in the subtitle of this year's Innovative Economic Symposium was not only the main topic of the submitted scientific articles but also the main obstacle in organizing this year's conference.

Due to the precarious situation all over the world caused by the COVID-19 pandemics, it was unclear until the very last minute whether it would be possible to host the conference in person at the venue in Samara, Russia. Eventually, the worsening epidemiological situation as well as the applied restrictions made it impossible for the participants to meet in person.

Therefore, this year's event took place online using the communication platform ZOOM. On 13th November 2020, academicians from China, Russia, Korea, Malaysia, Slovakia, Ukraine or the Czech Republic joined the virtual meeting in order to present their research and exchange expert knowledge.

The thematic focus of this year was Stable Development in Unstable World. The whole conference started with opening speech of prof. Ing. Marek Vochozka, MBA, Ph.D., Rector of the Institute of Technology and Business in České Budějovice (VŠTE) and dr. Valentina Mantulenko representing the Samara State University of Economics, the main host of this year's conference. The conference programme consisted of two plenary sections. All participants presented their interesting contributions and expressed their hopes that next year, it will be possible to meet in person again.

Expert guarantor

prof. Marek Vochozka
Rector of the Institute of Technology and Business (CZE)

prof. Taehee Lee
Chair of International School, Kookmin University (ROK)

prof. Zhao Jixin
Dean of the School of Economics and Management, North China University of Technology (CHN)

prof. Svetlana I. Ashmarina
Samara State University of Economics (RUS)

The Organizing Committee

prof. Marek Vochozka, Rector of ITB, School of Expertness and Valuation, The Institute of Technology and Business in Ceske Budejovice (CZE)

prof. Zhao Jixin, Dean of the School of Economics and Management, North China University of Technology (CHN)

prof. Taehee Lee, Chair of International School, Kookmin University (ROK)

Jakub Horák, MSc., MBA, ITB in Ceske Budejovice (CZE)

Veronika Machová, MSc., MBA, ITB in Ceske Budejovice (CZE)

Pavel Rousek, Ph.D., ITB in Ceske Budejovice (CZE)

Zuzana Rowland, PhD., MBA, ITB in Ceske Budejovice (CZE)

Tomáš Krulický, MSc., MBA, ITB in Ceske Budejovice (CZE)

Valentina Mantulenko, Dr., Samara State University of Economics (RUS)

Eva Kalinová, MSc., ITB in Ceske Budejovice (CZE)

Jiří Kučera, MSc., ITB in Ceske Budejovice (CZE)

Statement of Peer review

In submitting conference proceedings to *Web of Conferences*, the editors of the proceedings certify to the Publisher that

1. They adhere to its **Policy on Publishing Integrity** in order to safeguard good scientific practice in publishing.
2. All articles have been subjected to peer review administered by the proceedings editors.
3. Reviews have been conducted by expert referees, who have been requested to provide unbiased and constructive comments aimed, whenever possible, at improving the work.
4. Proceedings editors have taken all reasonable steps to ensure the quality of the materials they publish and their decision to accept or reject a paper for publication has been based only on the merits of the work and the relevance to the journal.

Title, date and place of the conference

Innovative Economic Symposium 2020 - Stable Development in Unstable World

13 November 2020

Samara State University of Economics, Samara, Russia

Proceedings editor(s):

Ing. Jakub Horak, MBA

Ing. Zuzana Rowland, MBA, PhD.

Mgr. Petr Suler, PhD.

Date and editor's signature

2020-12-17

17, avenue du Hoggar - PA de Courtabœuf – BP 112 - 91944 Les Ulis Cedex A (France)
Tél. : 33 (0)1 69 18 75 75 - Fax : 33(0)1 69 07 45 17 – www.edpsciences.org

Takeover as a Form of Corporate Conflict and the Process of Transfer of Stock Ownership`s Rights

I.Yu. Belayeva[1], *O.V.* Danilova[1], and *K.V.* Uskov[1,*]

[1]Financial University under the Government of the Russian Federation, Moscow, Russia

Abstract. The article considers the problem of corporate conflicts in the Russian practice of corporate relations in the joint-stock companies. Today, corporate relations are changing, the role of intangible resources is growing, and a 100% share in the company capital does not always guarantee control over the company. In this regard, the range of ways to acquire corporate control is expanding. The purpose of the article is to develop a classification of the methods of joint stock companies takeovers. The proposed classification differs from existing classifications by differentiating takeover methods according to the criterion "takeover tool used by the takeover initiator". The use of this classification by companies in the development of protective mechanisms against takeovers will make it possible to choose the most appropriate and effective protective measures in a particular situation and thereby avoid some corporate conflicts.
Keywords: Corporate conflicts; Corporate relations; Takeover; Property relations; Takeover methods; Property

1 Introduction

A corporation can be defined as a social system that operates in the economics field of and in the resource exchange field of. Corporate relations in such social systems arise on the initiative of potential participants in corporate relations (stakeholders) at the moment when they have the intention to obtain resources through interaction with the corporation.

The reason for the emergence of corporate relations is the subjective assessment of the stakeholder (potential stakeholder) about the possibility of obtaining benefits from interaction with the corporation, with its other stakeholders. The benefits of a stakeholder should be understood as the resources that the corporation provides him. They should be valuable to the stakeholder, should be able to directly or indirectly satisfy his interests. That is, corporate relations - relations between participants in the economic exchange of resources within the corporation in accordance with the rules and regulations in force in the corporation, which are determined by the institutional environment.

If the expected benefits of the stakeholder are equal to the benefits received, then in this case the activity associated with the impact on the corporation will be low. But if the real benefits and the expected ones do not match, then the stakeholder (or potential stakeholder) can begin to influence the corporation and create the conditions under which it will receive

[*]Corresponding author: KVUskov@fa.ru

the expected benefits, including, for example, attempting to redistribute control over property. This means that it will impact the corporation and / or its stakeholders. Thus, corporate relations arise as a result of the activity of potential participants in corporate relations, which, in turn, arises as a result of their interest in obtaining benefits from interaction with the corporation.

Of particular importance to the corporation are "key participants in corporate relations" (key stakeholders). Key stakeholders include individuals and legal entities that own key resources necessary for the smooth operation of the corporation, and also have the ability to have a significant impact on it. Key resources can be not only financial resources or the prevailing share in the authorized capital. Today, stakeholders influence corporations, even without ownership of shares or shares in the authorized capital. There are informal channels of influence, informal leaders who, although they have a great influence on the distribution of resources, are not taken into account by modern means of displaying the state of the company. That is, it is potentially possible to control the activities of the company without the right to property, it is enough to own key resources. In this regard, corporate relations are currently changing, they are being transformed, this affects the features of property management, and the transformation of property theory is also taking place.

In the case when stakeholders do not receive the necessary benefits from interacting with the company, this negatively affects corporate relations, which can lead to a redistribution of areas of influence on the corporation and lead to a redistribution of property rights.

One of the most common forms of corporate conflicts is a takeover, when a participant in corporate relations (a potential participant in corporate relations) seeks to acquire ownership rights. In other words, in the aspect of a takeover, corporate conflicts contribute to the transfer of rights to joint-stock property against the will of the current owners. These can be conflicts between large owners, when one of them seeks to gain corporate control, between current owners and external entities.

Analysis of scientific work in this area [1;2;3;4;5;6;7] allowed to form the following classification of external entities that implement acquisitions (takeover entities):

- competitors whose goal is to gain a larger market share;
- corporations, forming a vertically integrated holdings;
- investment companies whose goal is to resell the target corporation;
- specialized companies that provide takeover services;
- individuals (their motivatopn may differ, but, as a rule, their goal is to acquire a larger share in the authorized capital and get more control).

The implementation of a takeover can have both positive and negative consequences. This is always a negative moment for the current owner, who against his will loses control of the business and actually loses the company. In the case of an inefficient company that infringes on the rights of minority shareholders, a takeover can be a positive moment, so the new owner can improve the situation in the company and increase its capitalization.

The purpose of the article is to develop a classification of takeover methods of the joint-stock companies.

Tasks:

- analyze the main methods to acquire corporate control rights in Russia (analysis of scientific papers in this field);
- develop a general term related to the transfer of corporate control rights -«takeover»
- identify the typical takeover stages in the Russian practice.

2 Methodology

As research tools, such general scientific methods as systemic, factorial, functional, and comparative analysis were used. The study analyzed scientific papers in the field of organizing protection against threats from the mergers and acquisitions market; the practice of acquiring corporate control in Russia; practice of organizing protection against takeovers. An analysis of the main takeover methods in Russian practice was carried out. This allowed us to develop a classification of the takeover methods, which differs from the existing classifications by differentiating the takeover methods according to the criterion “takeover tool used by the takeover initiator”.

Initially, we analyzed the processes of transition of rights to corporate control (forms of corporate conflicts) that exist in Russian economic practice and are covered in the scientific literature. There are many processes that are similar in essence, but have different names, including "unfriendly takeover", "hostile takeover", "corporate takeover", and "corporate raid". These processes were analyzed in the works of Arutyunova K. G., Burykin D. V., Gomtsyan S. V., Iontsev M. G. Krichevsky N., Radygin A.D., Savchuk S., Sogrina D. V., Trifonova L. A., Etnov R. M. There was an idea to define a General term. The term "takeover" was chosen and the author's definition was given.

Takeover – actions of one and (or) several individuals (legal entities) aimed at acquiring rights to corporate control in the face of resistance from the majority of the company's owners. Various processes of transition of rights to corporate control were considered. We considered how the takeover subject gets control over the property, what algorithm of actions it performs. Similar points were identified when implementing different methods of acquiring corporate control rights, and a new classification was developed, which is described below.

3 Results

As a result of the analysis of modern Russian practice of acquiring control over corporations, as well as existing research in this area, a classification of takeover methods of joint stock companies has been developed (picture 1).

1. Takeover methods, the implementation of which is carried out in the framework of the right field.

1.1. The takeover method, the implementation of which is carried out through the shares of the target corporation:

1.1.1. Direct shares buying.

The condition for the implementation of the takeover method is the consent of the shareholders to sell their shares. However, such consent may be forced, and this must be taken into account.

There are several options for implementing this takeover method.

Option 1 – buying of a minimum controlling block of shares which is consolidated by a major shareholder or a group of shareholders of the target company. In this case, the takeover entity is able to make strategic decisions [1, p. 67].

This option provides 2 situations:

- the controlling shareholder (group of shareholders) voluntarily sells shares;

- the takeover entity forces the controlling shareholder (group of shareholders) to sell the shares.

Option 2 - buying of a minimum controlling block of shares which is consolidated by minority shareholders. This option is common in companies with dispersed share capital. The disadvantage of using this option can be considered a situation in which the demand for shares of the target company increases. Thus, obstacles are created to complete the transaction. The problem will be more significant when the threshold value of the control packet is high [1, p. 11-12]. Russian corporations are characterized by a low dispersed share capital, as a rule, a small number of owners.

Option 3 - buying of fewer shares than a controlling stake. Typically, this option is implemented with the aim of making decisions with the goal of forcing the remaining shareholders to sell their shares of the target corporation.

1.1.2. The purchase of votes of minority shareholders of the target corporation.

The purchase of votes is a way of gaining control over a corporation, used mainly in the United States. It is realized in the form of purchase of votes of minority shareholders and is an expensive takeover method. The purchase of votes takes place immediately before the general meeting of shareholders. The trustee receives from minority shareholders the right to vote on certain issues independently. The main purpose of the purchase of votes of shareholders is to change the composition of the board of directors of the target company [1, p. 11]

With the successful implementation of this takeover method, the takeover subject gets the opportunity to influence the decision-making process on the board of directors through its representatives. For example, this may be a change of management, sale of assets, changes in the internal documents of the company. The purchase of the votes of minority shareholders may also force the controlling shareholder to sell his shares.

A prerequisite for the effective implementation of this takeover method is the consolidation of 30 or 50 percent of the voting shares of the target company for voting.

In addition to acquiring the votes of minority shareholders, the acquisition entity may initiate shareholder agreements. Such agreements are concluded to achieve an agreed position in the voting process at the general meeting of shareholders.

1.2. Impact on stakeholders of the target corporation.

The aim of implementing this takeover method is to obtain the support of stakeholders in the process of acquiring control over the corporation. That is, the absorption entity in this case seeks to influence the actions and decisions of stakeholders.

Successful implementation of the takeover method will create conditions for influencing strategic decisions (if there is support from board members, managers, shareholders). In addition, the takeover entity may indirectly impede the company's activities (if there is support from creditors, suppliers, key customers, registrar, employees, etc.).

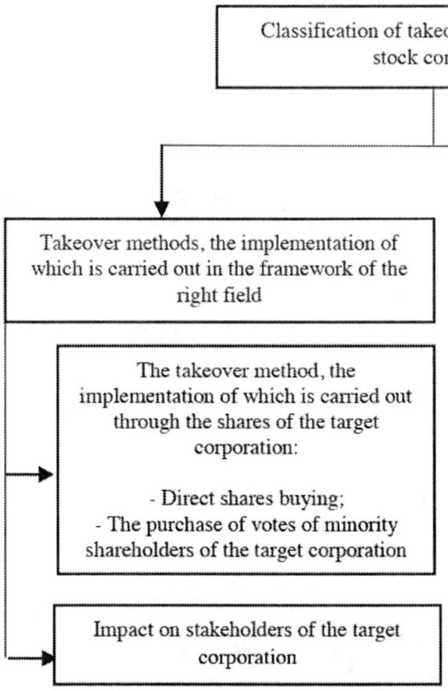

Fig. 1. Classification of takeover methods of joint stock companies
Source: Own processing.

Since the takeover is also realized through the bankruptcy of the target corporation, in this case the takeover entity either acquires the company's debts or stimulates the initiation of bankruptcy proceedings by its lender (that is, the lender acts in the interests of the takeover entity).

The existence of such takeover method forces corporations to organize high-quality interaction with stakeholders. They need to ensure the loyalty of at least key stakeholders, create conditions for long-term mutually beneficial cooperation. If this is achieved, the stakeholder will not contribute to the takeover entity and this will ensure the preservation of ownership for current shareholders.

Today, one of the most important factors affecting the modification of corporate relations is the loyalty of stakeholders.

There are many Russian and foreign studies of the concept of "loyalty", and it is understood differently [8]. Typically, loyalty is associated with the loyalty of only staff, customers, and the theoretical basis on this issue is mainly related to marketing.

In scientific papers, it is noted that loyalty has a positive effect on the quality of work of employees and their behavior, increases their responsibility, reduces the risks of their leakage, contributes to a more efficient functioning of the corporation, and increases its stability.

The presence of similar concepts of loyalty, difficulty in developing its diagnostic tools, problems with a clear definition of loyalty factors are the reasons that give loyalty to a corporation the status of a problem [9, p. 3].

Loyalty is associated with such characteristics of a stakeholder as benevolence, honesty, friendliness, openness, integrity, loyalty, devotion, acceptance of company values [8].

Loyalty should be considered not only in relation to internal, but also to external stakeholders of the corporation. Today, relations between companies and stakeholders in terms of social responsibility are becoming important, and external stakeholders play a special role here.

The analysis of scientific works in this subject area made it possible to develop a generalized interpretation of the term "stakeholder loyalty". Stakeholder loyalty is a characteristic of a stakeholder indicating a high probability of a positive attitude towards the corporation and a high probability of useful behavior within it.

A loyal stakeholder is ready to lose his resources (that is, to make concessions for the benefit of the company and to the detriment of himself) and to cooperate with the company for a long time.

Factors affecting stakeholder loyalty include "actual stakeholder satisfaction from engaging with the corporation"; "future satisfaction from engaging with the corporation"; "the impact on the stakeholder of the institutional environment of the corporation", as well as "the ability to obtain the same (or more valuable) resources outside the corporation" (for example, when interacting with other corporations) (picture 2).

The actual satisfaction of the stakeholder from interacting with the corporation is affected by expectations in obtaining resources, as well as the actual resources received by the corporation. Expectations are formed on the basis of the interests of a stakeholder, the share of his resources in a corporation, and the value of resources. With increasing stakeholder satisfaction from interacting with a corporation, its loyalty increases.

The stakeholder's satisfaction in the future is affected by the history of his relationship with the corporation (the results of relationships that can be both positive and negative), his attitude to the history of relations with the corporation of other stakeholders, and the duration of his cooperation with the corporation. The subjective assessment of the stakeholder about the high probability of obtaining the resources he needs from cooperation with the corporation in the future increases loyalty.

The influence of the institutional environment of a corporation on a stakeholder consists in the influence of social institutions that transmit loyal behavior, as well as in the actions taken by a stakeholder [10, p. 213]. The term "social institution" in the framework of this study is understood as an information object, which reflects the forms of behavior and their consequences. The stronger the influence of social institutions that broadcast loyal behavior, the higher the stakeholder loyalty.

The possibility of obtaining similar (more valuable) resources outside the corporation depends on factors such as the availability of cooperation proposals from the takeover entity, as well as the availability of cooperation proposals from other corporations [10, p. 213]. With such offers, stakeholder loyalty could potentially decrease.

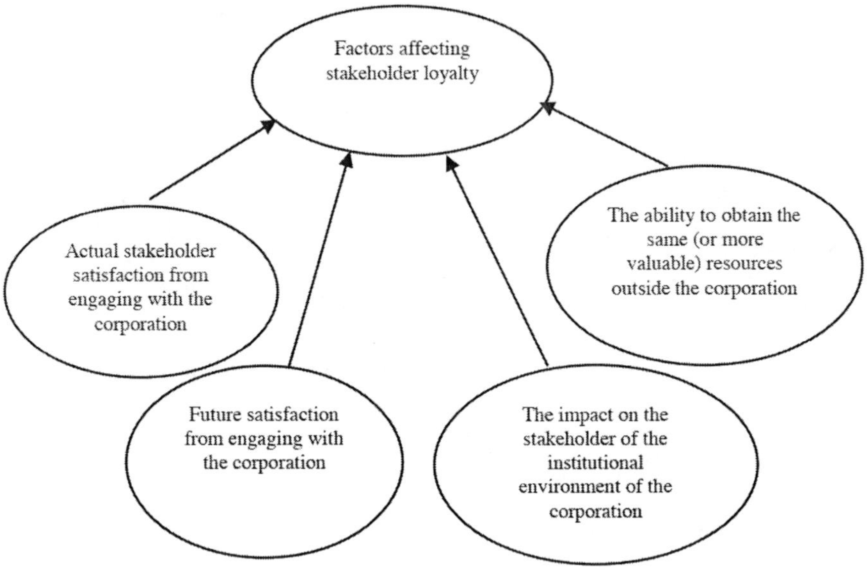

Fig. 2. Factors affecting stakeholder loyalty
Source: Own processing.

It is also worth noting that psychological characteristics affect loyalty. Next, we should consider the takeover methods, the implementation of which is carried out outside the framework of the right field.

2. Takeover methods the implementation of which is carried out outside the framework of the right field.

2.1. Takeover through stock theft.

When embezzlement of the shares of the target corporation, the entity takes actions outside the legal framework, and these actions are classified as a crime. All methods of stock theft can be generalized into two main groups. This is an imitation of transactions (the so-called illegal transactions) with shares, as well as the commission of illegal actions with the register of shareholders.

As a result of the implementation of the first option, the owner is deprived of the right to own shares without taking any actions to transfer them and receives information about changes in the register of shareholders after a long time. During this time, the subject of absorption manages to carry out transactions. Stolen shares are mixed with shares acquired legally and resold to bona fide buyers.

The second method of stealing shares is applied to corporations that maintain a register of shareholders without involving specialized organizations. In this case, the level of security is low. In case of registrar disloyalty, the subject of absorption can apply this method. The purpose of illegal actions with the registry is to write off shares from the personal accounts of some individuals and transfer them to the accounts of others without a proper transfer order [6, p. 81-82].

2.2. Influence on decisions of state bodies.

The algorithm for the implementation of this takeover method is as follows:

- purchase of shares from minority shareholders (or theft of shares);
- initiation of criminal cases against controlling shareholders, members of the board of directors, managers;
- fixing the rights to the assets of the target corporation through decisions of the courts (registration authorities);

- purchase of other shares of the target corporation;
- carrying out the takeover.

That is, the takeover entity, through the actions of corrupt officials (judges), implements the takeover, forcing the controlling shareholder to sell its shares [11].

Also, based on court rulings on securing a lawsuit, an arrest may be made on all significant assets of the target company, as well as a ban on management to enter into transactions. Under these conditions, it is impossible for current owners to withdraw assets or burden them with significant obligations.

If the takeover entity does not have enough resources to obtain the necessary court decision, the seizure of assets can be carried out as part of the criminal process on the basis of the actions of the inquiry officer, investigator or prosecutor during the investigation of criminal cases.

An implementation option for this takeover method is to transfer the registry to another registrar. This is based on the decisions of various courts to take measures to secure a claim. After this, the shareholding owned by the owner is stolen and resold to bona fide buyers [2].

4 Discussion

Based on the practice of acquiring corporate control in Russia, the takeover process is advisable to reflect the through three stages (picture 3).

1. Determination of the takeover motive.

There are many motives for the absorption. Individuals may seek to strengthen their influence in the corporation, the acquisition may be motivated by the acquisition of the right to use and disposal of its resources. The main motives for legal entities are entering new markets, increasing market share, obtaining contracts, tax benefits, increasing financial stability, diversification, acquiring product chains and others [12;13].

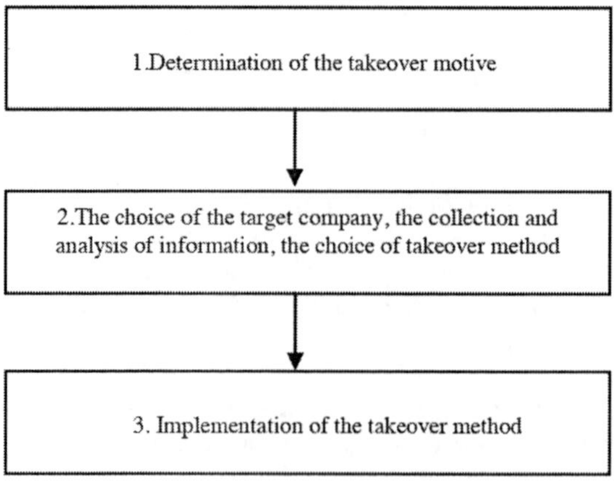

Fig. 3. Takeover process
Source: Own processing.

When a takeover motive is determined, the takeover entity searches for a suitable target company.

2. The choice of the target company, the collection and analysis of information, the choice of takeover method.

At this stage, the subject determines the appropriateness of the takeover. The takeover subject may be interested in inefficient companies with low capitalization, corporations that have an unstable dividend history and infringe on the rights of shareholders, in which corporate conflicts exist.

At this stage, the following information about the target company is analyzed:

- corporate information (history of privatization (creation) of the company; information on the degree of dispersion of share capital; composition of owners; information from the charter documents, regulations; reports and minutes of general meetings of shareholders; minutes of meetings of the board of directors (supervisory board), executive collegial body; information about activities of branches, subsidiaries and affiliates);
- information on financial and economic activities (information on receivables and payables; information on the availability of liquid assets; quarterly reports; balance sheets; information on tax history; features of interaction with partners, creditors, suppliers, consumers);
- information on shareholders (participants) and the management staff (including information on the informal interests of managers and shareholders, especially relationships, information on leaders, conflict history) [2; 5].

Also, the takeover entity may be interested in analyzing similar information about the key stakeholders of the target company.

After analyzing the information, the subject of absorption reveals the degree of difficulty of absorption and selects the method of absorption [2].

3. Implementation of the takeover method.

Many Russian scientists have studied the actions that the takeover subject performs in the takeover process. Radygin A.D. and Etnov R.M. distinguish six groups of takeover methods:

- purchase of shares from employees, investment institutions, brokers, banks in the secondary market,
- voluntary or administrative-compulsory entry into holdings,
- the purchase of assets that are liquidated in the bankruptcy process,
- purchase and transformation of debts into property and equity participation;
- the use of court decisions (for recognition of previously completed transactions (issues of securities) invalid;
- initiation of court orders, including falsification of court decisions [14; 7, p. 57-58].

Markov P.A. identifies several of the most common forms (methods) of absorption:

- acquisition of a controlling stake,
- abuse of shareholder rights,
- manipulation of preferred shares,
- “erosion” of shares (by issuing during the arrest of shares of the main shareholder),
- purchase of debt instruments,
- falsification of the transfer order, which is submitted to the registrar, as well as decisions of general meetings of shareholders;
- purchase of small blocks of shares,
- use of the "parallel" registry [15, p. 46-47].

Studying acquisition strategies, M. Besedin identifies two absorption methods: bankruptcy; the acquisition of shares (shares in the authorized capital) guaranteeing corporate control [16].

There are other classifications of absorption methods that are characteristic of Russian practice:

- the acquisition of small blocks of shares,
- creation of conditions for bankruptcy of the target company,
- challenging ownership of strategically important assets,

- actions aimed at reducing the value of the corporation and the subsequent acquisition of its assets;

- bribery of managers of the target company [4, p. 28].

Gomtsyan S.V. offers its own classification of takeover methods: purchase of votes, purchase of assets, direct purchase of shares, reorganization [1, p. 10].

The three methods of acquiring the companies that are most widely used are presented below:

- establishing control over the executive body;

- purchase of a controlling stake;

- initiation of bankruptcy followed by the purchase of assets [3].

Because a typical takeover process consists of three stages, it is necessary to build protection against such processes relative to these stages. Different types of protection may apply in different countries, and this is due to differences in legal systems. However, the general logic of protection, and this applies to a company from any country, may be identical, namely:

- organization of information influence on the takeover subject in order to cause it to refuse to acquire control over the company (the first level of protection against takeover);

- managing the loyalty of stakeholders in order to organize long-term mutually beneficial interaction between them and the company (the second level of protection against takeover);

- creating procedural barriers in the takeover process in order to prevent the company's corporate governance system from making decisions that contribute to the takeover (the third level of protection against takeover).

5 Conclusion

The article presents a classification of the takeover methods, which differs from the existing classifications by differentiating the takeover methods according to the criterion "takeover tool used by the takeover initiator", the application of which allows the target company to choose the most appropriate and effective protective measures in a particular situation and thereby avoid some corporate conflicts. It was revealed that the takeover entity can perform various actions to gain control over the corporation's activities, eroding corporate ownership, using corporate conflicts, especially corporate relations in a particular company. However, as already noted, even a 100% share in the company capital does not always guarantee control over the corporation if the resource base of the company includes more valuable resources. Then the owners of these resources control the company. This is due to the fact that property rights are not clearly defined.

References

1. S. V. Gomtsyan. Takeover rules for joint-stock companies: comparative legal analysis/ Gomtsyan Suren vardanovich / / volters kluver 136 p. (2010)
2. Yu. V. Koneva. *Unfriendly acquisitions and economic security of the enterprise. Legal guide of the head.* [online], Available at: http://www.klerk.ru/law/articles/59089 (2006)
3. A. Molotnikov. *Takeover of companies: an overview of methods and tools.* [online], Available at: http://www.lin.ru/document.htm?id=1249703346338427714 (2001)
4. L. L. Nikitin, D. V. Nurzhinsky, The strategy and tactics of takeover defense. *Mergers and Acquisitions*. (2), 28-34 (2003)

5. V. Pankratiev. Protection from hostile takeover. *Business Entropy*. [online], Available at: http://bizentropy.biz/articles/167-zashhita-ot-vrazhdebnogo-pogloshheniya-okonchanie.html
6. A. D. Osinovsky. *Shareholder vs. joint-stock company*. 2nd ed, 384 p. (2006)
7. A. D. Radygin, R. M. Etnov. *Problems of mergers and acquisitions in the corporate sector*. 154 p. (2002).
8. V. Dominyak. *Organizational loyalty: basic approaches*. [online], Available at: http://dominiak.ru/publ/publ15.html
9. S. S. Baranskaya. *Psychological factors of organizational loyalty*. 24 p. (2012)
10. K. V. Uskov. Managing the loyalty of company participants in the aspect of protection from absorption. *Vestnik universiteta (GUU)* (1), 210-214 (2014)
11. *The typical scheme of raiders ' seizure*. [online], Available at: http://zahvaty.net/tipovaya-sxema-rejderskogo-zaxvata.html/2 (2010)
12. V. G. Antonov, O. Yu. Kirillova, M. V Samosudov, H. P. Kharchilava. *Corporate management: Textbook for students of the specialty "Organization Management"*. State University of Management, Institute of Marketing of GUU. 287 p. (2010)
13. A. Ya. Cherenkov. Reasons and motives of corporate mergers and acquisitions. *Financial research* (14) [online], Available at: http://finis.rsue.ru/2007-1/Cher.pdf (2007)
14. E. S. Demidova. *Features of hostile takeovers in the corporate sectors of the Russian Federation and the European Union and the problem of their regulation*. [online], Available at: http://elibrary.finec.ru/materials_files/izv/IzvSPbUEF2007_1_c157_161_s.pdf (2007)
15. P. A. Markov. *Hostile takeover. Theory, practice, legal regulation: monograph*. 96 p. (2010)
16. M. Yu. Besedin. *Mechanism for ensuring economic security of economic entities from unfriendly acquisitions*. 177 p. (2007)
17. K. G. Arutyunova. *Ways to protect against "unfriendly acquisitions" in the civil law of Russia and the United States of America*. 222 p. (2010)
18. D. V. Burykin. *Unfriendly acquisitions as a threat to the economic security of joint-stock companies*. 216 p. (2006)
19. M. G. Iontsev. *Corporate takeovers: mergers, acquisitions, greenmail*. 224 p. (2006)
20. N. And. Krichevsky. *Raiding: the portrait of a phenomenon*. [online], Available at: http://www.ng.ru/ideas/2007-10-12/11_reiders.html (2007)
21. S. V. Savchuk. Analysis of the main motives of mergers and acquisitions. *Management in Russia and abroad* (5) [online], Available at: http://www.cfin.ru/press/management/2001-5/04.shtml (2001)
22. N. S. Sogrina. *Features of unfriendly acquisitions (raiding) in the transformed economy*. 189 p. (2010)
23. L. A. Trifonova. *Formation of a strategy for protecting joint-stock companies from unfriendly takeover*. 161 p. (2008)

The Relationship between Capital Structure and Financial Performance of the Company

V.V. Tretiakova[1,*], *M.S.* Shalneva[1], and *A.S.* Lvov[1]

[1]Financial University under the Government of the Russian Federation, Moscow, Russia

Abstract. The article examines and analyzes the relationship of key performance indicators (ROA, ROIC, change in market capitalization and price-to-book ratio) and the capital structure of the company based on the pharmaceutical industry in the UK for the 2009-2019 period. The study seeks to provide a practical evidence on the impact of external financing on company's financial performance and test applicability of the pecking order theory for the chosen companies. The research conducted uses panel data regression and Wald test to determine and analyze the effect of capital structure on the financial indicators of the company performance. The study used a sample of 185 UK companies from the pharmaceutical industry. The result of the research showed that equity has negative effect on price-to-book ratio and ROA and positive effect on change in market capitalization, while long-term debt has a positive relationship with price- to-book ratio and change in market capitalization. In addition, short-term debt has a negative effect on change in market capitalization, ROA and ROIC. The study also provides only partly coincidence of the results with the pecking order theory.
Keywords: Capital Structure; Financial Performance; Market Capitalization; Return On Assets; Return On Invested Capital.

1 Introduction

Since the 20th century, ways to evaluate company's development have been actively discussed and explored. One of the generally accepted financial measures is considered to be company's financial performance, which represents a subjective measure of the firm's effectiveness in using assets of primary occupation and generating revenues. In addition, financial performance also indicates the overall financial health of a company for a certain period of time [4]. To improve the financial performance internal and external sources of financing are chosen by the company, which subsequently affect its capital structure. The capital structure decision is one of the most significant for businesses as it may help to maximize returns of the company and function successfully in the competitive environment. Starting from the 50s of the last century various theories of capital structure are being developed, which are trying to find the best option for firms financing that will help them develop actively and improve their performance [1].

*Corresponding author: trevishka@gmail.com

Interestingly, very few research took the innovative industries of developed countries to study this relationship [8]. Thus, modern innovative companies remain unexplored for the relationship of their performance and capital structure. Hence, meticulous empirical study of this relationship can be relevant for a better understanding of the functioning of such companies in the modern world.

The aim of the article is to find out whether there is a relationship between capital structure and company performance indicators on the example of pharmaceutical companies in the UK. The aim is supported by the following objectives:

- To determine whether there is a relationship between capital structure and a firm's performance according to relevant literature;
- To discover the key financial indicators of the company performance;
- To examine the relationship between key financial indicators and capital structure;
- To identify the impact of external financing on the company financial performance;
- To test whether the financial resources allocation conforms one of the capital structure theories.

This study will attempt to close the previous researches' gaps and highlight any unexpected relations to get a better understanding of pharmaceutical companies' operational principles and ways to improve their financial performance, as well as allow investors to make more informed decisions regarding the financing of the above companies.

2 Methodology

The following research aims to empirically analyze the selected companies from pharmaceutical industry in the UK and test their relationship between capital structure and financial performance according to the hypotheses.

Since the external sources of financing of the capital structure is a combination of equity, long term and short term debt sources their impact on each indicator of financial performance will be considered separately. The following financial indicators of the company performance were selected: ROA, ROIC, DMC, PB ratio. The present research seeks to determine the impact of capital structure on the performance variables chosen and accept or reject the following hypotheses according to the following data sample analysis.

The present study examines the effect of capital structure on key financial indicators of the company's performance for UK companies in pharmaceutical industry. Therefore, based on the financial performance indicators chosen, the following hypotheses were formulated. Each hypothesis describes one of four financial indicators. The H1 states that external financing impacts negatively DMC as an indicator of company performance. The H2 claims that external financing impacts positively PB ratio as an indicator of company performance. The H3 verifies whether external financing impacts negatively ROA as an indicator of company performance. Finally, H4 examines whether external financing impacts positively ROIC as an indicator of company performance. Finally, based on testing the first four hypotheses, the H5 is formulated: financial resources allocation follows the pecking order theory.

The research uses secondary data only, since the necessary information about financial indicators of the company's performance as well as their capital structure cannot be obtained primary in the framework of this study.

The study collects data from different companies from the same industry over a long period of time. Since the data refer to the concept of "multidimensional data" and include measurements over a long period of time, it is decided to use panel study as a type of longitudinal design for econometric and statistical studies [2].

In order to measure the effect of independent variable on dependent one and test the hypotheses H0 and H1 panel data analysis techniques are applied. The hypothesis testing involves the use of several models: linear regression, pooled OLS panel regression model and Fixed and Random effects Panel regression model [11].

To test the relationship between capital structure and PB ratio, DMC, ROA and ROIC linear regression seems to be the best model to be applied. In a standard form, linear regression is represented by the formula:

$$Y_i = \beta_0 + \beta_1 X_1 + \ldots + \beta_n X_n + \varepsilon_i \qquad (1)$$

where: Y_i – dependent variable;
β_0– constant;
$\beta_1 - \beta_n$– regression coefficients;
$X_1 - X_n$– explanatory variable;
ε_i – statistical error.

In order to find betas and build regression model the method of ordinary least squares (OLS) is used as one of the most powerful methods to apply for the panel regression [11].

In order to estimate the relationship between capital structure and ROA and ROIC as dependent variables, the regression functions were created:

$$ROA_{i,t} = \beta_{0,i} + \beta_{1,i} * ST + \beta_{2,i} * LT + \beta_{3,i} * DSH + \varepsilon_{i,t} \qquad (2)$$
$$ROIC_{i,t} = \beta_{0,i} + \beta_{1,i} * ST + \beta_{2,i} * LT + \beta_{3,i} * DSH + \varepsilon_{i,t} \qquad (3)$$

The independent variables are represented by short-term debt (ST), long-term debt (LT) and growth rate of outstanding shares (DSH), $\varepsilon_{i,t}$ implies statistical error.

For estimating the relationship between capital structure and PB ratio, ROE is included into regression model as a control variable in order to control risk if increasing the leverage. Therefore, the regression formula for PB ratio as a dependent variable is:

$$PB\ ratio_{i,t} = \beta_{0,i} + \beta_{1,i} * ST + \beta_{2,i} * LT + \beta_{3,i} * ROE + \varepsilon_{i,t} \qquad (4)$$

By using the coefficient of determination or the coefficient of the R-squared, the study determines the possibility of applying a regression model. The R-squared as a selective correction between the predictor and the response can be measured according to the formula [11]:

$$R^2 = \frac{Explained\ variation}{Total\ variation} \qquad (5)$$

To avoid autocorrelation of random errors of the regression model and bias coefficient estimates, the autoregressive model is used in the current study to test PB ratio, ROA and ROIC.

To obtain the best results in using the OLS method for PB ratio, ROA and ROIC, the study uses the autoregressive model. Autoregression of the first and second orders is used, the processes of which are characterized by the formulas:

$$AR(1): X_t = c + rX_{t-1} + \varepsilon_t \qquad (6)$$
$$AR\ (2): X_t = c + a_1 X_{t-1} + a_2 X_{t-2} + \varepsilon_t \qquad (7)$$

To test how capital structure affects the DMC, Pooled OLS and Fixed and Random effects Panel regression models need to be applied:

$$DMC_{i,t} = \beta_{0,i} + \beta_{1,i} * DST + \beta_{2,i} * DLT + \beta_{3,i} * DSH + \varepsilon_{i,t} \quad (8)$$

In case of influence of factors which are not considered in the model and random deviations dependent on companies these regression models help to avoid heterogeneity bias.

The confidence interval of the study is 90%, therefore p-values at 10% level or lower propose a significant result.

Wald test is also used to verify that the impact of short-term debt and long-term debt is different when testing PB ratio, ROA and ROIC as dependent variables. The program EViews 11 was used for this task.

The analysis is based on pharmaceutical firms listed on the London Stock Exchange. Financial data was obtained by authors from the Bloomberg information system on the 22nd of May 2020 for the period between 2009 and 2019 and consists of company financial data: D/E ratio, the number of outstanding stocks, stock price, market capitalization, long-term and short-term debt, P/B ratio, ROE, ROA and ROIC. The primary sample consists of 203 companies. The data was exported to Excel. Due to lack of some company data, some companies were excluded from the sample. The final sample consists of 185 companies for 11 years (from January, 2009 to January, 2019) based on yearly observations.

In order to eliminate the omitted variable bias control variables are included into the model where it is necessary [11]. The Hausman test is also used to prevent false results where they are possible [3].

3 Results

In order to test the hypotheses H1-H5 and evaluate results of the research the econometric and statistical techniques mentioned in the previous chapter are applied. The hypotheses H1-H5 with a specification for DMC includes panel data regression with 132 cross-sections and a total number of 720 panel observations. Firstly, a Pooled OLS regression model is applied. The generalized formula of Pooled OLS regression presented in the Methodology subsection is changed according to the obtained values by EViews program:

$$DMC_{i,t} = 0.555374 * DSH - 0.00082 * DST + 0.0000931 * DLT \quad (9)$$

According to the results obtained, all independent variables have a significant factor at p-values of 0,0253 for short-term debt, 0 for long-term debt and 0 for growth rate of outstanding shares which is lower 10%. The R-squared is 0,085411 which is rather low. That means that the amount of variance in the dependent variables that is predictable from the independent variables is low. Based on these results, it can be concluded that the observed outcomes are not replicated by the model well. However, the result may be incorrect due to heterogeneity bias [12]. In order to test pooled OLS for the heterogeneity bias, fixed effects model is estimated where the R^2 should increase by adding dummy variables in the amount n-1. The formula mentioned in the Methodology chapter is written below according to the coefficients findings:

$$DMC_{i,t} = 0.634928 * DSH - 0.00075 * DST + 0.000126 * DLT \quad (10)$$

As seen from the table, the R-squared has increased and is now 0,282295, which is nearly 20% more than the previous result in pooled OLS regression model. In order to prove the significance of the model Redundant fixed effects test is conducted. According to the fixed effects model test, cross-sections fixed effects model is more significant than the pooled model as the p-value of cross-section F test is 0,0613 and is less than 10%. That means that the original data in the pooled OLS model includes heterogeneity bias. Therefore the fixed effects model is applied. In the fixed effects model all three independent variables impact significantly on DMC as a dependent variable.

Since it is always necessary to choose the application of a model of fixed or random effects, the random effects model is now applied. A Hausman test verifies the value difference between the estimates of two models in order to choose the correct one.

The following formula for the random effects model according to the results is:

$$DMC_{i,t} = -0.04195 + 0.572947 * DSH - 0.0008 * DST + 0.000101 * DLT \qquad (11)$$

With a probability result of 0,2656 which is more than 10% Hausman test shows that the difference between values of the estimators is not significant, therefore the null hypothesis is accepted and consequently the random effects model is appropriate for the current study.

As a result, the random effects regression model reveals that all independent variables have an impact on changes in the market capitalisation. Delta shout and delta long-term debt with a p-value both of 0 affects DMC positively, while delta short-term debt with a p-value of 0,0252 has a negative relationship with the dependent variable.

The analysis of the relationship between capital structure and PB ratio is presented below. The total number of observations is 521 with 104 cross-sections included. To avoid the autocorrelation of random errors of the regression model and bias coefficient estimates, the autoregressive model AR(1) and AR (2) is used. By substituting values the panel data regression for PB takes on the form:

$$PB\ ratio_{i,t} = 3.050119 + 0.031414 * DE - 0.03628 * ROE + 0.345918 * AR(1) + 0.442254 * AR(2) \qquad (12)$$

The results show that there is a significant positive impact of D/E ratio (p-value=0,044) and significant negative impact of ROE (p-value=0,01). However, it was decided to consider debt in particular by using such independent variables as long-term debt and short-term debt as it gives an opportunity to verify the difference between short-term and long-term debt impact. According to the results, the current formula is:

$$PB\ ratio_{i,t} = 2.850597 + 0.0463 * LT - 0.03928 * ROE + 0.348387 * AR(1) + 0.408909 * AR(2) \qquad (13)$$

The results from EViews 11 show that long-term debt with a p-value of 0,0106, which is lower than 10%, has a significant positive effect on PB ratio. As a control variable, ROE has a negative effect on PB ratio with p-value of 0,005 as well. Short-term debt is not considered as statistically significant factor as its probability is 0,1549, which is higher than 10% p-value taken for the study.

Since short-term and long-term debts are a part of D/E ratio as a percentage, in order to avoid test fallacy it is necessary to test if there is a difference of impact of long-term and short-term debt. For this reason Wald test is applied in the following study. With Wald test the current study checks whether long-term debt and short-term debt are equal in value and therefore can be excluded from the model due to insignificance. The test conducted shows that the p-value of t-statistics (other statistic tests are not considered since only one linear constraint is checked) is 0,0101 which is less than 10%, therefore the null hypothesis is rejected as there is a statistically significant difference. The two coefficients C(2) (short-term debt) and C(3) (long-term debt) are not simultaneously equal to zero or equal to each other, therefore the impact of short-term debt differs from the impact of the long-term debt, hence they are not removed from the model.

In order to measure the relationship between capital structure and ROA and ROIC, the analysis of two dependent variables is combined since both same techniques and statistical tests are conducted to determine the relation. To outline its relationship of ROA with short-term debt, long-term debt and delta shout panel data regression with 103 cross-sections and 470 observations is created. The formula for the ROA panel data regression according to the results is:

$$ROA_{i,t} = -35.2797 - 0.05902 * ST - 18.3895 * DSH + 0.556996 * AR(1) + 0.323419 * AR(2) \quad (14)$$

As can be seen from the table, the probability of short-term debt is 0,0017 and lower than 10% whereas long-term debt has a p-value of 0,645, which is higher than the probability determined for the current study. Therefore, short-term debt impacts ROA negatively, while long-term debt does not have any influence on the financial performance indicator mentioned. DSH has the most significant negative impact on ROA with the coefficient of -18,3895 and p-value of 0,0043. In order to test the difference in value of short-term and long term debt the Wald test is taken. The Wald test shows the p-value of t-statistics is at level of 0,0025 and is less than 10%, therefore the null hypothesis is rejected, where C(2) is short-term debt and C(3) is long-term debt, therefore the impact of short-term debt and the long-term debt is different and they do not need to be removed from the model.

The panel data analysis for ROIC as a dependent variable includes 104 cross-sections and 473 total panel observations. Since ROIC takes significantly negative values in the chosen sample of companies, it was also decided to bring the indicator to the exponential form, thus the dependent variable is EXP(ROIC/100)-1. Wald test is also conducted to exclude the identity between values of long-term and short-term debts. According to the results the formula for the ROIC panel data regression is:

$$ROIC_{i,t} = -0.40157 - 0.00054 * ST + 0.601243 * AR(1) + 0.31541 * AR(2) \quad (15)$$

The probability of t-statistic in Wald test is less than 10%, therefore there is a difference in impact between short-term debt and long-term debt, hence the panel data regression analysis is applicable for these independent variables.

Panel data regression for ROIC indicates the significant impact of short-term debt (with p-value of 0,0058) only. The p-values of long-term debt (0,5265) and delta shout (0,8378) show absence of statistical significance of the independent variables.

4 Discussion

In accordance with findings, the results were combined into the Table 2 in order to give a more demonstrative view. The first column represents investigated financial indicators of the company performance, the subsequent columns point out the relationship between dependent variables and equity, long-term and short-term debts. Positive impact is indicated by "+" symbol, negative impact – by "-" symbol and absence of any relationship is represented by "N/A".

Table 1. Analysis of results

Dependent variable	Equity	Long-term debt	Short-term debt
DMC	+	+	-
PB ratio	-	+	N/A
ROA	-	N/A	-
ROIC	N/A	N/A	-

Source: Own processing.

Based on the table above, certain conclusions need to be drawn. Firstly, all elements of capital structure follow approximately the same trend in the impact on the key indicators of financial performance of the company. This can be seen as a positive sign that the research is done correctly.

DMC has a significant positive relationship with equity and long-term debt and negative relationship with a short-term debt. This means that equity financing and long-term debt increases the level of market capitalization. Therefore, the hypothesis **H1** must be **rejected.**

The findings related to the relationship of PB ratio and capital structure components show that PB ratio corresponds positively with ROE which presents equity and negatively with short-term debt. This only partly corresponds with the hypothesis, thus it concludes that **H2** hypothesis is **rejected.**

It is found that ROA has a negative relationship with equity and short-term debt. This is partly consistent with Salim and Yadav (2012) who test Malaysian firms and find the negative relationship between short-term debt and ROA [9]. However, this contradicts the argument of Mehdi & et al., (2013) who also test pharmaceutical firms and state that there is no relationship between short-term debt and ROA and negative relationship between long-term debt and ROA, which are the opposite results to the current research [6]. The difference in results could be due to another time period and country chosen, as well as the difference in countries' economies and their level of the development. According to the findings results, the hypothesis **H3** should be **accepted.**

Surprisingly, ROIC shows no relationship with equity and long-term debt and negative relationship with short-term debt. This disagrees with Nasimi (2016) paper which states that ROIC has a positive relationship with debt to equity ratio. Hence, the **H4** hypothesis is **rejected** [7].

In conclusion, it should be noted that the impact of equity on financial performance indicators is worse than the impact of long-term debt. There is a negative relationship between equity and 2 out of 4 financial indicators, while long-term debt affects positively 2 out of 4 financial indicators and does not have any impact on the rest ones. This provides an evidence for the pecking order theory and acceptance of H5 [5]. However, short-term debt impacts even more negatively than equity does, which is not corresponded with the pecking order theory [10]. As a result, the findings partly follow the pecking order theory, hence the **H5** hypothesis is **rejected.**

5 Conclusion

The research has shown that long-term debt has a positive impact on such dependent variables as DMC and PB ratio, short-term debt negatively affects DMC, ROA and ROIC and equity negatively impacts PB ratio and ROA, while DMC positively. Thus, 4 of the 5 proposed hypotheses were rejected.

The current study suggests practical implications of the research conducted for the companies and investors. As it was stated in the introduction, company's performance is one of the fundamental financial indicators to measure the company's level of development and it also impacts the investment decisions. In this study information was obtained on the impact of capital structure on financial performance indicators in pharmaceutical companies in developed countries using the example of the UK. This allows companies to get correct estimation of their capital structure and optimize it better in order to improve the indicators of their performance. The study also provides management with the information on how debt impacts the financial performance of the pharmaceutical company. It could be used for making strategic decisions in order to achieve maximized profits, and lead to the rise of market capitalization. It is also important for making competent decisions regarding the activities of the company and raising funds in such a way as to avoid financial losses and improve their financial performance. Finally, understanding the optimal structure of external financing of pharmaceutical companies and evaluating the correctness of this financing may help investors make a decision about investing their own funds in a particular project. However, it also should be mentioned that the results may vary due to the country and industry specifics. Moreover, the size of the sample and the period taken may also change the results of the research as some important factors can be included. For example, the results may be different due to the financial crisis of 2008, the consequences of which may affect the financial performance of the company in subsequent years, but not all studies consider the same period. Secondly, there is a possibility of an insignificant result due to heteroscedasticity and, therefore, inefficiency of estimators [10].

Based on the results of this study, it is able to suggest some recommendations for the further research. Firstly, it is highly commended to use a wider sample of companies, as well as include more financial indicators of the firm performance in order to find more evident relationships between performance indicators and capital structure. Secondly, a greater number of developed countries should be taken into consideration to track the overall trend of the impact of external financing on the company's performance in these countries. It may give further development of the current research and help to explain more relations in the model. Finally, it is suggested to consider the effect of capital structure on company financial performance depending on the size of the company and the level of leverage, which can provide a basis for determining the differences in the impact on samples with different input data. Therefore, the current study can be developed in several areas of further research, which could deepen the scientific understanding of the relationship between key performance indicators of a company and capital structure.

References

1. Aggarwal, D., & Padhan, P. C. Impact of Capital Structure on Firm Value: Evidence from Indian Hospitality Industry. *Theoretical Economics Letters*, **07**(04), 982–1000. (2017)
2. Bryman, A., & Bell, E. *Business Research Methods 3e*. Oxford, United Kingdom: Oxford University Press. (2011)
3. Hausman, J. Specification Tests in Econometrics. Econometrica, **46**, 1251-1271. (1978)
4. Kenton, W. Financial Literacy. Retrieved from https://www.investopportunity.com/terms/f/financial-literacy.asp (2020)
5. Laisi, T. Pecking order theory in a bank-centered lending environment - Evidence from North European economies. *University of Vaasa*, 1. (2016)
6. Mehdi, M., Farimah, R., Forough , R., Seyed, M., & Jamshid, S. The Effect of Capital Structure on the Profitability of Pharmaceutical Companies The Case of Iran. Iran J Pharm Res., 573–577. (2013)

7. Nasimi, A. Effect of Capital Structure on Firm Profitability (An Empirical Evidence from London, UK). *Global Journal Of Management And Business Research,* **16**(4), 8-20. (2016)
8. Nemlioglu, I., & Mallick, S. K. Do innovation-intensive firms mitigate their valuation uncertainty during bad times? *Journal of Economic Behavior & Organization*, **177**, 913–940. (2020)
9. Salim, M., & Yadav, R. Capital Structure and Firm Performance: Evidence from Malaysian Listed Companies. *Procedia - Social and Behavioral Sciences*, **65**, 156–166. (2012)
10. Tucker, J., & Stoja, E. Industry membership and capital structure dynamics in the UK. *International Review of Financial Analysis*, **20**(4), 207–214. (2011).
11. Weisberg, S. *Applied Linear Regression*. Hoboken, NJ, United States: Wiley. (2013)
12. White, H. A Heteroskedasticity-Consistent Covariance Matrix Estimator and a Direct Test for Heteroskedasticity. *Econometrica*, **48**(4), 817. (1980)
13. Wooldridge, J. M. *Econometric Analysis of Cross Section and Panel Data*. Amsterdam, Netherlands: Amsterdam University Press. (2010)
14. Wooldridge, J. M. Control Function Methods in Applied Econometrics. *Journal of Human Resources*, **50**(2), 420–445. (2015)

Sources of Investment Support of Production and Business Activities in the Russian Economy

A.V. Streltsov[1], *G. I.*Yakovlev[1,*], and *I.V.* Kosyakova[2]

[1]Samara State University of Economics, Samara, Russia
[2]Samara State Technical University, Samara, Russia

Abstract. The problems of determining effective and available sources of investment for the purposes of technical and organizational development of enterprises and business activities are studied. An urgent requirement is to reverse the unstable dynamics of the industrial investments volumes implemented by Russian enterprises, the low degree of renewal of fixed assets of enterprises that go to a critically high level of wear and tear, which does not allow for high resource productivity and competitiveness of Russian enterprises of processing activities in the world market. The purpose of the study is to consider the indicators of the current state and the possibility of using certain available sources of investment in production, to analyze the characteristics of state regulation of investment activity in the economy. The article uses methods of statistical grouping, analysis of cause-and-effect relations, and meaningful economic interpretation of the phenomena of the investment process in the Russian and world economy conditions. A comparative analysis of the most popular sources of investment support for production and economic activities of enterprises is carried out, their advantages and disadvantages, and features of use are shown. As a result, recommendations were formed on the choice of investment sources in relation to various economically stable groups of manufacturing enterprises operating in markets of different regulatory severity.
Keywords: sources, investment, enterprises, business activity, competitiveness, development.

1 Introduction

Enterprises need various funds to carry out their production and sales activities, which are in practice mobilized from the totality of all available investment resources. Currently, in the context of the global socio-economic crisis caused by the ongoing viral pandemic, the growth of global uncertainties, trade and economic wars, and other adverse, restrictive factors, the economic authorities of many countries have begun to develop a national plan for economic recovery, identifying and attracting investment necessary for the development of productive forces, corresponding to the required size and sources. For the development of the scientific and industrial sector in Russia, it is important that the national plan for

*Corresponding author: dmms7@rambler.ru

long-term economic development contains not only measures to support individual industries, but also includes the development of end-to-end technologies, such as industrial Internet, robotics, radio electronics, automated production, etc.

It is clear that the absence of any well-thought–out policy regarding the development of productive forces, given the chronic underinvestment, expensive financial resources for enterprises of processing activities, with their current low efficiency and productivity, carrying out production on obsolete equipment, with periodically recurring crises, has led to the destruction and instability of many of them. This is especially noticeable in non-primary processing economic activities, primarily in mechanical engineering, which integrates advanced scientific and technological achievements.

2 Methodology

In this study, the authors rely mainly on the theoretical provisions of the production approach in the study of the content and process of industrial investment. They, according to the methodology of J. M. Keynes, have an exclusively industrial origin, while the income from the sale of products and services of the enterprise is divided into two components – consumption and investment. Despite the priority of the manufacturing sector over the financial sector, this approach does not deny the role of credit for business development, which is balanced by the obligations of industrial enterprises on the scale of the macro-economy.

This approach allows for various classifications of industrial investments in the study of enterprise development problems. As a rule, specialists distinguish internal and external sources of investment, according to Fig. 1. The first include retained earnings and depreciation. This includes a number of other less significant and stable components: investments of citizens and legal entities' money, unused part of social funds, etc. External loans include various forms of lending by commercial entities, government loans, funds raised on the stock market (consolidated funds for issuing shares, bonds, issuing bills, etc.). These sources of investment can be defined as traditional, widely used in a market economy. In addition, among the external ones, we can distinguish so-called "non-traditional" ones, which are less often used in enterprises, but have significant investment potential. These include leasing operations, cooperative interaction of enterprises, financial relations between subsidiaries and parent companies within the framework of international strategic alliances, etc.

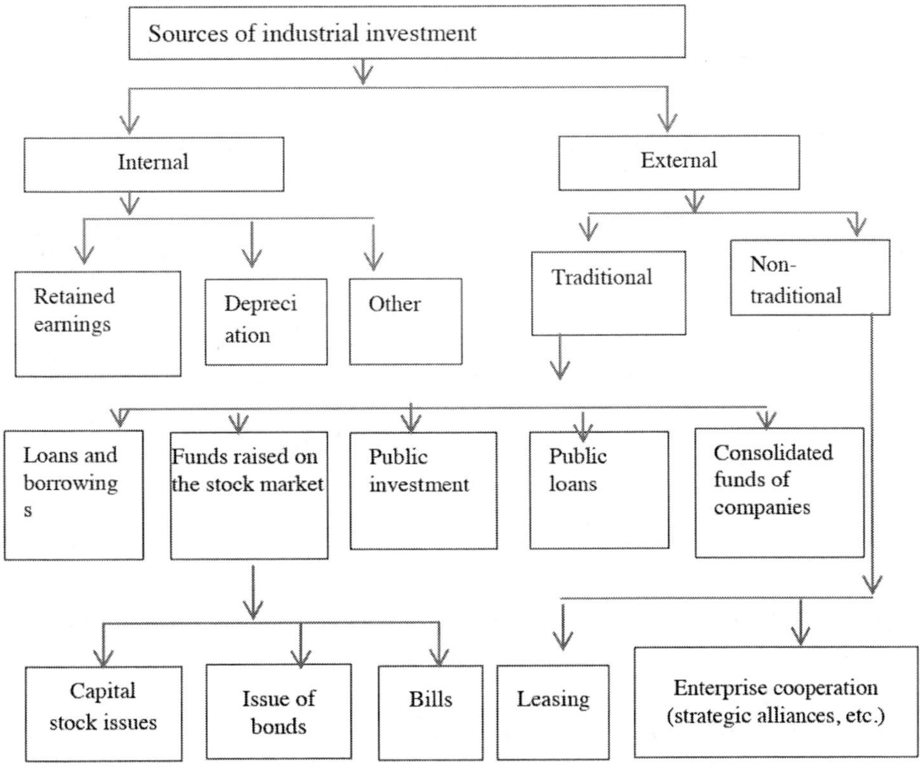

Fig. 1. Composition of investment sources of industrial enterprises
Source: Own processing.

Describing the resource capabilities of these forms of investment in relation to the subjects of the domestic Russian economy, we should note the following. The most significant source of own funds in machine-building enterprises is depreciation deductions, i.e. annual (monthly) deductions in proportion to the worn-out part of the fixed assets cost. However at present the use of depreciation for the purpose of updating the worn out elements of fixed assets is extremely insufficient and is limited by the following reasons:

1) Lack of incentives for targeted use of depreciation;
2) Using primarily the straight-line method of calculation;
3) Rather large actual rates of depreciation of the cost of cash in the case of their accumulation in the depreciation fund;
4) Insufficient motivation for updating fixed assets in general, since the current financial condition of many enterprises requires the immediate use of available funds, while the same unfavorable financial condition of consumers forces them to accept low-quality, but relatively cheap products.

We used methods of statistical grouping, comparative analysis, and meaningful economic interpretation of economic practice at the macro and micro levels. Experimental base of the study: results of activities of Russian and foreign enterprises in various industries, management bodies of national economies, especially in the formation and promotion of targeted industrial policy.

3 Results

The formation of future national champions in the production of innovative products and business activities which should be competitive on the international market. This formation requires a well-thought-out industrial policy and large-scale investment resources. However, the results of investment activity of domestic economic entities over the past nine years have been quite modest (Table 1).

Table 1. Key indicators of investment activity in the Russian economy

Indicator	Units	2010	2015	2016	2017	2018
Investment in fixed assets:	billion rubles	9152.1	13897.2	14748.8	16027.3	17595.0
per capita,	thousand rubles	64.1	94.9	100.6	109.1	119.8
as a percentage of the previous year	(in comparable prices)	106.3	89.9	99.8	104.8	104.3
as a percentage of gross domestic product	%	20.7	20.0	21.2	21.5	20.7
Depreciation of fixed assets	at the end of the year, percent	47.1	47.7	48.1	47.3	46.6
Coefficient of renewal of fixed assets	in constant prices, percent	3.7	3.9	4.4	4.3	5.1

Source: Own processing according to the Rosstat data of the Russian Federation [6].

The analysis of these tables shows quite unstable dynamics of investments implemented by Russian enterprises. Despite the fact that the degree of depreciation of fixed assets of enterprises in the vast majority is already approaching 50%, the coefficient of their renewal only slightly exceeds 5 percent. At this rate, the full renewal of fixed assets can only occur in twenty years, when the organizational and technical apparatus of production will be repeatedly changed throughout the world in accordance with the laws of changing technical patterns and cycles of modernization. In addition, outdated production capacities of domestic industrial enterprises will not be able to effectively support the digital transformation, increase the productivity of resources to the level of world-leading firms, thereby losing in ensuring the international competitiveness of their products.

In addition to depreciation, another important internal source of investment is the portion of net profit allocated for the accumulation purposes. In a number of large organizations, this can be quite a significant amount. However, in general, given the low level of profitability in the manufacturing industry and the need for funds to update fixed assets, it can be noted that this source of investment is an extremely small amount in comparison with the necessary need.

Moreover, considering the complex characteristics of the investment processes of security capital renewals, including, in addition to the above, this is an orientation when updating mainly on imported equipment and the low exchange rate of the domestic currency, significantly increasing the cost of procuring from abroad equipment and technological backwardness of the majority of enterprises, in addition to requiring new equipment to raise funds for the purchase of new technologies and corporate services, again mainly foreign, and generally continued for a fairly long period insufficient economic activity and lack of meaningful economic growth. All this determines the extremely low value and common set of internal sources – net profit and depreciation. In this regard, the role of using external sources, including borrowed ones, is increasing.

Leasing transactions can also be defined as a specific source of investment. Their specificity lies in the fact that the content of leasing is the provision of property for rent.

However, the income received by the leasing company is in the form of a certain percentage. Leasing companies are very closely connected with financial organizations (banks). A special feature of leasing operations is that they can not be considered as a source of investment for capital renovation of the enterprise. They are usually intended to financing the purchase of individual machines and units, most often transport vehicles. In modern conditions, state loans, direct state investment, as well as consolidated funds of integrated companies have become important for certain categories of machine-building enterprises. Despite their low availability, these sources have a very important characteristic for businesses: low cost and minimal risk compared to other options. However, the ability of enterprises to influence decision-making processes on their involvement is limited.

The possibility of increasing investment activity from the part of non-traditional sources of investment - the international cooperation and integration of enterprises now is quite promising [11, p.466]. Transnational strategic alliances and other forms of integration relations between enterprises of different jurisdictions allow not only to attract investment, but even in their absence, to reduce the overall need for them, reduce risks by dividing the range of works between partners in accordance with taken competencies.

Speaking about the state regulation of investment activity and its impact on the behavior of investors, we can not fail to note its relationship and, accordingly, the need to take into account the specifics of the economic activities of individual sub-sectors and machine-building enterprises. First of all, it is determined by the technological features of production and the functioning of the industry commodity market. In accordance with this, we can distinguish the following enlarged groups of enterprises and sub-sectors of mechanical engineering:

1. Oligopolistic sector, which includes enterprises whose target market of goods and services is controlled by large oligopolistic structures;
2. Marketing (market sector), with a market of goods and services based on competition;
3. A depressed sector where the market of goods and services is difficult for domestic enterprises to access for various limiting reasons;
4. The sector of military-industrial complex enterprises, the market of which is determined by the state order and regulated by the state.

Of course, to a certain extent, this division is somewhat conditional, since a diversified enterprise can operate in different markets. However, any enterprise has the main products that occupy the largest range, and the possibilities of production diversification in conditions of lack of investment are significantly limited by the existing technological scheme of production of the enterprise's products. In a number of major sub-sectors of mechanical engineering (automotive, shipbuilding, aerospace, railway engineering, etc.), expanding the production line requires significant time and resources, so that as a result, a diversified enterprise could carry out its activities in various segments of the commodity market.

Describing the enterprises of the oligopolistic sector, it can be noted that for them the possibility of attracting investment is closely linked with the establishment of partnerships with the relevant Russian oligopolies (UAC, USC, Russian Railways, etc.). At the same time, the state task is to coordinate and regulate cooperative relations and form specific projects for the development of machine-building enterprises, which are ultimately aimed at improving the efficiency of oligopolistic structures. Attracting the necessary investments for the implementation of the project should also include the participation of the latter in this process. Therefore, the state has a significant share in the capital of these structures, and its regulatory influence may well ensure an effective level of organization of these cooperative relations.

Marketing oriented enterprises (market type) are forced to rely mainly on the free market in their activities and, accordingly, form their investment projects more independently, in comparison with the previous type, on a market basis. These projects are significantly more sensitive to macroeconomic indicators, the state of the business environment, and the level of risk. Therefore, enterprises of this type are most dependent on various measures of state support.

Enterprises in the depressed sector of the economy, as a rule, can not independently ensure the proper level of efficiency of their investment projects, since the introduction of new products to the market for them is limited for various reasons. For example, because the market is busy with imported products or there are special technical requirements. In order to increase the investment activity of enterprises in this group, it is necessary to develop a comprehensive program for the development of sub-sectors based on restructuring, by creating large integrated holding companies, and organizing import substitution.

Investment activities of military-industrial complex enterprises are mainly regulated by the state. To bring it to a different qualitative level within the framework of state regulation, it is necessary not only to ensure and control the implementation of the state order, but also to set tasks and monitor their implementation for the diversification and development of production, which is also noted in the work of D. A. Nikolaev, who has considered in detail alternative existing and promising sources of financing for enterprises of the military-industrial complex [10, p.107].

4 Discussion

In most of the works of specialists, the prospects for investment support of enterprises and business activities are linked to the solution of problems of effective use of depreciation charges. Thus, Eroshevsky S. A. (2010) and co-authors considered the financial features of depreciation as a source of investment in the development of the production potential of an oil and gas producing enterprise [3, p.120]. Jodi L.Gissel (2016) notes that targeted reproduction of depreciation charges can provide a significant part of the financing of the costs of modernization of the enterprise [7, p.61].

Improving the efficiency of the depreciation process at enterprises requires the development and implementation of a targeted set of measures, the leading of which are legislative and tax measures. Miklyaeva Yu. S. supports a similar interpretation, noting the importance of depreciation in financing the reproduction of fixed assets in relation to enterprises in the oil refining industry [9, p.227].

However, among scientists-economists, mainly financial and accounting specialization, the view that depreciation should not be the main source of investment is becoming more widespread. It is no coincidence that Kurmakaeva E. Sh. (2006), considering the possibility of involving additional sources of financing in the system of reproduction of the fixed capital of Russian enterprises, focuses on the development of leasing relations, provided that appropriate favorable conditions are created for its development at the macroeconomic level [8, p.176].

Fedorov Yu. V. et al. (2016) identified the problems of insufficient investment level in the fixed capital of machine-building enterprises and incomplete use of the potential of external sources of capital investment financing. They suggested to strengthen the role of attracted and borrowed funds as sources of investment financing in the modernization and renewal of the fixed capital of machine-building enterprises [4, p.7].

Efimychev Yu. I. and co-authors (2012) analyzed the sources and volumes of financing of domestic enterprises at the stages of designing and putting products on the market with the participation of the state, business angels, and venture funds[2, p.239]. Vivek Ghosal and Usha Nair-Reichert (2009) examined the relationship of sources, mechanisms and financial instruments for attracting investment in the innovative activity of an industrial enterprise, conducted a comparative analysis of the most popular sources of attracting investment, with an assessment of advantages and disadvantages[14, p.536].

Rubtsova Yu. A. considered external sources of investment for enterprises, advantages and disadvantages of IPO as a source of investment financing for enterprises[12, p.431]. Gurieva I. V. (2014) analyzed the existing theoretical and practical approaches, procedures and algorithms for selecting sources of investment financing in the scientific economic literature and identified the minimum price of sources of financing, depending on the economic conditions, financial condition, goals and objectives of the enterprise that it solves as the main criterion[5, p.271].

Sotnikova L. N. and Kukush L. I. (2016) studied the main problems of selecting sources of financing for real investment projects by comparing the cost of attracting them for the enterprise and identified the main factors of pricing sources of real investment financing in enterprises [13, p.16]. With regard to investment problems, Dimitris Skuras et al. note that the size of a firm has a direct, inverted U-shaped influence on the probability of investment and an indirect influence through the linear influence of the size on the probability of innovation [1, p.1778]. Yuan Hu (2012) uses the example of Chinese enterprises to link investment in fixed assets with a contribution to reducing energy consumption, directed to improve energy efficiency of new investment projects in the field of fixed assets formation [15, p.327].

As a result, a brief comparative description of the sources of investment for the development of the enterprise is presented in Table 2.

Table 2. Comparative characteristics of investment sources for modernization of machine-building enterprises

Sources of investment		Characteristics			
№	Name of the source	Possibility of using	Value	Cost	Riskiness
1.	Profit	Max	Min	Low	Min
2.	Depreciation	Max	Changes depending on the start time of mobilization	Low	Min
3.	Government loans and public investments	Min	Max	Low	Min
4.	Commercial loans and borrowings	Average	Max	High	Max
5.	Issue of shares	Min	Average (limited by the capital ownership structure)	Low	Low
6.	Funds of the management company of the corporation	Limited (for a limited number of enterprises)	Max	Average	Low
7.	Strategic alliance	Min	Average (limited by the capital ownership structure)	High	Max
8.	Leasing	Average	Min	Low	Average

Source: Own processing.

This table describes the main characteristics of individual sources of industrial investment for the machine-building complex. This is determined by the number of generated estimates. For example, the possibility of using such a source of investment as "commercial loans" is estimated as average, although the banking sector's opportunities for financing seem to be maximum. However, in most domestic machine-building enterprises, the level of profitability of production is lower than the existing interest rate, which makes it difficult to attract loans.

5 Conclusion

Development and modernization is an integral stage of the life cycle of enterprise development and business structures. An urgent need for investment resources for its implementation occurs just when the financial and economic indicators of its activities deteriorate, and the investment attractiveness is minimal. The current order of financial relations in modern regulatory and methodological accounting and tax documents does not link depreciation with the need to create funds, reserves for updating fixed assets. It is no accident that many economists suggest making changes to the concept of accounting and ensuring the targeted reproduction use of depreciation charges. It is obvious that they are able to provide a significant part of the financing costs for the modernization of the enterprise. It should also be noted that it is the targeted use of depreciation for updating fixed assets that distinguishes the historical periods of our country, in which the development of industry was the most dynamic.

The investment justification of the technical and organizational development of an enterprise is largely influenced by both the characteristics of manufacturing enterprises and the specific characteristics of its current state. The most affordable and reliable sources of investment for enterprise modernization are profit and depreciation. The latter, as the practice of industrialized countries shows, can make a significant share in the investment justification for enterprise modernization. However, to achieve this in relation to the Russian business environment, it is necessary to implement a set of legislative and regulatory measures to improve the standards of accounting, tax accounting and planning.

References

1. D. Skuras, K. Tsegenidi, K. Tsekouras. Product innovation and the decision to invest in fixed capital assets: Evidence from an SME survey in six European Union member states. *Research Policy*, **37**(10), pp. 1778-1789. Available at: https://doi.org/10.1016/j.respol.2008.08.013 (2008)
2. Yu. I. Efimychev , Yu. O. Plekhova, A.V.Volkov. Analysis of sources and volumes of investment in domestic innovative enterprises. No.: 1-1, 2012, P.: 239-243Bulletin of the National Research Lobachevsky State University of Nizhni Novgorod. №:1-1, P.: 239-243 (2012)
3. S. A. Eroshevsky, A.V.Streltsov, K. A. Kalugina, T. S. Muravyova. Depreciation as a source of investment in the development of the production potential of an oil and gas producing enterprise. Problems of improving the organization of production and management of industrial enterprises: Intercollegiate collection of scientific papers.. Publishing house: Samara State University of Economics (Samara) №: 1-1, P.: 120-124.(2010)
4. Yu. V. Fedorov, O. A. Druzhkova. Analysis of investment in fixed assets and sources of their financing for machine-building enterprises. Socio-economic management:

theory and practice.. Publishing house: Izhevsk State Technical University University (Izhevsk) №:1(28), P. 4-12. (2016)

5. I. V. Gurieva Justification of the procedure for selecting sources of financing for investment in technical re-equipment of the enterprise Business inform. Publishing house: Research center of industrial development problems of the National Academy of Sciences of Ukraine (Kharkiv) № 3, P. 271-275 (2014)
6. Investment in Russia. 2019: Stat. sb. Rosstat. - M., 228 p. (2019)
7. Jodi L. Gissel. A case of fixed asset accounting: Initial and subsequent measurement. *Journal of Accounting Education* **37**, pp. 61-66. Available at: https://doi.org/10.1016/j.jaccedu.2016.10.001. (2016)
8. E. Sh. Kurmakaeva Additional sources of investment in the system of reproduction of fixed capital of enterprises. *Bulletin of the State Technical University of Saratov*. **3**(14), P. 176-180 (2006)
9. Yu. S. Miklyaeva Depreciation deductions as a source of investment in fixed assets of oil refining enterprisesScience of the XXI century: current directions of development. Publishing house of Samara State University of Economics (Samara) № 1-2, P. 227-230 (2017)
10. D. A. Nikolayev Sources of investment in enterprises of the military-industrial complex in modern economic conditions. Socio-economic and technical problems of the military-industrial complex: history, reality, innovations.. Interuniversity Collected papers on materials of V All-Russian scientific practical conference. 2018. Publishing house: Nizhny Novgorod State Technical University (Nizhny Novgorod) P.: 107-110 (2018)
11. E.M. Pimenova, A.V. Streltsov, G.I. Yakovlev. Development of International Production Cooperative Relations in the Digital Economy // Lecture Notes in Networks and Systems. – 84. - pp. 466-472. DOI: 10.1007/978-3-030-27015-5_56. URL: https://www.scopus.com/inward/record.uri?eid=2-s2.0-85070224756&doi=10.1007%2f978-3-030-27015-5_56&partnerID=40&md5=2764581e4c8f601bd1fef916eda9b52c (2020)
12. Yu. Rubtsova A.Capital issues are a source of investment financing for enterprises. *Topical issues of economic sciences*, (2) P. 431-435. (2008)
13. L. N. Sotnikova, L. I. Kukush. Comparison of prices of various financing sources of real investment in the enterprise. *Financial Bulletin*. Publishing house: Voronezh State Agrarian University named after Emperor Peter I (Voronezh). **4**(35), P. 16-23. (2016)
14. V. Ghosal, U. Nair-Reichert. Investments in modernization, innovation and gains in productivity: Evidence from firms in the global paper industry. *Research Policy*. **38**(3), P. 536-547. Available at: https://doi.org/10.1016/j.respol.2008.10.010. (2009)
15. Y. Hu. Energy conservation assessment of fixed-asset investment projects: An attempt to improve energy efficiency in China. *Energy Policy*, **43**, pp. 327-334. Available at: https://doi.org/10.1016/j.enpol.2012.01.009. (2012)

Project-based Learning Approach to Marketing Competencies Development

Radim Dušek[1,*]

[1]VŠTE, Faculty of Corporate Strategy, Department of Tourism and Marketing, České Budějovice, Czech Republic

Abstract. Due to the requirements for matching the needs of employers with the job applicants' skills, companies increasingly tend to use a key competency-based approach to select new marketers. The aim of this paper is to determine if the project-based learning method has helped developing economic university students' key competencies that are essential in the marketing related jobs. For testing and developing such key competencies, the method of project-based learning have been implemented into the Marketing communication course at the Faculty of Economics, University of South Bohemia in Czech Republic. Results show that innovated course had a positive effect especially on the development of abilities to work effectively in team, assessing and analysing information or work independently.
Keywords: project-based learning; marketing competencies; key competencies; development; pedagogy

1 Introduction and Literature Review

Marketing is everything, but it can be interpreted by various authors or experts with hundreds of different definitions. Concurrently this concept can be understood by entrepreneurs differently and inconsistently. For a small business, it is usually a paid advertisement on social networks or a new catalogue [1]. In the case of global corporations, this is already an almost scientific approach to how to tie the customer to their products, and marketing costs are among the highest.

With the growing importance of marketing, the demand of enterprises for more or less experienced employees who will prepare and implement their marketing strategies is also increasing, as well as the rising number of job applicants. Due to the requirements for matching the needs of employers with the individual job applicant's skills, HR managers increasingly tend to use a key competency-based approach to select the best candidates [2, 3].

Competencies can be defined as a generic complex of knowledge, values and attitudes, but primarily life skills such as problem solving, communication, collecting and analysing information, ability to work independently, leadership, creativity, ability to cooperate effectively and lifelong learning [4,5]. These competencies allow individuals to act

[*]Corresponding author: dusek@mail.vstecb.cz

adequately and efficiently within the various professional and personal life situations. They represent and incorporate a wide range of long term qualifications and requirements for a flexible response to the changing labour market demands and they also ensure maximum performance in the defined tasks [6].

In the developed countries, efforts have been put by universities for business or economic degree to prepare graduates who want to be marketers to meet actual labour market requirements. These requirements are no longer based only on theoretical knowledge, but especially on higher level of abilities which provides a framework to the behaviour and skills that are expected of competent marketers in various types of marketing specializations [7, 8]. Such competency-based approach consists of the desired framework of the individuals' key competencies for specific positions by means of a competency model as a basis for the selection of suitable applicants for a given position within an organization, for example in marketing department [9-11].

To be competent and capable in the marketing related jobs, authors refer mainly to this set of key marketing competencies [12-16]:

a) Creative thinking;
b) Communication;
c) Working independently;
d) Effective team collaboration;
e) Leadership;
f) Negotiation;
g) Assessing and analysing information;
h) Ability to manage a budget;
i) Public speaking;
j) Ability to work under pressure.

Most of these competencies are behavioural, so if students have certain basic individual capabilities, their competencies can be developed throughout higher education [17], so obvious next step in the competencies development is to find and implement suitable pedagogy methods enabling students to develop mentioned competencies. There are plenty of methods that have been proven to help in that, for example case studies [18], project-based learning methods [19, 20], experimental and problem-based learning [21] or business simulation games [22].

Project-based learning method is often considered as the one of the most suitable pedagogy for students' key competencies development. This approach can be characterized by creating the students' research teams to work on long-term projects (usually one or two semesters), which are connected to real world business practice and designed to solve the actual business tasks and problems [23, 24]. Advantages of the project-based learning method can be described as follows [25]:

1. Integrating wide range of skills and knowledge.
2. Connecting the students' skills and knowledge with the experience that will help to increase effectiveness of the learning process.
3. Activating students for lifelong learning.
4. Developing students' ability to communicate and cooperate in a team.
5. Developing students' organizational and time management skills.

2 Methodology

The aim of this paper is to determine if the project-based learning method has helped developing economic university students' key competencies that are essential in the marketing related jobs. For testing and developing chosen key marketing competencies, the method of project-based learning have been implemented into the Marketing communication course at the Faculty of Economics, University of South Bohemia in Czech Republic. As this course is in the final year of the study, it effectively serves to synthesing knowledge and skills from various marketing related courses from the five-year study in the field of trade and marketing. In lectures throughout the semester, students at first obtained enough relevant and actual theoretical background about marketing communication mix with examples of new trends effectively used by marketing experts. Then, on the first course seminar the teams (2-3 students per team) were created and each team had to choose one specific project. The concrete projects were specified by the global company that runs automotive factory in South Bohemia and its marketing managers were interested to cooperate with the faculty on this new project. At the remaining seminars, each team had to work on the project independently (but under the supervision of the course teacher - author of this paper), create a final report with the relevant recommendations and then introduce the project results to the company managers at the end of the semester.

The topics specified by the company were focused on:

a) marketing communication mix for new products,
b) HR marketing communication,
c) marketing communication strategy for company's events.

To find out effects of the project-based learning method on Marketing communication course's students, the research is aimed at determining whether and how this experience influenced the specific competencies development. So the following research question was determined:
Which marketing competencies, that are often required by marketing related jobs, have been developed in students by the project-based learning method experience?

Based on the course structure and project topics, these six marketing competencies were chosen and analysed based on the participants' self-evaluation data:

1. Assessing and analysing information
2. Communication
3. Creative thinking
4. Effective team collaboration
5. Leadership
6. Working independently

Data from all participants (76 students/28 teams) of Marketing communication course have been collected during both summer semesters 2018 and 2019 with the questionnaire survey right after they finished the course (quantitative, 25 questions, mainly five-point Likert scale, 76 responses).

Excel and R software have been used for the data analysis, namely nonparametric single sample Wilcoxon one-sided test for testing the hypotheses.

3 Results

The assumption was that the project-based learning method experience improved six marketing competencies mentioned above. After finishing the course, students self-evaluated the Marketing communication course's contribution on developing their competency framework on the five-point Likert scale *(where 5 = strongly agree; 4 = agree; 3 = undecided; 2 = disagree; 1 = strongly disagree)*.

Means are shown in Fig. 1. If mean is greater than 3, the course has helped to develop the particular competency. Based on the the students' self-evaluation, it can be argued that the Marketing communication course had a positive effect especially on the development of abilities to work effectively in team and assessing and analysing information. These results correspond to the project tasks and necessary activities, because at first students had to often communicate and meet company managers to obtain enough relevant and comprehensive information about the given problem, analyse it and then start to work on concrete creative solutions or ideas.

To a lesser extent, it helped to develop participants' creative thinking and communication skills. The suggestions and solutions of some teams were very creative, however, these interesting initial ideas were mostly suggested by few more creative participants, and then further developed by other team members. It was clear that in the case of some students, it was not possible to develop their creativity too much with this method, because they are just less competent in this area.

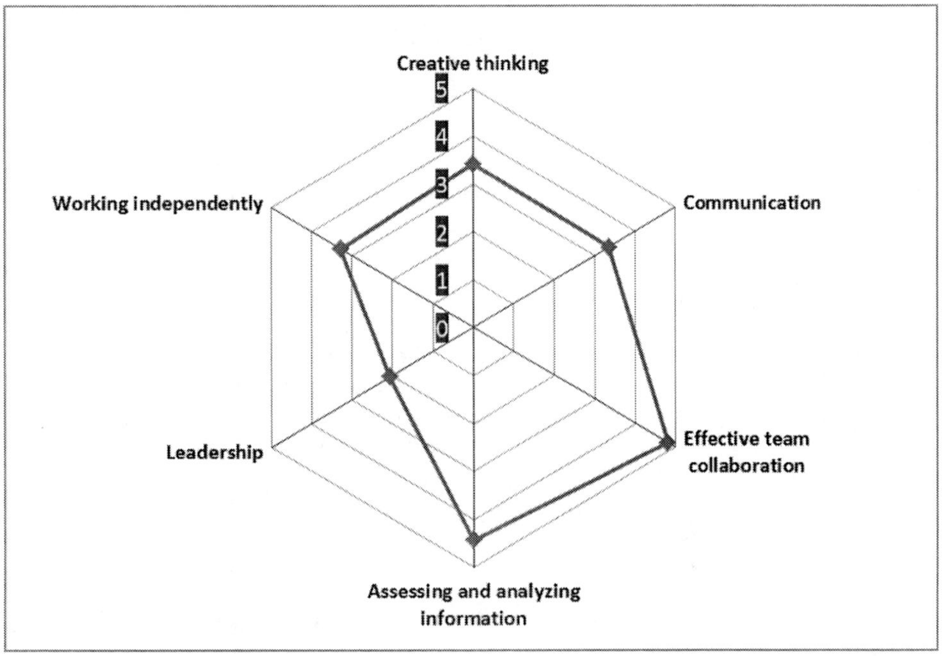

Fig. 1. Level of competencies' development (means, n = 76)
Source: Own processing.

Nonparametric single sample one-sided Wilcoxon test was used to test null hypotheses for all 6 analysed competencies as shown in Tab. 1.

H_0: Specific competency mean including 95% confidence interval is statistically significantly greater than 3.

H_A: non H_0

Table 1. Results of Wilcoxon test and accepted hypotheses at the 95 % confidence level

Marketing competencies	p-value	Accepted hypothesis
assessing and analysing information	1.753^{-07}	$\mathbf{H_0}$
communication	0.0002	$\mathbf{H_0}$
creative thinking	0.0004	$\mathbf{H_0}$
effective team collaboration	6.329^{-06}	$\mathbf{H_0}$
leadership	0.419	H_A
working independently	0.1736	H_A

Source: Own processing.

At the 95 % confidence level, the null hypothesis was accepted for these 4 of analysed 6 competencies: assessing and analysing information; communication; creative thinking; effective team collaboration. Mentioned competencies' means including 95% confidence interval are statistically significantly greater than 3, so it can be concluded that project-based learning method experience, according to students' self-evaluation, developed at least four marketing competencies in a certain way.

After only two semesters of Marketing communication course with the use of this alternative pedagogy, obtained findings are important in terms of the possible updating traditional approach to the education process and apply project-based learning methods into the other marketing courses such as marketing research, public relations or strategic marketing. These courses can be actualized to contribute more in terms of developing particular competencies that are demanded by labor market. At the same time, working on real, practical and actual projects helps to develop participants' ability to apply their theoretical background from other completed courses, for example trade, management or economy.

4 Conclusion

The aim of this paper is to determine if the project-based learning method has helped developing economic university students' key competencies that are essential in the marketing related jobs. For testing and developing chosen key marketing competencies, the method of project-based learning have been implemented into the Marketing communication course at the Faculty of Economics, University of South Bohemia.

To answer the research question, data from 76 participants of the course during 2018 and 2019 have been collected with the survey after they finished the course. Based on the course structure and project topics, six marketing competencies were chosen and analysed based on the participants' self-evaluation data - assessing and analysing information; communication; creative thinking; effective team collaboration; leadership; working independently.

Results show that innovated course had a positive effect especially on the development of abilities to work effectively in team and assessing and analysing information. These findings are important from the view of the possible updating traditional approach to the education process and apply project-based learning methods into the other marketing courses such as marketing research, public relations or strategic marketing. These courses can be actualized to contribute more in terms of developing particular key competencies demanded by the actual labour market situation.

References

1. P. Pillai, *Everything is Marketing and Marketing is Everything*. The Marketing Review (2012)
2. D. Liebenow, S. Uedelhoven, M. Stricker, J. Nachwei, *Quality in Competency Management: An Explorative Study to Develop a Benchmark for Competency Modeling and Competency Models.* Zeitschrift für Arbeits und Organisationspsychologie, **64**, 2, 120-148 (2020)
3. M. R. Weber, J. Lee, A. Crawford, *A suggested best practices for enhancing performance of soft skills with entry-level hospitality managers.* Anatolia-international journal of tourism and hospitality research, **31**, 1, 76-87 (2019)
4. M. Armstrong, *Armstong's Handbook of Human Resource Management Practice*. Philadelphia: Kogan Page (2011)
5. A. Wiek, L. Withycombe, C. L. Redman, *Key competencies in sustainability: a reference framework for academic program development.* Sustainability Science, **6**, 2, 203-218 (2011)
6. T. Weiss, S. Kolberg, *Coaching Competencies and Corporate Leadership*. Florida: CRC Press (2003)
7. V. Vojtko, R. Dušek, *Filling the gap between theory and practice by business simulation games.* Conference Proceedings of the 20th Annual International Conference Enterprise and Competitive Environment, March 9–10, 2017, Mendel University in Brno, 908-918 (2017)
8. M. Hay, *Business schools: a new sense of purpose.* Journal of Management Development, **27**, 4, 371–378 (2008)
9. D. D. Dubois, W. J. Rothwell, Competency-Based Human Resource Management. California: Davies-Black Publishing (2004)
10. C. J. Gonsalvez, F. L Calvert, Competency-based Models of Supervision: Principles and Applications, Promises and Challenges. Australian Psychologist. 49, 4, 200–208 (2014)
11. M. A. Campion, A. A. Fink, et al, Doing competencies well: best practices in competency modeling. Personnel Psychology. 64, 1, 225-262 (2011)
12. J. Hadrich, *New Key Competencies for Authorities, Companies and their Managers Part 2: Obstacles and Success Factors of modern Leadership.* Deutsche Lebensmittel-Rundschau, **116**, 5, 186-196 (2020)
13. J. M. Loveland, S. A. Thompson, J. W. Lounsbury, D. Dantas, *Is diffusion of marketing competence necessary for a market orientation? A comparative investigation of marketing managers and their defining traits*. Marketing Intelligence & Planning, **33**, 3, 469-484 (2015)
14. DQ. Xin, X. Zhao, *The Construction of College Students Self-marketing Competency Model.* Recent advance in statistics application and related areas, vols I and II – Proceedings from the 2nd Conference of the International Institute of Applied Statistics Studies, Qingdao, China (2009)
15. M. Rieckmann, *Future-oriented higher education: Which key competencies should be fostered through university teaching and learning?* Futures, **44**, 2, 127-135 (2012)
16. S. Nwankwo, A. Gbadamos. *Entrepreneurship Marketing: Principles and Practice of SME Marketing.* Oxon: Rouletge (2020)

17. M. Hay, Business schools: a new sense of purpose. Journal of Management Development, 27, 4, 371–378 (2008)
18. E. Anderson, B. Schiano, *Teaching with cases: a practical guide.* Harvard Business School Publ, Boston, Mass (2014)
19. W.N. Bender, *Project-based learning: differentiating instruction for the 21st centrury.* Thousand Oaks: Corwin (2012)
20. R. Maltese, *Project Based Learning: 25 Projects for 21st Century Learning.* Indianapolis: Dog Ear Publishing (2012)
21. S.R. Tiwari, L. Nafees, O. Krishnan, *Simulation as a pedagogical tool: Measurement of impact on perceived effective learning.* The International Journal of Management Education. **12**, 3, 260–270 (2014)
22. M. Pratt, S. Hahn, *Effects of Simulation on Student Satisfaction With a Capstone Course.* Journal of Hospitality & Tourism Education, **27**, 1, 39-46 (2015)
23. M. Almulla, *The Effectiveness of the Project-Based Learning (PBL) Approach as a Way to Engage Students in Learning.* Sage Open, **10**, 3 (2020)
24. J. Atkinson, *Developing teams through project-based learning.* Hampshire: Gower Publishing (2001)
25. R. Dušek, *Developing Student's Key Competencies through Cooperation with Enterprises.* SHS Web of Conferences - Eurasia: Sustainable Development, Security, Cooperation. **71** (2019)

Fuzzy Estimate of the Development of Passenger Cars Production in the Czech Republic

Simona Hašková[1,*]

[1]Institute of Technology and Business, School of Expertness and Valuation, Ceské Budejovice, Czech Republic

Abstract. The global outbreak of the COVID-19 and the measures taken, disrupted fundamentally economies around the world. Almost all sectors were affected. The experts have long emphasised the Czech economy's dependence on the automotive industry. Car producers and companies linked to them have been loaded by severe difficulties after the pandemic outbreak. The article shows one of the constructive ways how to forecast a change in the passenger cars production in the Czech Republic in 2020. Metodologically we lean on a procedure of the fuzzy approach. The prediction itself cannot be derived from the series of historical data of the variables that are related to the target output variable as shown in the fuzzy prediction of GDP for 2018 by this author. Due to the extreme situation caused by pandemic outbreak, the role of expert predictions come intensively into play with their outcomes becoming the set of input data to the fuzzy model. The result of the fuzzy forcast of a change in the cars production in CZ for 2020 shows a greater drop than the official statistical model claims.

Keywords: passenger car production, automotive industry, fuzzy approach, fuzzy prediction

1 Introduction

The global COVID-19 pandemic outbreak and, as a result, the containment measures taken, have profoundly disrupted people's lives and the economies. The private and public demand, industrial production, supply chains, labour market, commodity prices, capital flows and foreign trade have been hit. Given the deepness of these shock changes, it is obvious that the European Union was affected by severe economic recession. The COVID-19 outbreak trigged off a symmetric shock and due to tight economic connections of EU states the aggregate demand and supply shocks have intensified.

A current short time period forecast for 2021shows incomplete recovery in most of countries [1]. It is clear that crisis impact and the way the individual EU states get over is

*Corresponding author: haskovas@post.cz

going to be very uneven. How well and quickly the countries will pull together relies upon many external and internal factors [2]. There is not only the seriousness of the pandemic spread and the strictness of containment measures in play, but a great role belongs to particular economic exposures and pre-existing conditions as well as the policy responses.

The economic impacts of the current crises have been viewed from two basic levels [3] - the level of duration and the level of demand/supply side. For instance, the part time jobs and unemployment rise is expected in the short term occurring mainly on the demand side. Hysteresis occurrence in labour market has a potential of both supply and demand side and persisting in the medium term. Emerging market economy drop as a result of expected difficulties to rise external capital is rather perceived as a short-term effect affecting both the demand and supply side. Reorganization of cross border supply chain is expected from the long term affecting demand and supply side, etc.

Economists and politicians have often discussed the strong dependence of the Czech economy on the automotive industry and strong export focus on Germany. This dependence has been manifested most drastically during economic crises. The coronavirus outspread brought a significant decline in the production and car sales, automotive components and technologies. For example, Škoda Auto factory that is understood as the heart of the Czech industry stopped all lines. This factory creates on average for about five percent of GDP and accounts approximately 9% of exports. Some experts suggest to break out of the vulnerable car industry [4] and give more support to other domestic production and entrepreneurs. However, according to economists, such a scenario is not feasible; it is improbable that the Czech industry will be more radically reoriented in the future [5].

The aim of the paper is focused on a short term fuzzy prediction in the production of passenger cars in the Czech Republic. Due to external shock of the pandemic outbreak it is not possible to rely on a relevant sequence of historical data and the interventions of experts based on which the fuzzy prediction can be performed (see [6]). Therefore, we build the fuzzy prediction on the data based on expert scenarios of possible development of relevant variables. This procedure fully takes into account an extreme condition in which the economies currently appear.

2 Methodology: The Introduction to Fuzzy Approach

In the tasks of expert analysis or expert control of vaguely defined systems, the vague terms are usually presented by linguistic variables and their linguistic values (the terms), where the word „linguistic" is used in the sense of „ *communicated in natural language* ".

The relationship between combinations of terms of the input linguistic variables and the output linguistic variable are given by a set of rules. The tool to mathematically solve these kind of tasks is the fuzzy logic on which the fuzzy approach is based (more in [7]).

The fuzzy approach purposefully modifies the assumptions and therefore the conclusions of the „*principle of extension*" of fuzzy logic. This approach phases are visually demonstrated in Fig. 1 and concisely summarized as shown in Hašková [6]. and:

- The first phase is the fuzzification of the numerical vector $x = (x_1, \dots, x_n)$ entering the relation K, in which the vector x selects inferential rules relevant to its multi-criteria evaluation. These are the rules $((A_{1i}, \dots, A_{nj}), B_k) \in P$, in which terms′ formalizations $A_{1i}, \dots, A_{nj}$ of the left side are all nonzero numbers $\mu_{\underline{A}1i}(x_1), \dots, \mu_{\underline{A}nj}(x_n)$.
- Moving from the symbolic notation $((A_{1i}, \dots, A_{nj}), B_k)$ of the selected inference rule to its logic notation dependent on x in the form $(\min\{\mu_{\underline{A}1i}(x_1), \dots, \mu_{\underline{A}nj}(x_n)\}, \mu_{\underline{B}k}) = (\mu_{\underline{A}^*}(x), \mu_{\underline{B}k})$ we obtain a number $\mu_{\underline{A}^*}(x)$ indicating not only the degree of truthfulness that the situation characterized by the vector x actually occurred, and

therefore the degree of eligibility of the rule selection $((A_{1i},\ldots,A_{nj}), B_k)$ to its solution, but also the degree of significance of the term B_k in the multi-criteria evaluation of the vector x. $\mu_{\underline{B}k}$: $Y \rightarrow \langle 0,1\rangle$ is the standard membership function of the elements $y \in Y$ to the fuzzy set $\underline{B}_k$. On the other hand, $\mu_{\underline{B}k}^*: Y \rightarrow \langle 0,1\rangle$ in the form $\mu_{\underline{B}k}^*(y) = \min\{\mu_{\underline{A}^*}(x), \mu_{\underline{B}k}(y)\}$ limits from above the function $\mu_{\underline{B}k}$ by means of the number $\mu_{\underline{A}^*}(x)$.

- The components of the relation K will then be the elements of the set $Px = \{B_k: \exists y:(\min\{\mu_{\underline{A}^*}(x),\mu_{\underline{B}k}(y)\} > 0), ((A_{1i},\ldots,A_{nj}), B_k) \in P, y \in Y\}$. These are the partial intermediate results by which aggregation (logical sum) we obtain the formalized term $\underline{B}_{agg} = \{(y, \mu_{agg}(y)): y \in Y\}$ on Y, for which it applies (1)
$$\mu_{agg}(y) = max\left\{min\{\mu_{\underline{A}^*}(x), \mu_{\underline{B}k}(y)\}: \left((A_{1i},\ldots,A_{nj}), B_k\right) \in P, y \in Y\right\} \quad (1)$$
- The final phase of the fuzzy approach is defuzzification, in which last step we obtain the output value y_x of the multicriteria model of an alternative evaluation $(a_1,\ldots,a_c) \in A$ as the mean value of the elements $y \in \underline{B}_{agg}$ weighted by the values $\mu_{agg}(y)$ of their significance – see (2). Therefore
$$y_x = \int y \cdot \mu_{agg}(y)dy / \int \mu_{agg}(y)dy \quad (2)$$
where $\int$ is the symbol of a definite integral over the universe Y – more in detail in [8] and [9].

The process of the multi-criteria assessment of a short time prediction of the change in passenger cars production in the Czech Republic is illustrated in a solution to the specific task.

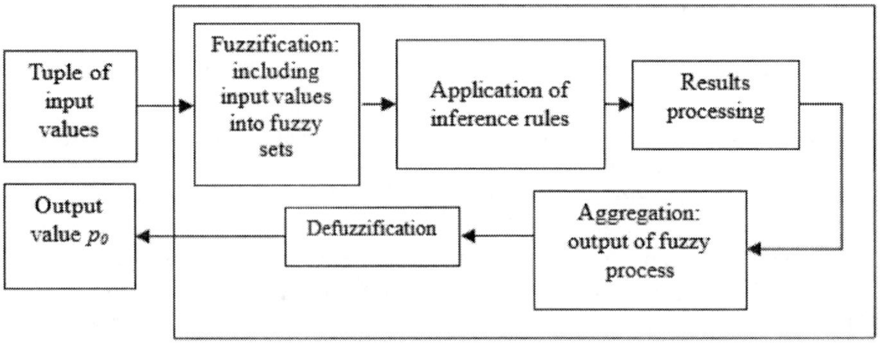

Fig. 1. Schematic representation of the multicriteria problem solving process within the fuzzy approach
Source: Own processing.

3 Application - Assumptions and Entry Data of the Fuzzy Predictive Model

The formulation of the fuzzy model of prediction of the next member of the time series has its specifics; we know a number of previous members of the resulting series of the target variable – these are historically measured data and data statistically predicted. Its

 https://doi.org/10.1051/shsconf/20219101005

consecutive values can be used in the prediction only if the economy has not appeared in a period of an extreme state, which deviates from the normal (expected) framework of its functioning.

In the case of a standard economic development, it would therefore be possible to estimate the current stage of development (decline, growth, stagnation), within which the prediction will fall when predicting the targeted variable – here, a year-on-year change in passenger cars production in the Czech Republic ($\Delta PS_{CZ}\%$). This estimate would be also supported by the development of the historical part of input values of basic linguistic variables, on which $PS_{CZ}\%$ depends (albeit to a large extent vaguely). These are then reflected mainly in the values of the extreme limits (intervals), in which the result of the prediction would be searched - see, for example, the work [10], which shows a short-term prediction of the development of the GDP growth rate.

Next Tab. 1 shows the basal values of the input linguistic variables $\Delta GD_{PG}\%$ (a year-on-year change in Germany's gross domestic product), $\Delta AP_W\%$ (a year-on-year change in the worldwide cars production), $\Delta AP_G\%$ (a year-on-year change in the cars production in Germany) and the output linguistic variable $\Delta PS_{CZ}\%$ (a year-on-year change in the production of passenger cars in CZ) between the years 2009 - 2019 and their econometric forecast for the year 2020.

Table 1. Basic macroeconomic data for fuzzy predicting models of the annual change in the production of passenger cars in CZ

ROK	**2009**	**2010**	**2011**	**2012**	**2013**	**2014**	**2015**	**2016**	**2017**	**2018**	**2019**	**2020**
$\Delta GDP_G\%$	-5,69	4,19	3,91	0,43	0,43	2,22	1,74	2,23	2,47	1,52	0,57	*-6,95*
$\Delta AP_W\%$	-12,7	24,2	3,4	5	3,4	3,4	1	4,4	2,1	0	-5,2	-22
$\Delta AP_G\%$	-9	12	5,4	-8,5	0	3,7	1,8	0	-1,8	-8,9	-7,8	-17,2
$\Delta PS_{CZ}\%$	**4,1**	**10,3**	**11,6**	**-1,6**	**-3,9**	**10,5**	**4,2**	**8**	**5,2**	**1,7**	**-0,68**	**-30**

Source: [11], [12], [13] / own processing.

From the course of $\Delta PS_{CZ}\%$ it can be concluded that in the period 2016 - 2019 the production of passenger cars in CZ was in a phase of decline, which, due to the „extreme“ – the COVID-19 pandemic, will deepen significantly in the following period. Therefore, the characteristics of this historical data in Tab. 1 cannot be used for the formulation of the fuzzy predictive model for 2020.

As already mentioned, the outbreak of the COVID-19 pandemic has led to an extreme exogenous economic shock, and the absence of a historically relevant precedent makes it impossible to capture the prediction using standard econometric models and fuzzy procedure as shown in [6]. As a result, any predictions are based on expert scenarios simulating possible shocks at a given set of input assumptions. It follows that the opinion of a knowledgeable expert becomes a major specificity of the fuzzy predictive model formulation. The experts take knowledge and available information into account through qualified scenario constructions, interventions in the model structure and the inference rule formulation.

It applies that the predictive model works with the *dimensionless* basal values of the output universe Y and the input universes U_{GDP_G}, U_{AP_W}, U_{AP_G} located within the interval $\langle 0, 100 \rangle$ and obtained by *converting the entry values* of the respective linguistic variables. The conversion of the inserted basal value x of the universe Y of the *output* linguistic variable to the dimensionless value $y \in Y$ is given by the formula $y = 100 \cdot (x - x_{min}) / (x_{max} - x_{min})$, where x_{max}, or respectively, x_{min} is the highest, or respectively, the lowest a priori (subjectively) expected result value. For the conversion in the opposite direction, $x = (y \cdot (x_{max} - x_{min}) / 100) + x_{min}$ applies.

If the predominant element strategy is chosen in formulating the system of inferential rules with three inputs, then for the conversion of the value x of the universe of the *input* linguistic variable having a positive or featureless influence on the output linguistic variable it applies that $u = 100 \cdot (x - x_{min}) \,/\, (x_{max} - x_{min})$. In the case of a negative influence on the output linguistic variable, $u = 100 - \left(100 \cdot (x - x_{min}) \,/\, (x_{max} - x_{min})\right)$ holds.

In our case, the output linguistic variable is a year-on-year change in the production of passenger cars in CZ in the year immediately following the historical period. The percentage increase in gross domestic product in Germany (GDP) indicates a *positive effect* – an increase in the growth of production of personal cars in CZ; a decrease of automotive production worldwide and in Germany ($AP_W\%$, $AP_G\%$) tend to decelerate the car production in CZ (a *negative impact*).

4 Results

In order to estimate the targeted $\Delta PS_{CZ}\%$, the expert knowledge published in (Europian Economic Forecast from spring 2020 [14]) is used with three global scenarios for 2020 being considered. Scenario i. assumes a six-week "lockdown" with a global GDP decline of 8%. A longer-term "lockdown" scenario ii. means a 15.5% decrease in GDP and scenario iii. works with the second wave of the pandemic, which would require additional measures and restrictions in autumn and winter 2020 – this assumes a 10.5% decline in GDP (the forecast in Tab. 1 for 2020 represents the scenario iii). The scenarios include assumptions such as duration of restrictive measures, sector demand, uncertainty, tourism activity, preventive savings, public distancing, automatic stabilizers, discretionary fiscal policy and liquidity support.

The formulation of the predictive fuzzy model for the year 2020 is based on the characteristics of these scenarios. Scenario i. reaches almost 76% of scenario iii (the baseline); scenario ii. accounts for approximately 148% of scenario iii. This delimitation allows the formulation of the limits in which the resulting forecast $\Delta PS_{CZ}\%$ for 2020 will apear: the result of the prediction for 2020 will be searched within the limits of the given values $x_{max} = -44$, $x_{min} = -23$, i.e., the maximum decrease – the baseline prediction in 2020 x 1.48, the minimal decrease – the baseline prediction in 2020 x 0.76.

The conversion of the $\Delta PS_{CZ}\%$ x value given by scenario iii. $x_{20} \approx 33$ $|(100 \cdot (-30 - (-23)) \,/\, (-44 - (-23))| \approx 33)$. Analogously, for the recalculated vector input values (x_{GDP_G}, x_{AP_W}, x_{AP_G}) for the year 2020 we get an interval according to the formula (MAX [maximum decrease – baseline prediction 2020 x 1.48]; MIN [minimum decrease – baseline prediction 2020 x 0.76]). The conversion of basic values in 2020 of the given inputs is driven as follows: $u = 100 \cdot (x - x_{min}) \,/\, (x_{max} - x_{min})$ in the case of a positive influence on the output linguistic variable and $u = 100 - \left(100 \cdot (x - x_{min}) \,/\, (x_{max} - x_{min})\right)$ in the case of a negative influence on the linguistic output.

- $x_{GDP\_G} = 100 \cdot |(-6{,}95 - (-5{,}3)) \,/\, (-10{,}3 - (-5{,}3))| = 33$;
- $x_{AP\_W} = 100 - |(100 \cdot (-22 - (-16{,}7)) \,/\, (-32{,}6 - (-29))| = 47$;
- $x_{AP\_G} = 100 - |(100 \cdot (-17 - (-13)) \,/\, (-33 - (-25))| = 50$.

Thus, the triple $(33; 47; 50) \in U = U_{GDP\_G} \times U_{AP\_W} \times U_{AP\_G} = \langle 0, 100\rangle^3$ of internal (converted) input values has been created (see point 1 in section 2).

A coarser variant of this model with one inner and two boundary fuzzy sets for the terms low (*L*), normal (*M*) and high (*H*) value well suits our purpose. The interval U is divided into five sections by points a, b, c, d with the courses of membership functions (3):

• $\mu_{\underline{L}}(x) = 1$ for $x < a$, $\mu_{\underline{L}}(x) = (b - x) / (b - a)$ for $a \le x < b$, $\mu_{\underline{L}}(x) = 0$ otherwise; • $\mu_{\underline{M}}(x) = (x - a) / (b - a)$ for $a \le x < b$, $\mu_{\underline{M}}(x) = 1$ for $b \le x < c$, $\mu_{\underline{M}}(x) = (d - x) / (d - c)$ for $c \le x < d$, $\mu_{\underline{M}}(x) = 0$ otherwise; • $\mu_{\underline{H}}(x) = 0$ for $x < c$, $\mu_{\underline{H}}(x) = (x - c) / (d - c)$ for $c \le x < d$, $\mu_{\underline{H}}(x) = 1$ otherwise.	(3)

Because of the high *external* uncertainty (e.g., the uncertain impact of the pandemic on economies) and *internal* uncertainty of the model we choose the points a, b, c, d evenly distributed within the interval $U \in \langle 0, 100 \rangle$, in which a = 20, b = 40, c = 60, d = 80. Therefore, the courses of the membership functions of the converted output and all the converted inputs are identical (Fig. 2, in which these courses are plotted above the domain of universe Y of converted values of the output linguistic variable):

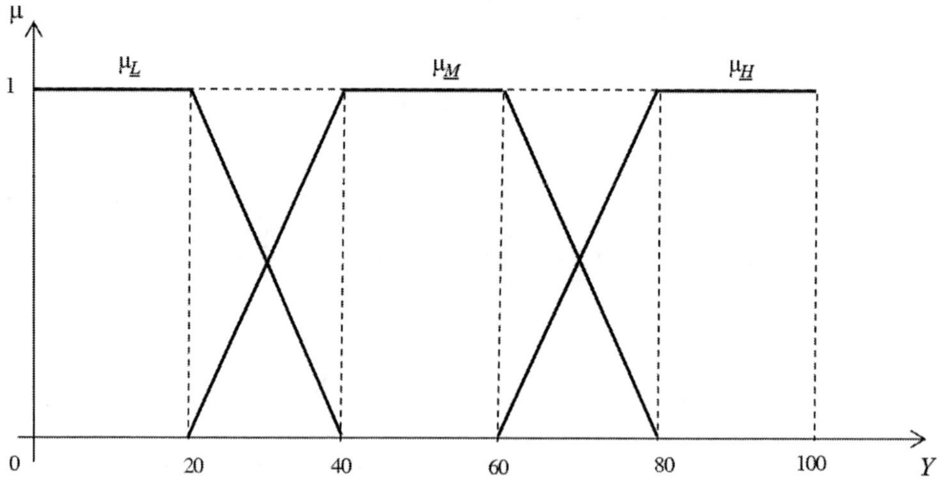

Fig. 2. Courses of the membership functions with even distribution of the points a, b, c, d within the interval values $y \in Y = \langle 0, 100 \rangle$
Source: Own processing.

The following *fuzzification table* Tab. 2 valid for i = $GDP_G\%$, $AP_W\%$, $AP_G\%$, whose elements are the values $\mu_{\underline{A}i}(u_i)$, where index $\underline{A} \in \{\underline{L}, \underline{M}, \underline{H}\}$, is derived from the above stated equations and inequalities given by (3):

Table 2. Fuzzification table valid for i = $GDP_G\%$, $AP_W\%$, AP_G, whose elements are values $\mu_{\underline{A}i}(x_i)$, where $\underline{A}_i \in \{\underline{L}_i, \underline{M}_i, \underline{H}_i\}$

Interval	$x_i < 20$	$20 \le x_i < 40$	$40 \le x_i < 60$	$60 \le x_i < 80$	$x_i \ge 80$
$\underline{L}_i$	1	$(40 - x_i) / 20$	0	0	0
$\underline{M}_i$	0	$(x_i - 20) / 20$	1	$(80 - x_i) / 20$	0
$\underline{H}_i$	0	0	0	$(x_i - 60) / 20$	1

Source: Own processing.

In the fuzzification table, only non-zero elements are taken into account and with its help the set $X = \{(\underline{L}_{GDP\_G}; 0.35), (\underline{M}_{GDP\_G}; 0.65), (\underline{M}_{AP\_W}; 1), (\underline{M}_{AP\_G}; 1)\}$ is created.*

* Comment on creating a set X using a fuzzification table:

- The value $x_{GDP\_G} = 33$ (for i = GDP_G%) falls within the interval of the second column; there are two non-zero elements in it - one $((40 - x_i) / 20)$ in the line $\underline{L}_{GDP\_G}$, gives after substituting (40 - 33) / 20 = 0.35, and therefore $(\underline{L}_{GDP\_G}; 0.35)$ - the other $((x_i - 20) / 20)$ in the line $\underline{M}_{GDP\_G}$, gives after substituting (33 - 20) / 20 = 0.65, and therefore $(\underline{M}_{GDP\_G}; 0.65)$.
- The value $x_{AP\_W} = 47$ (for i = AP_W%) falls within the interval of the third column; there is one non-zero element (1) in it in the $\underline{M}_{AP\_W}$ line, and therefore $(\underline{M}_{AP\_W}; 1)$.
- The value $x_{AP\_G} = 50$ (for i = AP_G%) falls again within the interval of the third column; there is one non-zero element (1) in it in the $\underline{M}_{AP\_G}$ line, and therefore $(\underline{M}_{AP\_G}; 1)$.

A two-element set $LF = \{(\underline{L}_{GDP\_G}, \underline{M}_{AP\_W}, \underline{M}_{AP\_G}), (\underline{M}_{GDP\_G}, \underline{M}_{AP\_W}, \underline{M}_{AP\_G})\}$ is formed from the set X. After applying the inference rules $F(\underline{L}_{GDP\_G}, \underline{M}_{AP\_W}, \underline{M}_{AP\_G}) = \underline{M}$ and $F(\underline{M}_{GDP\_G}, \underline{M}_{AP\_W}, \underline{M}_{AP\_G}) = \underline{M}$ to triplets from LF, we can convert the general expression $\min\{\min\{\mu_{\underline{A}1}(x_1),\ldots,\mu_{\underline{A}n}(x_n)\}, \mu_{F(\underline{A}1,\ldots,\underline{A}n)}\}$ to a concrete form $\min\{\min\{0.35; 1; 1\}, \mu_{\underline{M}}\} = \min\{0.35, \mu_{\underline{M}}\} = \mu_{\underline{M}}^*$: $Y_{\underline{M}} \to \langle 0, 0.35\rangle$ in the first case and $\min\{\min\{0.65; 1; 1\}, \mu_{\underline{M}}\} = \min\{0.65, \mu_{\underline{M}}\} = \mu_{\underline{M}}^*$: $Y_{\underline{M}} \to \langle 0, 0.65\rangle$ in the second case. For the membership function $\mu_{\underline{AGG}}$: $\langle 33 - 100\rangle \to \langle 0, 1\rangle$ to the fuzzy set $\underline{AGG}$ we get: $\mu_{\underline{AGG}} = \max\{(\mu_{\underline{M}}^*; 0.35), (\mu_{\underline{M}}^*; 0.65)\}$, where $\mu_{\underline{M}}^*$: $\langle 33 - 80\rangle \to \langle 0; 0.65\rangle$, i.e. $\mu_{agg}(y) = 0$ for $y < 33$, $\mu_{\underline{AGG}}(y) = 0.65$ for $33 \le y \le 67$, $\mu_{\underline{AGG}}(y) = (80 - x) / 20$ for $67 \le y \le 80$, $\mu_{\underline{AGG}}(y) = 0$ for $y > 80$ (see the strongly drawn contour in Fig. 3).

By an approximate numerical approximation of the values of the integrals in the formula for calculating the horizontal coordinate of the position of the center of gravity, see part 2 relation (2), the value p_0 representing the estimate of the value PS_{CZ}% for 2020 can be estimated as $p_0 = (0.65 \cdot (40 + 50 + 60 + 70) / (4 \cdot 0.65) \approx 52.5$ %. After recalculation x_0 to ΔPS_{CZ}%, we get the predicted value of $\Delta PS_{CZ}\% = (52.5 \cdot (-44 - (-23) / 100) -23) \approx -34$.

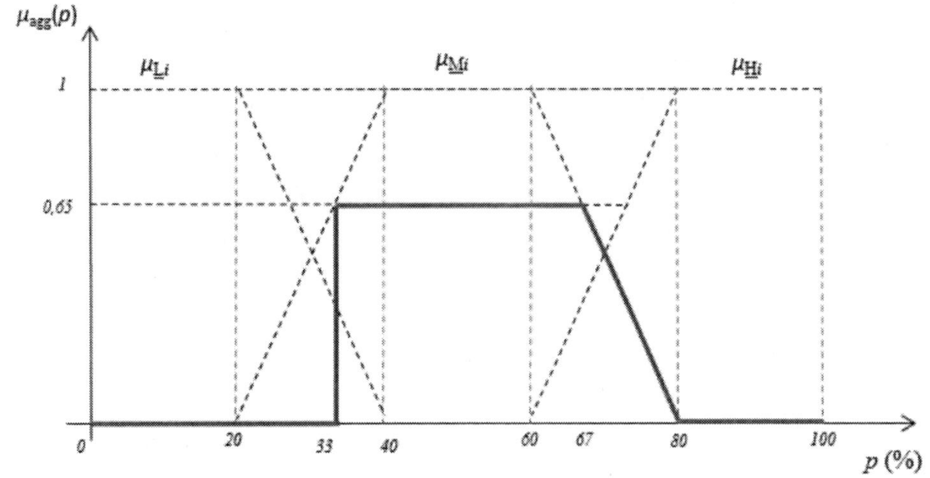

Fig. 3. The course of the output aggregate function $\mu_{\underline{AGG}}$ for PS_{CZ}%
Source: Own processing.

5 Discussion and Conclusive Remarks

The aim of the paper was to present a general fuzzy algorithm formulated within the Zadeh's fuzzy approach for solving a predictive economic problem under conditions of internal uncertainty of the model and external uncertainty relating to the entry data used. Such procedure offers an expected (predicted) objective result as an alternative to statistically expected values.

The fuzzy algorithm is presented and justified in view of the transformation of one of the important concepts of fuzzy logic, which is the extension principle, at the level of linguistic variables and their terms, within which the problem is solved. Furtheron, this algorithm is used to predict the change in the production of passanger cars in CZ for 2020 (the comparison with the official econometric forecast).

Under „normal", i.e., an expected behaviour of economy, the entry data are partial components of the historical development of the indicators influencing, although vagualy, the trend of a target variable (the influencing indicators here are a change in Germany's gross domestic product, a change in the worldwide cars production and a change in the cars production in Germany, the target variable is a change in the production of passenger cars in CZ). Due to the COVID-19 outbreak the procedure of the fuzzy prediction following the expected trends is excluded. Instead, the expert scenarios of the possible variables′ (indicators′) development are taken into account.

The data of the indicators are used in the fuzzy model formulation phase, within which an expert (through the qualified interventions in the structure of the model and the formulation of inference rules) gets the opportunity to take into account his/her knowledge and experience, as well as in the process of solving a problem that is purely mechanical and independent of the human factor and in which these indicators play the role of external inputs to the fuzzy algorithm.

It consists of five basic steps, which are *fuzzification*, application of *inference rules*, further *processing of intermediate results*, *aggregation* and *defuzzification*. The contribution to the fuzzification process is the formulation of a fuzzification table, which makes the solution much more transparent and easier. Regarding the inference rules in the function of tools for partial transformations of inputs to an output, only two of their total number of 27 were used: $((\underline{L}, \underline{M}, \underline{M}), \underline{M})$, $((\underline{M}, \underline{M}, \underline{M}), \underline{M})$. The right-hand sides of the two rules correspond with the predominant element strategy in formulating the system of inferential rules.

The result of the solution (the course of $\mu_{\underline{AGG}}$ function) at the level of terms of the linguistic variable $\Delta PS_{CZ}\%$ is shown in Fig. 3, in which the designation $\Delta PS_{CZ}\%$ is a notation abbreviation for the predicted change in the production of passenger cars in CZ expressed as a percentage of the highest drop $\Delta PS_{CZ}\%$ estimate derived from the official pessimistic GDP scenario iii. decrease wordwide. By recalculation of the distances of the centers of gravity of strongly delimited areas of the vertical coordinate axes to a value of the original scale, the prediction of $\Delta PS_{CZ}\%$ value for 2020 was obtained as -34%, which is slightly worse result than the 2020 econometric estimate that gives -30% prediction.

References

1. Eurostat *Newsrelease – euro indicators*.[on-line] https://ec.europa.eu/eurostat/documents/2995521/10294804/4-13052020-AP-EN.pdf/dfa765ad-4a32-8135-f98f-69e565347750 (2020)

2. Evgenidis, A., Papadamou, S., & Siriopoulos, C. The yield spread's ability to forecast economic activity: What have we learned after 30 years of studies?. *Journal of Business Research*, *106*, 221-232. (2020)
3. del Rio-Chanona, R. M., Mealy, P., Pichler, A., Lafond, F., & Farmer, D. Supply and demand shocks in the COVID-19 pandemic: An industry and occupation perspective. *arXiv preprint arXiv:2004.06759*. (2020)
4. Gasiorek, M., Serwicka, I., & Smith, A. Which manufacturing sectors are most vulnerable to Brexit. *UK trade policy observatory. Briefing paper*, *16*. (2018)
5. Srivastava, M., & Tyll, L. The effect of industry-specific networking behaviour on the internationalization performance of Czech SMEs. *European Business Review*. (2020)
6. Hašková, S. An alternative approach for estimating GDP growth rate: fuzzy prediction model. *ACC Journal*. (2019)
7. Zadeh, L. A. Fuzzy logic and approximate reasoning. *Synthese*, **30**(3-4), 407-428. (1975)
8. Grabisch, M. The application of fuzzy integrals in multicriteria decision making. *European journal of operational research*, **89**(3), 445-456. (1996)
9. Běhounek, L., & Cintula, P. From fuzzy logic to fuzzy mathematics: A methodological manifesto. *Fuzzy Sets and Systems*, **157**(5), 642-646. (2006)
10. Hašková, S., & Fiala, P. A fuzzy approach for the estimation of foreign investment risk based on values of rating indices. *Risk Management*, **21**(3), 183-199. (2019).
11. Statista Germany: Growth rate of the real gross domestic product (GDP) from 2009 to 2021. [on-line] https://www.statista.com/statistics/375203/gross-domestic-product-gdp-growth-rate-in-germany/(2020)
12. Statista Estimated worldwide automobile production from 2000 to 2019. [on-line] https://www.statista.com/statistics/262747/worldwide-automobile-production-since-2000/ (2020)
13. Statista Automobile production in Germany from 2007 to 2019. [on-line] https://www.statista.com/statistics/589254/car-production-germany/ (2020).
14. Europian Commission European Economic Forecast, Spring 2020. [on-line] https://ec.europa.eu/info/sites/info/files/economy-finance/ip125_en.pdf (2020).

Prediction of Financial Health of Business Entities of Selected Sector Using Balance Analysis II. by Rudolf Doucha and Verification of Its Predictive Ability through ROC

Dusan Karpac[1,*] and *Viera* Bartosova[1]

[1]University of Zilina, The Faculty of Operation and Economics of Transport and Communications, Department of Economics, Zilina, Slovakia

Abstract. Forecasting business failure is a worldwide known term, in a global notion, and there is a lot of prediction models constructed to compute financial health of a company and, by that, state whether a company inclines to financial boom or bankruptcy. A healthy financial management of a business entity is very important for the proper operation of the business, and it is therefore very important to know how to assess financial health and to anticipate possible problems that will be easier to eliminate in advance. Globalized prediction models compute financial health of companies, but the vast majority of models predicting business failure are constructed solely for the conditions of a particular country or even just for a specific sector of a national economy. Predictive models can indicate whether an entity tends to prosper or bankruptcy, and so we can assess the financial health of the business. This paper provides a description of the balance analysis II. by Rudolf Doucha, discusses its application to a sample of 266 Slovak subjects and points to its prediction in the given field. The verification of the ability to forecast bankruptcy or financial stability has been evaluated through ROC analysis.
Keywords: Specificity, Financial prediction, ROC curve, AUC

1 Introduction

The term "enterprise financial health forecasting" is widely known in the world, but the intensity of how different countries are addressing this issue is uneven. Business entities in our country forecast their financial stability only in a minimalist way, mainly because the conditions for the Slovak market have so far not designed a general model of prediction that is fully applicable to any entity, regardless of the sector in which it operates. In view of this, it is necessary at least to verify which models are suitable for use in a particular sector, and that fact was the main reason for us to write this study. Since the Slovak and Czech economies are very similar to each other, we have chosen to use the balance analysis II. by Rudolf Doucha.

[*]Corresponding author: dusan.karpac@fpedas.uniza.sk

Evaluation of efficiency is an integral part of rational behavior of units that aims to survive in a competitive environment in a long-term [1].

Every corporation has an economic and moral responsibility to its stockholders to perform well financially [2].

However, the number of bankruptcies in Slovakia has been growing for several years without an apparent macroeconomic cause. To prevent a rapid denigration and to prevent the outflow of foreign capital, various efforts are being zealously implemented. Robust analysis using conventional bankruptcy predictive tools has shown that existing models are adaptable to local conditions, particularly local legislation [3].

Financial hardship is the financial state of an enterprise when an enterprise has serious payment problems that cannot be resolved otherwise than by radically changing its operating or financial activity [4, 5].

Bankruptcy is a situation where an enterprise is unable to pay its debts and ultimately fails to meet its obligations. Bankruptcy is a condition that ends all business and economic activities, so it is also referred to as economic death [6].

Financial analysis represents a basic aspect of financial management of a business entity and points to the operation of the company. The state of financial stability of each entity is most reflected in the different forms of the profit indicator. Profit, within the result of management, represents one of the most important flow quantities. This explains how effectively corporate capital is used in the entity [7].

This study shows how the balance analysis II. indexes sample entities to bankrupt and prosperous businesses, and through the ROC analysis we verify the ability to properly redistribute these businesses.

2 Literature review

In order to ensure the best financial health of an enterprise, we can predict it, i.e. to predict how the funds and the total assets of an enterprise should be treated in the short or long term for their most efficient use. In order to identify the level of financial health of the company, prediction models, also called early warning systems, are used. It assesses the health of the company on the basis of complex characteristics, both in terms of past and current developments, as well as in terms of the future development of the entity [8].

Paul Joseph Fitzpatrick, in 1932, first published a study comparing the development of indicators in solvent and insolvent businesses. He pointed out that the development of selected business indicators - indicators long before the serious economic difficulties that usually result in insolvency - begin to differ in the endangered enterprises. Later, Merwin, Beaver and Altman dealt with this issue. These are considered to be the founders of forecasts and predictions of the financial health of business entities [9].

In a more recent study, the interdependence of sovereign default risk and banking fragility in Russia was investigated, using credit default swaps as a proxy for default risk.

Predictive models are constructed around the world by authors from different countries. In Slovakia, Chrastinova and Hurtosova are the most known authors [10, 11, 12]. Also in Slovakia very important author in the area of predictions is Gundova. In the neighboring Czech Republic, there are also several representatives of the area of financial crisis predictions. Absolutely best known are the spouses Inka and Ivan Neumaiers or Kislingerova. In the Czech Republic, there is also significant author, doing research and working in the prediction of bankruptcy area, Ruckova or as well Rudolf Doucha, whose model we used for evaluating in this study.

3 Methodology

ROC analysis was developed as a standard methodology to quantify the ability of the signal receiver to correctly distinguish objects of interest from the background noise in the system. In the area of financial risk forecasting, ROC analyzes the correct distribution of subjects and compares two categorization algorithms, one of which is considered to be a comparative element [13].

The base reallocation we used is based on the Commercial Code and evaluates financial health under the following conditions: Equity / Liabilities <0.04; Current assets / short-term liabilities <1; Negative EAT. Business entities that met all three conditions simultaneously were included in the class of non-prosperous subjects and others were designated as prosperous businesses.

The sample of entities we have selected consists of 266 businesses in the same area. Data analysis reported 236 prosperous and 30 non-prosperous companies, and this was considered to be the standard benchmark in verifying the selected prediction model.

Balance analysis II. provides the system evaluating the company in four basic directions (stability, liquidity, activity, profitability) and subsequently one aggregate indicator. The shape the aggregate indicator of this model is as follows [14]:

$$\mathbf{BA\ II. = (2S + 4L + A + 5P) / 12} \tag{1}$$

Where:

S = stability, L = liquidity, A = activity, P = profitability

Each of these areas consist of several indicators and have their own aggregate eventual indicators as following:

<u>*Stability aggregate indicator*</u> $\mathbf{S = (2S1 + S2 + S3 + S4 + 2S5) / 7}$ (2)

Where:

S1 = equity/fixed assets; S2 = (equity/total liabilities)·2; S3 = equity/external sources; S4 = total assets/ (short-term debts·15); S5 = total assets/(stocks·15)

<u>*Liquidity aggregate indicator*</u> $\mathbf{L = (5L1+ 8L2 + 2L3 + L4) / 16}$ (3)

Where:

L1 = (2·financial assets)/short-term debts; L2 = ((financial assets + receivables)/short-term debts)/2,17; L3 = (current assets/short-term debts)/2,5; L4 = (work capital/total liabilities) · 3,33

<u>*Activity aggregate indicator*</u> $\mathbf{A = (A1+ A2 + A3) / 3}$ (4)

Where:

A1 = (total sales/2)/total liabilities; A2 = (total sales/4)/equity; A3 = (value added · 4)/total sales

<u>*Profitability aggregate indicator*</u> $\mathbf{P = (3P1 + 7P2 + 4P3 + 2P4 + P5) / 17}$ (5)

Where:

P1 = (10 · EAT)/value added; P2 = (8 · EAT)/equity; P3 = (20 · EAT)/total liabilities; P4 = (40 · EAT)/sales; P5 = (1,33 · operational profit)/operational profit + financial profit

The classification of the resulting values: BAII. > 1 symbolizes a healthy financial situation of the company; 0,5 < BAII. ≤ 1 neutral zone of unallocated results and if BAII. ≤ 0,5 unhealthy financial situation of the company. Since ROC analysis compares two sorting algorithms of subjects into two categories, we have modified the distribution of the model as follows: if BAII. > 0,75 symbolizes a healthy financial situation of the company; and if BAII. ≤ 0,75 unhealthy financial situation of the company.

The class of enterprises, which according to the prediction model should be labeled as average, has been redistributed exactly in half, which means that we have put its middle value in the scoring interval for businesses in the gray zone, and it has been used as the threshold for determining the prosperity of entities.

4 Results

For the purpose of the paper we calculated the financial standing of each company through the selected prediction model. The results were compared to the definition of the company in crisis according to the Slovak legislation and by the confusion matrix of the prediction model we got the parameters evaluating the predictability of the selected model. These data are presented in the following Table 1.

Table 1. Ranking success rate ratings for Balance analysis II. by Rudolf Doucha

Type I error	0,00%
Type II error	20,76%
Sensitivity	79,24%
Specificity	100,00%
Accuracy	81,58%
AUC	89,60%

Source: Own processing.

In Table 1 we can see that error of II. type, consisting of the false denomination of 49 enterprises for non-prospering, amounts to 20,76%. This is due to the fact that the 49 subjects that were mistakenly identified were in the section of 236 healthy business entities and therefore the number of incorrectly marked enterprises ultimately did not cause such a large percentage of error of second kind. The overall accuracy of the model has been reduced and evaluated to 81,58%, but despite the overall accuracy of the model, the standard element evaluating its classification capability is the AUC area.

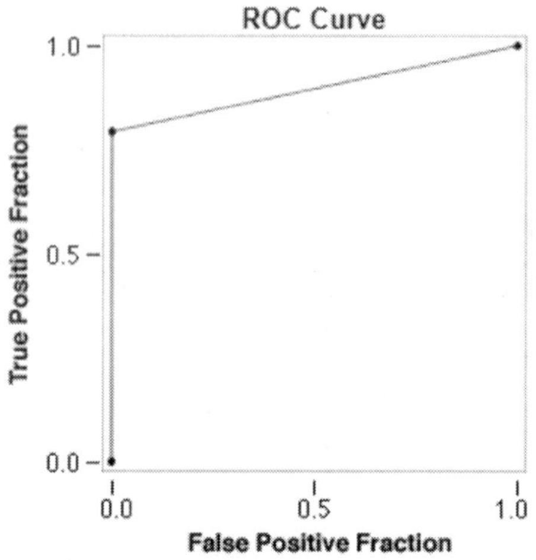

Fig. 1 Representation of the area under the ROC curve for Balance analysis II. by Rudolf Doucha
Source: Own processing.

The AUC parameter reaches out 89,6% for the model application, and thus it can be demonstrated that the selected prediction model has a good classification ability if we reflect to the classification rule resulting from the theoretical knowledge obtained from Klepac and Hampel [15].

5 Conclusion

In all companies, the importance of enterprise evaluation keeps growing within today's ever changing economic environment. Enterprise evaluation is the basic element for understanding the sources of company competition and, at the same time, it is a source for company strategy implementation support.

Forecasting the financial health of business entities is very important. On the one hand, this is important for the businesses themselves because it makes it possible to recognize the adverse effects on development and also provides opportunities for their removal. On the other hand, it is important for investors, creditors and other entities in contact with the business that they want to know about their future development in order to protect their deposited resources.

For this contribution, 266 entities from selected sector were redistributed to bankrupt and prosperous businesses under the terms of the Commercial Code. We then processed financial analysis and enterprise prediction through Balance analysis II. by Rudolf Doucha. We compared the results of the classification of these two algorithmic methodologies and based on that comparison we worked with the first and second type error parameters, sensitivity and specificity. Subsequently, using the results of these variables, we processed the ROC curve analysis, from which the resulting AUC value under the curve, was a clear indication, and according to that, the chosen predictive model was able to clearly predict the financial health of the enterprise in the selected sector of Slovakia.

Verification analysis showed that the model has a good classification capability in the area, achieving 89,6% success

This is paper is an outcome of project VEGA: 1/0544/19 Formation of the methodological platform to measure and assess the effectiveness and financial status of non-profit organizations in the Slovak Republic.

References

1. K. Valaskova, P. Durana, P. Adamko, J. Jaros, Financial compass for Slovak enterprises: modeling economic stability of agricultural entities. *Journal of Risk and Financial Management*, **13**(5), 92 (2020)
2. J. Salaga, V. Bartosova, E. Kicova, Economic Value Added as a measurement tool of financial performance. *Procedia Economics and Finance*, **26**, 484-489 (2015)
3. K. Valaskova, T. Kliestik, M. Kovacova, Management of financial risks in Slovak enterprises using regression analysis. *Oeconomia Copernicana*, **9**(1), 105–121 (2018)
4. L. Vagner, Facility Management as a Partner of Cost Controlling at Costs Optimization in the Selected Enterprise. *17th Annual Conference on Finance and Accounting*, pp. 621-629 (2017)
5. M. Durica, I. Podhorska, P. Durana, Business failure prediction using cart-based model: A case of Slovak companies. *Ekonomicko-manazerske spektrum*, **13**(1), 51-61 (2019)
6. L. Svabova, M. Durica, Being an outlier: a company non-prosperity sign?. Equilibrium. *Quarterly Journal of Economics and Economic Policy*, **14**(2), 359-375 (2019)
7. M. Durica, J. Frnda, L. Svabova, Decision tree based model of business failure prediction for Polish companies. *Oeconomia Copernicana*, **10**(3), 453-469 (2019)
8. M. Kovacova, T. Kliestik, K. Valaskova, P. Durana, Z. Juhaszova, Systematic review of variables applied in bankruptcy prediction models of Visegrad group countries. *Oeconomia Copernicana*, **10**(4), 743-772 (2019)
9. T. Kliestik, K. Valaskova, G. Lazaroiu, M. Kovacova, J. Vrbka, Remaining financially healthy and competitive: the role of financial predictors. *Journal of Competitiveness*, **12**(1), 74 (2020)
10. P. Kral, K. Janoskova, Condition of acceptability of project risk in management of the project portfolio. *15th International Scientific Conference on Globalization and its Socio-Economic Consequences*, pp. 345-352 (2015)
11. T. Kliestik, M. Misankova, K. Valaskova, L. Svabova, Bankruptcy prevention: new effort to reflect on legal and social changes. *Science and Engineering Ethics*, **24**(2), 791-803 (2018)
12. P. Gundova,. Verifikácia vybraných predikčných metód na vzorke slovenských podnikov. *Acta Academica Karviniensia*. **4**, 26-38 (2014)
13. T. Fawcett, *An introduction to ROC analysis*. Pattern Recognition Letters (2006)
14. R. Doucha, *Finanční analýza podniku: praktické aplikace*. Praha: VOX Consult, pp. 224 (1996)
15. V. Klepac, D. Hampel, Prediction of bankruptcy with SVM classifiers among retail business companies in EU. *Universitatis Agriculturae et Silveculturae Mendelianae Brunensis*, 627-634 (2016)

Jones' Model and Its Modifications in the Conditions of the Slovak Republic

Lenka Strakova[1,*] and *Lucia* Svabova[1]

[1]The University of Zilina, Faculty of Operation and Economics of Transport and Communications, Department of Economics, Zilina, Slovakia

Abstract. One of the key tasks of financial accounting from its beginnings to the present day is to determine the performance of the company. The financial statements should provide users with a true and fair view of the financial position and financial performance of the entity during the period. At present, profit represents the most frequently accepted measure of a company's financial performance. An important prerequisite for profit as a reliable measure of performance is its quality, which can be influenced by various factors or techniques resulting from earnings management. This paper aims to compare the detection capability of the Jones model and its modifications for assessing the occurrence of earnings management in the conditions of the Slovak Republic. We use the regression analysis and comparison method, based on which we compare the detection capability of the Jones model and its modifications for assessing the occurrence of earnings management in the conditions of the Slovak Republic. The contribution of the paper lies in the observation of the Jones model and its modifications to determine a suitable model for assessing the existence of earnings management in companies in Slovakia, which will be the subject of future research.
Keywords: Earnings relevance, earnings management, Jones model, Modifications of the Jones model

1 Introduction

One of the key tasks of financial accounting from its beginnings to the present day is to determine the performance of the company. [19] The financial statements should provide users with a true and fair view of the entity's financial position and financial performance during the period. [10] At present, the company's managers and analysts use a wide range of tools, algorithms, and methods to determine the value of profit representing the most commonly accepted measure of a company's financial performance, to predict the future development of the company's financial health. [12] The relevance of the accrual principle applied in accounting is undeniable. [25] Accrual accounting is, thanks to the time and material allocation of costs and revenues, a good system for evaluating the effectiveness of management, based on which it is possible to measure and detect earnings management. [14] Earnings management is one of the most challenging, debated, controversial and at the

*Corresponding author: lenka.strakova@fpedas.uniza.sk

same time the most promising topics in finance and financial management. [11] Nevertheless, it is still difficult to provide a uniform definition of earnings management, as evidenced by the fact that the authors of various professional studies very often perceive earnings management differently. The main reasons for the lack of consensus on earnings management are its inconsistency, ambiguity, and problematic measurability. Studies aimed at detecting and measuring earnings management are very often based on discretionary accruals and work with models that estimate the discretionary share of reported profit, starting with simple models that identify discretionary accruals with total accruals to more sophisticated models that try to divide the total accruals into a discretionary and non-discretionary part. However, systematic evidence regarding the relative performance of these models in detecting earnings management does not yet exist.

1.1 Earnings relevance

The question of why profit is such an important indicator of a company [15] that it has become the subject of management and manipulation is answered by Ronen and Yaari [6] explaining three approaches based on the separation of shareholder ownership and management in publicly traded companies. This creates a conflict of interest between the parties, as management decisions may not always be in line with the views of business owners. Depending on the awareness of existing third parties about the true profit and the power to implement the decisions taken, these authors distinguish between a costly contracting approach, a decision-making approach, and a legal approach. The costly contracting approach emphasizes the importance of profit information in creating effective contracts. In the event of unforeseen situations, the parties remain bound by the old contracts and use the earnings management application to fulfill their contractual contracts. In terms of informing third parties about the true profit and the power to make decisions, the first approach presupposes complete information and the power of shareholders to control the management of the company. The decision-making approach describes profit as a valuable source of information for decision-making in a given company. If the shareholders are rational, it is not possible to carry out earnings management without the implicit or explicit consent of investors. The decision-making approach presupposes power, but not shareholder awareness. The legal approach is based on the lack of tools for business owners to effectively control management. However, profit allows owners to use limited tools more effectively, helping to eliminate information asymmetries between owners.

1.2 Earnings management

The phenomenon of earnings management belongs to the integral and fundamental part of their business finance. [16] Earnings management is a legal and very effective method of accounting techniques and may be used to obtain specific objectives of the enterprises involving the manipulation of accruals. [23] However, despite the unequivocal legality of earnings management, the creative nature of these activities, which show the financial situation and performance of the company according to management's ideas and not according to the needs of users of financial statements, cannot be denied, as McKee [26] confirms. The first of the definitions relating to earnings management by Degeorge, Patel and Zeckhauser [4] reads Strategic use of managerial judgment in influencing profit values reported by external users of financial statements. It is achieved primarily by timing reported or actual economic events to shift between periods. Other authors define EM as the use of flexibility in the selection of accounting methods that influence managers' decisions about future cash flows. [13] Sosnowski [24] explains that reducing financial limitations of enterprise can cause earnings management because enterprise tries to obtain higher external capital.

1.3 Earnings management testing

The issue of measuring and detecting earnings management is a relatively frequently researched problem. Despite a large amount of professional literature, there is no uniform opinion on the most effective testing methods. In the researched literature, the method using residues from aggregated accrual models, which is based on the classical Jones model, is the most used. The second most used method is models based on specific accrual accounts. In research, the most used residual model from aggregated accrual accounts in most cases provides results that indicate evidence or confirmation of the use of earnings management. This model is also cited by Dechow, Sloan and Sweeney [17] as the model with the strongest and best results. Although he is not talking exactly about residues from aggregated accruals, he is talking about the modified Jones model, which is the basis for it, so they can be considered identical. On the other hand, Dechow, Richardson, and Tuna [18] use the modified Jones model to test accrual accounts as explanatory variables for the skewed, non-standard profit distribution that is identified, for example, by Burgstahler and Dichev [2].

1.4 Jones model and its modifications

The present review of the scientific literature related to earnings management points to diversity [1] because the Jones model and its modifications use the different intensity of earnings management testing. [7] Jones's model mitigates the assumption of non-discretionary accrual constant. This model takes into account the impact of changes in the economic situation of the company on the non-discretionary accrual with the help of tangible fixed assets and changes in sales as independent variables of the model. Jones's model divides the time series of companies' profits into two stages, which are estimation periods and event periods. The estimation period represents zero discretion accrual. The event period assumes that the discretionary accrual is not zero, resulting in earnings management. The main assumption of Jones' model is that companies do not implement earnings management during estimation periods. This assumption is very unlikely to be met, which may result in bias in the measurement and detection of earnings management. [5] Jones' model includes the assumption of non-discretion in sales. If a company manages profit through discretionary revenues, then Jones's model removes part of the profit from the discretionary accrual estimate, which the author herself acknowledges as a limitation of her model. [8] The consequence of the limitation is the fact that the given model detects earnings management, even in its absence. [27] Dechow, Sloan, and Sweeney [17] sought to streamline Jones' original model and therefore proposed to adjust the variable change in sales (ΔREV) to the variable change in receivables (ΔREC). The purpose of such a modification was to reduce the measurement error of the discretionary accrual if the discretionary accrual was generated through revenues. Jones' modified model points to the assumption that any change in sales revenue from a supplier credit is the result of earnings management. However, it is uncertain in the literature whether a given modification improves the original Jones model. The authors dealing with this issue, Jeter and Shivakumar [3] consider such an assumption to be unrealistic and therefore tried to design their model. As part of the creation of their model consisting in the modification of the original Jones model, they proposed to include cash flow from operating activities in the non-discretionary part of earnings management, due to its more effective detection within a group of companies with extreme cash flow reporting. The author Kasznik [20] also tried to improve Jones' original model, incorporating cash flow (FOCFO) as the third explanatory variable in his model. Another of the authors on the subject, Key [9] added a new variable to the original Jones model, which is intangible assets, and justified this transformation by using the relationship between intangible assets and depreciation, which are one of the components of the non-discretionary accrual. Teoh, Welch, and Wong [21] also attempted to modify Jones' model and focused on short-term accruals. They defined short-term accrual as the difference between changes in non-monetary current assets and changes in operating current liabilities. They determined the non-discretionary part of the short-term accrual by increasing the company's sales. Kothari et al. [22] presented a modification of the original Jones model, which is also known as "performance matching" while offering 2 approaches. The first is the pairing of similar companies, which alleviates the need for the least-squares DA estimation. The second, linear approach, is to consider the return on assets (ROA) to regulate the performance of the organization.

2 Methodology

The conference contribution aims to compare the detection capability of the Jones model and its modifications for assessing the occurrence of earnings management in the conditions of the Slovak Republic. For achievement relevant data we use the financial database, Amadeus. In our research, we analyze the Jones model and its modifications in the conditions of the Slovak Republic. Our analysis is applied to a sample of 1253 enterprises in the condition of the Slovak republic. In research, we use a selection method and comparison method, based on which we compare the detection capability of the Jones model and its modifications for Slovak enterprises according to the adjusted coefficient of determination, standard deviation, and the level of statistical significance of the earnings model as such and individual independent variables entering the analyzed earnings models. Within the analysis of statistical significance, we use the F - the test by the medium of setting hypotheses H0 and H1.

3 Results and discussion

We analyze profit models using criteria proposed by the literature, which are adjusted coefficient of determination, standard deviation, and the level of statistical significance of the profit model as such and individual independent variables entering the analyzed profit models. We apply individual profit models to a sample of 1253 Slovak companies, within which we selected companies according to the following criteria.

Table 1. Selected criterion

Criterion	Values
Sector	the automotive industry, mechanical engineering, electrical engineering, energy, mining, construction
Sales	> 3 000 000 EUR
Earnings	> 100 000 EUR
Assets	> 5 000 000 EUR
State of the company	uncontested
Type of ownership	private domestic enterprises, private international enterprises, private foreign enterprises

Source: Own processing.

3.1 The adjusted coefficient of determination

The adjusted coefficient of determination expresses the explanatory power of the profit model. We performed a statistical regression analysis to determine the values of the adjusted coefficient of determination. Using the adjusted coefficient of determination, we then explain the variability and relevance of the analyzed profit models, by comparing the obtained results over time and between models, with higher values indicating a higher explanatory power of the model and vice versa.

Table 2. The adjusted coefficient of determination

Earnings model	The adjusted coefficient of determination			
	2015	**2016**	**2017**	**Average**
Jones model	0.1093	0.0380	0.0329	0.0601
Modified Jones model	0.1155	0.0509	0.1255	0.0973
Jeter and Shivakumar model	0.1228	0.0371	0.0345	0.0648
Kasznik model	0.1217	0.0478	0.0513	0.0736
Key model	0.1060	0.0387	0.0324	0.0590
Teoh, Welch and Wong model	0.0085	0.0095	0.0968	0.0383
Kothari model	0.1132	0.0475	0.1292	0.0966

Source: Own processing.

Based on the table, we can see that in 2015, the two highest values of the adjusted coefficient of determination were recorded. These are cash flow models by Jeter and Shivakumar with a coefficient of determination of 0.1228 and Kasznik with a coefficient of determination of 0.1217. Both of these profit models thus explain the variability of the total accrual of more than 12%. The achieved values for the remaining profit models are as follows: Modified Jones model with determination coefficient value 0.1155, Kothari model with determination coefficient value 0.1132, Jones model with determination coefficient value 0.1093, and Key model with determination coefficient value 0.1060. These profit models explain the variability of the total accrual in the range from 10 to 12%. The relatively low explanatory power represented by the adjusted coefficient of determination was achieved by the Teoh, Welch, and Wong model, with a coefficient value of 0.0085, which indicates the variability of the dependent variable to less than 1%. In the examined the year 2016, it reached the highest value of the adjusted coefficient of determination 0.0509 Modified Jones model. This model is followed by the Kasznik model with the value of the coefficient of determination 0.0478 and Kothari model with the value of the coefficient of determination 0.0475, Key model with the value of the coefficient of determination 0.0387, Jones's model with the value of the coefficient of determination 0.0380 and cash flow model by Jeter and Shivakumar with the value of the coefficient 0.0371. Relatively low values of the adjusted coefficient of determination were achieved similarly to 2015 by Teoh, Welch, and Wong model with the value of the coefficient of determination 0.0095. In the observed period of 2017, the Kothari model reached the highest explanatory power represented by the adjusted coefficient of determination 0.1292, which explains the variability of the total accrual to almost 13%. Then follows the Modified Jones model with the value of the coefficient of determination 0.1255, Teoh, Welch and Wong model with the value of the coefficient 0.0968, Kasznik model with the value of the coefficient of determination 0.0513, Jeter and Shivakumar model with the value of the coefficient of determination 0.0345. Jones's model reached the value of the coefficient of determination 0.0329, Key model reached the lowest value of the coefficient of determination at the level of 0.0324.

3.2 Standard deviation

Through the analysis of the standard deviation, we explain the variability of the values of the variables of the selected profit models. If the standard deviation becomes low, it means that the given profit model more effective measures and detects earnings management in a selected sample of companies operating in the conditions of the Slovak Republic. The following tables provide a summary of each profit model.

Table 3. Overview of Standard Deviations

Jones model		**Intercept**	**ΔREV**	**PPE**	
2015	Coefficient	0.0232	0.0194	-0.1794	
	Standard deviation	0.0138	0.0190	0.0302	
2016	Coefficient	-0.0019	0.0172	-0.1063	
	Standard deviation	0.0143	0.0209	0.0307	
2017	Coefficient	-0.0078	-0.0500	-0.1064	
	Standard deviation	0.0191	0.0217	0.0397	
Average	Coefficient	0.0045	0.0045	0.1307	
	Standard deviation	0.0157	0.0205	0.0335	
Modified Jones model		**Intercept**	**ΔREV - ΔREC**	**PPE**	
2015	Coefficient	0.0252	-0.0342	-0.1764	
	Standard deviation	0.0137	0.0200	0.0303	
2016	Coefficient	0.0069	-0.0490	-0.1094	
	Standard deviation	0.0141	0.0235	0.0305	
2017	Coefficient	0.0102	-0.1280	-0.1192	
	Standard deviation	0.0181	0.0216	0.0377	
Average	Coefficient	0.0141	0.0704	0.1350	
	Standard deviation	0.0153	0.0217	0.0328	
Jeter and Shivakumar model		**Intercept**	**ΔREV**	**PPE**	**CFO**

2015	Coefficient	0.0034	0.0070	-0.1910	0.2149
	Standard deviation	0.0163	0.0195	0.0305	0.0945
2016	Coefficient	-0.0099	0.0150	-0.1119	0.0840
	Standard deviation	0.0171	0.0210	0.0314	0.0987
2017	Coefficient	-0.0248	-0.0539	-0.1151	0.1575
	Standard deviation	0.0236	0.0220	0.0402	0.1287
Average	Coefficient	-0.0104	0.0106	0.1393	0.1521
	Standard deviation	0.0190	0.0208	0.0340	0.1073
Kasznik model		**Intercept**	**ΔREV**	**PPE**	**ΔCFO**
2015	Coefficient	0.0253	-0.0022	-0.1833	0.2904
	Standard deviation	0.0137	0.0213	0.0300	0.1323
2016	Coefficient	-0.0028	-0.0011	-0.1081	0.3084
	Standard deviation	0.0143	0.0228	0.0305	0.1592
2017	Coefficient	-0.0132	-0.0710	-0.1035	0.4589
2017	Standard deviation	0.0190	0.0229	0.0393	0.1830
Average	Coefficient	0.0031	0.0248	0.1316	0.3526
	Standard deviation	0.0157	0.0223	0.0333	0.1582
Key model		**Intercept**	**ΔREV**	**PPE**	**IA**
2015	Coefficient	0.0232	0.0193	-0.1792	-0.0082
	Standard deviation	0.0141	0.0190	0.0304	0.5720
2016	Coefficient	0.0009	0.0169	-0.1042	-0.6589
	Standard deviation	0.0146	0.0209	0.0308	0.6123
2017	Coefficient	-0.0117	-0.0501	-0.1063	0.6763
	Standard deviation	0.0196	0.0216	0.0394	0.7160
Average	Coefficient	0.0041	0.0046	0.1299	0.0031
	Standard deviation	0.0161	0.0205	0.0335	0.6334
Teoh Welch and Wong model		**Intercept**		**ΔSALE-ΔREC**	
2015	Coefficient	-0.0410		-0.0390	
	Standard deviation	0.0080		0.0210	
2016	Coefficient	-0.0345		-0.0460	
	Standard deviation	0.0083		0.0240	
2017	Coefficient	-0.0363		-0.1209	
	Standard deviation	0.0105		0.0219	
Average	Coefficient	-0.0373		0.0686	
	Standard deviation	0.0089		0.0223	
Kothari model		**Intercept**	**ΔREV - ΔREC**	**PPE**	**ROA**
2015	Coefficient	0.0209	-0.0330	-0.1742	0.0003
	Standard deviation	0.0156	0.0200	0.0301	0.0008
2016	Coefficient	0.0091	-0.0500	-0.1100	-0.0001
	Standard deviation	0.0164	0.0239	0.0305	0.0009
2017	Coefficient	0.0272	-0.1337	-0.1243	-0.0019

	Standard deviation	0.0214	0.0220	0.0380	0.0012
Average	Coefficient	0.0191	0.0722	0.1362	0.0006
	Standard deviation	0.0178	0.0220	0.0329	0.0010

Source: Own processing.

Based on the table, we can see that the Jones model achieves relatively high values of the average standard deviations of individual variables. The first variable change in sales (ΔREV) represents the average standard deviation of 0.0205. The average standard deviation of the second independent variable on tangible fixed assets (PPE) reaches the average standard deviation of 0.0335. In the modified Jones model, the first variable expressed as the difference between the change in sales and the change in receivables (ΔREV - ΔREC) causes the average standard deviation to decrease to 31% of the average value of the coefficient and the average standard deviation of the second variable tangible fixed assets (PPE) reaches 25% of the average value of the coefficient. The Jeter and Shivakumar model also achieves relatively high values of standard deviations. The mean standard deviation of the variable ΔREV is 0.0208. The average standard variable PPE reaches 25% of the average value of the coefficient and CFO reaches 71% of the average value of the coefficient. The Kasznik model is again one of the models achieving relatively high values of the mean standard deviation. The mean standard deviation ΔREV is 90% of the mean value of the coefficient. The average standard deviation of the PPE is 25% of the value of the coefficient. The last variable ΔCFO reaches the average value of the standard 45% of the average value of the coefficient. The average standard deviations in the case of the Key model reach extreme values. The average standard deviation of the variable ΔREV reaches the value 0.0205, which represents an extreme proportion of the average value of the coefficient. The average standard deviation of the PPE is 26% of the value of the coefficient. The standard deviations of long-term intangible assets (IA) reach high values in the individual years 2015-2017, which is also reflected in the high shares of the average standard deviation from the average value of the coefficient. The Teoh, Welch, and Wong model with a single independent variable ΔSALE - ΔREC achieves relatively low values of the standard deviation and the average standard deviation reaches 33% of the average value of the coefficient. The Kothari model for the variable ΔREV - ΔREC achieves an average standard deviation of more than 31% of the average value of the coefficient and the second variable PPE reaches more than 24%. The average standard deviation of the ROA compared to the other variables of this model reaches a high share of the average value of the coefficient.

3.3 The significance level of models

As part of the analysis, we test profit models using the F - the test of statistical significance, while setting hypotheses H0 and H1. Hypothesis H0 tells us that the studied idol is not statistically significant, hypothesis H1 tells us that the studied idol is statistically significant. The following table provides an overview of the accepted hypotheses and a percentage expression of the statistical significance of the investigated profit models.

Table 4. Statistical significance of models

Model	**2015**	**2016**	**2017**	**Statistical significance %**
Jones model	H_1	H_1	H_1	100
Modified Jones model	H_1	H_1	H_1	100
Jeter and Shivakumar	H_1	H_1	H_1	100
Kasznik model	H_1	H_1	H_1	100
Key model	H_1	H_1	H_1	100

Teoh Welch and Wong model	H_0	H_0	H_1	33.33
Kothari model	H_1	H_1	H_1	100

Source: Own processing.

Based on the table, we can see that the Jones model, the Modified Jones model, the Jeter and Shivakumar model, the Kazsnik model, the Key model, and the Kothari model represent statistically significant models because we have accepted hypothesis H1. The percentage of their statistical significance reaches 100%. The Teoh, Welch, and Wong model is statistically insignificant in 2015 and 2016 because we adopted the H0 hypothesis. The percentage expression of the statistical significance of a given profit model reaches 33.33%.

3.4 The significance level of variables

We test the variables contained in the individual analyzed profit models by comparing the p-value with the significance level $\alpha = 0.05$. Again, we set the hypotheses H0 and H1. Hypothesis H0 states that the variables contained in the analyzed profit models are not statistically significant and hypothesis H1 states that the variables contained in the analyzed profit models are statistically significant. If the comparison shows that the p-value is less than the significance level α, we reject hypothesis H0 and accept hypothesis H1. A model whose variables are statistically significant explains earnings management correctly, and conversely, a model whose variables are statistically insignificant indicates that its variables do not explain earnings management, ie the manipulation of discretionary accrual. The following table provides an overview of accepted and rejected hypotheses for all variables contained in the analyzed profit models, together with a percentage expression of their statistical significance.

Table 5. Statistical significance of variables

Variables	**2015**	**2016**	**2017**	**Statistical significance %**
Jones model				
ΔREV	H_0	H_0	H_1	33.33 %
PPE	H_1	H_1	H_1	100.00 %
Modified Jonesovej model				
ΔREV – ΔREC	H_0	H_1	H_1	66.67 %
PPE	H_1	H_1	H_1	100.00 %
Jeter and Shivakumar				
ΔREV	H_0	H_0	H_1	33.33 %
PPE	H_1	H_1	H_1	100.00 %
CFO	H_1	H_0	H_0	33.33 %
Kasznik model				
ΔREV	H_0	H_0	H_1	33.33 %
PPE	H_1	H_1	H_1	100.00 %
ΔCFO	H_1	H_0	H_1	66.67 %
Key model				
ΔREV	H_0	H_0	H_1	33.33 %
PPE	H_1	H_1	H_1	100.00 %
IA	H_0	H_0	H_0	0.00 %
Teoh Welch and Wong model				

ΔSALE – ΔREC	H_0	H_0	H_1	33.33 %
Kothari model				
ΔREV – ΔREC	H_0	H_1	H_1	66.67 %
PPE	H_1	H_1	H_1	100.00 %
ROA	H_0	H_0	H_0	0.00 %

Source: Own processing.

Based on the results from the table, we see that the Jones model achieves a high percentage expression of the statistical significance of the variables contained, especially in the case of the PPE variable, which appears to be statistically significant throughout the analyzed period. However, the variable ΔREV is recorded for statistical significance only in the last year of 2017, and therefore the percentage expression of the statistical significance of this variable is at the level of 33.33%. The modified Jones model achieves a high percentage of the statistical significance of all independent variables. The variable ΔREV - ΔREC is statistically insignificant only in the first year of 2015, the variable PPE is statistically significant in the whole analyzed period. The Jeter and Shivakumar model also achieves a high percentage of the statistical significance of the variable only in the case of PPE. The remaining variables appear to be statistically significant and their percentage of statistical significance reaches the level of 33.33%. The Kasznik model has a statistically significant variable, which is PPE in all three years, and a ΔREV variable, which is statistically significant in only one of the three analyzed periods. The variable ΔCFO is statistically significant only in two of the three analyzed periods, while its percentage of statistical significance reaches the level of 66.67%. The Teoh, Welch, and Wong model contains only one variable ΔSALE - ΔREC, which, however, is statistically significant only in the last year of 2017, while its percentage of statistical significance reaches the level of 33.33%. The variables contained in the remaining models appear to be statistically insignificant for all three years, with the result that they do not explain the manipulation of the discretionary accrual. Although the key model achieves a high percentage expression of the statistical significance of the PPE variable, the ΔREV variable is statistically insignificant, reaching a level of statistical significance of 33.33% and the IA variable is statistically significant, while its statistical significance is at 0%. Kothari's model also contains a statistically insignificant variable ROA, whose level of statistical significance reaches 0%.

3.5 Final evaluation of the achieved results

We evaluated selected analyzed profit models based on three criteria, which adjusted the coefficient of determination, standard deviation, the level of statistical significance of individual profit models, and statistical significance of variables contained in the analyzed profit models. The following table provides an overview of the criteria met by the given profit models for the examined period of three years.

Table 6. The final summary of selected models

Earnings management models	The adjusted coefficient of determination	Standard Deviation	Statistical		Fulfilled criteria
			model	variables	
Jones model	x	x	✓	x	1
Modified Jones model	✓	✓	✓	✓	4
Jeter, Shivakumar model	x	x	✓	x	1
Kasznik model	x	x	✓	✓	2
Key model	x	x	✓	x	1

Teoh, Welch, Wong	x	✓	x	x	1
Kothari model	✓	x	✓	x	2

Source: Own processing.

Based on the table, we see that the Modified Jones model meets all the criteria, because this model achieved the highest value of the adjusted coefficient of determination, the lowest values of average standard deviations of independent variables contained in the model, high percentage of statistical significance not only of the model but also variables. Kasznik's model and Kothari's model meet two of the four criteria, with Kazsnik's model meeting the criterion of statistical significance of the model itself and its variables, and Kothari's model meeting the criterion of the adjusted coefficient of determination and statistical significance of the model itself. The remaining models such as Jones' model, Jeter and Shivakumar model, Key model and Teoh, Welch, and Wong model meet one of the four criteria. The Jones model did not meet the criteria of the adjusted coefficient of determination, standard deviation, and statistical significance of the variables contained in the model. Jeter and the Shivakumar model did not meet the criteria of the adjusted coefficient of determination, the standard deviation, and also the statistical significance of the variables contained in this model. The key model did not meet the same criteria. The Teoh, Welch, and Wong model did not meet the criterion of the adjusted coefficient of determination and statistical significance of the model itself, but also its variables. Based on the results of the analysis, the Modified Jones model appears to be the most effective profit model for measuring and detecting earnings management in the conditions of the Slovak Republic. Compared to the other analyzed profit models, this model achieves a higher value of the adjusted coefficient of determination and the lowest average standard deviation. It is a statistically significant model, while almost all of its variables reach statistical significance.

4 Conclusion

The analysis was focused on the measurement and detection of earnings management in the conditions of the Slovak Republic using the Jones model and its modifications. To find out whether the Jones model and its modifications can capture and detect earnings management in the conditions of the Slovak Republic, we set four criteria, based on which we assessed the effectiveness of selected models. The four main criteria included the adjusted coefficient of determination, the standard deviation, the statistical significance of the profit models, and the statistical significance of the variables contained in the analyzed profit models. The highest explanatory power represented by the average value of the adjusted coefficient of determination was achieved by the Modified Jones model at the level of 0.0973, which means that the model explains the variability of the total accrual to 9.73%. The second in order is Kothari's model with a value of 0.0966. The following are Kasznik's model with a coefficient value of 0.0736, Jeter, and Shivakumar with a coefficient value of 0.0648, Jones's model with a value of 0.0601, Key model with a value of 0.0590. The Teoh, Welch and Wong models show the lowest explanatory power expressed by the average value of the adjusted coefficient of determination at the level of 0.0383. The second criterion was the standard deviation of the variables contained in the analyzed profit models. In terms of this criterion, the Modified Jones model was the most effective, mitigating the high standard deviation of the first variable of the original Jones model. Its modification into the variable ΔREV - ΔREC caused a decrease in the average value of the standard deviation to 31% of the average value of the coefficient. The average standard deviation of the PPE variable reached almost 25% of the average value of the coefficient. The Teoh, Welch, and Wong models also achieved low values of the mean standard deviation. The statistical significance of the profit model itself was met by several models, including the Jones model, the Modified Jones model, both cash flow models, the Key model, and the Kothari model. The percentage expression of statistical significance in the examined period reaches the level of 100% for all analyzed models. In terms of the statistical significance of individual variables contained in the analyzed profit models, the Modified Jones model was the most effective. This model achieves a high percentage of statistical significance for both. The Kasznik model satisfies the last criterion. Based on the above findings, we consider the Modified Jones model to be the most effective in the conditions of the Slovak Republic.

Acknowledgment

This paper was prepared with the support of VEGA 1/0210/19 Research of innovative attributes of quantitative and qualitative fundaments of the opportunistic earnings modeling which authors gratefully acknowledge.

References

1. A. Novak Sedlackova, D. Svecova, A. Tomova, Slovak Airports as Profit/Loss Making Entities: Consequence for Transport Policies. *Journal of Tourism and Services*, **10**(19), 153-163 (2019)
2. D. Burgstahler, I. Dichev, Earnings management to avoid earnings decreases and losses. *Journal of Accounting and Economics*, **24**(1), 99-126 (1997)
3. D.C. Jeter, L. Shivakumar, Cross-sectional estimation of abnormal accruals using quarterly and annual data: effectiveness in detecting event-specific earnings management. *Accounting and Business Research*, **29**(4), 299-319 (1999)
4. F. Degeorge, J. Patel, R. Zeckhauser, Earnings Management to Exceed Thresholds. *Journal of Business*, **72**(1), 1-33 (1999)
5. H. Hoglund, Detecting earnings management with neutral networks. *Expert Systems with Applications*, **39**(10), 9564-9570 (2012)
6. J. Ronen, V. Yaari, Earnings Management. Emerging Insights in Theory, Practice and Research. New York: Springer (2008)
7. J.C. Hollowell, K. Boris, J. Vrbka, E. Kovalova, Cognitive Decision-Making Algorithms for Sustainable Manufacturing Processes in Industry 4.0: Networked, Smart, and Responsive Devices. *Economics, Management, and Financial Markets*, **14**(4), 9-15 (2019)
8. J.J. Jones, Earnings management during import relief investigations. *Journal of Accounting Research*, **29**(2), 193-228 (1991)
9. K.G. Key, Political cost incentives for earnings management in the cable television industry. *Journal of Accounting and Economics*, **23**(3), 309-337 (1997)
10. K. Valaskova, T. Kliestik, M. Kovacova, Management of financial risks in Slovak enterprises using regression analysis. *Oeconomia Copernicana*, **9**(1), 105-121 (2018)
11. K. Valaskova, T. Kliestik, M. Kovacova, Assessment of selected models of earnings management in economic conditions of Slovakia. *Proceedings of the 33rd International-Business-Information-Management-Association*, pp. 3922-3931 (2019)
12. M. Kovacova, T. Kliestik, K. Valaskova, P. Durana, Z. Juhaszova, Systematic review of variables applied in bankruptcy prediction models of Visegrad group countries. *Oeconomia Copernicana*, **10**(4), 743-772 (2019)

13. M.R. Sankar, K.R. Subramanyam, Reporting Discretion and Private Information Communication through Earnings. *Journal of Accounting Research*, **39**(2), 365-386 (2001)
14. M.T.H. Khanh, P.A. Thu, The effect of financial leverage on real and accrual-based earnings management in Vietnamese firms. *Economics and Sociology*, **12**(4), 299-333 (2019)
15. P. Durana, A. Zauskova, L. Vagner, S. Zadnanova, Earnings drivers of Slovak manufacturers: Efficiency assessment of innovation management. *Applied Sciences*, **10**(12), 4251 (2020)
16. P. Durana, K. Valaskova, D. Chlebikova, V. Krastev, I. Atanasova, Heads and tails of earnings management: quantitative analysis in emerging countries. *Risks*, **8**(2), 57 (2020)
17. P. Dechow, R. Sloan, A.P. Sweeney, Detecting earnings management. *The Accounting Review*, **70**(2), 193-225 (1995)
18. P.M. Dechow, S.A. Richardson, I.A. Tuna, Why are earnings kinky? An examination of the earnings management explanation. *Review of Accounting Studies*, **8**(6-9), 355-384 (2003)
19. Q. L. Huynh, Reputation to the vicious circle of earnings quality and financial performance. *Economics and Sociology*, **12**(2), 361-375 (2019)
20. R. Kasznik, On the Association between Voluntary Disclosure and Earnings Management. *Journal of Accounting Research*, **37**(1), 57-81 (1999)
21. S.H. Teoh, I. Welch, T.J. Wong, Earnings management and the underperformance of seasoned equity offerings. *Journal of Financial Economics*, **50**(1), 63-99 (1998)
22. S.P. Kothari, A.L. Leone, CH.E. Wasley, Performance matched discretionary accrual measures. *Journal of Accounting and Economics*, **39**(1), 163-197 (2005)
23. T. Kliestik, K. Valaskova, E. Nica, M. Kovacova, G. Lazaroiu, Advanced methods of earnings management: monotonic trends and change-points under spotlight in the Visegrad countries. *Oeconomia Copernicana*, **11**(2), 371-400 (2020)
24. T. Sosnowski, Earnings management in the private equity divestment process on Warsaw Stock Exchange. *Equilibrium. Quarterly Journal of Economics and Economic Policy*, **13**(4), 689–705 (2018)
25. T. C. Hoang, D. M. Joseph, The effect of new corporate accounting regime on earnings management: evidence from Vietnam. *Journal of International Studies*, **12**(1), 93-104 (2019)
26. T.E. McKee, Earnings Management: An Executive Perspective, Indiana: Thomson (2005)
27. V.L. Bernard, D.J. Skinner, What motivates managers' choice of discretionary accruals?. *Journal of Accounting and Economics*, **22**(1), 313-325 (1996)

An Empirical View on the Determinants of Sustainable Economic Development: Evidence from EU(28) Member States

Beata Sofrankova[1,*], *Dana* Kiselakova[1], and *Erika* Onuferova[1]

[1]University of Presov in Presov, Faculty of Management, Presov, Slovak Republic

Abstract. Sustainable development represents a comprehensive approach that focuses on economic, social and environmental aspects in order to meet the needs of the present without compromising the ability of future generations. The main aim of the paper is to reveal the impact of selected global multicriteria indices (through Global Innovation Index, Doing Business Index, Economic Freedom Index, Corruption Perception Index and Human Development Index) on sustainable development (expressed by Sustainable Economic Development Assessment) within EU(28) countries for the period of 2011 to 2018. Secondary data were processed using panel data regression analysis. Results of average score in the field of sustainable economic development showed that Finland (84.5) achieved the best-performed, while the worst position was recorded by Romania (56.5). Looking at the selected indices average scores, the leading position was indicated in the case of Denmark (80.5) and the worst results was obtained by Greece (57.6). Based on the results of the panel regression analysis, to the key determinants influencing the sustainable economic growth of the EU(28) countries belong such variables as innovation activity, business environment, corruption issues and human resources. The presented findings provide valuable insight toward to strategic priorities review, as well as, it can also help governments move from plans to action.

Keywords: sustainable growth, innovation, entrepreneurship, economic freedom, corruption, human resources, EU(28) countries

1 Introduction

Nowadays, the concept of sustainable development has become a widely recognized aim for human society. At national and global level, the adoption of sustainable development strategies represents the key driving force in achieving the long-term economic growth, in creating conditions for environmental protection, as well as, in guaranteeing a healthy life and social inclusion. For sustainable development to be achieved, it is necessary to harmonize and interconnect all these core elements for the well-being of individuals and societies.

*Corresponding author: beata.sofrankova@unipo.sk

The worldwide globalization led to the adoption of the Sustainable Development Goals. From international perspective, it is a key moment for building and implementing of urgent and inclusive action plans. As a part of a new sustainable development roadmap, the 2030 Agenda was approved, which contains the 17 Sustainable Development Goals, a call to action to protect the planet, to create an environment of well-being for people and to achieve the economic prosperity [1]. At present, many international institutions and organizations publish recognized annual reports focus on assessment of countries concerning to various range of factors expressing sustainability potential.

The study was aimed to examine the impact of global multi-criteria determinants (namely, the Global Innovation Index, the Doing Business Index, the Economic Freedom Index, the Corruption Perception Index and the Human development Index) on sustainable growth (measured through the Sustainable Economic Development Index) using the panel data regression analyses. The basic dataset included EU(28) member states. The research was carried out for the years from 2011 to 2018. The purpose of this research was to identify key determinants influencing the sustainable economic growth of countries analyzed.

The paper is organized into following sections. At first, it was carried out the literature review concerned to selected global indices in relation to sustainability of economies. In the next section, the methodology, data, aim of paper and research question were introduced briefly. The following empirical section was focused on performing research and presenting revealed results. The last section summarized the significant findings and the main recommendations for supporting sustainable development of EU member states.

2 Literature review

An increasing growth of globalization and internationalization of economic, environmental and social system determine the quality of human well-being. However, many analysts claim this exponential growth has already exceeded the capacity of resources. Deciding where and how to deploy their effort and resources is a central challenge. Boston Consulting Group's ***Sustainable Economic Development Assessment (SEDA)*** can be a powerful tool in this effort. The issue of sustainable development assessment was examined by many researchers. Huttmanova and Valentiny [2] analyzed the actual situation in the case of economic and environmental pillar of sustainable growth within EU member states, using indicators as SEDA, HDI, and Environmental Performance Index (EPI) and so on. Paper findings revealed that an achieving a positive result in one dimension of sustainability cause negative results in other dimensions. As reported by Strezov, Evans and Evans [3], global indices enable to quantify all dimensions of sustainable prosperity (including economic, environmental and social aspects). In this context, authors applied selected various indices (for example Change in Wealth Index (CWI), Environmental Sustainability Index (ESI), Global Well-Being Index (GWI), Happy Planet Index (HPI), Sustainable Society Index (SSI) and SEDA. Based on research results was proposed a normalized average sustainability index (NASI), as an integration of all above-mentioned indices. The literature on evaluation of sustainability has largely focused on investigating environmental and economic potential. In this regard, Bassetti, Blasi and Sedita [4] researched interconnection among environmental conditions and economic performance. The empirical evidence confirmed that well-advised environmental orientation positively affects economic situation of nations and contribute to future well-being.

An assumption that sustainable development of economies is supported by innovation activities was proved by many research papers. In this backdrop, Zhylinska, Balan and Andrusiak [5] suggested new methodological approach to assess the level of innovation towards to sustainable economic development. Authors applied three indices, namely the

Global Innovation Index (GII), the European Innovation Scoreboard (EIS) and the Global Competitiveness Index (GCI). Innovation potential determining sustainability of the economies was calculated with the multicriteria methods (for example Analytic Hierarchy Process). The performed analysis confirmed a significant problem in implementing the sustainable development concept and innovation activities worldwide. According to Kastrinos and Weber [6], leading the world towards the Sustainable Development Goals (SDGs) is an ambition of the European Union (EU) with special attention on innovation potential. The success of EU in achieving SDGs depend on the effectivity of R&D and innovation policy. In this context, authors emphasis the issue of foresight in innovation development which will contribute to sustainability transitions in Europe. Mozas-Moral, Bernal-Jurado, Fernandez-Ucles and Medina-Viruel [7] investigated the Sustainable Development Goals (SDG) in relation to the Information and Communication Technologies (ICT). Specifically, study examined factors for the active use of technological innovation. In order to achieve the paper aims, the fuzzy set Qualitative Comparative Analysis (fsQCA) was used. The results showed that ICT contributes significantly to sustainability of firms and economies.

The countries cannot achieve sustainable economic development without a favourable business environment. In this context, Cervello-Royo, Moya-Clemente, Perello-Marin and Ribes-Giner [8] analyzed the economic and sustainable development factors of countries that may be related to opportunity-driven entrepreneurship. Research was realized based on data of 57 countries for 2017. In order to achieve settled goal, a cross-national analysis using the fsQCA methodology was implemented. Data were retrieved from four databases: Global Entrepreneurship Monitor, Country Risk Score, ***Doing Business Index (DBI)***, and Sustainable Development Goals Index. The findings revealed the positive interconnection between above-mentioned variables. According to Corrales-Garay, Mora-Valentin and Ortiz-de-Urbina-Criado [9] entrepreneurship belong to the key element in the sustainable development field. Presented study suggested a conceptual model for analyzing entrepreneurship through open data. For his purpose, a descriptive analysis and a co-word analysis were performed. Paper results showed how business agents used information published by public administrations to generate new entrepreneurial initiatives, especially novel business models associated with new mobile applications. Thereafter, these activities contribute to creation of a value added in building initiative toward to sustainable prosperity. In the context of sustainable growth, Babajide, Lawal, Asaleye, Okafor and Osuma [10] examined the relationship between financial stability and entrepreneurship conditions in connection to the selected SDGs. The pooled ordinary least squares (OLS) and random effects techniques were applied in research paper. The findings revealed that financial stability of economies, as well as, business environment have a significant positive effect on achieving sustainable economic goals.

In recent years, the growing number of studies examine the impact of economic freedom and democracy on sustainable of economies. In this backdrop, Economou [11] examined the impact of economic freedom from the point view of foreign direct investment (FDI) inflows on economic sustainability and competitiveness of the European Union economies for the last 20 years. Findings confirmed that economic freedom have a positive impact on FDI inflows, which lead indirectly to sustainable economic growth. The other research devoted to economic freedom and sustainable development was presented by Mushtaq and Ali Khan [12]. They examined the effect of economic freedom, expressed by ***Economic Freedom Index (EFI)*** on sustainable development, measured by Sustainable Development Index (SDI) of countries. The paper findings revealed that economic freedom positively influences sustainable economic growth. Iacobuta and Gagea [13] state that institutional quality lies at the bottom of achieving sustainability level. For this reason, authors examined the relationship between economic freedom (as a measure of institutional quality), and sustainable development within EU27 countries using EFI and Sustainable Society Index (SSI). The empirical findings confirmed that components of economic freedom (property rights or business freedom) significantly correlate with sustainable development.

The issue of corruption is regarded as a one of the key barriers for sustainable economic development. In this context, sustainable Development Goals are committed to measuring aspects of corruption over time. The role of corruption in public sector was evaluated by Zouaoui, Qudah, and Arab [14]. Authors focused on analyzing the ***Corruption Perception Index (CPI)*** over 2000-2014 for all countries included in the CPI using hierarchical classification methods. The aim of paper was to group countries according level of corruption and investigate the volatility of corruption. As reported by Fanea-Ivanovici, Muşetescu, Pană and Voicu [15], fighting corruption and enhancing tax compliance through digital public services represent key factors for increasing sustainable development. Armeanu, Vintilă and Gherghina [16] empirically investigated the drivers of sustainable economic development within EU(28) countries related to economic indicators, corruption, business environment, infrastructure and technology. By means of panel data regression models were examined above-mentioned factors of sustainability growth. In terms of economies analyzed, the CPI confirmed a positive association with sustainable economic growth.

In every economy and society people have an enormous impact on creating a better future. Human resources are belonged to the most important elements of competitiveness and sustainable performance. In this backdrop, Aguiar, Tome, Farias, Machado and Ribeiro [17] claim that the sustainable growth and competitive advantages of countries are influenced by human development level. Authors devoted to examining the relations between government actions and ***Human Development Index (HDI)***. The paper findings revealed the existence of a significant positive correlation between the HDI and government action plans. Liu, Brown and Casazza [18] pointed to fact that various sustainability indicators tend to reflect different or even converse outcomes. In this context, authors investigated and compared five sustainable development indicators (HDI, EPI, ESI, and Emergy-based Sustainability Index and Surplus Biocapacity). The research study confirmed a significant correlation among variables.

3 Data and Methodology

The main aim of paper is to reveal the impact of selected global multicriteria indices (GII, DBI, EFI, CPI and HDI) on sustainable development (representing by SEDA indicator) for the period of 2011 to 2018. To perform the analysis, we choose the aggregation of EU(28) countries. In this context, the following research question (RQ) was formulated:

RQ: *Is there a statistically significant relations between the sustainable economic growth and selected multicriteria determinants of sustainable development within the EU(28) countries for the period analyzed?*

The secondary data concerning to individual indices was collected on the basis of annually online reports published by international organizations, such as [19 - 23, 1]. Among other things, the following research is also focused on deeper analysis of the selected six multi-criteria indices within EU member's states, so a brief overview of their framework is given in the Table 1 below.

Table 1. Brief structure framework of the indices analysed

Index	Brief index overview
SEDA Sustainable Economic Development Assessment	SEDA is a proprietary diagnostic tool for measuring well-being and identifying priorities for sustainable growth. The assessment relies on a total 40 indicators and the overall SEDA score represents the aggregation of the score for the 10 dimensions integrated into three main subindices: ▪ 1st Subindex – *Economics* (1st – 3rd dimension) – Income, Economic Stability and Employment; ▪ 2nd Subindex – *Investments* (4th – 6th dimension) – Education, Health and Infrastructure; ▪ 3rd Subindex – *Sustainability* (7th – 10th dimension) – Income equality, Civil Society, Governance and Environment.
GII Global Innovation Index	GII was launched in order to measure innovation progress of society. The final GII score is the simple average of two main subindices. Each of them consists of pillars and overall GII indicator examines approximately 80 partial indicators: ▪ 1st Subindex – *Innovation Input* (1st – 5th pillar) – Institutions, Human capital and research, Infrastructure, Market sophistication and Business sophistication; ▪ 2nd Subindex – *Innovation Output* (6th – 7th pillar) – Knowledge and technology, Creative outputs.
DBI Doing Business Index	DBI measures aspects of business regulation affecting small and medium-size domestic firms. The methodology of DBI calculates the percentile for the individual indicators of the economy. The overall DBI is calculated on the basis of 11 different areas (dimensions) of business regulation (indicator set), whereas the first ten DBI dimensions are included in the DBI score and the last dimension (*11th Labor market measures*) only provides information on the state and development in this area: ▪ 1st Dimension – *Starting a business;* 6th Dimension – *Protecting minority investors;* ▪ 2nd Dimension – *Dealing with permits;* 7th Dimension – *Paying taxes;* ▪ 3rd Dimension – *Getting electricity;* 8th Dimension – *Trading across borders;* ▪ 4th Dimension – *Registering property;,* 9th Dimension – *Enforcing contracts;* ▪ 5th Dimension – *Getting credit;* 10th Dimension – *Resolving insolvency.*
EFI Economic Freedom Index	Economic freedom is the fundamental right of every human to control his or her own labor and property. The overall EFI score consist of twelve aspects of economic freedom that covers own formulas and methodology used to compute the overall score which is grouped into four broad categories: ▪ 1st Pillar – *Rule of law* (1st – 3rd component) – Property rights, Government integrity and Judicial effectiveness; ▪ 2nd Pillar – *Government size* (4th – 6th component) – Government spending, Tax burden and Fiscal health; ▪ 3rd Pillar – *Regulatory efficiency* (7th – 9th component) – Business freedom, Labor freedom and Monetary freedom; ▪ 4th Pillar – *Market openness* (10th – 12th component) – Trade freedom, Investment freedom and Financial freedom.

CPI Corruption Perception Index	CPI assess countries based on how corrupt a country's public sector is perceived to be by experts and business executives. The composite CPI is based on the hundreds specific question used to collect data capturing the six main aspects of corruption: ▪ 1st Dimension – *Bribery*; ▪ 2nd Dimension – *Diversion of public funds*; ▪ 3rd Dimension – *Prevalence of officials* using public office for private gain; ▪ 4th Dimension – *Ability of governments to contain corruption* and enforce effective mechanisms; ▪ 5th Dimension - *Red tape*; ▪ 6th Dimension - *Excessive bureaucratic burden*.
HDI Human Development Index	HDI quantify a different elements of inequalities in various aspects of their lives. HDI is a summary measure of achievements in three key dimensions of human development: ▪ 1st Dimension – *Long and healthy life* – Life expectancy at birth (Index); ▪ 2nd Dimension – *Knowledge* – Expected years of schooling and Mean years of schooling (Education Index); ▪ 3rd Dimension - *A decent standard of living* – GNI per capita index.

Source: Own processing.

The basic output of descriptive statistics within the selected global multicriteria indices is presented in the following Table 2. The analysis was realized for the period of 2011 to 2018.

Table 2. Descriptive statistics of the selected global multicriteria indices

Descriptive statistics	SEDA	GII	DBI	EFI	CPI	HDI
Mean	72.48	49.51	74.30	69.07	65.40	87.62
Median	71.25	49.20	74.88	69.00	64.50	87.65
Std. Deviation	8.22	7.55	5.75	5.65	15.23	3.84
Skewness	-0.02	0.15	-0.22	-0.39	-0.08	-0.26
Kurtosis	-0.95	-1.07	-0.68	-0.45	-1.05	-0.83
Minimum	55.40	34.20	60.04	53.20	33.00	78.20
Maximum	85.40	64.80	85.66	80.40	94.00	94.20
Percentiles 25	67.00	42.85	70.15	65.10	54.00	84.70
Percentiles 75	81.40	56.35	78.90	73.95	78.00	90.80

Source: Own processing.

In order to compare data, one simple adjustment was realized. In the case of HDI indicator was indicated different score range, so indices comparisons were not possible. For this reason, we decided to transform the original HDI scores (range from 0 to 1) to the same range (from 0 to 100), as in the case of other indices, as follows:

$$I_{adjusted} = I_{basic} * 100 \quad (1)$$

To investigate the relations between the selected independent variable Y (SEDA) and other dependent variables X_1 (GII), X_2 (DBI), X_3 (EFI), X_4 (CPI) and X_5 (HDI), panel data regression analyses was used. Panel data regression analysis presenting one of the multidimensional statistical methods. Within presented research, a ***Fixed Effect Model*** (FEM) was applied, which can be expressed by the following model [24]:

$$y_{it} = \beta_1 x_{it1} + \beta_2 x_{it2} + \dots + \beta_k x_{itk} + \alpha_1 z_{i1} + \alpha_2 z_{i2} + \dots + \alpha_{1q} z_{iq} + u_{it} \quad (2)$$

where *y_{it} - i^{th} value of variable Y in the basic file; X_1-X_k - explanatory variables; i - cross-sectional dimension; t - time dimension; β_0 - intersection of the y-axis with the regression line; Z_1-Z_q - individual effects; β_n - regression coefficient in the basic file; u_{it} - i^{th} random error of variable.*

If the individual effects of Z_1 and Z_q are unobservable but correlated with explanatory variables, then the FEM model has the following form:

$$y_{it} = \alpha_i + \beta_1 x_{it1} + \beta_2 x_{it2} + \dots + \beta_k x_{itk} + u_{it} \quad (3)$$

where *α_i - specific constant for each cross-sectional unit.*

To process the above-mentioned data using selected statistical method, the STATISTICA software (13th ed.) was used.

4 Empirical results

To achieve settled paper aims, the following research analysis was performed. To begin with, the global multicriteria assessment of EU member states was made based on selected global indices. In the next section of paper are presented the results of panel data regression analysis in order to examined the impact of selected global indices (GII, DBI, EFI, CPI and HDI) on sustainable development. All analyses were realized for the years 2011-2018.

4.1 Global multicriteria assessment of EU member states

The first part of the presented research results is devoted to an in-deeper analysis of selected input global indices development categorized into two groups - SEDA indicator as the main variable of sustainability growth evaluation and others multicriteria indices as key determinants of sustainable potential. The average SEDA score development within EU member states over the period analyzed is presented in the Figure 1 below.

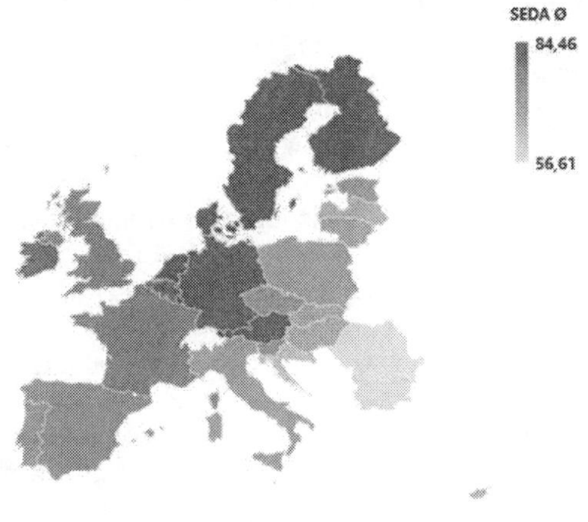

Fig. 1. The average SEDA score of the EU member states over the years 2011-2018
Source: Own processing based on Sustainable Economic Development Assessment interactive guide.

The average SEDA score of EU(28) countries (72.48) ranged from a minimum value of 55.40 recorded by Romania, up to the maximum value of 85.40, indicated in the case of Sweden. Over the period analyzed was revealed an increasing trend of sustainable development (an increase about 0.97%). Despite of this positive findings, it was found the decrease of the score almost in all EU countries during the 2013, 2014 and 2017.

The next paper subsection was focused on an in-depth development analysis of the five selected global multicriteria indices (GII, DBI, EFI, CPI and HDI). In the following Figure 2 are presented the average indices score development within EU member states.

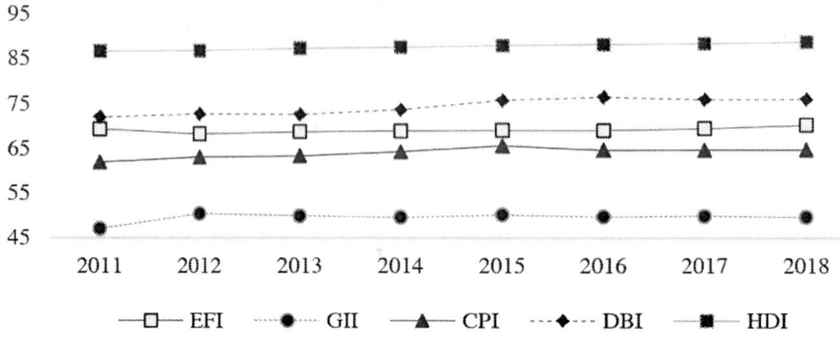

Fig. 2. The average global indices score of the EU member states over the years 2011-2018
Source: Own processing based on the selected global multicriteria indices reports.

Over the period analyzed, the ***GII*** score achieved value at 49.73 per year, on average. In 2010 was recorded the lowest values score (46.87) and during the analyzed years the innovation level increased about 5.85%. To the most innovative countries within EU were included Netherlands, United Kingdom and Sweden. On the other hand, Greece, Romania and Croatia were included among the countries with the lowest achieved score of GII. Furthermore, the overall GII ranking showed that exactly 14 countries ranged above and 14 below of EU average score.

The assessment of EU member states on the basis of indicator ***DBI*** showed unstable development of business environment. Despite of the positive increase trend (about 5.58%; from 2011 to 2018), in 2013 and 2017 was revealed the significant year-on-year decrease. The leading position in the creating of suitable business environment belonged to Denmark, United Kingdom and Sweden (score at level of 82.92 on average), while the lowest innovation score was found in Malta, Greece and Croatia. Moreover, the overall DBI ranking showed that 12 countries ranged above and 16 below of EU average score.

When evaluating the average scores of ***EFI*** for the period of 2011 to 2018, an increasing development trend is visible, but only about 1.39% on average. The EFI score ranged from the lowest average value of 69.25 in 2011, to the highest value reached in 2018 (70.21). Over the period analyzed, on average, the EU member states achieved the higher score than value 70 only in the last analyzed year. The best conditions for economic freedom were created in Ireland, Estonia and Denmark. On the other hand, the lowest level of economic freedom was revealed in Greece, Croatia and Italy. In this regard, the proportion of countries in EFI ranking was balanced toward to EU average score (14 countries above and 14 below EU average score).

One of the other determinants of sustainable development of countries was corruption, expressed through ***CPI*** indicator. At the beginning it is important to emphasise that score scale range from 0 (highly corrupt) to 100 (very clean). Looking at the development trend of this index for the EU(28) countries, the average values increased significantly from 61.89 (in 2011) to 64.68 (in 2018). By comparison of CPI score range (max to min) was revealed significant findings. In the case of Bulgaria was found the lowest score at level of 33.00 (highly corrupt), while the maximum score was indicated for Finland. Both these results was reached in 2010. The proportion of countries in CPI ranking in respect of EU average was identical as in the case of DBI.

The last determinants of sustainable economic assessment was indicator ***HDI***. Over the period analyzed, the HDI score achieved the positive trend (about 2.66%). The lowest score was revealed in 2010, while the highest score in 2018. In the case of this index only was recorded an increasing development in every single analyzed year. To the countries with the best conditions for life were belonged Denmark, Netherlands and Germany. On the other hand, Bulgaria, Romania and Croatia were indicated as the worst in creating suitable conditions for society. Furthermore, up to 15 countries achieved score above EU average, while only 13 ranked below EU average.

After considering the average score of selected multicriteria indices from 2011 to 2018, we also compiled a two-dimensional figure, whereby the results are presented in below (Figure 3). In order to assess indices overall, the indicator SEDA (as a key element of sustainable potential assessment) was compared with selected global multicriteria indices (on average) during the period analyzed. Findings confirmed research expectations. To the best-performed countries in field of sustainability, innovation, business environment, economic freedom, corruption and human development belong Denmark, Finland and Sweden. On the other side, to the worst-performed were included Romania, Bulgaria and Greece. Looking at the Figure 3, it can be compiled three clusters of countries according to above-mentioned scores of global indices. This classification can be useful tool for policy-makers in building nation action plans and country-to-country comparison.

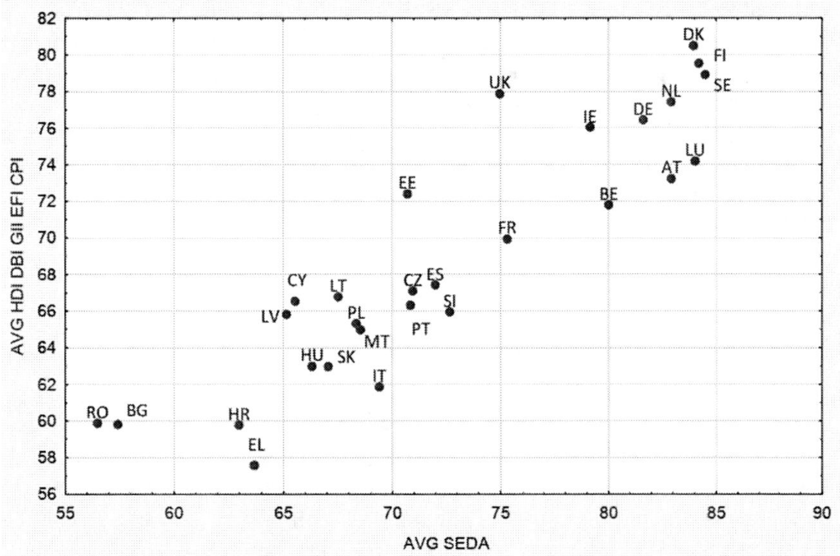

Fig. 3. The scatterplot of average SEDA against other global indices over the years 2011-2018
Source: Own processing.

4.2 Results of panel data regression analysis

The last subsection of presented study was devoted to the quantitative assessment of impact the selected determinants of sustainable growth, expressed through selected global multicriteria indices, such as GII, DBI, EFI, CPI and HDI on sustainable economic development (SEDA index) within the EU(28) member states over the period of 2011-2018. For purpose of achieve settled research aim, the panel data regression analyses were performed. In the following Table 3 are presented the results of regression analysis. In presented research, a ***Fixed Effect Model*** (FEM) was used.

Table 3. Panel data regression analysis between SEDA and selected multicriteria indices

Panel Data Regression Summary for Dependent Variable: **SEDA**
Fixed Effects Model (**FEM**)
R= .0.9376; R^2= .0.8790; Adjusted R^2= .0.8763;
F(5.218)=317.06; p<0.0000; Std. Error of estimate: 2.8908
N=224 observations included 28 cross-sectional units; Time series length=8

	Coefficient	**Std. Err.**	**t-ratio**	**p-value**
Intercept	**-46.1436**	**7.9777**	**-5.7841**	**0.0000***
GII	**0.1718**	**0.0578**	**2.9740**	**0.0030***
DBI	**-0.0918**	**0.0469**	**-2.8224**	**0.0052***
EFI	0.0622	0.0524	1.1879	0.2362
CPI	**0.1641**	**0.0254**	**6.4488**	**0.0000***
HDI	**0.5637**	**0.1032**	**11.6778**	**0.0000***

* indicates significance level at 0.05 level.

Source: Own processing.

Based on the appropriateness analysis of the regression model (Table 3), the regression equation is appropriate with regard to Fisher's test criterion at the selected significance level $\alpha = 5\%$ (p = 0.0000). The coefficient of determination (adjusted R^2), which explains the variability of the dependent variable (SEDA), achieved value 87.63%. The regression analysis confirmed the statistically significant impact (p=0.0000) of four determinants of sustainability (GII, DBI, CPI and HDI) on sustainable economic growth of EU(28) countries (measured by indicators SEDA) over the period analyzed. Moreover, findings showed that three global indices (GII, CPI and HDI) influenced the SEDA positively, while DBI indicator have negative impact (-0.0918). The most significant impact on sustainable prosperity was indicated in the case of indicator HDI (0.5637). The relations between variables can be expressed by the following panel data regression model:

$$SEDA = -46.1436 + 0.1718GII - 0.0918DBI + 0.1641CPI + 0.5637HDI$$

5 Discussion

It is more important than ever for policy-makers to develop and implement a complex strategies to achieve the sustainable development. As Gu, Wand, Hua and Liu [25] stated, continued economic development has become an inevitable requirement for sustainable growth. Authors applied triple bottom line of sustainable development in order to measure the level of economic prosperity. Findings confirmed a close relation between entrepreneurship and sustainability. According to Sira, Vavrek, Vozarova and Kotulic [26], human resources, knowledge and its creation, transfer and preservation becoming a key element in ensuring a country's sustainable position in a competitive environment. By modelling of macroeconomic context of current economic sustainable development through indicators as inflation, unemployment and corruption was dealt by Litavcova, Jencova and Vasanicova [27]. The results confirmed a significant positive impact of above-mentioned macroeconomic indicators on sustainable growth. As reported by Stefko, Bacik, Fedorko, Olearova and Rigelsky [28], knowledge of the options that digital technology and innovation bring, leads to the right set of strategies and will follow the effective achievement of sustainability aims. Dobrovic, Gallo, Mihalcova, Stofova and Szaryszova [29], examined the relations between innovation, economic sustainability and competitiveness, considering the results of performance assessment models within the EU countries. Paper results revealed a positive relationship among composite indicators.

All above-mentioned studied (including literature review) are in line with this paper findings and lead to the conclusion that innovation potential, conditions for entrepreneurship, corruption problems and human resources belong to the key determinants of sustainable economic growth.

6 Conclusion

Understanding the progress countries make over time, both overall and at the level of specific dimensions, is critical to setting more-effective national priorities and strategies. Ongoing economic, social, political changes, and many other factors, cause the differences in the global sustainability of economies. In this context, Kiselakova, Sofrankova, Onuferova and Cabinova [24] investigated the relations between various multicriteria macroeconomic indices. This paper was elaborated as a continuation of the previous research.

The main aim of paper was to reveal the impact of selected global multicriteria indices (GII, DBI, EFI, CPI and HDI) on sustainable economic development (representing by SEDA indicator) for the years of 2011 to 2018. To perform the analysis, the aggregation of EU(28) countries was chosen. The research aim assigned was met.

Based on the achieved average values of the selected global multicriteria indices and by their comparison within the European Union area was found:

- In the case of the Sustainable Economic Development Assessment (SEDA), the Finland achieved the best-performed (84.51), while Romania was ranked to the worst-performed countries (56.52).
- In the case of the multicriteria determinants (represented by global indices such as GII, DBI, EFI, CPI an HDI), to the best-performed country belong Denmark (80.48) and, as the worst-performed country was identified Greece (57.58).
- By comparison of EU(28) member states, the most significant score range (on average) was recorded within CPI assessment (49.88) and the lowest score range was found within the HDI assessment (13.14).
- Results of the panel data regression analysis confirmed that sustainable economic development is influenced significantly by such determinants as innovation and technology, business conditions, corruption environment and potential of human resources.

Sustainable development is every country's goal, and the ability to convert wealth into well-being varies widely among countries, reflecting different societal choices and policy decisions. Governments grapple every day to balance competing priorities. Deciding where and how to deploy their effort and resources is a central challenge. In this regard, presented results could be useful in fulfilling the already established EU priorities for the period 2019- 2024 [30]: A European Green Deal, A Europe fit for the digital age, An economy that works for people, A stronger Europe in the world, Promoting European way of life, and A new push for European democracy.

This article is one of outputs of project VEGA No. 1/0279/19 "Model approaches to increase performance and competitiveness in the European area in the context of sustainable development".

References

1. United States [online], Available at: https://repositorio.cepal.org/bitstream/handle/11362/40156/S1801140_en.pdf?sequence=25 (2019)
2. E. Huttmanova, T. Valentiny, Assessment of the economic pillar and environmental pillar of sustainable development in the European Union. European Journal of Sustainable Development, 8(2), 289-298 (2019)
3. V. Strezov, A. Evans, T. J. Evans, Assessment of the economic, social and environmental dimensions of the indicators for sustainable development. *Sustainable Development*, **25**(3), 242-253 (2017)
4. T. Bassetti, S. Blasi, S. R. Sedita, The management of sustainable development: A longitudinal analysis of the effects of environmental performance on economic performance. *Business Strategy and the Environment*, (2020)
5. O. I. Zhylinska, V. H. Balan, I. V. Andrusiak, A comparative assessment of the level of innovation support for sustainable economic development. *Marketing and Management of Innovations*, **3**, 354-364 (2017)
6. N. Kastrinos, K. M. Weber, Sustainable development goals in the research and innovation policy of the European Union. *Technological Forecasting and Social Change*, **157**, (2020)
7. A. Mozas-Moral, E. Bernal-Jurado, D. Fernandez-Ucles, M. J. Medina-Viruel, Innovation as the backbone of sustainable development goals. *Sustainability*, **12**(11), (2020)
8. R. Cervello-Royo, I. Moya-Clemente, M. R. Perello-Marin, G. Ribes-Giner, Sustainable development, economic and financial factors, that influence the opportunity-driven entrepreneurship. An fsQCA approach. *Journal of Business Research*, **115**, 393-402 (2020)
9. D. Corrales-Garay, E. M. Mora-Valentin, M. Ortiz-de-Urbina-Criado, Entrepreneurship through open data: An opportunity for sustainable development. *Sustainability*, **12**(12), (2020)
10. A. Babajide, A. Lawal, A. Asaleye, T. Okafor, G. Osuma, Financial stability and entrepreneurship development in sub-Sahara Africa: Implications for sustainable development goals. *Cogent Social Sciences*, **6**(1), 1-25 (2020)
11. F. Economou, Economic freedom and asymmetric crisis effects on FDI inflow: The case of four South European economies. *Research in International Business and Finance*, **49**, 114-126 (2019)
12. S. Mushtaq, R. E. Ali Khan, Economic freedom and sustainable development: A panel data analysis. *The Pakistan Journal of Social Issues*, 89-97 (2018)
13. A. Iacobuta, M. Gagea, Institutional quality, economic freedom and sustainable development. A comparative analysis of EU countries. *17th International Economic Conference on The Economic World Destiny - Crisis and Globalization?*, (2010)
14. A. Zouaoui, A. A. Qudah, M. B. Arab, World Corruption Perception Index Analysis. *Research Journal of Finance and Accounting*, **8**(24), (2017)
15. M. Fanea-Ivanovici, R. C. Muşetescu, M. C. Pană, C. Voicu, Fighting corruption and enhancing tax compliance through digitization: Achieving sustainable development in Romania. *Sustainability*, **11**, 1-24 (2019)

16. D.Ş. Armeanu, G. Vintilă, Ş.C. Gherghina, Empirical study towards the drivers of sustainable economic growth in EU-28 countries. *Sustainability*, **10**(1), 1-22 (2018)
17. A. Aguiar, M. A. Tome, A. D. Farias, C. Machado, D. N. M. Ribeiro, Relationship between country governance indicators and Human development index. *Gestao e Desenvolvimento*, **16**, 53-70 (2019)
18. G. Liu, M. T. Brown, M. Casazza, Enhancing the sustainability narrative through a deeper understanding of sustainable development indicators. *Sustainability*, **9**(6), 1-19 (2017)
19. Boston Consulting Group [online], Available at: http://hdr.undp.org/sites/default/files/hdr2019.pdf (2019)
20. Heritage Foundation [online], Available at: https://www.heritage.org/index/pdf/2019/book/index_2019.pdf (2019)
21. INSEAD and the World Intellectual Property Organization (WIPO) [online], Available at: https://www.globalinnovationindex.org/gii-2019-report (2019)
22. Transparency International [online], Available at: https://www.transparency.org/files/content/pages/2019_CPI_Report_EN.pdf (2019)
23. World Bank Group [online], Available at: https://www.doingbusiness.org/content/dam/doingBusiness/media/Annual-Reports/English/DB2019-report_web-version.pdf (2019)
24. D. Kiselakova, B. Sofrankova, E. Onuferova, V. Cabinova, The evaluation of competitive position of EU-28 economies with using global multi-criteria indices. *Equilibrium-Quarterly Journal of Economics and Economic Policy*, **14**(3), 441-462 (2019)
25. W. T. Gu, J. Y. Wand, X. Y. Hua, Z. D. Liu, Entrepreneurship and high-quality economic development: based on the triple bottom line of sustainable development. *International Entrepreneurship and Management Journal*, (2020)
26. E. Sira, R. Vavrek, I. K. Vozarova, R. Kotulic, Knowledge economy indicators and their impact on the sustainable competitiveness of the EU countries. *Sustainability*, **12**(10), 1-22 (2020)
27. E. Litavcova, S. Jencova, P. Vasanicova, On modelling of macroeconomic context of current economic development. *34th International Conference Mathematical Methods in Economics (MME)*, 506-511 (2016)
28. R. Stefko, R. Bacik, R. Fedorko, M. Olearova, M. Rigelsky, Analysis of consumer preferences related to the use of digital devices in the e-commerce dimension. *Entrepreneurship and Sustainability Issues*, **7**(1), 25-33 (2019)
29. J. Dobrovic, P. Gallo, B. Mihalcova, L. Stofova, P. Szaryszova, Competitiveness measurement in terms of the Europe 2020 Strategy. *Journal of Competitiveness*, **10**(4), 21-37 (2018)
30. European Commission [online], Available at: https://ec.europa.eu/info/strategy/priorities-2019-2024 (2019)

Value Added Tax and Its Influence on Pricing and Price Decision Making of Companies – A Case Study

Anna Kubjatkova[1,*], *Anna* Krizanova[1], and *Veronika* Jurickova[1]

[1]University of Zilina, The Faculty of Operation and Economics of Transport and Communications, Department of Economics, Zilina, Slovakia,

Abstract. Pricing is a complex process that is influenced by many factors, which companies must constantly take into account when creating prices. The final selling price is formed by various influences, including the impact of value added tax (VAT). This tax affects all final consumers, but it also affects the businesses themselves in setting their sales prices. The main goal of this paper is to analyze the impact of VAT on pricing and to prove the fact that VAT values affect both the payer and the non-payer of VAT, if they buy from the VAT payer. The purpose is to point out the difference in the selling price between these two entities and thus to prove the influence of VAT on their pricing. In this paper, the methods of induction and deduction were used, as well as methods of analysis and synthesis. Based on the performed analyzes, we proved the influence of VAT on the pricing of both the payer and the non-payer of VAT, while the difference is observable in the amount of their selling price and tax liability.
Keywords: pricing, VAT, VAT payer, VAT non-payer, selling price

1 Introduction

It is in the interest of each state to control the tax compliance of all taxable persons, whether legal or natural persons. Taxes are the most important item of state budget revenue. Building an economically sustainable tax system is key to securing government activities [1]. An effective tax system creates the conditions in the country for a quality business environment necessary for the growth of a market economy [2]. All taxpayers are therefore equally important, and it is not just a sector of trade that is considered to be the power of the economy, especially in developing countries [3].

The tax system of the Slovak Republic consists of direct and indirect taxes. Direct taxes include income tax, real estate tax, or motor vehicle tax. In the case of direct taxes, the person is obliged to pay them and pay them to the state budget, while in the case of an employee-employer relationship; the employer is obliged to pay this tax on behalf of the employee. The principle of payment of indirect taxes is that one person pays them and the other person makes payment to the state budget. Indirect taxes consist of value added tax

*Corresponding author: anna.kubjatkova@fpedas.uniza.sk

(VAT) and other excise taxes, while indirect taxes especially VAT, are, according to Kufelova [4] a more important component of the state budget than direct taxes. Therefore, tax audits are also carried out more frequently under indirect taxes. The reason is also that with VAT the possibility of tax evasion is much greater because VAT significantly more affects the solvency of individual tax subjects, which also affects the solvency of the state itself.

The risk of tax evasion is a constant threat in all countries, with taxpayers often aware of the fact that control authorities are severely limited in certain situations. According to Hybka, the risk of VAT evasion is also increased by the fact that the system and procedure for applying VAT to countries are often confusing and complicated [5].

VAT is an effective tool for the economy for several reasons. Author Cnossen [6] argues that, in general, VAT reduces the marginal cost of public finances and is thus an effective tool for increasing the tax ratio of countries that have included VAT in their tax system. According to author Adhikari [7], the effectiveness of VAT is also based on the assumption that this tax is designed effectively and is well enforceable. Loss of VAT effectiveness can occur, for example, in the exemption of small companies from VAT, or the exemption from VAT of problem-taxing agricultural and service companies in developing countries. The author notes in his research that most studies aimed at increasing the efficiency of VAT focus on high-income countries and therefore there is little evidence to assess the increase in VAT efficiency across rich and poor countries. Poorer countries have a much weaker VAT collection system, which means that the efficiency gains from VAT are questionable for them.

The pricing of a company depends on several factors, while its process is formed by a many pricing decisions, while the prosperity of not only Slovak companies is influenced by various important determinants [8]. The inherent influence of psychological aspects in pricing cannot be denied [9]. According to Valaskova, Kliestikova, and Krizanova [10] and Vatankhah, Zarra-Nezhad, and Amirnejad [11], not only knowledge of legislation but also consumer preferences is key to pricing. According to Chetthamrongchai and Saengchai [12] monitoring customer loyalty and loyalty and the factors that influence them also increase the efficiency of the pricing process and its outcome. The result of the company's pricing is an efficient pricing system and the selling price itself, in which all pricing decisions are reflected. The influence of VAT on the process of pricing and pricing of the company is clear in the final phase of creating and shaping the selling price. It is, therefore, necessary for companies to apply the VAT system correctly and to know its current rate.

The final customer and his satisfaction is the key to the long-term survival of the company [13] and as the final link in the retail chain, it is influenced by the amount of the selling price, while VAT is a part of it. However, this customer is often unaware of the presence of this component of the selling price and this affects his behavior. Several authors deal with the issue of the shopping behavior of customers in the Slovak Republic from different perspectives. The author Majerova [14] conducted a survey of the specifics of the behavior of Slovak customers on the Internet. The authors Gajanova, Nadanyiova, and Moravcikova [15] analyzed customer segments based on demographic and psychographic aspects. Researches by different authors from the environment of consumer behavior focused on green marketing is also a benefit in this area [16, 17]. There is, therefore, a large amount of research in the field of consumer behavior in the Slovak Republic, but in the environment of this country, there is a great lack of studies dealing with the impact of VAT on the behavior of consumers or companies themselves.

2 Methodology

The current amount of VAT in the Slovak Republic is regulated by Act no. 222/2004 Coll. on value added tax. This legal regulation recognizes 2 types of VAT, namely:

- basic VAT rate of 20%,
- reduced VAT rate of 10%.

A company needs to be able to assign the correct VAT rate to the goods or services it offers. According to the law, the basic VAT rate applies to all goods and services that are not subject to a reduced tax rate. According to legal regulation, the reduced tax rate applies to the following groups of goods and services:

- foods such as specific meat, fish or healthy foods,
- medicines and selected medical supplies,
- books, magazines, and periodicals,
- accommodation services [18].

2.1 VAT payer vs. VAT non-payer

Companies must distinguish between the payer and the non-payer of VAT, as this affects the selling price as a result of the company's pricing. A VAT payer in the Slovak Republic is an entity that:

- exceeded the turnover of 49 790 € during 12 consecutive months of business,
- purchased goods or services from the EU whose total value in a calendar year exceeds 14 000 €.

A non-VAT payer is an entity that does not pay value added tax but does not use the right to deduct it. The most frequent non-payers of the tax are small self-employed persons, ie self-employed persons. Companies, non-profit and other organizations can also be non-payers of value added tax [18].

An important issue in setting the price and thus in determining the selling price is not only whether the company itself is a payer or non-payer of VAT, but also whether the seller (supplier) is a payer or non-payer of VAT. It very often happens that a company, as a VAT payer, buys goods or a service from a non-VAT payer, and when it sells those goods or services to the final customer, the company no longer counts the sales amount with VAT but without VAT. It can be argued that the relationship that if a payer buys from a non-payer, he also sells without VAT no longer applies. This error affects the selling price to a large extent, with the payer not paying value added tax to the State and thus committing tax fraud.

2.2 Tax liability calculation method

The value added tax return compares output tax (tax on the invoice issued to the supplier) and input tax (tax on the invoice received by the customer from the supplier). If the company has a higher amount of output, it means that it invoiced more for VAT than it has the amount on invoices received from the supplier, it means that the company has incurred tax liability. In this case, the company must pay to the state the amount it has calculated in the tax return. The opposite is the case of excessive deduction when the company has a higher amount of VAT on supplier invoices than the amount of VAT invoiced by the customer. In this case, the state will reimburse the company for the amount of the excess deduction [4].

To calculate the tax liability, the company must have have information such as:

- purchase price without VAT - purchase price of inputs without VAT,
- added value - operating costs and profit,

- VAT rate - 20% (unless it is a reduced tax rate).

Tax liability calculation formula [18]:

$$\mathbf{TL = OVAT - IVAT} \tag{1}$$

Where:
TL = tax liability, OVAT = output VAT, IVAT = input VAT

The input VAT item is the relevant percentage (20%) of the purchase price excluding VAT. The company calculates the output VAT item as an appropriate percentage (20%) of the tax base [18]:

$$\mathbf{TB = PPW + AV} \tag{2}$$

Where:
TB = tax base, PPW = purchase price without VAT, AV = added

The effect of VAT on a company's pricing is reflected in the selling price, which the company determines according to whether it is a VAT payer or a non-payer. Most business entities purchase material from another VAT payer, so the procedure is as follows. The VAT payer determines the selling price as follows [18]:

$$\mathbf{SPV = TB + OVAT} \tag{3}$$

Where:
SPV = selling price with VAT

The VAT non-payer will determine his selling price according to the formula [18]:

$$\mathbf{SP = PPV + AV} \tag{4}$$

Where:
SP = selling price, PPV = purchase price with VAT

The difference between a payer and a non-payer is therefore obvious. Unlike the non-payer, the VAT payer is also obliged to file a tax return for value added tax. Of course, companies determine the selling price as the optimal price to obtain the highest possible profit [19].

Classical procedures or even intelligent technologies can be used to calculate the tax liability. Their development and the creation of an integrated system can bring significant benefits not only in this area [20]. These systems can help make big data-based decisions to achieve results that can make the process itself time-saving [21].

3 Results

The impact of value added tax can be illustrated by a model example of a company, which we will mention once as a payer and once as a non-payer of VAT. The company buys material for the production of the product from the VAT payer. The purchase price of the material without VAT is 1 250 €. The company assumes a cost and profit of 700 € for this product. The rate of value added tax is basic, i.e. 20%.

The calculation of the selling price of a product where the company is a **VAT payer** is shown in Table 1:

Table 1. Calculation of the sales price of the VAT payer

Item name	Amount in €
PPW	1 250
IVAT	250
AV	700
TB	1 950
OVAT	390
Selling price	**2 340**

Source: Own processing.

Tax liability = 390 - 250 = 240 €

The calculation of the selling price of the product in case the company is a **VAT non-payer** is shown in Table 2:

Table 2. Calculation of the sales price of the VAT non-payer

Item name	Amount in €
PPV (PPW + IVAT)	1 500
AV	700
TB	2 200
OVAT	0
Selling price	**2 2000**

Source: Own processing.

If the company is not a VAT payer, its tax base is the purchase price as a whole, in which case it is the purchase price including VAT, as the company purchased from the VAT payer. If the company is a non-VAT payer, it has no tax liability, ie it does not file a VAT return.

We would like to emphasize that it is not possible to draw general conclusions from this case study, this research is aimed at providing an overview of the issue of the impact of value added tax on the payer and non-payer of VAT in specific conditions. This case study aims to provide information primarily to the general public through the analysis and processing of expertise.

4 Conclusions

In today's competitive environment, the importance of the company pricing process and its application on the market is growing. Properly set selling price is the key to gaining a competitive advantage and confirming market leadership. Therefore, companies must take into account all the factors that influence the creation of the selling price.

Based on the performed analysis, the difference in the sales price is visible, so it is appropriate to claim that VAT affects the company's pricing. Some companies also make pricing decisions based on VAT, which they have to take into account if they are VAT payers. However, it is clear from the example that non-VAT payers must also take into account the amount of VAT in the case of purchase from a VAT payer, so when purchasing material, non-VAT payers decide between VAT payers or non-payers as suppliers. Of course, to draw general conclusions, it is necessary to create additional simulations, which will be the subject of further research.

Pricing decisions and pricing of the company is a complex process, which is influenced by various factors, including the amount of VAT. Companies, whether or not they are VAT payers, must take their amount into account when creating the selling price of their products and services. In addition to being required to control the VAT calculation system, companies must also correctly determine the level of tax rate applied to specific products and services. Also, the control of the business relationship between the payer and the non-payer of VAT is in practice a necessary part of functioning in the market.

This is paper is an outcome of project VEGA no. 1/0718/18: The impact of psychographic aspects of pricing on the marketing strategy of companies across products and markets.

References

1. M. Schratzenstaller, *Sustainable tax policy: Concepts and indicators beyond the tax ratio*. Obs. Diagn. Econ. **141**(5), 55–77 (2015)
2. M. Buno, M. Nadanyiova, D. Hraskova, *The comparison of the quality of business environment in the countries of Visegrad group*. Procedia Economics and Finance, **26**, 423-430 (2015)
3. J. Vrbka, E. Nica, I. Podhorska, *The application of Kohonen networks for identification of leaders in the trade sector in Czechia*. Equilibrium. Quarterly Journal of Economics and Economic Policy, **14**(4), 739-761 (2019)
4. I. Kufelova, *Ceny a cenove rozhodovanie*. Bratislava: Wolters Kluwer. (2017)
5. M. M. Hybka, *Collection dilemmas and performance measures of the value-added tax in Germany and Poland*. International Journal of Management and Economics **54**(2), 110–121 (2018)
6. S. Cnossen, *VAT and agriculture: Lessons from Europe*. International Tax and Public Finance **25**(2), 519–551. (2018)
7. B. Adhikari, *Does a value-added tax increase economic efficiency?* Economic Inquiry. **58**(1), 496-517, (2020)
8. K. Valaskova, T. Kliestik, M. Kovacova, *Management of financial risks in Slovak enterprises using regression analysis*. Oeconomia Copernicana, **9**(1), 105–121 (2018)
9. K. Valaskova, V. Bartosova, P. Kubala, *Behavioural aspects of the financial decision-making*. Organizacija, **52**(1), 22-32 (2019)
10. K. Valaskova, J. Kliestikova, A. Krizanova, *Consumer perception of private label products: an empirical research*. Journal of Competitiveness, **10**(3), 149-163 (2018)
11. S. Vatankhah, M. Zarra-Nezhad, G. Amirnejad, *An empirical assessment of willingness to accept "low-cost" air transport services: Evidence from the Middle East*. Journal of Tourism and Services, **10**(18), 79-103 (2019)
12. P. Chetthamrongchai, S. Saengchai, *The impact of perceived service quality, customer perception and price strategy on pharmacy customer devotion. Polish* Journal of Management Studies, **20**(1), 139-148 (2019)
13. M. Nadanyiova, *The Possibility of Introducing the CRM System in Railway Company Cargo Slovakia*. 17th International Conference on Transport Means, 87-92 (2013)

14. J. Majerova, *Analysis of Specifics in Buying Behavior of Slovak Customers in Internet Environment.* 2nd International Conference on Social Sciences Research, **5**, 172-178(2014)
15. L. Gajanova, M. Nadanyiova, D. Moravcikova, *The Use of Demographic and Psychographic Segmentation to Creating Marketing Strategy of Brand Loyalty.* Scientific Annals of Economics and Business, **66**(1), 65-84 (2019)
16. D. Moravcikova, A. Krizanova, J. Kliestikova, M. Rypakova, *Green marketing as the source of the competitive advantage of the business.* Sustainability, **9**(12), 2218 (2017)
17. J. Majerova, *Analysis of Slovak Consumer's Perception of the Green Marketing Activities.* 4th World Conference on Business, Economics and Management, **26**, 553-560 (2015)
18. Act no. 222/2004 Coll. Value Added Tax Act as amended.
19. V. Oliinyk, O. Kozmenko, I. Wiebe, S. Kozmenko, *Optimal control over the process of innovative product diffusion: the case of Sony corporation.* Economics and Sociology, **11**(3), 265-285 (2018)
20. N. A. Zhuravleva, E. Nica, P. Durana, *Sustainable smart cities: networked digital technologies, cognitive big data analytics, and information technology-driven economy.* Geopolitics, History, and International Relations, **11**(2), 41–47 (2019)
21. T. D. Meyers, L. Vagner, K. Janoskova, I. Grecu, G. Grecu, *Big Data-driven algorithmic decision-making in selecting and managing employees: advanced predictive analytics, workforce metrics, and digital innovations for enhancing organizational human capital.* Psychosociological Issues in Human Resource Management, **7**(2), 49–54 (2019)

The Importance of Probabilistic Analysis in the Application of the Marketing Strategy of the Selected Brand

David Vrtana[1,*] and *Anna* Krizanova[1]

[1]University of Zilina, Faculty of Operation and Economics of Transport and Communication, Department of Economics, Zilina, Slovak Republic

Abstract. Entering foreign markets can be an opportunity for some brands. Expansion contributes to developing competitiveness, building a brand and increasing profits, which will be reflected in overall business performance in a market environment. In this article, we elaborate on the identification data of the Starbucks brand. We identify the international environment and the current marketing strategy of the brand. We also focus on defining its marketing performance with regard to the application of the current strategy in the international environment when entering the target foreign market. Based on the analyzed data on the brand and performance, we approach the probabilistic analysis in order to identify the right foreign market. The performed analysis consists of two basic stages. The first step is a rough selection analysis and a fine selection analysis. In the discussion part, we define the advantages and limitations of the brand's entry into a new foreign market. We support the individual statements with the results of a coarse and fine selection analysis. We also describe the use and limitations of the brand's cooperation strategy in the future. In the final part of the article, we identify the identified statements with the current marketing strategy and suggest the direction of its development when entering a new foreign market.
Keywords: Starbucks, marketing strategy, probabilistics analysis, foreign markets, brand

1 Introduction

Starbucks is a globally known brand with an international environment. It currently operates approximately 31,256 cafes in 65 countries around the world. From a marketing standpoint, Starbucks' mission is to support social development. Majerova attaches importance to social development. He argues that with a marketing orientation towards social development, a company can expand its customer base and thus contribute to the formation of society as a whole. [1] This view is shared by Buno et al. [2] A café should be a space where people can meet, socialize, make new friends and relationships. Therefore, the design of the interior of cafes is adapted to the main message of the brand. The mission

*Corresponding author: david.vrtana@fpedas.uniza.sk

seeks to promote the brand to the international environment and create a certain type of integration component among people. In order to fulfill its mission, the brand's goal is to constantly acquire new customers and retain existing ones. In addition to customer orientation, Starbucks also focuses on environmental policy. By making coffee obtained from soil, the brand strives to reduce the adverse effects on the environment. In order to promote the environment, the brand strives to reduce overall water, energy and recycle various types of materials. Combining customer focus with an orientation to protect and support the environment is a good example of how to apply a green marketing strategy. By simply focusing on the environment, the brand proves that it is globally socially responsible, which can ultimately contribute to changing the shopping behavior of customers of competing companies. This view is also supported by Hollowel et al. [3.4]

Since 2014, Starbucks has seen a growing trend in net income. In 2017, Starbucks' net income was $ 2.885 million. In 2018, net income was $ 4.518 million. In 2019, the brand's net income was $ 3.599 million. In 2020, there was a decrease in net income by an average of 9% compared to the previous period. The decline in revenue recorded in 2019 and 2020 can be attributed to various factors such as the economic crisis caused by the SARS 2 COVID-19 virus. [3] The growing revenues from 2014 can be attributed to the expansion of the brand to foreign markets, the creation and implementation of a new mission in conjunction with a new marketing strategy. Kliestikova et al. they attribute customer loyalty as an important part of a marketing strategy. [6] For this reason, as the target markets expand, the brand will also expand its customer base. Valaskova et al. [7]

Expansion into foreign markets and a change in the brand's marketing policy can make a significant contribution to building a customer base and thus to achieving higher revenues. In connection with the ongoing global economic crisis, we decided in the article to identify a new target market for the Starbucks brand. In order to correctly identify, we followed the rules of probabilistic analysis. Within it, we performed a rough and fine selection analysis.

1.1 Materials and methods

The research problem in our article was defined as follows: Based on the known theoretical and empirical basis to identify what are the new potential markets with a certain degree of economic maturity for the Starbucks brand. Based on the name-defined problem, we focused our research on the detailed possibilities of Starbucks' expansion into new international markets. We consider research to be important for further analysis of the possibility of developing the economy of a particular state. The main goal of our research is to identify potential countries that, to a certain degree of maturity, can be attractive to the globally stable Starbucks brand.

In this article, we have used several methods in order to obtain the required information correctly. We used the method of analysis, synthesis and comparison in the theoretical evaluation of the researched issues. Within these methods, we compared the views and opinions of different authors on the researched issues. Based on induction, deduction and description, we analyzed existing research and case studies in the field, which were secondary sources for us. In order to identify potential foreign markets for the Starbucks brand, we considered them justified and important. Subsequently, in the analytical part of the work, we performed a probabilistic analysis, which we divided into two stages. The first stage was a rough selection analysis and the second part was a fine selection analysis. We used individual methods in order to select the most suitable foreign market for the entry of the Starbucks brand.

We used the FOE indicator in the probabilistic analysis. It is an indicator that compares the foreign trade turnover of a given country with the economy created by GDP. According to the authors, the indicator is suitable for examining the economic openness of a given country and according to it it is possible to identify the extent to which foreign entities can participate in the domestic economy, or how strongly they can participate in it. [8] Based on the above, we decided to use the FOE indicator as a way to identify the various degrees of involvement of the economies of selected countries in the globalization process and the importance of the impact of growing foreign trade in these countries. We calculated the indicator using the following relationship:

$$FOE = ((Export + Import) / GDP) * 100$$

We also needed to determine the size of the country's export performance and import intensity. We found the export performance from the relationship:

$$\text{Export performance} = (Export / GDP) * 100$$

We also needed to identify the import intensity. To determine it correctly, we used the following calculation:

$$\text{Import intensity} = (Import / GDP) * 100$$

We decided to choose this indicator due to its importance for defining the degree of openness of our assessed potential foreign markets for the entry of an international brand.

When selecting and defining individual countries, we took into account material and system factories. Among material factories, we have identified the economic dimension of a given country and the maturity of a given country. In addition to material things, we have also identified system factories, which are influenced by legislative changes, economic instruments, as well as customs and traditions. The reason for the selection of individual countries was their defined development strategy, which is a key factor in achieving economic growth of the country in order to increase the living standard of the country.

Philippines: In 2016, the country's economic growth was 6.5%, in 2019 it was 6.7%. There is also a growing trend in foreign direct investment, which was worth USD 2.8 billion in 2016 and was worth 201 in 2019., Despite the fact that the Philippines is considered to be the poorest part in the region, it is recording its economic growth to a lesser extent, but for foreign companies, this may herald building a more efficient internal economy. There is also an increasing turnover of foreign trade, which in 2015 amounted to 129.8 billion. USD and in 2018 to 182.15 billion USD. Another reason for choosing this country was the fact that the most developed sector in the country is the services sector, which accounted for 56.22% of GDP in 20118.

Kenya: The reason for choosing this foreign market is a market-oriented system of economic functioning. In percentage terms, there is a growing trend of foreign direct investment from 2015, when they reached 0.98%, until 2019, when they reached 1.4%. Among the East African region with a liberalized foreign trade system, the Republic of Kenya is considered to be the most developed. This is also the reason for choosing the country. Also, according to statistics from the Ministry of Foreign Affairs, there is an assumption of economic growth of 5% on average in the following period.

Cambodia: The reason for choosing this country is the increase in the country's annual GDP and economic growth potential. The development of annual GDP is on average around 7%.

Madagascar: This is a country that has seen GDP growth since 2014, from 3.3% in 2014 to 5.2% in 2019. The reason for choosing the country is the fact that economic growth caused by the services sector, which also includes business services.

The reason for choosing the markets of Slovenia and Croatia was their significant economic growth. Slovenia has seen GDP growth from € 37.6 billion since 2014. € to 48 billion € in 2019. There is also a growing trend of foreign direct investment from 2014, when they reached a value of 2%, until 2019, when their value is 2.6%. In the case of Croatia, GDP growth has also been recorded since 2015, when it reached 1.6% and in 2019

2.9%. In 2019, for comparison with the Slovak Republic (where the Starbucks brand already operates), the value of GDP was at the level of 2.3%. Also the most important sector of the economy of both countries is the services sector with a share of about 66%. Revenues from the services sector of both countries form a significant part of Croatia's GDP.According to the data from individual analyzes, we found that the markets of Slovenia, Croatia, Madagascar, Kenya, the Philippines and Cambodia are expected to increase their economic growth in the near future. [9]

2 Probability analysis

As mentioned at the beginning of the article in this chapter, we will focus on performing a probabilistic analysis in order to select the target foreign market of the Starbucks brand.

In the first phase of the analysis, we define possible target markets. In order to properly identify potential target markets, it is necessary to determine where the Starbucks brand currently operates and where it does not. Currently, the brand operates most intensively in Europe, Asia. It has the smallest operations in Southeast Asia, Southeast Africa and Southern Europe. [10] These areas are relevant when deciding a brand to choose a new foreign target market. The following table provides an overview of the areas where the brand could operate.

Table 1. Potential Starbucks brand areas and countries [10]

South Europe	Southeast Africa	Southeast Asia
Croatia	Madagascar	Philippines
Slovenia	Kenya	Cambodia

Source: Own processing.

In order to choose the right foreign market, we will "marginally" examine the political situation, economic risk, cultural aspects, product suitability and other significant factors that could positively or negatively affect the brand's entry into one of these markets.

The first factor we will examine is the product. From a marketing point of view, the product is the basic object of purchase and sale. For the international coffee trade, it is necessary to pay attention to its ever-growing popularity. The International Coffee Organization (ICO) was established to control the coffee trade. All coffee-growing countries are members of this organization. According to an international organization, 110 million bags of coffee are currently grown annually. Two thirds of this is exported. The most important coffee growers are countries located in South and Central America. Countries operating in Africa are also important growers. These countries include, for example, Togo, Ghana, Liberia, Congo, Ethiopia, Tanzania, Madagascar and the like. There are also countries in Asia that are international coffee growers. These countries include Vietnam, the Philippines, Indonesia, Yemen and the like. [11]

The secret of Starbucks coffee is in the quality of coffee beans and the way they are roasted. Certain types of beans are used for individual types of coffee, which come mainly from Latin America, Kenya and Indonesia. Therefore, we decided to consider some countries in more detail, in order to enter the Starbucks brand into a new foreign market.

2.1 Rough selection analysis

Within the probabilistic analysis, we focused more deeply on selection analyzes. The first is a rough selection analysis, in which we analyze in more detail the information on political, legal, economic and socio-cultural factors of the countries that we have chosen in Table 1.

As part of a rough selection analysis, we examined the functional openness of the economy (FOE). It is an indicator that compares the turnover of foreign trade (exports + imports) with the economy generated by GDP. [12] For evaluation, we selected the countries listed in Table 1 that have good economic prospects for Starbucks entry. The values of the indicator are shown in Table 2. Individual data are obtained from 2017 in millions of USD.

Table 2. Criteria for assessing the functional openness of the economy [14]

Kritériá		**Croatia**	**Slovenia**	**Madagascar**	**Kenya**	**Philippines**	**Cambodia**
Economics	Export	183 319	35 752	12 688	1 079	4 904	54 510
	Import	181 047	31 943	14 140	1 973	6 460	57 586
	Nominal GDP	65 019	42 987	12 871	61 746	340 302	19 542
	FOE	560,399	157,477	208,437	4,942	3,339	573,615

Source: Own processing.

The table shows the values related to nominal GDP, export and import information. After analyzing them, we decided that the markets to which the Starbucks brand could expand are Croatia, Madagascar and Cambodia, as they reach the highest values of the indicator of functional openness of the economy.

2.2 Fine selection analysis

After evaluating the functional openness of the economy in the previous chapter, we decided that the potential market for the entry of the Starbucks brand could be Croatia, Madagascar or Cambodia. In order to choose the right target market for the brand, we decided to proceed with a fine selection analysis. In it, we assign weights (assuming that the sum of weights is equal to 1) and values based on the performed analysis. As part of the analysis, we compare individual factors. In them, we identify areas that have been available from analysis and research and compare them. In the individual columns, we identify those areas that we consider important for the analysis. For example, coffee consumption in Croatia is an important factor. In Madagascar, for example, we consider high unemployment and so on to be a significant factor. In the individual columns, the essential areas from the assessed factor are identified. We assigned individual values on a scale from 1 to 5, where 1 means the least suitable environment and 5 the most suitable environment. The overall summary and rationale for the selection are described in Table 3.

In the above analysis, we have identified the following criteria with the greatest weight: economic situation, geographical location, market conditions, market size. We gave the least weight to the membership of countries in international groupings, the environment and the competitive influence of other brands.

Table 3. Fine selection analysis for the assessed areas [13]

Criterion	Weight	Value	Croatia	Madagaskar	Cambodia
Economic situation	0,20	C: 3 M: 1 K: 1	Optimal environment - unemployment 0.1%, open economy and conditions for good economic growth.	Satisfactory economic growth - textile industry. Increase in GDP, decrease in inflation. Decline in tourism.	High dynamics of economic growth, economic potential to be among the 30 largest economies in the world.
Geographical distance	0,20	C: 4 M: 2 K:1	Well-built infrastructure of expressways and motorways.	The importance of air transport, low quality of roads and railways. 70% of foreign trade flows through the port of Toamisina.	Low infrastructure. The advantage is tourism.
Market Conditions	0,15	C: 3 M: 2 K: 2	The average annual coffee consumption is 5.3 kg per capita.	High unemployment, cheap labor, lowest wage costs.	Accelerating macroeconomic growth, increasing the level of transport, tourism, digital infrastructure.
The population	0,15	C: 2 M: 3 K: 2	4 437 460	20 653 556	15 000 000
Environment	0,10	C: 2 M: 3 K: 2	Optimal location, higher wages, intensive tourism in the summer months.	Industrial centers around the capital, infrastructure development, energy and mining. The potential is ecotourism.	Population growth, low inflation, stability of public finances, greater transparency.
Country membership in the international groupings.	0,10	C: 4 M: 3 K: 2	EU, NATO, COE, OPCW, OSCE, WTO, OECD and the like.	UN (since 1960), UNESCO, UNIDO, UNCTAD, ECA, International Monetary Fund (IMF), WTO, African Union (AU) and the like.	WTO, APEC, OSN and the like.
Competition	0,10	C: 1 M: 2 K:2	Higher competition and market openness.	Weak competitive brands.	Weak competitive brands.

Source: Own processing.

After evaluating the individual criteria, we calculated the points assigned to each country. The allocation of points depended on the evaluation of the current situation belonging to the evaluated criterion. As a result, the country gained the most points in Croatia with 19 points, Madagascar with 16 points and Cambodia with 12 points. Based on the above, it can be argued that Croatia should be the new foreign market for the Starbucks brand.

Despite the lower population compared to the other countries under comparison, the country has a better built infrastructure, well-established conditions for economic growth, and an optimal location for setting up operations. [14]

3 Discussion

Based on the evaluation of individual countries, we would recommend the Starbucks brand to expand to Croatia for several reasons. As a country, Croatia has a well-established infrastructure in the form of expressways and motorways, which can be used to ensure the supply of coffee to individual plants. Furthermore, the country has membership in several international groups, which can be an advantage in expanding the brand to this country. The optimal environment or the openness of the economy are a good factor for opening new operations in the country.

In addition to the analysis of the market economy of selected countries, it is also necessary to analyze the future development of the Starbucks brand. The advantages include a global operation with more than 200,000 employees, 31,256 operations and operations in 65 countries around the world. The Starbucks brand currently ranks 460th in the global brand rankings. Another advantage of the future operation is the added value, which can be seen in a long tradition (since 1971), in the product and the overall image of the brand. In addition to global operations, added value, the overall marketing strategy can also be viewed positively. It is currently focused on creating new business opportunities with the acceptance of global environmental policy. The strategy also includes cooperation to develop and improve the product portfolio (Arabica coffee) around the world by raising the status of Indian coffee. In order to enhance the quality and profile of Indian coffee, Starbucks uses the combined strategies of two leading beverage companies, Starbucks Coffee Company and Tata Global Beverages. The alliance focuses on a narrow product range and mainly on product quality.

Starbucks is looking to expand into new target markets, reach new customers there, which is matched by various interactive campaigns and constant adaptation to customer requirements. By generalizing the strategy, it can be argued that it is driven by customers and their behavior, not by the competition. However, product quality and brand orientation must be important elements of this strategy. Its importance in the marketing strategy is emphasized by Kliestiková and Janosková. [15,16] Nadanyiova puts another importance on brand perception, arguing that companies should ensure quality marketing communication, which should include brand consideration. [17]

There may also be some restrictions on expansion into new markets. An example is 2007, when the Starbucks brand sought to expand into the Indian market. The brand wanted to assert itself together with the Future Group, but without success. The reason for the failure to enter this market was the political restriction by the Indian government. It was a failure due to ambiguity about the foreign property structure. The Indian government cited the reason for exceeding the foreign investment limit - Starbucks and Future Group owned more than 50% of separate branded retailers, thus exceeding the set limit in India. Subsequently, after five years, the brand managed to enter the Indian market with a new strategy and the abolition of the existing cooperation partnership. Another problem was the unusualness of drinking coffee. According to the International Coffee Organization, it was only 80 g per capita, while in Croatia it is 5.3 kg per capita and in the USA it is 4.11 kg per capita. People in India prefer drinking tea, and the tea industry is an important element in supporting and developing the economy in this country. [18,19]

Based on the above, it is possible to perceive some restrictions on entering the Croatian market. On the one hand, it may be a good precondition for the country's economic growth, but on the other hand, there may be restrictions that will prevent the brand from expanding. One of them may be the ongoing economic crisis, people's preferences, ignorance of quality coffee, political constraints of the country, strong competitiveness of other brands and the like. [20] These restrictions are only general as they have not been examined in this article.

We mentioned above the benefits of the future development of the Starbucks brand. In addition to the benefits, the negatives must also be assessed. One is the strength of competition, suppliers and the market environment in the form of oversaturation. That's why Starbucks needs to be able to respond effectively to every market change. This should be part of an already established and implemented marketing strategy.

By expanding into new target markets, a company can increase competitiveness, support brand development, expand target customer segments and, ultimately, increase business profits. [21,22] The importance of customer loyalty is also emphasized by Hollowell et al., Who argue that the primary goal of the company is customer orientation. This may also apply to expansion into the international environment. [23] However, expansion needs to take into account ongoing internal marketing and management processes. Their importance is emphasized by Mira et al. [24] A plan for new foreign markets with an assessment of individual factors must be an essential part of the marketing strategy. Rosnerova and Majerova also agree with this opinion. [25,26]

4 Conclusion

Based on an evaluation of all important factors, it can be argued that the Starbucks brand has built up good potential for the future. It has created space for expansion with its products to new foreign markets. Subsequently, we generalized the strategy from the analyzes, the primary goal of which is customer orientation and the expansion of the existing customer base in new target markets. As this is the primary goal of the strategy in the article, we considered the importance of three different areas for Starbucks' entry into the new market. The results of the analyzes show that the potential market for the brand's entry into foreign Croatia market is Southern Europe, due to the assumption of economic growth in the future. The subject of our further research will be a detailed examination of the possibility of entering and applying the Starbucks brand on the Croatian market.

Acknowledgement

The paper is an output of the science project VEGA no. 1/0718/18: The impact of psychographic aspects of pricing on the marketing strategy of companies across products and markets.

References

1. J. Majerova, Analysis of Specifics in Buying Behavior of Slovak Customers in Internet Environment. *2nd International Conference on Social Sciences Research (SSR 2014)*, **5** (2014)

2. M. Buno, M. Nadanyiova, D. Hraskova, The comparison of the quality of business environment in the countries of Visegrad group. *4th World Conference on Business Economics and Management (WCBEM),* **26** (2015)
3. Starbucks Company Information [online], Available at: https://www.starbucks.com/about-us/company-information (2020)
4. J. C. Hollowell, B. Kollar, J. Vrbka, E. Kovalova, Cognitive decision-making algorithms for sustainable manufacturing processes in Industry 4.0: networked, smart, and responsive devices. *Economics, Management, and Financial Markets*, **14**(4), 9–15 (2019)
5. Starbucks Net Income 2006 – 2020, SBUX [online], Available at: https://www.macrotrends.net/stocks/charts/SBUX/starbucks/net-income (2010)
6. J. Kliestikova, P. Durana, M. Kovacova, Naked consumer's mind under branded dress: case study of Slovak Republic. *Central European Business Review*, **8**(1), 15-32 (2019)
7. K. Valaskova, J. Kliestikova, A. Krizanova, Consumer perception of private label products: an empirical research. *Journal of Competitiveness*, **10**(3), 149-163 (2018)
8. C. Gräbner, P. Heimberger, J. Kapeller, F. Springholz, Understanding economic openness: a review of existing measures. *Review of World Economics,* 82, 152-166 (2020)
9. Economic information on the territory 2016 – 2019. [online], Available at: https://www.mzv.sk/documents/724626/620840/Chorv%C3%A1tsko+-+ekonomick%C3%A9+inform%C3%A1cie+o+terit%C3%B3riu+2020 (2019)
10. Number of Starbucks locations worldwide 2003-2019. [online], Available at: https://www.statista.com/statistics/266465/number-of-starbucks-stores-worldwide/ (2019)
11. International Coffee Organization [online], Available at: http://www.ico.org/mission07_e.asp (2020)
12. World Data atlas [online], Available at: https://knoema.com/atlas (2020)
13. Y. T. Lo, A. Jusoh, A. Mardani, D. Streimikiene, Restaurant branding matters: a quantitative report on how brand image can moderate relationship. *Journal of International Studies*, **11**(4), 270-281 (2018)
14. L. Gajanova, M. Nadanyiova, J. Kliestikova, et. al., The Potential of Using Bluetooth-Based system as a Part of Proximity Marketing in the Slovak Republic. *Marketing and Management of Innovations,* **2,** (2019)
15. J. Kliestikova, K. Janoskova, Branding with understanding: how national profile of consumer influences brand value perception. *Marketing and Management of Innovations*, **3**, 149-157 (2017)
16. J. Kliestikova, M. Kovacova, A. Krizanova, P. Durana, E. Nica, Quo vadis brand loyalty?: comparative study of perceived brand value sources. *Polish Journal of Management Studies*, **19**(1), 190-203 (2019)

17. M. Nadanyiova, Neuromarketing - An opportunity or a threat?. *Communications - Scientific letters of the University of Zilina*, **19**(4), 90-94 (2017)

18. Tata Global Beverages and Starbucks Form Joint Venture to Open Starbucks Cafes across India [online], Available at: https://stories.starbucks.com/stories/2012/tata-global-beverages-and-starbucks-form-joint-venture-to-open-starbucks-ca/ (2012)

19. Starbucks Faces One Enormou Problem in India [online], Available at: https://www.businessinsider.com/starbucks-opens-in-india-2012-10 (2012)

20. M. Poradova, Content marketing strategy and its impact on customers under the global market conditions. *19TH International Scientific Conference Globalization and Its Socio-Economic Consequences 2019 – Sustainability in the Global-Knowledge Economy*, **74** (2020)

21. E. Princes, Boosting impulse buying behavior in marketing management: customer satisfaction perspective. *Polish Journal of Management Studies*, **20**(2), 403-413 (2019)

22. M. Nadanyiova, K. Kramarova, Green Marketing and its impacts on Consumer's Green Purchasing Behaviour. *International Scientific COnference on Marketing Identity* (2019)

23. J. C. Hollowell, Z. Rowland, T. Kliestik, J. Kliestikova, V. V. Dengov, Customer loyalty in the sharing economy platforms: how digital personal reputation and feedback systems facilitate interaction and trust between strangers. *Journal of Self-Governance and Management Economics*, **7**(1), 13–18 (2019)

24. M. R. C. Mira, A. F. A. Moura, L. D. S. Mónico, A new measure of the quality of tourism product. *Journal of Tourism and Services*, **9**(17), 1-22 (2018)

25. Z. Rosnerova, D. Hraskova, The impact of globaliization on the business position of European Union. *19TH International Scientific Conference Globalization and Its Socio-Economic Consequences 2019 – Sustainability in the Global-Knowledge Economy*, **74** (2020)

26. J. Majerova, T. Kliestik, Brand valuation as an immanent component of brand value building and managing. *4th World Conference on Business, Economics and Management (WCBEM)*, **26** (2015)

On the Relationship between Bank Lending Indicators and General Economic Indicators

Yury Korobov[1,*], *Sergey* Bogomolov[1], *Larisa* Ilyina[1], and *Marina* Plotnikova[1]

[1]Yuri Gagarin State Technical University of Saratov, Socio-Economic Institute, Department of Banking, Money & Credit, Saratov, Russia

Abstract. One of the most important tasks of any state is to ensure stable economic growth. Banks can play an important role in performing this task, primarily by providing loans. The purpose of the study is to identify the relationship between indicators of banks' lending activity and general indicators of economic development. Index of physical volume of GDP and index of physical volume of fixed capital investment were selected as resultant economic indicators, and growth rate of debt on bank loans (overall and by loan types), the share of loans in fixed capital investment, and the ratio of debt on bank loans to GDP were used as factor variables. The study of the dynamics of these indicators showed that the trajectory of economic indicators has a general tendency to decrease their values; the dynamics of economic indicators depends more on bank lending to legal entities than on lending to individuals, and often reflects the change in the share of loans in fixed capital investment with a time lag; economic growth is more strongly influenced by bank lending to legal entities than by lending to individuals. The revealed patterns indicate the need to develop a monetary policy aimed at stimulating corporate lending and moderate curbing consumer lending.
Keywords: bank, lending, economic indicators, investment, gross domestic product.

1 Introduction

Ensuring stable economic growth and at the same time creating conditions for a positive role of banks in its implementation is one of the most important tasks of any state, including Russia. In the President's message to the Federal Assembly of Russian Federation in 2020, the idea of the need for active participation of banks in solving large-scale socio-economic problems facing the country was voiced several times. In particular, banks are encouraged to implement preferential mortgage programs. In addition, one of the measures proposed to launch a new investment cycle, which implies an annual increase in investment of at least 5% and achieve a ratio of investment to gross domestic product (GDP) at 25% by 2024, is to ensure the availability of long-term money by increasing the availability of credit in the real sector of the economy. It is worth noting that in *Russian*

*Corresponding author: yurikor1965@gmail.com

Financial Market Development Program for 2019-2021 [1], ensuring access to financial services and capital is also identified as one of the main tasks of the *Bank of Russia*. At the same time, within the framework of this document, it is advisable to form a development strategy for each component of the financial market, including the credit market as its most significant sector [2]. In another strategic document – *Monetary Policy Guidelines for 2020–2022* [3], *Bank of Russia* emphasizes that its monetary policy creates conditions for accelerating economic growth, but cannot affect the economic potential itself, which depends primarily on productivity of labor and capital, and the speed of introduction of new technologies.

Considering these guidelines for the near term, it is important to identify the relationship between the rate of economic growth and development of bank lending. In this regard, the purpose of this study is to identify the relationship between indicators of banks' lending activity and general indicators of economic development in Russia.

The objectives of the study are: 1) to assess the trajectory of the dynamics of economic indicators in Russia for the period 2000-2018; 2) to analyze the evolution of bank lending indicators for the same period; 3) to compare the dynamics of general economic indicators and bank lending indicators in order to identify the relationship between them; 4) to determine the specific patterns of this relationship; 5) to formulate guidelines for monetary policy, taking into account the revealed patterns.

2 Literary Research

Relationship between banks' lending activity and macroeconomic indicators is one of the most important topics of economic research. In our opinion, the theoretical basis of specific empirical studies based on quantitative data in this area can be considered scientific works in which the question of existence of any connection between credit and economic development was raised in the most general form. Thus, the classics of economic thought (*Smith*, *Ricardo*, etc.) attached great importance to production, but at the same time diminished the role of credit. On the contrary, *Schumpeter*, *Hahn* and others, overemphasized credit as the primary engine of economic development [4]. In accordance with the ideas of *Marx* [5], credit should not be put forward as the primary engine of economic development, but it can contribute to it. And, therefore, excessive lending is not the root cause of crises, but it can develop and exacerbate accumulated production problems. *Keynes* [6] emphasized that one of the consequences of the increase in bank lending is the expansion of production based on investment growth. *Hayek* [7], who investigated how credit expansion affects the production structure, came to the conclusion that credit expansion, which is not supported by a previous increase in voluntary savings, distorts the production structure, artificially making it excessively capital intensive, and ultimately leads to recessions [8].

Modern scientists also continue to study the relationship between lending activity and macroeconomic indicators. While some authors emphasize mutual causal relationship between lending dynamics and economic development [9], others detect influences of the macroeconomic environment on bank lending [10], but most researchers prove the dependence of economic growth on bank lending, focusing on lending to companies [11, 12, 13, 14, 15, 16, 17, 18]. In particular, *Lavrushin* [19] examines in detail possible positive results of lending for the economy and negative consequences in case of illiterate handling of credit, and proposes a number of indicators to assess the role of credit in the economy.

Research in this area is also conducted by regulatory financial institutions. Thus, *International Monetary Fund (IMF)* representatives periodically prepare relevant reviews, and, in particular, in 2015, it was revealed to what extent credit growth is reflected in GDP growth in developing countries [20]. *Bank of Russia* experts found that an increase in loans to the real economy before reaching the threshold of 86% of GDP contributes to economic growth, while after passing this threshold, on the contrary, can lead to macroeconomic instability [21].

Thus, taking into account the results of these and other similar studies, we consider it necessary to study the peculiarities of the relationship between indicators of banks' lending activity and macroeconomic indicators on the basis of an extended list of factor variables, which is what this article is focused on.

3 Data and methods

Agreeing with the opinion that initially banks' lending activity influences economic development, we have selected as resultant (dependent) following economic indicators:

- Index of physical volume of GDP (hereinafter – GDP index);
- Index of physical volume of fixed capital investment (hereinafter – investment index).

As factor variables were chosen following bank lending indicators:

- Growth rate of debt on bank loans – overall and separately on loans to resident legal entities and to resident individuals (hereinafter – growth rate of bank loans);
- Share of loans in sources of fixed capital investment financing (hereinafter – share of loans in investment);
- Ratio of debt on bank loans (overall and by loan types) to GDP (hereinafter – ratio of bank loans to GDP).

The dynamics of these indicators was studied for the period 2000-2018 in Russia. Main data sources for the study were analytical materials of *Bank of Russia*, and *Federal State Statistic Service*.

The identification of the peculiarities of the relationship between selected resultant and factor variables required the use of graphical models, systems approach, comparative approach, dialectical method, deduction and induction.

4 Results and discussion

Comparative analysis of the dynamics of the GDP index and the investment index shows that their changes practically coincide and have a general tendency to decrease the values (Fig. 1). So, while for the periods 2000-2002, 2003-2005, 2006-2008 average values of these indices did not drop below 106%, the periods 2009-2011, 2012-2014, 2015-2018 not even reached the level of 105% (Table 1). At the same time, the investment index is characterized either by a higher growth or by a larger decrease in comparison with the GDP index, which indicates its more noticeable sensitivity to both crisis phenomena and the emergence of favorable conditions for economic development.

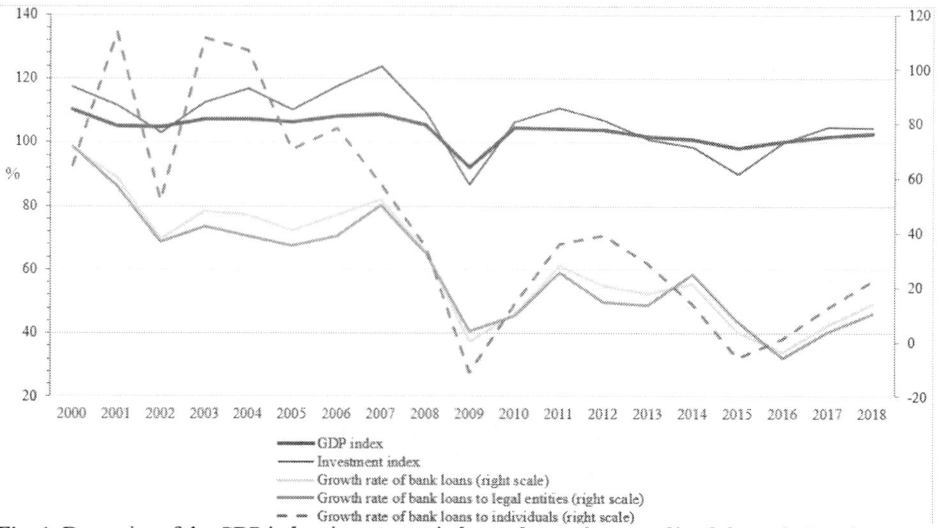

Fig. 1. Dynamics of the GDP index, investment index and growth rates of bank loans in Russia
Source: Own processing based on data of *Bank of Russia* [22] and *Federal State Statistic Service* [23].

Table 1. Average values of general economic indicators and bank lending indicators in Russia

Figure	Average values for the periods:					
	2000-2002	2003-2005	2006-2008	2009-2011	2012-2014	2015-2018
GDP index	106.6	106.9	107.3	100.3	102.1	100.7
Investment index	110.7	113.2	117.0	101.2	102.0	99.7
Growth rates of bank loans	56.4	45.2	44.0	12.9	20.0	5.0
Share of loans in investment	3.7	7.1	10.6	9.3	9.7	10.2
Growth rate of bank loans to legal entities	55.2	38.8	40.6	13.0	17.8	4.0
Growth rates of bank loans to individuals	76.6	96.2	57.1	13.1	27.3	7.6
Ratio of bank loans to GDP	14.3	22.8	35.6	40.8	46.8	48.3
Ratio of bank loans to legal entities to GDP	13.3	19.3	27.1	31.5	33.4	35.0
Ratio of bank loans to individuals to GDP	1.0	3.5	8.5	9.3	13.4	13.3

Source: Own processing based on data of *Bank of Russia* [22] and *Federal State Statistic Service* [23].

Assuming that the lending activity of banks is one of the factors of economic development, we analyzed to what extent the growth rate of loans to legal entities and individuals is reflected in the GDP and investment indices. The analysis shows that the dynamics of lending to legal entities almost coincide with the dynamics of total volume of lending. Trends in the growth rate of lending to individuals, on the contrary, often differ from this trajectory. Thus, in 2003, 2006, and 2010-2011, there was an increase in lending to individuals similar to the general dynamics of bank lending, but the actual growth rates of loans to individuals significantly exceeded the overall values of the corresponding indicators, which is primarily due to the more intensive development of consumer lending in comparison with corporate lending in these years. In other cases, trends in bank lending to individuals repeated trends in bank lending in general, but with some delay: while the decline in the overall growth rate of bank loans was observed in 2001-2002 and 2012-2013, the decline in the growth rate of loans to individuals occurred only in 2002 and 2013-2015.

However, in certain periods of time, the dynamics of the consumer loans segment, on the contrary, outpaced the overall trajectory. Thus, the slowdown in the growth rate of loans to individuals in the pre-crisis period already began in 2007, while the overall rate of bank loans significantly decreased only a year later. Or, the observed gradual increase in the growth rate of loans to individuals during 2016-2018 was accompanied by a similar trend in the corporate loans segment only starting from 2017.

Comparing dynamics of the growth rates of bank loans with dynamics of GDP and investment indices, one can see that their trajectories are similar and are characterized by a decrease in values. The slowdown in lending dynamics is associated, in particular, with an increase in concentration in the banking sector [24]. The trajectory of changes in macroeconomic indices is more similar to the trajectory of growth rate of loans to legal entities, which indicates a closer dependence of economic development on dynamics of corporate lending than on dynamics of consumer lending.

As a follow-up to this idea, we will evaluate the impact of the ratio of bank loans to GDP on the dynamics of the studied macroeconomic indicators (Fig. 2). It is obvious that there is no similarity in trends between the growth rate of loans and the ratio of loans to GDP [25]. During the analyzed period 2000-2018, there was a gradual increase in the value of the ratio of loans to GDP from 11.6% at the end of 2000 to 46.5% at the end of 2018. At the same time, over these years, the contribution of lending to individuals in this indicator has noticeably increased: if at the end of 2000 ratio of this segment to GDP was only 0.6%, then at the end of 2018 it jumped to 14.3%.

However, the GDP and investment indices in 2000-2018 did not show similar significant growth, moreover, the development of consumer lending was sometimes accompanied by a parallel decrease in the values of the GDP and investment indices. In this regard, it should be noted that some authors emphasize the risk of domination of short-term consumer loans in our country over long-term [26], and there are studies that prove that corporate lending has a more beneficial effect on economic growth than consumer lending. [27, 28].

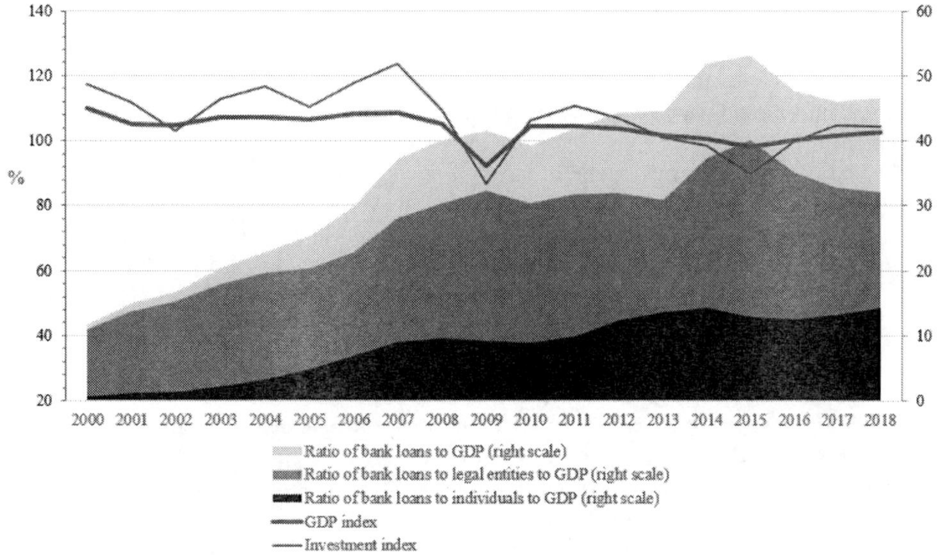

Fig. 2. Dynamics of GDP index, investment index and ratio of bank loans to GDP in Russia
Source: Own processing based on data of *Bank of Russia* [22] and *Federal State Statistic Service* [23].

In support of this idea, we compared GDP and investment indices with the dynamics of share of loans in investment (Fig. 3).

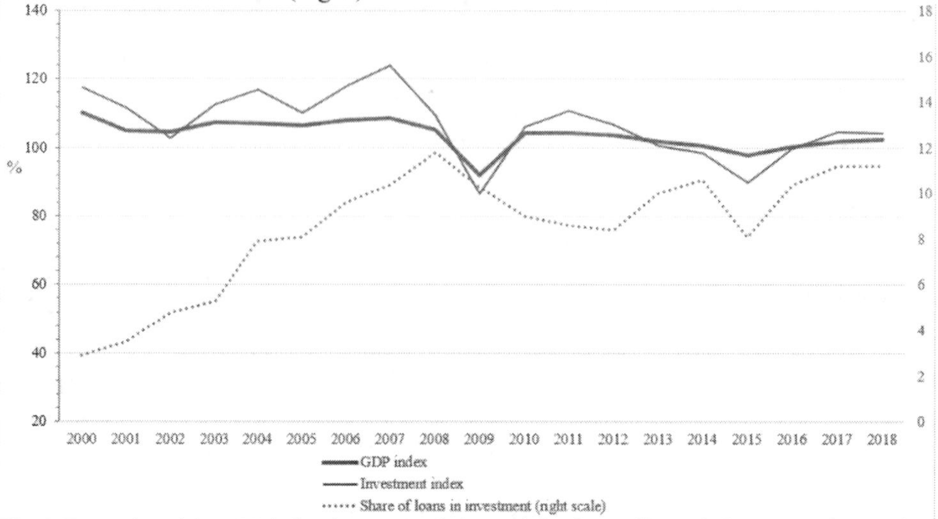

Fig. 3. Dynamics of the GDP index, investment index and the share of loans in investment in Russia
Source: Own processing based on data of *Bank of Russia* [22] and *Federal State Statistic Service* [23].

The results of this analysis showed that changes in the share of loans in investment often affect growth rates of macroeconomic indices in about 2-3 years. For example, a noticeable increase in the share of loans in investment in 2000-2002, 2004, 2006-2008, and 2013-2014 was accompanied by an increase in macroeconomic indices in 2003-2004, 2006, 2010-2011, and 2016-2018, respectively. Conversely, the downward trend in the share of loans in investment in 1998-2000 and 2009-2012 may have influenced the decline in the values of macroeconomic indices in 2001-2002 and 2012-2015, respectively. A similar relationship is shown by a comparison of the dynamics of the average values of these indicators presented in Table 1. Thus, the increase of the average values in 2003-2005, 2006-2008, 2012-2014 compared to the previous three-year periods was observed for the macroeconomic indices, and the share of loans in investment. At the same time, in 2015-2018, the average value of the share of loans in investment increased in comparison with the previous period, while the average values of macroeconomic indices decreased. In our opinion, the fact that in some cases an increase in the share of loans in investment is reflected in economic growth, and in others it is not, can be attributed primarily to the fact that this share itself, even despite the increase, may remain insignificant. This indicator would certainly have a more noticeable impact on economic growth if it reached the level of more than 20%, as in developed countries, where it is equal to 30-50% [29].

5 Conclusion

Summarizing the results of the study, we can draw the following conclusions:

1. The trajectory of dynamics of GDP and investment indices in most cases coincides and has a general tendency to decrease the values, while the investment index is more sensitive to both crisis phenomena and the emergence of favorable conditions for economic development, including lending factors.

2. The dynamics of lending to legal entities almost coincides with the dynamics of total volume of lending, but trends in the growth rate of lending to individuals often differ from this trajectory.

3. Economic indicators are more dependent on the development of bank lending to legal entities than on lending to individuals.

4. In most cases, there is a dependence of the GDP and investment indices on the share of loans in investments with a time lag of 2-3 years, however, there are facts that contradict this conclusion, which may be due to the low values of the share of loans in investment in Russia, which are insufficient to obtain a more noticeable impact on the economic growth.

5. The revealed patterns of interrelation between bank lending indicators and general economic indicators indicate the need to develop a monetary policy aimed at stimulating corporate lending and moderate curbing consumer lending.

References

1. Russian Financial Market Development Program for 2019-2021 [online], Available at: https://www.cbr.ru/Content/Document/File/87952/fm_development_program_2019-2021.pdf (2019)
2. Н.Е. Бровкина, *Развитие финансового рынка в контексте его кредитной составляющей*. Банковские услуги, **10**, с.10-16 (2019) [N.E. Brovkina, *Development of the financial market in the context of its credit component*. Banking services, **10**, pp.10-16 (2019)].
3. Monetary Policy Guidelines for 2020–2022 [online], Available at: https://www.cbr.ru/eng/publ/ondkp/on_2020_2022/page/ (2019)
4. Э.Я. Брегель, А.С. Цаголов, *Теория кредита*. Москва: Госфиниздат (1933) [E.Ya. Bregel, A.S. Tsagolov, *Theory of credit*. Moscow: State Financial Publishing (1933)].
5. K. Marx. *Capital. A critique of political economy, vol. 3: The process of capitalist production as a whole*. New York: International Publishers (1894).
6. J.M. Keynes, *The general theory of employment, interest and money*. London: Macmillan (1936).
7. F.A. Hayek, *Prices and production*. London: Routledge (1935).
8. J. Huerta de Soto, *The Austrian School: market order and entrepreneurial creativity*. Cheltenham: Edward Elgar (2008).
9. Д.В. Бураков, *Управление цикличностью движения кредита*. Москва: Русайнс (2015) [D.V. Burakov, *Credit cycle management*. Moscow: Ru-Science (2015)].
10. D. Hancock, J.A. Wilcox, *Has there been a "capital crunch" in banking? The effects on bank lending of real estate market conditions and bank capital shortfalls*. Journal of housing economics, **3** (1), pp.31-50 (1993).
11. А.Г. Аганбегян, *Инвестиционный кредит – главное звено преодоления спада в социально-экономическом развитии России*. Деньги и кредит, **5**, с.11-18 (2014) [A.G. Aganbegyan, *Investment loan as a main link to overcome recession in Russia's socio-economic development*. Money and credit, **5**, pp.11-18 (2014)].
12. М.В. Антонова, Е.В. Кутищева, *Роль банковского кредита в развитии экономики Липецкой области*. Пермский финансовый журнал, **1**, с.48-59 (2014) [M.V. Antonova, E.V. Kutischeva, *Role of bank lending in economic development of Lipetsk region*. Perm financial magazine, **1**, pp.48-59 (2014)].
13. А.П. Вожжов, Е.Л. Гринько, С.П. Вожжов, Д.В. Черемисинова, *Архитектоника финансового обеспечения и регулирования устойчивого и сбалансированного экономического роста*. Москва: КНОРУС (2017) [A.P. Vozhzhov, E.L. Grinko,

S.P. Vozhzhov, D.V. Cheremisinova, *Architectonics of financial support and regulation of sustainable and balanced economic growth*. Moscow: KNORUS (2017)].

14. Г.В. Кобылинская, *Финансовое обеспечение инвестиционной деятельности и ее влияние на региональное развитие*. Север и рынок: формирование экономического порядка, **3**, с.134-144 (2018) [G.V. Kobylinskaya, *Financial support to investment activities and its influence on regional development*. The North and the Market: Forming the Economic Order, **3**, pp.134-144 (2018)].
15. В.В. Маганов, *Корреляционный анализ зависимости банковского кредита, сбережений и динамики валового внутреннего продукта*. Научное обозрение, **8** (1), с.307-313 (2014) [V.V. Maganov, *Correlation analysis of the dependence of bank credit, savings and the dynamics of gross domestic product*. Scientific Review, **8** (1), pp.307-313 (2014)].
16. Б.В. Мдавар, *Роль кредитования банковской системы Иордании в экономической деятельности страны*. Экономика и предпринимательство, **11** (2), с.175-179 (2015) [B.V. Mdawar, *The role of lending to the banking system of Jordan in the country's economic activity*. Journal of Economy and Entrepreneurship, **11** (2), pp.175-179 (2015)].
17. Н.Н. Столбовская, В.А. Максименко, Д.Г. Кулакова, Е.М. Коликова, *Оценка взаимосвязи динамики объемов банковского кредитования и объемов ВВП в России на основе применения метода регрессионного анализа*. Финансовые исследования, **4**, с.58-70 (2017) [N.N. Stolbovskaya, V.A. Maksimenko, D.G. Kulakova, E.M. Kolikova, *Evaluation of the relationship between the dynamics of the volume of bank lending and GDP in Russia on the basis of the method of regression analysis*. Financial Research, **4**, pp.58-70 (2017)].
18. M.N. Ibrahim, M.E. Shah, *Bank lending, macroeconomic conditions and financial uncertainty: evidence from Malaysia*. Review of Development Finance, **2** (3-4), pp.156-164 (2012).
19. О.И. Лаврушин, *Эволюция теории кредита и его использование в современной экономике*. Москва: КНОРУС (2016) [O.I. Lavrushin, *Evolution of credit theory and its use in modern economy*. Moscow: KNORUS (2016)].
20. M. Garcia-Escribano, F. Han, *Credit expansion in emerging markets: propeller of growth?* IMF Working Paper, **WP/15/212** (2015).
21. М. Мамонов, А. Пестова, В. Панкова, Р. Ахметов, О. Солнцев, *Финансовый сектор, экономический рост и макроэкономическая стабильность*. Банк России. Серия докладов об экономических исследованиях, **21** (2017) [M. Mamonov, A. Pestova, V. Pankova, R. Ahmetov, O. Solncev, *Financial sector, economic growth and macroeconomic stability*. Bank of Russia. Economic Research Report Series, **21** (2017)].
22. Bank of Russia [online], Available at: http://cbr.ru (2020)
23. Federal State Statistic Service [online], Available at: http://rosstat.gov.ru (2020)
24. С.М. Богомолов, Л.В. Ильина, Ю.Е. Копченко, *Консолидация банковского сектора России: характеристика процессов*. Вестник Саратовского государственного социально-экономического университета, **3**, с.101-105 (2018) [S.M. Bogomolov, L.V. Ilyina, Yu.E. Kopchenko, *Consolidation of banking sector in Russia: characterization of processes*. Bulletin of Saratov State Socio-Economic University, **3**, pp.101-105 (2018)].
25. С.М. Богомолов, М.В. Плотникова, *Банковская политика в области обеспечения качества кредитных услуг и направления ее сбалансированности*. Интеллект.

Инновации. Инвестиции, **6**, с.25-31 (2019) [S.M. Bogomolov, M.V. Plotnikova, *Banking policy of ensuring quality of credit services and the direction of its balance*. Intellect. Innovation. Investments, **6**, pp.25-31 (2019)].

26. О.Е. Кузина, Н.А. Крупенский, *Перекредитованность россиян: миф или реальность?* Вопросы экономики, **11**, с.85-104 (2018) [O.E. Kuzina, N.A. Krupenskiy, *Over-indebtedness of Russians: myth or reality?* Economic Issues, **11**, pp.85-104 (2018)].
27. T. Beck, B. Büyükkarabacak, F. Rioja, N.N. Valev, *Who gets the credit? And does it matter? Household vs. firm lending across countries*. The B.E. Journal of Macroeconomics, **12** (1), pp.1-46 (2012).
28. L. Florian, *Enterprise credit, household credit and growth: new evidence from 126 countries*. Luxembourg: University of Luxembourg (2016).
29. А.Г. Аганбегян, *Размышления о финансовом форсаже (по мотивам книги «Финансовые стратегии модернизации экономики: мировая практика»)*. Деньги и кредит, **8**, с.5-10 (2015) [A.G. Aganbegyan, *Reflections on financial boosting (on the motives of the book "Financial strategies for economic modernisation: global practice")*. Money and credit, **8**, pp.5-10 (2015)].

A Bibliometric Analysis of Branding Trends: Does Irrationality Really Matter?

Jana Majerova[1,*], *Lubica* Gajanova[1], and *Margareta* Nadanyiova[1]

[1]University of Zilina, Faculty of Operation and Economics of Transport and Communications, Department of Economics, Slovak Republic

Abstract. Irrationality and its transposition into theory of consumer behaviour is generally considered to be one of the leading trends in contemporary managerial research. It becomes to be notoriety without need of its verification in conditions of specifics of managerial practice across products and markets. However, there are still platforms for discussion about its impact on current state of knowledge. One of these platforms is brand management. On the one hand – the most dominant brand value source is its image creating attribute where irrationality has its undisputable position but on the other hand – qualitative attributes of brand are still considered as significant. Thus, the rationality is highlighted in scope of branding. Therefore, the schism between rationality and irrationality is visible. The aim of this paper is to identify the position of irrationality in the branding trendology. Methodologically, VOSviewer has been used as a basic research platform for the bibliometric analysis of contemporary scientific literature. We have found out, that there is a significant discrepancy between trends in branding theory worldwide and in Slovak Republic. Thus, we create platform to discuss about the applicability of branding patterns which have been modified in scope of irrationality and questioning the existence of Homo Oeconomicus.
Keywords: brand, branding, irrationality, bibliometric analysis

1 Introduction

Consumption is an integral part of human existence. It can be considered as a driving force for the sustainable development of national economies. As early as 1776, Adam Smith stated that the wealth of nations comes from the production and consumption of goods and services. Consumption has thus become the object of research, not only from an economic but also from a sociological or psychological perspective. This fact is also stated by the fact that the Nobel Prize in Economics was awarded in 2017 to Richard H. Thaler, who researches the psychological factors behind the economic decisions of individuals. By emphasizing the need to take into account the psychological attributes of decision-making of individuals as economic entities, Thaler became one of the key representatives of the school, the so-called behavioural economics, the projection of which into corporate reality across the various areas of its management, becomes an imperative of the effective

*Corresponding author: jana.majerova@fpedas.uniza.sk

operation of the company in the market aimed at achieving defined strategic goals. While at the macro level the behavioural approach to consumption is a relatively new line of thought, at the micro level the application of behavioural principles is experiencing a renaissance. In the context of corporate governance, there has been a deviation in the recent past from the original emphasis on the need to take them into account in favour of implementing quantitative methods, even in traditionally behavioural areas such as marketing [1,2,3]. However, the fundamental divergence of the essence of problem areas of marketing and such methods of solution has ultimately proved that the behavioural approach has the status of condition sine qua non not only in marketing but also in the complex management of the company. On the one hand, consumption, which accelerates business activity and the development of national economies, is the subject of macroeconomic research, on the other hand, it transcends at the micro level through almost all business activities. Since behaviour as such is a cross-cutting sociological-psychological category, its examination in the context of consumption as a purely economic category explicitly requires the application of a behavioural approach. Consumer behaviour is traditionally defined as goal-oriented behaviour in which individuals (consumers), resp. their groupings, acquire, use, hold or dispose of goods and services with the intention of increasing their satisfaction and saturating their living needs while maximizing the usefulness gained [4,5]. The fact that these processes are not always efficient is in favour of rejecting the theory of rationality. According to Wansink, a significant part of the products purchased (more than 12%) is never used [6].

2 Theoretical background

Table 1 summarizes the genesis of the theory of rationality in consumer behaviour.

Table 1. Genesis of the theory of rationality in consumer behaviour

Author	Date	Nature
Smith, A.	1776	The first rationally conceived theory of consumption
Mill, J.S.	1836	The first recorded simplification of man for the needs of economics
Jevons, W.S.	1865	The mechanical concept of economic man, which formed the basis of neoclassical economics
Menger, C.	1871	An alternative approach to rationality in economics based on border conditions and the establishment of a tradition on which the marginal current of economic thinking close so far, the so-called classical economics
Becker, G.	1992	A shift from the simplification of man as an economic subject to greater complexity while preserving the mathematical nature of the created models

Source: Own processing according to Zalega [7,8].

An unambiguous definition of rationality in the consumer behaviour of an individual is still absent in the professional literature. In fact, the only agreement between the authors is that a person behaves rationally if he always chooses the one that is best for him from the variants available to him. This is done by satisfying these basic axioms:

- axiom of comparison complexity - it is necessary for the consumer to be able to decide whether product A brings him higher, lower or equivalent benefit than product B,
- axiom of transitivity - assuming that the benefit of product A is higher than the benefit of product B and at the same time the benefit of product B is higher than

the benefit of product C, the benefit of product A is higher than the benefit of product C,

- the axiom of greed - a larger number of products is always considered more beneficial than a smaller amount, provided that the product provides the consumer with a positive satisfaction of his needs,
- the axiom of diversity - the consumer prefers to achieve average consumption over extreme differences in consumption [9].

The fact that these axioms are not fully fulfilled in every consumer decision-making was admitted by neoclassical economists, but did not consider this fact in their models, which is a significant shortcoming [10]. They also disagreed on what is best for the consumer. In this sense, many variants come into consideration - from a very narrow definition, where the criterion is understood own benefit, which can be quantified in monetary units as the highest profit, or the smallest possible loss (absolutely rationalist), to a very broad definition of own benefit. , which may include such factors as sincere interest in other people, prestige, evaluation of oneself as a good person, etc. (partially behaviouralistic). However, irrational factors, the existence of which the proponents of the theory of rationality were aware of on the one hand, were considered inexplicable redundancies of economic events, which resulted in the belief that their inclusion in models to improve their predictive abilities is not possible. The theory of irrationality in consumer behaviour is based on the assumption that human behaviour is not sufficiently explainable using the assumptions of rationality and that we can achieve a more accurate result only if we assume some systematic lack of rationality in the model. The basic stream of ideas based on the denial of rationality in consumer behaviour is the school of so-called behavioural economics, which is elaborated in the professional literature by several theories. These vary among themselves due to changes in assumptions and in the approach to modifying standard (usually neoclassical) models. These theories are: 1) theory of limited rationality; 2) prospect theory; 3) theory of mental accounting and 4) theory of self-control.

The theory of limited rationality originated in 1957, when H. Simon first defined the term "limited rationality" as a means of proposing a rational choice that takes into account the cognitive limits of the economic entity - the decision-maker, i.e. limits arising from his knowledge and subjective possibilities. This theory is based on the assumption that the decision maker selects from the given alternatives with a known probability distribution the result that maximizes the expected value of a given utility function. What is best for the consumer is based on his subjective judgment.

The theory of prospects, formulated by D. Kahneman and A. Tversky in 1979, is an alternative theory to the theory of limited rationality. Its essence lies in the fact that it is a natural aversion to loss for the consumer, as a result of which the probability of unlikely events is overestimated and the probability of normally probable events is underestimated. The aversion to loss is always assessed against a certain reference point, which in practice means that the same situation can be assessed differently by consumers. A specific feature of this theory is that it does not assume the consumer's orientation according to the resulting usefulness, but according to the subjectively perceived value. The difference between these approaches is that while the utility function displays the benefit of various possible states, the value function describes the relationship between individual changes and their subjectively perceived value [11].

The theory of mental accounting was developed by Thaler in 1985 to explain irrationality in consumers' financial decisions. In principle, it is based on the value function characteristic of the theory of prospects and begins to work with the concept of hedonic efficiency, which defines the consumer's effort to ensure that the transactions he considers bring him maximum satisfaction with minimal psychological harm. It assumes that the revenues and losses of the implemented decision are evaluated with respect to a subjectively perceived reference point, which is derived from the nature of the sources of financing and the approach to the perception of the risk of the transaction. The basis of this theory is therefore the analysis of subjectivism in the perception and evaluation of realized transactions, in the division of sources of their financing into the so-called mental accounts and in the periodicity of evaluating the efficiency of management on individual accounts [12].

The theory of self-control is based on the assumption of the existence of a discrepancy between the intended and actual actions. It assumes that current consumption has a higher added value for the consumer than deferred consumption. This implies the need to reconsider the exponential discount approach to quantify future consumption benefits. Instead of an approach based on discounting equally large periods of time at the same discount rates, a modified approach is needed, where periods closer to the present need to be discounted at higher discount rates. The phenomenon manifests itself in the fact that if the consumer plans a future purchase, his plans may change at the moment the future merges with the present, without any other change in circumstances [13]. Solomon states that, in addition to traditional psychological categories, irrationality in consumer behaviour is also influenced by categories resulting from other social sciences [14]. This theory identifies micro and macro-consumer behaviour; giving it a predominantly psychological (micro) and sociological (macro) dimension. The theory follows from the concept of individual freedom of decision and the degree of influence of endogenous, resp. exogenous factors on his shopping behaviour. However, the current literature rejects such an approach and returns to emphasizing the behavioural nature of the phenomenon of consumer behaviour explored through the prism of its dualistically conceived psychological-sociological aspects in the context of consumption as a key economic variable [15,16]. This approach modifies the traditional concept of shopping behaviour formulated in the context of the so-called triads of purchasing situations (ordinary or repeated purchase, modified purchase and new purchase) with respect to the degree of rational consumer involvement (cognitive, habitual and affective shopping behaviour). This concept follows the models of consumer behaviour formulated by Blackwell-Miniard-Engel [17]. Although these models state the presence of sociological and psychological influences, they abstract from their concretization across products and markets, thus actually evaluating the significance of these aspects in the context of neoclassical postulates and thus do not reflect the scientific shift from rationalism to behaviouralism. Unlike models constructed in other areas of economic research, which have a strong predictive power, for this reason, the models of consumer behaviour constructed so far have a persistent descriptive character and mostly serve as a support apparatus for ex post analysis. The above also applies to brand management models, which were paradoxically built on behavioural pillars when they were created. There are several reasons why the behavioural-oriented approach is currently experiencing a renaissance within them. In the context of corporate governance, the most important of these are:

- short-term trend of transition from behaviouralism to the implementation of the principles of so-called marketing engineering and emphasis on quantitative methods,
- the need to revise the originally conceived models with regard to regional disparities of behavioural aspects of brand management.

In the context of theoretical concepts of brand management with a significant immediate impact on corporate governance, it is the fragmentation of theoretical approaches leading to the decline of the strategic dimension of brand management at the corporate level. The implementation of fragmented theoretical approaches to the behavioural aspects of brand management leads to an undesirable spiral effect. It has the following course: domestic companies apply inappropriate concepts of brand management - brand value does not grow further or decreases - managers prefer to abandon the strategic dimension of brand management activities - domestic brands lose their competitive potential compared to foreign - the market is distorted - remain in the market only strong foreign brands using their original management concepts - the impression of the so-called "good practice" - domestic brand managers apply seemingly effective foreign concepts and the cycle repeats itself.

3 Methodology

The idea of visualizing bibliometric networks, often referred to as science mapping, has received serious attention since the early days of bibliometric research. Visualization has turned out to be a powerful approach to analyze a large variety of bibliometric networks, ranging from networks of citation relations between publications or journals to networks of co-authorship relations between researchers or networks of co-occurrence relations between keywords. Over time, researchers have started to analyze larger and larger networks, leading to the need for more advanced visualization techniques and tools. At the same time, professional users of bibliometrics, for instance research institutions, funding agencies, and publishers, have become more and more interested in bibliometric network visualizations. To make bibliometric network visualizations available to a wider public, both inside and outside the bibliometric research community, researchers have developed a number of software tools, most of which are freely available. Nowadays, widely used are these two: VOSviewer and CitNetExplorer. VOSviewer is a tool that we have developed over the past few years and that offers in a relatively easy way the basic functionality needed for visualizing bibliometric networks [18, 19]. The VOSviewer provides distance-based visualizations of bibliometric networks. By default, VOSviewer therefore displays only the nodes in a bibliometric network and does not display the edges between the nodes. In the visualizations provided by VOSviewer, the distance between two nodes approximately indicates the relatedness of the nodes. By providing distance-based visualizations rather than graph-based ones, VOSviewer is especially suitable for visualizing larger networks. Because of its strong focus on visualization, VOSviewer offers less functionality for analyzing bibliometric networks than other tools. However, VOSviewer does have some special text mining features. We´ve applied VOSviewer on the data obtained in Web of Science database since year 2016 to be sure that only hot topics are included to detect trends in scope of brand value building and management. We didn´t focus only on top journals in this field, but we´ve applied key words approach regardless source of the paper. Thus, brand management has been investigated while to criteria to the inclusion of the topic obtaining this key words has been its appearance in more than twelve sources. Based on this, clustering could be provided. The basic dataset has been formed by more than 3000 papers.

4 Results

The aim of the paper was to identify the position of irrationality in the branding trendology. Methodologically, VOSviewer has been used as a basic research platform for the bibliometric analysis of contemporary scientific literature. We have found out, that there is a significant discrepancy between trends in branding theory worldwide and in Slovak Republic. The reason for studying the trends in brand value building and management separately worldwide and in Slovak Republic was that in specific conditions of Slovak Republic, quality is still perceived as main brand value source while in worldwide perspective, image is dominating brand value source. Thus, it has been at the beginning of the research expected that irrationality and behaviourism would be relevant to trendology worldwide where image is perceived as main brand value source. Similarly, in scope of Slovak Republic, irrationality wouldn´t be perceived as leading trend in branding due to the fact that quality is traditionally considered as main brand value source.

As it is shown in Fig. 1 where the spheres of interest of scientific community in scope of brand management worldwide are graphically expressed, the irrationality resp. behaviourism is not detected as individual area of interest. It is included in "green group" considered as brand attributes where significant attention is paid to behavioural loyalty. However, it is obvious that irrationality is indirectly included in all the created groups of interest of contemporary branding science. These groups are: 1) retail branding (red); 2) digital branding (blue); 3) brand attributes (green); 4) brand value sources (yellow) and 5) regional specifics of branding (violet).

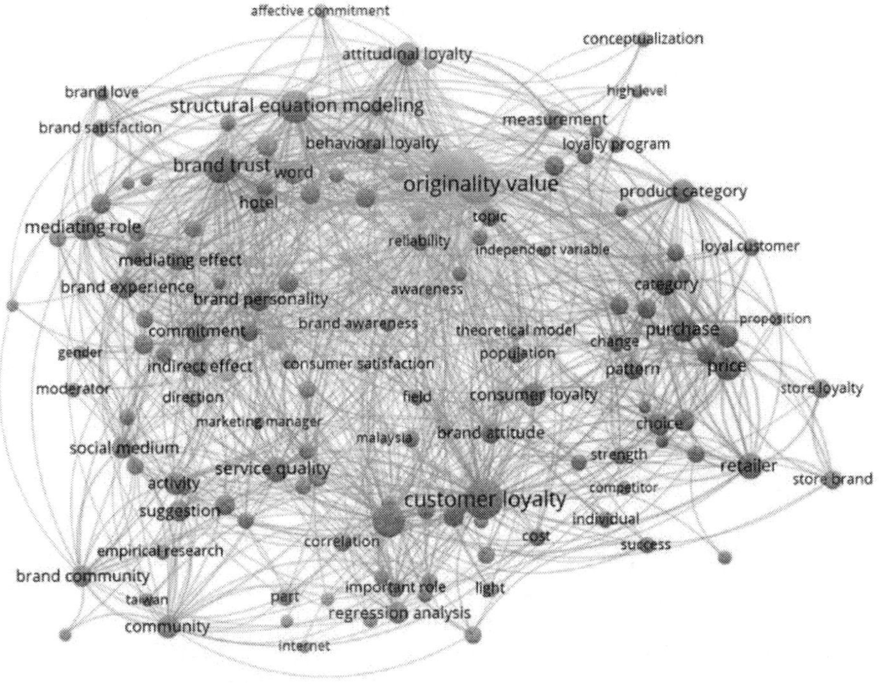

Fig. 1. Spheres of interest of scientific community in scope of brand management worldwide
Source: Own processing in VOSviewer platform.

As it is shown in Fig. 2 where the spheres of interest of scientific community in scope of brand management in Slovak Republic are graphically expressed, the irrationality resp. behaviourism is not detected as individual area of interest. When comparing the bibliometric map and identified areas of interest of contemporary branding science worldwide and in Slovak Republic, the explicit appearance of attribute "behavioural" is not present in specific conditions of Slovak Republic. It can be also stated, that irrationality is neither indirectly included in the created groups of interest of contemporary branding science. These groups are: 1) consumer perception (red); 2) Slovak specifics of branding (green) and 3) financial aspects of branding (blue). Thus, the original presumption that irrationality and behaviourism is relevant to trendology worldwide where image is perceived as main brand value source, has been partially verified. Similarly, in scope of

Slovak Republic, it has been proven that irrationality is not perceived as leading trend in branding due to the fact that quality is traditionally considered as main brand value source.

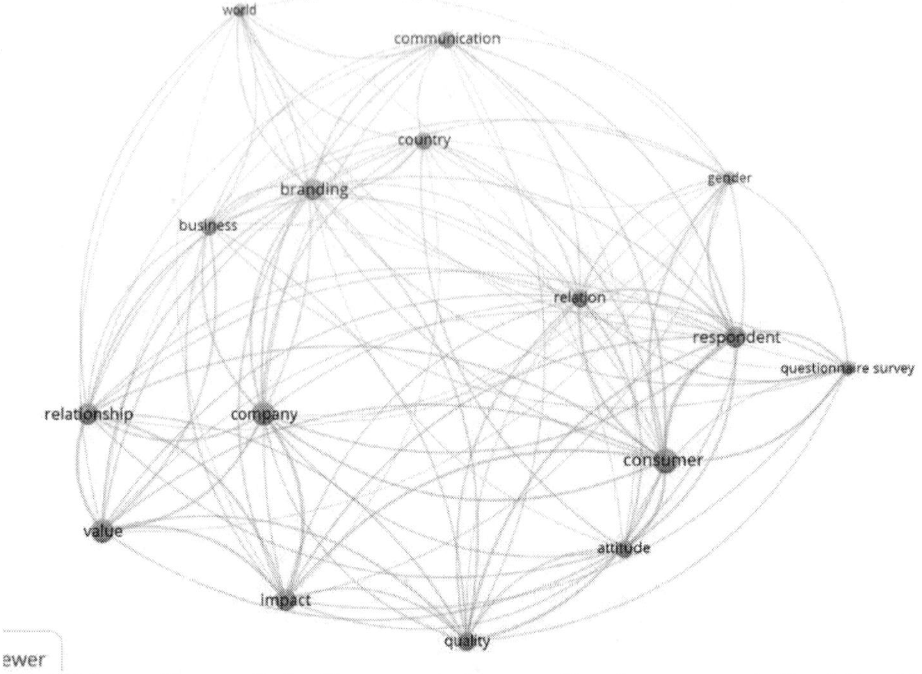

Fig. 2. Spheres of interest of scientific community in scope of brand management in Slovak Republic
Source: Own processing in VOSviewer platform.

5 Conclusion

The aim of the paper was to identify the position of irrationality in the branding trendology. We´ve assumed that there will be significant difference between importance of irrationality in branding worldwide and in specific conditions of Slovak Republic. The reason for such a presumption was that the irrationality and behaviourism would be relevant to trendology where image is perceived as main brand value source (worldwide) while it wouldn´t be perceived as leading trend in branding due to the fact that quality is traditionally considered as main brand value source (Slovak Republic). Methodologically, VOSviewer has been used as a basic research platform for the bibliometric analysis of contemporary scientific literature. We have found out that branding trends worldwide are: 1) retail branding; 2) digital branding; 3) brand attributes; 4) brand value sources and 5) regional specifics of branding. In specific conditions of Slovak Republic these trends are: 1) consumer perception; 2) Slovak specifics of branding and 3) financial aspects of branding. Thus, we have verified the original assumption about significant difference in branding trendology based on the dominant brand value source (image generally vs. quality in Slovak Republic). This fact indicates the need of systematic evaluation of trends in contemporary brand management provided in the light and shadow of bibliometric overview.

This contribution is an output of scientific project VEGA no. 1/0718/18: The impact of psychographic aspects of pricing on the marketing strategy of companies across products and markets.

References

1. S. Parihar, Actors Affecting Consumer Involvement in Purchase of Garments: A Quantitative Analysis for Marketing Strategies. *The Journal of Indian Management & Strategy*, **22**(2), pp. 24-31 (2017)
2. M.E. David, F.R. David, F.R. David, The Quantitative Strategic Planning Matrix: A New Marketing Tool. *Journal of Strategic Marketing*, **25**(4), pp. 342-352 (2017)
3. P. Ceniga, V. Sukalova, Future of logistics management in the process of globalization. *Procedia Economics and Finance*, **26**, 160-166 (2015)
4. M. Kienzler, C. Kowalkowski, Pricing Strategy: A Review of 22 Years of Marketing Research. *Journal of Business Research*, **78**, pp. 101-110 (2017)
5. K. Matusinska, M. Stoklasa, The state of strategic marketing application in Czech small and medium-sized enterprises. *Scientific Annals of Economics and Business*, **66**(2), 193-212 (2019)
6. B. Wansink, A. Brasel, S. Amjad, The Mystery of the Cabinet Castaway: Why We Buy Products We Never Use. *Journal of Family and Consumer Sciences*, **92**(1), pp. 104-107 (2000)
7. T. Zalega, Rationality and Methods of Research into Consumer Market Behaviour. *Equilibrium-Quarterly Journal of Economics and Economic Policy*, **7**(4), pp. 77-99 (2012)
8. J. Olah, E. Krisan, A. Kiss, Z. Lakner, J. Popp, PRISMA statement for reporting literature searches in systematic reviews of the bioethanol sector. *Energies*, **13**(9), 2323 (2020)
9. G. Barros, H. A. Simon, The Concept of Rationality: Boundaries and Procedures. *Revista de Economia Política*, **30**(3), pp. 455-472 (2010)
10. M. Klepek, The role of demographic factors in consumer perception of value from brand communication on Facebook. *Central European Business Review,* **9**(3), 56-73 (2020)
11. D. Kahneman, A. Tversky, Prospect Theory: An Analysis of Decision Under Risk. *Econometrica: Journal of the Econometric Society*, **47**(2), pp. 263-291 (1979)
12. R.H. Thaler, Mental Accounting and Consumer Choice. *Marketing Science*, **4**(3), pp. 199-214 (1985)
13. R.H. Thaler, Behavioral Economics. *Journal of Political Economy*, **125**(6), pp. 1799-1805 (2017)
14. M.R. Solomon, Consumer Behaviour: Buying, Having and Being. Essex: Pearson (2015)
15. H. Willman-Iivarinen, The Future of Consumer Decision Making. European *Journal of Futures Research*, **5**(1), 14 (2017)
16. E. Giampietri et al., A Theory of Planned Behaviour Perspective for Investigating the Role of Trust in Consumer Purchasing Decision Related to Short Food Supply Chains. *Food Quality and Preference*, **64**, pp. 160-166 (2018)
17. R.D. Blackwell, P.W. Miniard, J.F. Engel, Consumer Behaviour. 9th edition. Mason, Ohio: South-Western College Publishing (2001)
18. T. Chankoson, T. Thabhiranrak, The moderating role of brand awareness between the relationship of emotional attachment, brand relationship and positive word of mouth. *Polish Journal of Management Studies*, **20**(1), 129-138 (2019)

19. L. Lizbetinova, The quality of communication in the context of regional development. *DETUROPE-The Central European Journal of Regional Development and Tourism*, **6**(3), 22-38 (2014)

Impact of Dividend Policy on the Market Value of the Company

Ivan Eryomin[1,*], *Olga* Likhacheva[1], and *Lyudmila* Chernikova[1]

[1]Financial University under the Government of the Russian Federation, Moscow, Russia

Abstract. The article is devoted to the analysis and assessment of the impact of dividends on the market value of the company. The study is based on data from company reports and statistics from the Moscow Exchange. This paper presents a regression analysis of a choice consisting of 1) 20 systems of Russian companies and 2) 5 largest companies in the oil and gas sector for the period 2013-2019. The regression results show that dividends have a positive effect on capitalization only if the policy is based on the residual principle. The conclusions made in the work, which provide recommendations and a statistical assessment of dependence, can find practical application in determining the dividend policy of an organization and predicting changes in the capitalization of companies.
Keywords: dividend policy, capitalization, discount rate, dividends, regression analysis, capital structure, shares.

1 Introduction

An effective dividend policy is extremely important for a company in its quest to maximize the wealth of its shareholders. In recent decades, dividend policy has attracted the attention of financial analysts around the world. It has also been a source of controversy and the subject of intense theoretical and empirical research. The link between dividend policy and share prices remains one of the controversial aspects of dividend policy.

At the same time, despite the significant amount of research carried out, practically none of them concerned Russian companies and, in particular, the oil industry, so this work differs from previous studies in several respects. First, the analysis is fully focused on the Russian economy, which means that it takes into account its specific features. Secondly, the period covered by this study (2013–2019) is unique and largely covers the latest trends in the development of the market and economy. Finally, recent studies of dividend policy in Russia focus on the analysis of factors influencing the development of dividend policy and the size of dividend payments [1], rather than its impact on the value of the firm.

To achieve this goal, the following tasks were set: to study the results of theoretical and empirical studies of the problem under consideration, to review the literature on the topic, to analyze the legislative aspects, to conduct a regression analysis of the sample, to check

*Corresponding author: ivan-erx@yandex.ru

the statistical significance of the model, to determine practical recommendations for using the analysis results.

1.1 Concept of dividends and dividend policy

In general, the concept of dividends (from the Latin word dividendum - what is to be divided) means the share of the company's net profit distributed among shareholders (participants) as a return on their investment in the company. They can be paid to shareholders either in cash (English cash dividend) or by new shares (English stock dividend).

A similar definition is given in article 43 of the Tax Code of the Russian Federation, however, in this definition, the key features are used to calculate the tax base, therefore it is separately specified that payments to shareholders (participants) of an organization in the form of transferring shares of the same organization into ownership are not dividends [2]. At the same time, Article 42 of the Federal Law "On Joint Stock Companies" states that dividends can be paid in money, and in cases stipulated by the charter of the company, other property [3]. Thus, if the company decides to pay dividends in shares, such payment is not taxed (which has been 13% since 2015), since legally this is a simple change in the structure of the authorized capital (or its increase) (see Fig. 1).

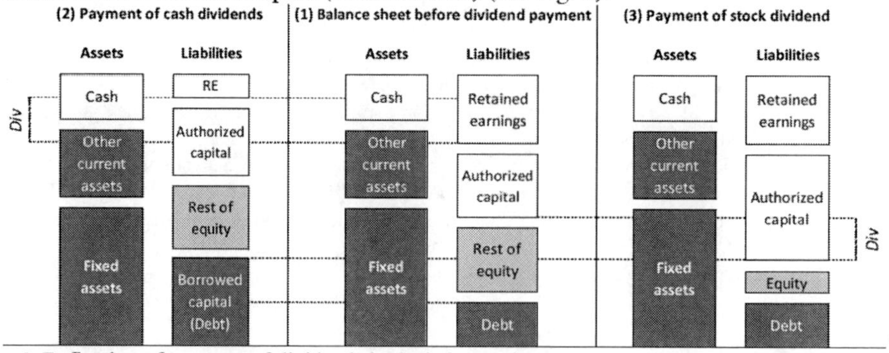

Fig. 1. Reflection of payment of dividends in the balance sheet
Source: Own processing.

In practice, dividends are not rarely paid in the form of shares. Archer Daniels Midland, an international agro-industrial corporation (ADM), sometimes called the "dividend aristocrat", has regularly paid dividends in shares of 5% of the number of shares held for more than 20 years [4].

The decision of the organization as to how much profit can be paid as dividends and how much should be retained by the firm is the task of the dividend policy.

1.2 Market value of the company

The value of the enterprise is broader than the concept of market capitalization, which reflects only the market value of equity (Formula 1).

$$EV = MV_{com} + MV_{pref} + D - I_{subsidiary} - Cash \quad (1)$$

where MV_{com} *- the market value of common shares*
MV_{pref} *– market value of preferred shares*
D *– market value of debt*

$I_{subsidiary}$ – *value of investments in associated companies [5]*

Cash is deducted because the total debt can be reduced (paid off) at the expense of cash and cash equivalents (Formula 2).

$$Net\ debt = total\ debt - cash\ and\ cash\ equivalents \quad (2)$$

All other things being equal, the dividend policy affects only the cost of equity. The market value of debt mainly includes loans that are not quoted on the stock exchange, are paid in full accordance with the agreement to a specific bank and are sold to them only in extreme cases. Bond loans, the value of which is obtained by discounting the expected cash flows of the bond to date using an appropriate discount rate, are debt securities traded in the bond market. Their value may change when dividends are declared, since they will demonstrate the company's solvency (in accordance with the signal theory of dividends), but only in the short term. In a study by J. Hanjinikolaus and A. Calay, for example, it is confirmed that with a decrease in dividend payments, bonds lost in value, i.e. they became less attractive, as did stocks. At the same time, in the opposite situation, the bonds remained unchanged.

At the same time, studies show that credit availability and cost are mainly influenced by "credit information" and legal regulations, as this reduces risks. This is supported by the results of Quoc Trung Tran, which conducted a regression analysis of a sample of 37,673 observations collected in 39 countries between 2013 and 2015, with the relationship being stronger in countries with strong regulations [6].

It is also assumed that if there is a relationship between the payment of dividends and the market value of borrowed capital, then it is reflected in the cost of equity, moreover, both managers and shareholders are guided by the cost of equity. Thus, when assessing the impact of dividend policy on the market value of a company, it is advisable to take into account the change in the value of equity capital only.

1.3 Effectiveness of dividend policy

The main source of financing for the organization's activities is its own capital, an important part of which is retained earnings. The payment of dividends reduces the volume of reinvested profits, therefore, the dividend policy affects the volume of own sources of financing and, indirectly, the volume of attracted debt capital. In other words, the payment of dividends, on the one hand, can increase the attractiveness of shares, but on the other hand, it will not allow increasing the book value of capital and, even worse, it will not provide an opportunity to invest these funds in profitable projects.

In this regard, there are still discussions in the scientific literature about the ability of the dividend policy to influence the value of shares and about the optimal size of the share of profit distributed among shareholders. A significant number of points of view and a large number of studies have accumulated, each of which confirms one or another opinion, but they are all based on several theories of dividend policy:

- The theory of the preference of dividends (Bird-in-the-hand theory)
- Tax differential theory
- Signal theory of dividends
- The theory of conformity of the dividend policy to the composition of shareholders or the theory of clientele (clientele theory)
- Classical dividend irrelevance theory

1.4 Theories of dividend policy

The theory of the preference of dividends (bird-in-the-hand theory) is based on the fact that investors, proceeding from the principle of minimizing risk, always prefer current dividend payments to potential benefits in the future.

The positive effect on the market value is also explained by the fact that current income (paid in the form of dividends) is more expensive than income deferred for the future, due to its inherent risk [7] and it is not just about discounting. Since the amount of dividends paid indicates the stability and sustainability of the company's results, the minimum required shareholder return (discount rate) decreases, causing an increase in the present value of cash flows.

In addition, in some countries, due to legislation, shareholders prefer the payment of dividends to the reinvestment of profits due to the different tax rates on exchange rate gains and dividends [8].

This fact is taken into account by the **tax differential theory**, according to which, in order to take into account the impact of tax expenditures, Brennan (1970) adjusted the CAPM model to make the rate more attractive for companies with lower dividends (for the case when tax the rate on dividend yield is higher), thus weakening Miller and Modigliani's assumption that there are no taxes (see The Theory of Dividend Independence).

Despite extensive empirical research, this hypothesis does not appear to be supported by the data. Moreover, Poterba (1987) documented the stability of dividend payments during periods of extensive US tax changes [9].

Although the impact of taxes remains inconclusive, increasing attention is being paid to the problem of information asymmetry. Subsequent research by Watts (1973) and others has confirmed that the accrual of dividends increases the price of a stock, while a reduction in dividends tends to lower prices. Information asymmetry also made it possible to take into account the cost of paying dividends [10]. In addition, managers are often faced with weakened oversight in the event of a stable dividend payout, as high dividends can increase shareholder confidence in management.

Signaly theory pf dividents, built on these principles, asserts that when assessing the fundamental value (or assessing the attractiveness of shares by shareholders), significant information is carried by the change in the level of dividends - a signal that the company is on the rise and expects an increase in profits in the next periods.

If the amount of dividends is dictated from the outside, then, for example, in Russian practice [11], this measure in most cases has a negative effect on the value, since the political factor is involved.

The theory of conformity of dividend policy to the composition of shareholders or **clientele theory** challenges all previous provisions and argues that dividend policy is influenced by the interests of the majority of the company's shareholders, some of whom prefer the current yield, while others favor the growth of share prices in the future. In other words, the impact on value is a secondary concern.

Ciaran Driver, Anna Grosman, and Pasquale Scaramozzino show that investor pressure on businesses is more than just overseeing management activities (which have opposing interests according to agency theory). However, it is argued that such pressures arise in poorly performing financial markets and are responsible for the "increased appetite" for liquidity and short-term income [12].

The classical dividend irrelevance theory says that dividends do not affect the value of stocks. This position is substantiated by Merton Miller and Franco Modigliani in a theoretical article in 1961. They reasoned as follows.

During each period, the firm determines its investment policy, attracts financing, pays dividends, which can be secured by new loans in the capital market (issue of shares) or by increasing borrowing. Proving the first theorem about the independence of the company's value from the financing structure, the authors conclude that two identical firms (with equal EBIT), differing only in the volume of borrowings, are worth the same [13].

At the same time, the first theorem does not take into account the effect of risk on investor behavior, because a company with a leverage due to the fact that *r* (interest rate on loans) is lower than WACC, there is a lower minimum required return. The second theorem on the weighted average cost of capital proves the opposite. In other words, both the rate and the income are equal for both firms, which, in accordance with the income approach, gives equal value to the companies, which means that the value of the company is mainly determined not by the dividend policy and not the amount of borrowed funds, but by the efficiency of activities (profitability, etc.) [13].

Thus, in a perfect capital market, dividend policy matters only because it affects the ratio of the return on investment and the required rate of return (i.e. IRR> WACC). Therefore, the dividend policy should be based on the residual principle. The sequence of actions of the company's management should be as follows:

1) If investment projects provide a level of profitability that exceeds the required rate of return, then they are accepted.

2) Dividends are paid if not all profits are used to invest in these projects.

In other words, the reinvestment of profits increases the welfare (market value of the company) only if it is spent efficiently: if the generated income covers the cost of raising capital. In fact, the validity of MM's arguments is recognized by almost everyone, and the dispute is about whether taxes, emission costs and other market imperfections affect the situation. These assumptions are built and confirmed by some practical research.

1.5 Foreign experience

In the Russian economy, the capital market is less developed than the European or American, so the theory of Miller and Modigliani may not work in its pure form.

First of all, in Russia, as elsewhere, there is tax differentiation, although dividends on shares and income from their sale are taxed at an equal rate of 13% (for residents of the Russian Federation), the fact is that income tax on the sale of securities is paid in the moment money is withdrawn from the account, while dividends are taxed at the moment the money is credited to the account, even though they may be reinvested in shares. As a result, shareholders should choose not to receive dividends (all other things being equal).

Presumably, given the importance to the economy of the oil and gas sector, Russia is somewhat reminiscent of Nigeria, where some of the problems are most pronounced. From this point of view, the analysis of Nigeria's oil industry is of interest.

The article "The Impact of Dividend Policy on the Shareholders Value of Listed Firms in the Nigerian Petroleum Marketing Industry" (Alhaji Ali Tijjani, Mohd Norfian Alifiah and Gana Kafiya Wakil) assesses the impact of dividend policy on the value of Nigerian oil companies. Regression analysis of a sample of six firms listed on the Nigerian Stock Exchange (NSE) for the period 2008-2017 refutes the hypothesis that "dividend payments by oil firms in Nigeria do not significantly affect share prices."

The result showed a positive significant relationship between dividend policy and the market value of the firms in question. [14] This study confirms the literature on the relevance of dividends by Gordon (1962) and Walter (1963) and rejects Miller and Modigliani (1961), given that their underlying assumptions may not hold true in the Nigerian context.

 https://doi.org/10.1051/shsconf/20219101013

However, the authors explained this dependence by the weak protection of minority shareholders and the underdevelopment of the capital market, respectively, to analyze the situation in Russia, additional research is needed.

2 Methodology

The analysis methods used by the study include regression. The hypothesis being tested is the presence of the influence of dividend policy on the value of firms (market value of equity capital).

As a dependent variable (y_i), capitalization in absolute terms or its growth rate can be taken (since its value varies quite strongly depending on the firm). The independent variables (x_{ij}) under consideration include the factors mentioned in the first chapter, the influence of which is presumably the strongest: the amount of dividends (or dividend yield), the share of profit allocated to dividend payments, return on equity ROE, leverage $\left(\frac{D}{E}\right)$, as well as the Moscow Exchange's IMOEX index (since during recessions, as noted earlier, companies keep the amount of dividends at the same level).

3 Results

3.1 Regression analysis of the largest companies

As already mentioned, the founders of the classical dividend theory based their assumptions on some premises associated with a perfect capital market:

- free and equal accessibility of information for all investors, no transaction costs, rationality of shareholders' behavior
- guaranteed placement of additionally issued shares

Other theories are based on criticism of these premises, therefore, in order to get as close as possible to the conditions of a perfect market and to minimize the influence of other variables, it is advisable to consider the largest Russian companies in terms of capitalization, the so-called blue chips.

These are the most stable and liquid firms, so there is an opportunity to sell or buy a significant amount of shares of these companies at any time during a trading session on the stock exchange without a significant loss in price. In addition, a significant number of minority shareholders and investors ensure the rationality of the players' behavior, minimizing the impact of "deviations" of individual cases and irrational actions, and due to the fact that the largest companies are always in the spotlight, information spreads rather quickly and is practically equally accessible. The prominence and size of these issuers allows them to place additionally issued shares with virtually no loss.

In view of the above explanations, the 20 largest companies in terms of capitalization were selected (as of December 31, 2019) [15]. Data were obtained from firms' annual reports for the period 2013-2019. [16] (altogether 140 observations, see Table 1). Let's note right away that market capitalization is the total value of the company's shares currently in circulation (both common and preferred) [15].

Table 1. Summary table of random sample of empirical observations

№	Ticker	Period	ROE	D/E	Dividend per share, RUB	Change in capitalization, %	Capitalization, rub.	Dividend yield, %
1	GASP	31.12.2013	12,1%	0,39	7,20	-3,82%	3 286 120 325 649	5,19%
2	SBER	31.12.2013	19,2%	8,68	3,20	9,12%	2 260 945 225 920	3,06%
-/-	-/-	-/-	-/-	-/-	-/-	-/-	-/-	-/-
121	GASP	31.12.2019	8,7%	0,50	0,00	67,46%	6 077 464 231 688	0,00%
122	SBER	31.12.2019	20,4%	5,68	18,70	36,51%	5 709 855 444 600	7,07%
123	ROSN	31.12.2019	15,6%	1,51	33,41	5,29%	4 775 538 924 340	7,41%
124	LKOH	31.12.2019	15,8%	0,41	192,00	17,82%	4 405 472 500 000	3,12%
125	NVTK	31.12.2019	53,0%	0,21	32,33	11,72%	3 833 639 955 600	2,56%
126	GMKN	31.12.2019	139,2%	3,54	1488,02	48,17%	3 050 339 795 376	7,72%
127	SIBN	31.12.2019	19,1%	0,73	37,96	21,71%	1 994 901 823 109	9,02%
128	SNGS	31.12.2019	-0,4%	0,18	0,00	67,04%	2 107 381 083 926	0,00%
129	TATN	31.12.2019	25,6%	0,65	64,47	6,74%	1 776 863 479 240	7,90%
130	PLZL	31.12.2019	105,5%	3,35	162,98	31,34%	944 878 136 366	2,32%
131	YNDX	31.12.2019	4,9%	0,27	0,00	40,43%	791 769 254 430	0,00%
132	NLMK	31.12.2019	22,5%	0,76	17,36	-9,02%	858 829 463 492	12,11%
133	CHMF	31.12.2019	50,8%	1,36	115,88	-0,13%	788 293 259 060	12,31%
134	MTSS	31.12.2019	151,4%	21,64	42,50	35,06%	641 980 080 969	13,23%
135	ALRS	31.12.2019	24,7%	0,69	2,63	-13,79%	624 475 435 768	3,10%
136	VTBR	31.12.2019	12,2%	8,38	0,00	35,36%	596 573 717 758	0,00%
137	IRAO	31.12.2019	14,8%	0,36	0,20	29,35%	525 393 000 000	3,90%
138	MAGN	31.12.2019	15,5%	0,50	3,83	-1,84%	471 165 624 450	9,08%
139	POLY	31.12.2019	25,1%	1,20	0,82	35,84%	454 483 915 087	0,08%
140	MGNT	31.12.2019	5,1%	4,03	304,19	-2,44%	349 861 681 715	8,86%
Standard deviation			**45%**	**3,50**	**254,25**	**38,23%**	**1 293 895 108 782**	**4,11%**
The coefficient of variation			**232,8%**	**169%**	**275,1%**	**204,8%**	**97,1%**	**83,5%**

Source: Own processing on the basis of reporting data and information from the Moscow Exchange website.

To find the most accurate regression model, consider several options for the arrangement of variables:

1. y_i – capitalization, rub.; x_{i1} – ROE; x_{i2} - $\frac{D}{E}$; x_{i3} - dividend per share, rub.
2. y_i – capitalization growth, %; x_{i1} – ROE; x_{i2} - $\frac{D}{E}$; x_{i3} - dividend per share, rub.
3. y_i – capitalization growth, %; x_{i1} – ROE; x_{i2} - $\frac{D}{E}$; x_{i3} - dividend yield, %
4. y_i – capitalization, руб.; x_{i1} – ROE; x_{i2} - $\frac{D}{E}$; x_{i3} - dividend yield, %

The multiple linear regression model for individual observations in general looks like (formula 3):

$$y_i = \beta_0 + \beta_1 ROE_i + \beta_2 \left(\frac{D}{E}\right)_i + \beta_3 Div_i + \varepsilon_i \tag{3}$$

The results of the given calculations show that all equations describing the dependence are statistically insignificant (see Table 2). In addition, checking the significance of the coefficients shows that they are all statistically close to zero, therefore, there is no statistical relationship between the explained variable Y and the explanatory variables X_j) (j = 1,2,3). This may be due to the fact that there are other factors affecting the cost. On the other hand, the results obtained may be due to significant differences between the companies in question. The considered firms, although quite similar in size, still operate in different industries, which affects financial results, profitability of production, and other indicators.

It is worth noting that in those models where R^2 is higher (in models 1, 2, 3), the coefficients before x_3, showing the relationship between this variable and the effective factor y_1, are the lowest relative to other coefficients. In model 2 and 3, they are 0 (or very close to it). In model 4, the coefficient is negative. Analysis of factors using paired regression (the relationship between capitalization and dividends) does not change the result, this model is also statistically insignificant.

In other words, as was proved by Miller and Modigliani, the market value of companies depends mainly on market conditions, the state of the economy, management efficiency, profitability of production, and other features, and not on dividend policy.

At the same time, this cannot yet be asserted, since it is likely that the low quality of the model is caused by strong differences between the companies. This is confirmed by the too large value of the coefficient of variation. It is found by the following formula (formula 4):

$$V = \frac{\sigma}{\bar{x}} \cdot 100\% \quad (4)$$

σ – standard deviation, $\bar{x}$ – mean

This coefficient shows the degree of variability in relation to the sample mean [17], in other words, how much the x_i deviates from the mean among the observed observations. In the sample under consideration, some indicators deviate from the average by more than two times (see Table 1, row "Coefficient of variation").

Table 2. Regression analysis calculation table

№	Parameter estimates	*t-statistics*	Statistical significance $b_i(t = 1,98)$	Checking the Significance of an Equation		
	b0	1 259 053 868 517	9,469	Yes	R^2	0,0128
	b1	- 47 984 848 897	-0,160	No	F	0,5875
	b2	14 810 214 675	0,401	No	$F_{\alpha;\, m;\, n-m-1}$	2,6712
1	b3	561 642 082	1,232	No	Equation significance	No
	b0	0,2013	5,107	Yes	R^2	0,0064
	b1	-0,0199	-0,224	No	F	0,2926
	b2	-0,0070	-0,641	No	$F_{\alpha;\, m;\, n-m-1}$	2,6712
2	b3	0,0000	0,302	No	Equation significance	No
	b0	0,2013	4,270	Yes	R^2	0,0088
	b1	-0,0199	0,022	No	F	0,4013
	b2	-0,0070	-0,714	No	$F_{\alpha;\, m;\, n-m-1}$	2,6712
3	b3	0,0000	-0,645	No	Equation significance	No
	b0	0,2284	6,784	Yes	R^2	0,0039
	b1	0,0019	0,088	No	F	0,1768
	b2	-0,0078	0,289	No	$F_{\alpha;\, m;\, n-m-1}$	2,6712
4	b3	-0,5300	0,536	No	Equation significance	No

Source: Own processing.

Thus, it is premature to conclude that the relationship between dividends and changes in the company's value is insignificant. To minimize the impact of differences in company characteristics, it is advisable to consider the largest companies in the same industry, for example, oil.

3.2 Analysis of Russian companies in the oil and gas sector

Based on the same rating of the largest companies on the Moscow Exchange website, companies engaged in the production and processing of oil were selected: Rosneft (ROSN), LUKOIL (LKOH), Gazprom Neft (SIBN), Surgutneftegaz "(SNGS), Tatneft (TATN).

All collected data from the reports and from the Moscow Exchange website are collected in Table 3: **ROE** is the ratio of net profit to equity, **D/E** is borrowed funds on equity, **Dividend yield** = Dividend / market value of one share * 100%, **IMOEX** – the value of the Moscow Exchange index at the end of the year and the others. You can see that the range of values has greatly decreased – the coefficients of variation are lower.

The considered model looks as follows (formula 5):

$$y_i = b_0 + b_1 ROE_i + b_2 \left(\frac{D}{E}\right)_i + b_3 Div_i + b_4 Div(\%)_i + b_5 d_i + b_6 IMOEX_i + e_i \quad (5)$$

where $b_0, b_1, b_2, b_3, b_4, b_5, b_6$ – empirical regression coefficients (estimates of the parameters $\beta_0, \ \dots, \ \beta_6$. e_i – error terms (deviation estimates ε_i). The results of the calculations are shown in Table 4.

Table 3. Calculation table of indicators (x_{ij} and y_i)

№	Ticker	Period	ROE	D/E	Dividend per share, rub.	Dividend yield, %	Share of dividends in Net profit	IMOEX	Capit-n, billion rubles
1	ROSN	31.12.2013	17,4%	1,38	12,85	5,14%	24,72%	3045,87	2650,0
2	LKOH	31.12.2013	9,9%	0,39	110,00	4,56%	31,50%	3045,87	1723,6
3	SIBN	31.12.2013	18,7%	0,57	9,38	6,36%	23,82%	3045,87	699,8
4	SNGS	31.12.2013	13,5%	0,16	0,60	1,77%	7,68%	3045,87	1209,1
5	TATN	31.12.2013	15,5%	0,34	8,23	3,82%	22,89%	3045,87	468,9
6	ROSN	31.12.2014	12,1%	2,03	8,21	4,20%	24,86%	2369,33	2071,4
-/-	-/-	-/-	-/-	-/-	-/-	-/-	-/-	-/-	-/-
30	TATN	31.12.2018	27,2%	0,55	84,91	11,11%	87,45%	1396,61	1664,7
31	ROSN	31.12.2019	15,6%	1,51	33,41	7,41%	43,99%	1504,08	4775,5
32	LKOH	31.12.2019	15,8%	0,41	192,00	3,12%	21,38%	1504,08	4405,5
33	SIBN	31.12.2019	19,1%	0,73	37,96	9,02%	42,64%	1504,08	1994,9
34	SNGS	31.12.2019	-0,4%	0,18	0,00	0,00%	0,00%	1504,08	2107,4
35	TATN	31.12.2019	25,6%	0,65	64,47	7,90%	72,85%	1504,08	1776,9
Standard deviation			7%	0,62	72,94	2,47%	22,87%	533,89	1212,3
The coefficient of variation			47,7%	83,6%	144,8%	52,1%	77,1%	25,92%	62,0%

Source: Own processing based on reporting data and information from the Moscow Exchange website.

In the model under consideration, the coefficient $R^2 = 0,7110$, which indicates good predictive properties of the equation. Hypothesis H_0 ($R^2 = 0$) is rejected in favor of the alternative hypothesis, since $F > F_{\alpha;\, m;\, n-m-1}$, in other words, the equation is statistically significant. Parameters b_2 (financial leverage), b_3 (dividend per share) and b_6 (Moscow Exchange index) are significant, since $|t_{b_i}| > t = 2{,}045$, which indicates a positive linear relationship between capitalization and the amount of financial leverage and dividend per share.

Thus, Theorem 3 (on the ratio of prices of a company with 100% equity capital and a company with debt obligations), taking into account taxes, was confirmed: with an increase in the size of loans $\left(\frac{D}{E}\right)$, the value of the company also increases.

It should be noted that the influence of the dividend yield on capitalization is negative. This is probably due to the fact that after the payment of dividends there are drawdowns, the value of which is equal to the size of the dividend per share. These downturns, or dividend gaps, take some time to recover, and the larger the drawdown, the longer it takes for the price to recover, which is certainly reflected in the equity market value. However, after the corresponding calculations, the parameter b_4 turned out to be statistically insignificant, therefore, there is no effect of this indicator on capitalization.

An interesting fact is that during a market fall (a decrease in the IMOEX index), the capitalization of the companies in question is growing, perhaps this is due to the fact that investors view these companies as a more stable investment object during the crisis.

Table 4. Checking the Significance of the Regression Equation

Parameter estimates		t-statistics	Statistical significance b_i ($t = 2{,}048$)	Проверка значимости уравнения	
b0	2588,645	3,7407	Yes	R^2	0,7110
b1	1274,748	0,5768	No	F	11,4838
b2	1373,949	5,9139	Yes	F_{test}	2,4453
b3	9,540	4,8684	Yes	Equation significance	Yes
b4	-14090,53	-1,6121	No	-/-	
b5	-417,913	-0,4404	No		
b6	-0,738	-3,0311	Yes		

Source: Own processing.

Thus, for the companies under consideration, there is a positive relationship between the size of the dividend per share and the market value of capital. The implication is that the payment of dividends allows shareholders to get a return on their investment in the company, while potential investors are inspired by the ownership of shares in the companies. Investors are responding to the dividend declaration as a sign of financial prosperity, removing uncertainty about firms' ability to operate profitably. However, we must not forget that the amount of dividends must be reasonable; it is good if the company allocates "extra" funds for dividends, i.e. those that did not go to unprofitable projects. Shareholders usually respond favorably when companies distribute excess money instead of "wasting it on unprofitable projects." [4].

Table 5. Dynamics of the amount of dividend per share and capitalization, 2011-2019

№	Ticker	Period	Dividend per share	Capitalization, billion rubles
1	LKOH	31.12.2011	75	1445,5
2	LKOH	31.12.2012	90	1706,1
3	LKOH	31.12.2013	110	1723,6
4	LKOH	31.12.2014	154	1886,0
5	LKOH	31.12.2015	177	2001,0
6	LKOH	31.12.2016	195	2916,2
7	LKOH	31.12.2017	215	2823,4
8	LKOH	31.12.2018	250	3739,1
9	LKOH	31.12.2019	192	4405,5
Standard deviation			59,46	1025,46
Coefficient of variation			32,2%	36,8%

Source: Own processing.

3.3 Assessment of the impact of dividend policy on the market value of PJSC "Lukoil"

Lukoil is one of the few private companies with minimal government influence on policy making. An interesting fact is that Lukoil adheres to the policy that Miller and Modigliani spoke about: the total amount of dividends on outstanding shares minus shares held by LUKOIL Group entities is at least 100% of the adjusted free cash flow, which is defined as net cash. received from operating activities, net of capital expenditures, interest paid, repayment of lease liabilities, and expenses for the acquisition of shares in LUKOIL. In other words, the funds remaining after their distribution to profitable projects (for which

NPV> 0) go to dividend payments. Thus, the analysis of the dependence of capitalization on dividend payments for this company is of considerable interest.

The results of evaluating the relationship between dividend per share and capitalization confirm a positive relationship, however, it should be remembered that they give a result if certain rules are observed: dividends must be paid on a leftover basis. The equation and coefficient b_1, reflecting the relationship between the factors under consideration, are statistically significant.

Table 6. Checking the Significance of Equation and parameters

Parameter estimates		t-statistics	Statistical significance b_i (t = 2,045)	Checking the Significance of an Equation	
b0	318,445	0,4617	No	R^2	0,6189
				F	11,3690
b1	13,567	3,3718	Yes	F_{test}	3,6767
				Equation significance	Yes

Source: Own processing.

To assess the long-term impact on capitalization, you can consider a regression model with a time lag, i.e. "Shift" the variable x_i (dividend per share) by several years. The results of this bias are shown in Table 7.

Table 7. Checking the significance of the equation and coefficients with a time lag

Lag, years	Parameter estimates		t-statistics	Statistical significance b_i (t = 2,3646)	Checking the Significance of an Equation	
1	b0	383,706	1,3027	No	R^2	0,8987
	b1	14,485	7,8808	Yes	Equation significance	Yes
3	b0	616,036	2,8054	Yes	R^2	0,9299
	b1	17,778	9,6390	Yes	Equation significance	Yes
5	b0	585,500	2,3474	No	R^2	0,9479
	b1	26,096	9,5421	Yes	Equation significance	Yes

Source: Own processing.

Thus, the larger the "offset", the higher R^2, i.e. the quality of the model, which suggests that dividends can play a role even after a few years. At the same time, this is true only for Lukoil, other companies may have different results.

4 Conclusion

Thanks to statistical data and an extensive methodological base of the study, an analysis was made of the influence of dividend policy on the market value of the company. Based on the main results of the study, the following conclusions were made

The payment of dividends by oil companies in Russia has a positive effect on the price of their shares. In addition, the theorem on the growth of the firm's value with an increase in financial leverage was confirmed, which is explained by the effect of the tax shield.

The effect of stable dividend payments on capitalization growth in the following periods was confirmed. So, using the example of observations for the period 2011-2019 for the indicators of Lukoil, an increase in the accuracy of the explanatory model, taking into account the time effect, was statistically revealed.

In addition, using the example of Lukoil, it is shown that despite the positive relationship between dividends and growth in value, the optimal dividend policy is to pay dividends on a residual basis - first of all, it is necessary to pay attention to rational investments. Exactly as described in the conclusions of Miller and Modigliani.

At the same time, the study confirms the arguments that dividend policy, when adhering to the listed rules, positively affects the value of shares, as argued in the work of Lintner (1956) and Gordon (1962).

The management of companies engaged in oil production and refining must first of all effectively invest retained earnings, and not withhold the remaining funds, but direct them to dividends or to buy back shares. This is the most preferred policy as it is usually perceived favorably by shareholders.

References

1. A. Abramov, A. Radygin, M. Chernova, R. Entov, The “dividend puzzle” and the Russian stock market. Part 2. Voprosy Ekonomiki, **2020**(2), 59-85 (2020)
2. "Налоговый кодекс Российской Федерации (часть первая)" от 31.07.1998 N 146-ФЗ (ред. от 01.04.2020) НК РФ Статья 43. Дивиденды и проценты [online], Available at: http://www.consultant.ru/document/cons_doc_LAW_19671/fc207b30c9a64882be531927d7fbe7053769d6ca/ (10.05.2020)
3. Федеральный закон от 26.12.1995 N 208-ФЗ (ред. от 04.11.2019, с изм. от 07.04.2020) "Об акционерных обществах" (с изм. и доп., вступ. в силу с 01.01.2020) Статья 42. Порядок выплаты обществом дивидендов [online], Available at: http://www.consultant.ru/document/cons_doc_LAW_8743/9f4fbb73c39e25049b6c7c98bbbec1b66f83a9f2/ (10.05.2020)
4. R. Breyli, S. Mayers, *Principles of corporate finance*. ЗАО «Олимп-Бизнес», Moscow (2008)
5. Enterprise value [online], Available at: https://en.wikipedia.org/wiki/Enterprise_value (10.05.2020)
6. Quoc Trung Tran, *Creditors and dividend policy: Reputation building versus debt covenant*. European Research on Management and Business Economics **25**(3) (2019)
7. Теория независимости (иррелевантности) дивидендов [online], Available at: https://studme.org/263776/finansy/teoriya_nezavisimosti_irrelevantnosti_dividendov (2020)
8. М. Эскиндаров, М. Федотова, *Корпоративные финансы*; под ред. М.А. Эскиндарова, М.А. Федотовой. – Москва : КНОРУС (2020)
9. V. Aivazian, L. Booth, S. Cleary, *Dividend policy and the organization of capital markets*. Journal of Multinational Financial Management. **13** (2) (2003)
10. R. Watts, The information component of dividends. Journal of Business 46, 191/211 (1973)
11. РБК. Путин одобрил выплату госкомпаниями дивидендов в 50% от прибыли [online], Available at: https://www.rbc.ru/economics/27/04/2017/59020eaa9a794712f125072c (2020)
12. C. Driver, A. Grosman, P. Scaramozzino, Dividend policy and investor pressure. Economic Modelling [online], Available at: https://ez.el.fa.ru:2052/science/article/pii/S0264999318309623 (2020)
13. Modigliani-Miller Theorem [online], Available at: http://faculty.haas.berkeley.edu/parlour/teaching/corp_intro.pdf (2020)

14. A.Tijjani, M. Alifiah, G.Wakil, The Impact of Dividend Policy on the Shareholders Value of Listed Firms in the Nigerian Petroleum Marketing Industry. International Journal of Recent Technology and Engineering (IJRTE) ISSN: 2277-3878, **8** (2S9) (2019)
15. Московская биржа. Рыночная капитализация [online], Available at: https://www.moex.com/s26 (2020)
16. Conomy. Эмитенты [online], Available at: https://www.conomy.ru/search (2020)
17. Коэффициент вариации [online], Available at: https://ru.wikipedia.org/wiki/Коэффициент_вариации (2020)

Benefit's Portfolio in a Company in Connection to the Generation of Managerial Applicants

Kateřina Jančíková[1] and *František* Milichovský[1,*]

[1]Department of Economics, Faculty of Business and Management, Brno University of Technology, Brno, Czech Republic

Abstract. The topic of remuneration of executive managers (included CEO) is considered such a key factor of personal management (PM). The issue of remuneration of high-ranking managers is a key area of personnel management and corporate governance. This is still a much-discussed topic, which seeks to answer the most important question: How to set management rules in companies so that all stakeholders behave effectively? It is high ranking managers who are a specific group in which the creation of a remuneration contract must ensure the optimal connection of their performance with measurable indicators of the company. The consequences of their decisions can be clear for up to several years, so setting optimal benchmarks is more than important. The main aim of the research paper was to find a dependence between the type of respondent generation and the sought-after benefits in the managerial position. The hypothesis was defined as follows: the type of respondent generation influences the choice of benefit. IBM SPSS Statistics 25 was used to process and evaluate the data obtained from the structured questionnaire survey. The mathematical-statistical methods used to verify the dependence between selected variables were used the Chi-square test of independence of two variables.
Keywords: benefits and remuneration; motivation; corporate identity; human resource management

1 Introduction

Today, the company's reputation and reputation is not only related to its business practices and is not only influenced by customer satisfaction. Nowadays, great emphasis is placed on employees and their satisfaction, as employees are bearers of the brand. Each employee is unique, has specific characteristics, skills, and abilities that influence his / her behavior, values, and attitudes. These can be a source of high performance for the organization. Therefore, to maximize employee performance and ensure that it proudly presents the brand of the organization, it is necessary to motivate the employee correctly. The structure of workers has also changed considerably recently and the labor market is dominated by Y and Z generation. They are young people whose perception of work performance, motivation, and benefits differ significantly from the traditional conception

[*]Corresponding author: milichovsky@fbm.vutbr.cz

recognized by Generation X. freedom, the possibility of self-realization and the benefits they demand are primarily non-monetary and intangible. The following chart shows the change in the population structure in the Czech Republic over the next decade. More and more Z-workers will accumulate in the labor market, and it is now necessary to change the minds of company representatives and move from traditional forms of motivation and remuneration to new ways. According to Forbes (2019), generation X includes people born 1968-1982, generation Y includes people born 1983-1997, and generation Z includes people born after 1998.

The purpose of this article is to create a conceptual framework that will help HR managers and business owners to positively influence corporate identity through employee motivation and benefits provided. The article first defines the corporate reputation, then motivation and last but not least defines the specific characteristics of the Z generation, which is becoming new blood for the labor market. The article will present data and results of the research conducted in the Czech work environment. The second part of the paper describes the methodology and employed methods. The third part presents gained results with relevant discussion and finally, there is a conclusion of the paper.

2 Theoretical background

Every organization, regardless of size, has its own corporate identity (Balmer, Gray, 2003; Olins, 1989) and we understand an organization as any form of the profit-making institution (Boulton, Allen, Bowman, 2015). As a result, the concept of corporate identity can be applied very widely. The notion of corporate identity is not precisely defined (Kitchen et al., 2013; Leitch, Motion, 1999; Melewar, 2003). In general, corporate identity can be seen as the "essence of society" (Olins, 1978). The core values of the company are a central aspect of the corporate identity and initially come from the founders (Urde, 2003). Over time, however, corporate identity and company values change under the influence of several stakeholders, including employees (Iglesias et al., 2013). Human resources are becoming increasingly important for the impact of the social reputation of the organization. World studies show the impact of the ability to recruit talented employees on improving the company's reputation (Clardy, 2005; Turban, Cable, 2003), further improved corporate reputation through fair treatment (Koys, 1997; Shaughnessy, 2005). A conceptual framework that combines the role of HR and the company's reputation appears in its culture. The human resources function plays an important role in organizations where intellectual capital represents a competitive advantage for the company. Examples of such companies are Google or Intel, where great emphasis is placed on employee motivation and satisfaction (Friedman, 2009).

The role of employee champion maximizes employee benefit by responding to workers' concerns and allowing them to increase competencies in line with the organization's goals (Ulrich, 1997). Employee champion addresses a wide range of working conditions that increase employee loyalty, such as pay, pay, safety and job satisfaction, effective training, career planning and mentoring, etc. Employee champions at Wegmans (a private food chain based in New York) indirectly improved the company's reputation and financial performance. Wegmans ranked in the 1st position in the Fortune Magazine's 100 Best Companies in 2005, ranked 2nd in 2006 and 3rd in 2007. This company provides generous employee benefits, including a scholarship program for universities. This assistance is provided to more than 19,000 employees in a lot of $ 59 million. The program encourages all employees to pursue their educational goals, promote upward career mobility, a good community, and improve Wegmans' reputation as the number one employer for all people (Wegmans, 2007; Cesari, Gonenc, Ozkan, 2016). For this company, employees come first, and customers come second. Another company that takes great care of its employees is

Google. It is regularly ranked worldwide among the best employer rankings. In this company, they are committed to a work environment that is designed to maximize employee productivity, loyalty, and creativity. Google is characterized by low employee turnover because work and the environment are appealing to employees and want to stay in the company (Friedman, 2009). This company is also an attractive employer for the youngest generation of Z.

So far, Generation Z has been viewed "only" as young customers. Now all of this generation is entering the labor market, raising awareness of this generation as colleagues and employees. Each generation is specific and has its basic characteristics. Generation Z members demand respect and want to be treated as partners, not as traditional employees. A typical characteristic is the requirement of freedom to perform their work, which has so far usually been reserved for senior positions. They also have a choice of working time and contact with people outside their work. If the employer grants them such working conditions, then these workers will be rewarded with a great work commitment, unexpected independence, and purposefulness (Volek, 2018).

According to the latest surveys, up to 9 out of 10 employees occasionally monitor job offers, which means that if an employer wants to retain their employees, they must do their utmost to understand their values, needs, and remuneration, as well as the incentive to adapt (Clifford, 2014). Money is important to the new generation Z, but that's not all. Therefore, working conditions, good relationships with colleagues, praise or appreciation, teamwork, or meaningfulness of work also play an important role in the decision-making process.

Both Y and Z are aware of the challenges they face. They assume that it will not be easy to find employment at the time of emerging Industry 4.0, but they believe they will be better off than their parents. Thus, Z can be even more optimistic than the Y generation (Deloitte, 2018). According to a survey by Ozkan and Solmaz (2015), the social environment is very important for the Z generation. This is also an important factor in building a corporate culture. The survey also showed that this generation is very confident and understands work as a means to achieve their dreams. If they are not satisfied at work, they can easily leave the job. They are not afraid of uncertainty about losing their job.

The purposefulness of the Z generation directs these workers to high positions in companies that can reach them very quickly. The negative of this generation is lower loyalty to their employers and more frequent change of employer. All the more this fact will manifest itself when the company motivation program is not aligned with the priorities of these employees. For example, if a company is only fixed to generate its profits, then it loses loyalty to its employees (Deloitte, 2018). Figure 1 shows the factors for which the respondents of the performed research should sort the individual aspects according to their importance when considering work in the selected company. The survey was attended by 10,455 millenniums and 1,844 workers from the Z generation. However, this is in line with the idea that business owners should share their wealth with their employees. It also reflects the fact that the Y and Z generation must think more about their future and the security of their future family (Deloitte, 2018).

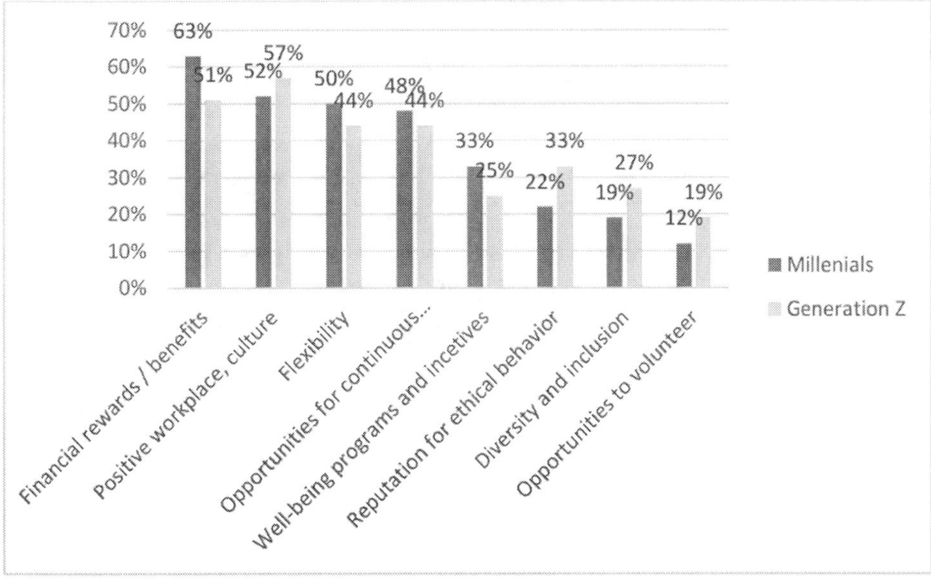

Fig. 1. Important aspects of work for the Y and Z generations
Source: Own processing.

According to a PWC survey (2019b), CEOs are very aware that there is a shortage of skilled workers in the labor market, so they need to be interested in something. However, each employee has his / her specific needs and offering each employee a special incentive program is not very realistic in the business environment. Therefore it is necessary to focus on groups of employees, e.g. according to generational classification, as described above in the article. To retain employees, managers and CEOs will have to take on non-traditional forms of work. These will not only reflect flexible cooperation models but will also reflect the new population structure.

The new forms of benefits currently appearing in the labor market, as well as wage increases, reflect the growth of the economy and CEO's expectations. According to a PWX survey (2019a), more than 50% of CEOs expect the economy to grow in the coming period. When comparing Czech CEOs and world CEOs, the Czechs are more optimistic about the future outlook for the world economy (PWC, 2019c), 70% of CEOs believe globally in the growing economic development, while 85% of CEOs believe in the Czech environment.

The attractiveness of the company in the labor market is also ensured by the efforts of employers to respect the needs of their employees. These needs (priorities) change over time and their hierarchy also changes. The general breakdown includes monetary (material) and value (non-material). Employee satisfaction consists of 3 components and the wage addresses only the basic needs of the employee. The second part of the satisfaction is the feeling of belonging, which is very closely connected with the corporate culture. The last part is the work itself (Chrenko, 2019). If all 3 components are set correctly, then employee satisfaction is greatly increased. According to a survey conducted by the Gallup Institute in 2017 on employee satisfaction, the 2nd and 3rd components have an impact of up to 87% on employee engagement and business productivity growth of up to 30% (Chrenko, 2019).

It can be expected that specific characteristics of a business, such as size or level of risk, will strongly affect performance compensation. Many authors have already empirically demonstrated the relationship between the size of a business and the level of management's remuneration and CEOs (Rosen, 1992; Murphy, 1985; Zhou, 2000; Ryan, Wiggins, 2001). According to Murphy (1999) and Song and Wan (2017), it is possible to read that the sensitivity between performance and manager's salary is weaker among larger American companies.

3 Methodology

The main aim of the research paper was to find a dependence between the type of respondent generation and the sought-after benefits in the managerial position and CEO. The chosen benefits are in Table 1.

Table 1. The dividing of chosen benefits as individual variables

Tangible benefits	company car	company laptop	
	financial contribution	company mobile phone	
Non-tangible benefits	proportionate ownership	home office	sick days
	extra holidays	employer brand	

Source: Own processing.

The hypothesis was defined as follows:

- H_0: the type of respondent generation does not affect the choice of benefit;
- H_1: the type of respondent generation influences the choice of benefit.

IBM SPSS Statistics 25 was used to process and evaluate the data obtained from the structured questionnaire survey. The statistical methods were used to verify the dependence between selected variables by Pearson Chi-square test of independence. This test is used to monitor the relationship between selected variables. To confirm the alternative hypothesis (that is, there is a relationship between defined variables) we have chosen a significance level of 95 %.

Data for this research were obtained through a structured questionnaire survey, which included both open and closed questions. The questionnaire survey was realized in January 2020. The basic population included 925 students of the Brno University of Technology, who have the potential to become senior managers in companies. Number of all students, which are divided into individual study level refer to individual generation. Master students could be included to generation Y and master students are consist in generation Z. Individual generations have different abilities, which are required at managerial positions. The difference between both groups are based on study programs, because bachelors' programs are made as professional programs and graduates are prepared for managerial work in various areas (e.g. accounting and taxes, production, ITC area, medium managerial positions in companies). The master programs includes parts, connected to managerial work and also prepare graduates for scientific work. In the long-term, both groups of bachelor and master students are almost similar. 522 respondents were randomly selected from the sample, with 50% of the respondents being Y population and 50% were respondents from the Z population. The reason for this dividing is that both generations are equal as the workforce with relevant offer to employer. Generation Y provides experiences is the most important thing, generation Z gives IT knowledge as its nature.

The return rate of the questionnaire survey was 34.48%, i.e. 184 fully completed questionnaires. Due to missing data, 5 questionnaires were excluded from the evaluation. Respondents of population Y represented 41% (73 respondents) and respondents of population Z represented 59% (112 respondents). The study among graduates of Brno University of Technology, which declares that within two years after graduation, will reach 3 out of 10 graduates in managerial positions (Brno University of Technology, 2018), proves that this is a correctly selected group of respondents.

4 Results

The relevance of benefits as part of the remuneration is based on several studies such as Castilla (2011), Dobbin et al. (2011), or Tomaskovic-Devey (2014). Test for independence (Pearson's chi-square) was employed to examine the stated hypothesis. The main base for verification is created a pivot table, showing frequencies of answers from individuals in connection to their generation. In the labor market, there are two main generations, which are prepared to become CEO soon. It is obvious from table 1, that tangible benefits are more preferred by both generations of workers in comparison to intangible benefits. All tangible benefits are acceptable by more than three-quarters of respondents in both generations. The main reason could be to "touch" the benefit. For intangible benefits, it is possible to corroborate the relevance of extra holiday and sick-days such an important part of the reward in an organization. According to the values in table 2, it is interesting, that both generations require the correct brand of their company.

Table 2. Pivot table of chosen benefits and individual generations

		Generation	No answer	Interest of chosen benefits				Relative positive interest of all respondents
				Absolutely disagree	Rather disagree	Rather agree	Absolutely agree	
Tangible benefits	Company car	Z	24	2	14	41	31	81,82 %
		Y	16	3	8	30	15	80,36 %
	Financial contribution	Z	26	2	21	45	18	73,26 %
		Y	16	0	11	29	16	80,36 %
	Company laptop	Z	26	2	18	38	28	76,74 %
		Y	16	2	6	19	29	85,71 %
	Company mobile phone	Z	26	2	18	38	29	76,74 %
		Y	16	2	6	19	29	85,71 %
Non-tangible benefits	Extra holidays	Z	25	0	13	39	35	85,06 %
		Y	15	0	4	32	21	92,98 %
	Sick days	Z	24	0	9	50	29	89,77 %
		Y	16	0	3	24	29	94,64 %
	Proportionate ownership	Z	28	7	43	31	3	40,48 %
		Y	18	10	20	13	11	44,44 %
	Employer brand	Z	24	2	14	41	31	81,82 %
		Y	16	3	8	30	15	80,36 %
	Home office	Z	28	7	43	31	3	40,48 %
		Y	18	10	20	13	11	44,44 %

Source: Own processing.

According to respondents' answers (see Table 2), it is obvious, that millennials prefer such benefits, which can support their families and provide them more time to be with family (mainly financial contribution, extra holidays, and sick-days). Generation Z prefers these benefits, which offer the kind of freedom for company financial sources (company car, extra holidays, sick days). On the contrary, for proportionate ownership and home office usually, both generations do not look for. Possibility of a home office is more preferred by employers in a stressful era such is during the COVID-19 situation, where a lot of companies used home-office instead of reducing staff (Papagiannidis, Marikyan, 2020; Zhang et al., 2020).

In the results of Pearson's chi-square test of independence, there are proved nine benefits in connection to the generation of applicants. From these nine connections, there are found only three relations, which could be considered as statistical dependencies – reach a significant p-value under 0,05. In the results of the Pearson's chi-square test of independence (see Table 3), there are found out dependencies between (these results match the required level of significance of potential error under 5%):

- generation and proportionate ownership (p-value = 0,005);
- generation and employer brand (p-value = 0,001);
- generation and home office (p-value = 0,033).

The p-value, which expresses the probability of validity of the null hypothesis. The lower the p-value, the lower the probability of a null hypothesis. It can, therefore, be concluded from the above data that we can only accept alternative hypotheses about the validity of the relationship between gender and selected sources of job offers only for job servers and social networks. Furthermore, we should consider the level of significance chosen. We are working with a 5% significance level in the paper, so the p-value should be lower than this significance level. Only for social networks the p-value $<\alpha = 0.05$, therefore we reject the null hypothesis and accept the alternative hypothesis about the dependence of selected elements.

Other benefits, which were verified for dependence with the generation, did not meet dependence with defined generations of applicants. Their p-values were over the required 5% error. There are two benefits, which significances are close to the relevant value (in comparison to others); company laptop has p-value 0,175 and sick days reach p-value for 0,152. Both values are high, but they could be put under the monitoring process and revised. From others point of the view, did not prove relevant p-values, which are accepted on the level of confidence 95% because they are higher than 0,4.

The intensity of gained dependencies is defined by the contingency coefficient, which can be in interval <0;1>. Gained coefficients for observed dependencies are as follow (all values are shown in Table 3):

- generation and proportionate ownership has intensity 0,275, which is described as rather low value;
- generation and employer brand reaches intensity 0,310, which is described as low value;
- generation and home office have intensity 0,232, which is described as a rather low value.

These three relationships could steady the planning of human resources, benefit offering, and hire relevant applicants, the fittest for a job position.

Table 3. Pearson's test of the relationship between chosen benefits and generations (Y and Z)

			value	p-value
Tangible benefits	Company car	*Pearson Chi-Square*	2,110	0,716
		Contingency Coefficient	0,106	
	Financial contribution	*Pearson Chi-Square*	2,506	0,644
		Contingency Coefficient	0,116	
	Company laptop	*Pearson Chi-Square*	6,336	0,175
		Contingency Coefficient	0,182	
	Company mobile phone	*Pearson Chi-Square*	1,888	0,756
		Contingency Coefficient	0,101	
Non-tangible benefits	Extra holidays	*Pearson Chi-Square*	2,896	0,408
		Contingency Coefficient	0,124	
	Sick days	*Pearson Chi-Square*	5,289	0,152
		Contingency Coefficient	0,167	
	Proportionate ownership	*Pearson Chi-Square*	15,051	0,005
		Contingency Coefficient	0,275	
	Employer brand	*Pearson Chi-Square*	19,322	0,001
		Contingency Coefficient	0,310	
	Home office	*Pearson Chi-Square*	10,459	0,033
		Contingency Coefficient	0,232	

Source: Own processing.

5 Conclusions

Benefit offer helps companies to care about current staff and due to its portfolio could catch relevant applicants from various generation type (Bolton et al., 2013). In general, both generations Y and Z look mainly for tangible benefits to reach certitude. From non-tangible benefits become more important company brand in connection to internal corporate culture and work environment, and to help create corporate identity supporting the brand. In the case of a high-quality company brand, employees change their mindset and their loyalty to the company have become much stronger and higher (Benraïss-Noailles, Viot).

The main aim of the research paper was to find a dependence between the type of respondent generation and the sought-after benefits in the managerial position and CEO. According to our results, we found out that there are three dependencies between analyzed generations Y and Z, and (1) proportionate ownership, (2) employer brand, and (3) home office. Their intensities are at low levels. Due correct setup of benefits portfolio it is possible to create relevant competitive advantage because the importance of human capital corroborates corporate strategies on the way of making good internal culture with linking to benefits' portfolio, corporate values, and strategic vision, and mission (Schnidman, Hester,

Plunkte, 2017). A similar situation confirms the work of Gašková (2020) under the condition of a Czech university alumni condition.

The main limitation of the study is focused on the Czech environment, where we chose university students and graduates as potential applicants of managerial jobs. Our results could be adequate for alike countries as Slovakia, Poland, Hungary or Ukraine are.

References

1. J. M. T. Balmer, S. Greyser, *Revealing the Corporation: Perspectives on Identity, Image, Reputation, Corporate Branding, and Corporate-level Marketing: An anthology*. London: Routledge (2003)
2. L. Benraïss-Noailles, C. Viot, Employer brand equity effects on employees well-being and loyalty. *Journal of Business Research*, (2020) 10.1016/j.jbusres.2020.02.002 (IN PRESS)
3. R. N. Bolton, A. Parasuraman, A. Hoefnagels, N. Migchels, S. Kabadayi, T. Gruber, Y. Komarova Loureiro, D. Solnet, Understanding Generation Y and their use of social media: a review and research agenda. *Journal of Service Management*, 24(3), 245-267. (2013) 10.1108/09564231311326987
4. J. G. Boulton, P. M. Allen, C. Bowman, *Embracing Complexity: Strategic Perspectives for an Age of Turbulence*. Oxford: Oxford University Press. (2015)
5. A. Clardy, Reputation, goodwill and loss: Entering the employee training audit equation. *Human Resource Development Review*, 4(3), 279-304. (2005) 10.1177/1534484305278243
6. C. Clifford, *What Young People Want From Work*. [online] Available at: https://www.entrepreneur.com/article/234934 (2014)
7. A. De Cesari, H. Gonenc, N. Ozkan, The effects of corporate acquisitions on CEO compensation and CEO turnover of family firms. *Journal of Corporate Finance*, 38, 294-317. (2016) 10.1016/j.jcorpfin.2016.01.017.
8. Deloitte. *Deloitte Millennial Survey: Millennials disappointed in business, unprepared for Industry 4.0*. [online] Available at: https://www2.deloitte.com/content/dam/Deloitte/global/Documents/About-Deloitte/gx-2018-millennial-survey-report.pdf (2018)
9. D. J. Flanagan, K. C. O'Shaughnessy, The effect of layoffs on firm reputation. *Journal of Management*, 31(3), 445-463. (2005) 10.1177/0149206304272186.
10. B. Friedman, Human resource management role implications for corporate reputation. *Corporate Reputation Review*, 12, 229-244. (2009) 10.1057/crr.2009.17.
11. J. Gašková, Servant leadership and its relation to work performance. *Central European Business Review*, 9(3), 24-37. (2020) 10.18267/j.cebr.236
12. O. Iglesias, N. Ind, M. Alfaro, The organic view of the brand: A brand value co-creation model. *Journal of Brand Management*, 20(8), 670-688. (2013) 10.1057/bm.2013.8.
13. P. J. Kitchen, M. Tourky, D. Dean, A. S. Shalaan, Corporate identity antecedents and components: Toward a theoretical Framework. *Corporate Reputation Review*, 16(4), 263-284. (2013) 10.1057/crr.2013.18.
14. D. J. Koys, Human resourced management and fortune's corporate reputation survey. *Employee Responsibilities and Rights Journal*, 10(2), 93-101. (1997) 10.1023/A:1025632711742.

15. S. Leitch, J. Motion, Multiplicity in corporate identity strategy. *Corporate Communications: An International Journal*, 4(4), 193-199. (1999) 10.1108/13563289910299319.
16. T. C. Melewar, Determinants of the corporate identity construct: A review of the literature. *Journal of Marketing Communications*, 9(4), 195-220. (2003) 10.1080/1352726032000119161.
17. K. J. Murphy, Corporate performance and managerial remuneration: an empirical analysis. *Journal of Accounting and Economics*, 7(1-3), 11-42. (1985) 10.1016/0165-4101(85)90026-6.
18. W. Olins, *Corporate identity-making business strategy visible through design*. London: Thames and Hudson. (1989)
19. W. Olins, *Corporate personality: An inquiry into corporate identity*. New York: Mayflower Books. (1978)
20. M. Özkan, B. Solmaz, The changing face of the employees- generation z and their perception of work (A study applied to university students). *Procedia Economics and Finance*, 26, 476-483. (2015) 10.1016/S2212-5671(15)00876-X
21. S. Papagiannidis, D. Marikyan, Smart offices: A productivity and well-being perspective. *International Journal of Information Management*, 51, 1-11. (2020) 10.1016/j.ijinfomgt.2019.10.012
22. S. Rosen, *Contracts and the market for executives in contract economics*. Blackwell Publishers. (1992)
23. H. E. Ryan, R. A. Wiggins, The influence of firm and manager-specific characteristics on the structure of executive compensation. *Journal of Corporate Finance*, 7(2), 101-123. (2001) 10.1016/S0929-1199(00)00021-3
24. A. Schnidman, L. Hester, P. Plunkte, (2017). Global recruiting trends 2017 - What you need to know about the state of talent acquisition. [online] Available at: https://hrs.wsu.edu/wp-content/uploads/2015/09/linkedin-global-recruiting-trends-report.pdf (2017).
25. W.-L. Song, K.-M. Wan, Explicit employment contracts and CEO compensation. *Journal of Corporate Finance*, 44, 540-560. (2017) 10.1016/j.jcorpfin.2014.11.002
26. D. B: Turban, D. M. Cable, Firm reputation and applicant pool characteristics. *Journal of Organizational Behavior*, 24(6), 733-750. (2003) 10.1002/job.215
27. D. Ulrich, *HR Champions*. Boston, MA: Harvard Business Press. (1997)
28. M. Urde, Core value-based corporate brand building. *European Journal of Marketing*, 37(7/8), 1017-1040. (2003) 10.1108/03090560310477645
29. M. Volek, *Mladí Češi jako zaměstnanci: Cílevědomější, než se zdají*. [online] Available at: https://www.ipsos.com/cs-cz/mladi-cesi-jako-zamestnanci-cilevedomejsi-nez-se-zdaji (2018)
30. S. X. Zhang, Y. Wang, A. Rauch, F. Wei, Unprecedented disruption of lives and work: Health, distress and life satisfaction of working adults in China one month into the COVID-19 outbreak. *Psychiatry research*, 288, 1-6. (2020) 10.1016/j.psychres.2020.112958
31. X. Zhou, CEO pay, firm size, and corporate performance: evidence from Canada. *Canadian Journal of Economics*, 33(1), 213-251. (2000) 10.1111/0008-4085.00013

Normality and Its Verification as a Basic Prerequisite for the Application of VaR

Jan Chutka[1,*] and *Ladislav* Vagner[1]

[1]University of Zilina, Department of Economics, Zilina, Slovakia

Abstract. If we start to deal with the topics of investing or trading in the financial markets, sooner or later we will encounter the topic of risk. Risk is one of the basic input variables in assessing the suitability and profitability of an investment and therefore there are a number of procedures and methods for its quantification. In our article, we address the issue of portfolio risk quantification through the VaR method and verification of its basic assumption, namely the normal distribution of values. The aim of the article is to compile an overview of procedures and methods for verifying the normal distribution and compare their specifics. In the first chapter of our paper, we focused on the definition of risk, its types. The second chapter focuses on methods of risk quantification and description of individual methods. In the third chapter, we will describe in detail the possibilities of verifying the normality of the distribution of values. In the last chapter, we will briefly interpret the information obtained and identified the advantages, disadvantages and other specifics of individual methods of verifying the normal distribution. We consider the aim of the article to be fulfilled and we believe that it will be a valuable contribution in this area of research.
Keywords: normality, normal distribution, Value at Risk, risk quantification

1 Introduction

In today's global world of finance, with the rapid development of modern technologies and investment approaches, the issue of risk identification and measurement is still relevant. Even today, risk remains one of the main input factors in assessing the suitability, adequacy and profitability of investments and transactions worldwide. Therefore, there are many procedures and methods for its quantification. In our article, we will try to focus on one of the most used tools for measuring portfolio risk, namely Value at Risk. The basic premise of the application of this method is the assumption of a normal distribution of the values. And the normal distribution is a very interesting and discussed assumption of this method.

*Corresponding author: jan.chutka@fpedas.uniza.sk

2 Methodology

The basic methods of formal logic are used in the work as an analysis and synthesis of theoretical knowledge on the basis of which we have come from general information to identify the specifics of individual methods of verifying the normal distribution. We have further described these methods in detail with their mathematical formulas and identified their specifics necessary for their use. We were also interested in the Value at Risk method in detail, which is currently one of the most widely used methods for calculating portfolio risk. The aim of the article is to compile an overview of procedures and methods for verifying the normal distribution and compare their specifics.

3 Risk

After very fundamental macroeconomic changes, which were identified in the 1970s, two basic trends emerged, which determined the increasing level of competition and uncertainty. The first is the deregulation of economic policy with a focus on market-oriented instruments, the second is globalization, which has forced companies to face the true nature of global competition. These trends have caused individuals and financial institutions to face a wide range of risks today. We can also define financial risk as what unites companies from possible losses on the capital and money markets due to possible shifts in interest rates, prices of securities held, losses from foreign exchange transactions, insolvency of the bond issuer and many others. The position that companies risk losing can be carefully optimized with many financial market instruments. Therefore, manufacturing companies and non-financial service providers focus on business risk management. However, when looking at financial market institutions such as banks, insurance companies, funds, I see that they focus on managing, predicting, arranging protection and advising on financial risks. The aim and main interest of such institutions is to try to measure risks and subsequently control the price of these risks. Today, there is no uniform doctrine of financial risk, which implies that the terminology is not precisely specified, but nevertheless in the following subchapter we present commonly used terminology in risk management [5,7].

3.1 Risk typology

In general, financial risk is classified into the following relatively general groups:

- **Market risk**
- **Credit risk**
- **Liquidity risk**
- **Operational risk**
- **Legislative risk**

We define market risk as the risk of loss of the value of the portfolio held caused by sharp changes in the prices of individual assets held. As its existence is linked to the valuation of assets in the market. The factor that most significantly affects the amount of market risk is the volatility of assets. In other words, the size of price movements over time around their average or around the reference price of owned assets in the portfolio. Credit risk arises when a responsible counterparty is unwilling or unable to meet its obligations. The amount of risk is then defined by the amount of exposure of such a claim that the relevant counterparty should have paid. More generally, we can also define credit risk as the potential loss of market value suffered by a company providing funds to counterparties due to a credit event. Such a credit event can be understood as a change in the counterparty's ability to meet its obligations, which is usually reflected in a change in the market value of such loans, a possible change in the counterparty's rating, or a change in the probability of client default. Liquidity is the ability to quickly monetize an asset so that there is no significant difference between the current selling price and the current selling price of that asset. In financial institutions, let us distinguish several types of liquidity risk. The most serious are Product / market liquidity risk, which arises due to the inability of a bank or other institution to sell its assets due to the high volumes that are in its portfolio. We define operational risk or the risk of direct or indirect losses as risk that arises due to inadequate or incorrect internal processes, systems, human resources behaviour, or due to external events. In the past, operational risks were calculated on the basis of expert considerations and there were no comprehensive quantitative tools to measure these risks. A significant step forward in the quantification of operational risks and their prevention was the progress in the field of computer technology, which enabled the effective recording of these events and their evaluation. Legislative risk is closely linked to credit risk and operational risk. This type of risk arises when a counterparty that incurs a loss in a particular transaction seeks to avoid paying for the transaction by finding an error in the contract on a legal basis. Legislative risk also arises when there are indications that a counterparty has no competence to enter into an agreed transaction [11,13,15].

4 Risk measurement

The increasing volatility of exchange rates, interest rates and commodity prices has created the need to create new financial instruments and analytical tools for risk management. When we begin to study risk management in detail, we identify two main streams of this development. The first basic stream of development in this area is the development of information technologies. The second is the development of financial theory, the emergence of new products to mitigate risks or to eliminate volatility from the owned portfolio [12, 16].

4.1 Value at Risk

The definition, although not in a strictly mathematical spirit, can be drawn directly from The 1996 Risk Metrics Technical Document: *"Value-at-risk measures the maximum potential change in the value of a portfolio of financial instruments with a given probability within a predetermined time horizon. VaR answers the question: how much can I lose with x% probability in a given time horizon?* [6].

For the purposes of our article, we will use the calculation formula for calculating normal linear VaR and scaling VaR published by Carlo Alexander in 2008 in his book: "Market Risk Analysis Vol. IV. Value-at-Risk Methods" [1].

4.2 Normal linear VaR

$$\mathrm{VaR}_\alpha = \Phi^{-1}(1-\alpha)\sigma - \mu \qquad (1)$$

Where Φ is the standard normal distribution function. However xα = −VaRα by definition, and Φ-1(α)= −Φ-1 (1−α) by the symmetry of the standard normal distribution, Carol Alexander (2008) substitute these into the above mentioned formula and he created an analytic formula for the VaR for a portfolio with an i.i.d. normal return [1,8,9].

4.3 Scaling VaR

VaR is often measured in a short-term risk horizon such as 1-day VaR and then scaled to represent VaR in the longer term risk horizon. The way how should VaR, that is estimated over one risk horizon be scaled to a VaR that is measured over a different risk horizon describes Carol alexander in detail as follows. We will use the formula for calculating normal linear VaR.

$$\mathrm{VaR}_{h,\alpha} \approx \Phi^{-1}(1-\alpha)\sqrt{h}\,\sigma_1 - h\,\mu_1 \quad (2)$$

Where μ1 and σ1 are the expectation and standard deviation of the normally distributed daily returns. We now can use a log approximation to the daily discounted return. We now can approximate the h-day log return with the ordinary h-day return, and deduce that this is (approximately) normally distributed. Then the h-day VaR is given by the approximation as follows [1].

5 Normal distribution

Each quantitative method has its specific limitations and assumptions that are necessary for the very possibility and correctness of the application of these methods. One of the basic assumptions of applying the mentioned VaR calculation to a specific data set is that this data must come from a normal distribution of values. The reason for such an assumption is the fact that the calculation itself is based on the distribution function of the normal distribution of values, and if this assumption were not verified and the data thus obtained by the VaR calculation were met, they would have no explanatory power. This is the reason for the need to verify the existence of a normal distribution of values before the actual application of this method [10,14, 17, 18].

5.1 Pearson's test

Pearson's test is among the most famous tests. It is a test which is based on the comparison of distributions between observed (empirical) and expected (theoretical) frequencies. This the test is generally recommended for random testing on a large scale. The test can be briefly described as follows. We have a random selection, with an unknown distribution function (empirical distribution) and also a known distribution function (theoretical normal distribution). At a preselected level of significance, we test the null hypothesis H_0: $F(x) = F_0(x)$ against the alternative hypothesis H_1: $F(x) \neq F_0(x)$. The test formula is as follows:

$$\chi^2 = \sum_{i=1}^{k} \frac{(n_i - n \cdot p_i)^2}{n \cdot p_i} \quad (3)$$

We reject the H_0 hypothesis at the level of significance of alpha if $X^2 > X^2_{1-\alpha,\ k-r-1}$. Prerequisite for the correct use of the Pearson test is the following assumptions [3]:

$n * p_i \geq 4$ and $k - r - 1 \geq 6$

5.2 Shapiro Wilk test

The Shapiro-Wilk test is based on determining whether points are constructed quantile of a quantile graph (Q-Q fence) significantly differ from the regression line intersected by these points. We test the null hypothesis H_0: $F(x) = F_0(x)$ at a preselected level of alpha significance against the alternative hypothesis H_1: $F(x) \neq F_0(x)$. The Shapiro Wilk test is mainly used for selections of smaller ranges where the number of this file does not exceed 50 units. The formula for testing is as follows:

$$W = \frac{\left(\sum_{i=1}^{m} a_{i,n} \cdot \left(x_{(n-i+1)} - x_{(i)}\right)\right)^2}{\sum_{i=1}^{n}\left(x_{(i)} - \bar{x}\right)^2} \tag{4}$$

The more the value is measured W approaches 1, thus there is agreement between the theoretical and better empirical distribution. Hypothesis H0 about the normal distribution of the basic set from which the random selection was made, we reject the level of significance of alpha, if applicable [2]:

$W \leq W_\alpha(n)$ where $W_\alpha(n)$ is critical value.

5.3 Kolmogorov Smirnov test

The Kolmogorov-Smirnov test is based on a comparison of the distribution function of the assumed (theoretical) distribution of the continuous type with the empirical (selective) distribution function. It should only be used to verify hypotheses that determine the theoretical distribution function unambiguously, i. not only in terms of shape, but also in terms of parameter values. Unlike the Person's test, this test can also be performed for random selections of relatively small ranges. At a preselected level of alpha significance, we test the null hypothesis H_0: $F(x) = F_0(x)$ against the alternative hypothesis H_1: $F(x) \neq F_0(x)$, where F is the empirical distribution function of the sample and F0 is the considered theoretical distribution function. Empirical distribution function:

$$F(x) = \begin{cases} 0 & \text{for} \quad -\infty < x < x_{(1)} \\ \frac{i}{n} & \text{for} \quad x_{(i)} \leq x < x_{(i+1)},\ i = 1, \ldots, n-1 \\ 1 & \text{for} \quad x_{(n)} < x < \infty \end{cases} \tag{5}$$

The test calculation formula is based on the expression of the value of D which is defined as a maximum distance between values F and F0:

$$D = sup\ |F_0(x) - F(x)| \tag{6}$$

Hypothesis H0 is rejected at the level of significance alpha, if the calculated value of D exceeds the critical value $D > D_\alpha(n)$ [4].

6 Results

Through tests on the verification of normal distribution, we verify the assumption that the data (sample file) come from the base file with normal distribution. Based on the theoretical advantages of individual normality tests, we can identify the following specifics pros and cons of individual tests.

Table 1. Description of the specifics of individual tests

Tests	Specifics
Pearson´s test	• One of the most famous and widespread tests. • Designed for testing large files. • It is necessary to follow the Cochrane rule, if these conditions are not met, it is recommended to merge adjacent intervals with a small number.
Shapiro Wilk test	• Based on verification of whether the points of the Q-Q graph differ significantly from the regression line intersected by these points. • Designed for testing selections with a smaller range ($n \leq 50$).
Kolmogorov Smirnov test	• It is used only to verify hypotheses that determine the theoretical distribution function unambiguously. • A very specific model. • To test files with a range $n \leq 30$, we will use the exact value $D_{\alpha}(n)$ from the table of critical test values.

Source: Own processing.

7 Conclusion

The aim of the article was to compile an overview of procedures and methods for verifying the normal distribution and compare their specifics. In the first chapter of our paper, we were focused on the definition of risk, its types. The second chapter focuses on methods of risk quantification and description of individual methods. In the third chapter, we described in detail the possibilities of verifying the normality of distribution of values. In the last chapter, we briefly interpreted the information obtained and identified the advantages, disadvantages and other specifics of individual methods of verifying the normal distribution. We consider the aim of the article to be fulfilled and we believe that it will be a valuable contribution in this area of research.

8 Discussion

Our paper dealt with the issue of testing and verifying the normality of the distribution as a basic prerequisite for the application of the VaR method. If we consider the application of the Value at Risk method after considering the application of the method, we must also ensure all the conditions that are necessary for the proper functioning of the method itself. The reason for the need to verify the normality of the distribution at the Value at Risk method is that the VaR calculation itself is based on the distribution function of the normal distribution. In practice, we cannot successfully apply the data obtained by this method if the set of values for which we calculate VaR does not have a normal distribution. Also, when it comes to identifying and verifying the normal distribution, it is not always clear how to implement it. There are a number of procedures and tests that, in various ways, directly or indirectly try to identify whether or not the file we are researching comes from a normal distribution. These tests are then selected based on the specifics of the examined group, but most often based on its scope. We have given the specifics of individual methods in the previous chapter, but in the case of real use of these methods it is necessary to proceed mainly from the structure and essence of verification of normality in terms of computational complexity, which is now replaced by a number of powerful computing software.

Acknowledgment

The paper is an output of the science project VEGA 1/0210/19 Research of innovative attributes of quantitative and qualitative fundaments of the opportunistic earnings modelling.

References

1. C. Alexander, Market Risk Analysis Vol. IV. Value-at-Risk Methods, (2008).
2. D. Markechova, A. Tirpakova & B. Stehlikova, Basics of statistics for teachers. UKF in Nitra, pp. 405, ISBN 978-80-8094-899-3, (2011).
3. E. Ostertagova, Applied statistic. Elfa, Kosice, pp. 161, ISBN 978-80-8086-171-1, (2011).
4. E. Ostertagova, Applied statistics in the computer environment of MATLAB. Equlibria, Kosice, pp. 193, ISBN 978-80-8143-006-0 (2012).
5. G. Holton, Value-at-Risk: Theory and Practice, Academic Press, New York, (2003).
6. J.P. Morgan, RiskMetrics Technical document, Reuters, New York, (1996).
7. J. Vrbka, E. Nica, I. Podhorska, The application of Kohonen networks for identification of leaders in the trade sector in Czechia. Equilibrium. *Quarterly Journal of Economics and Economic Policy*, **14**(4), 739-761 (2019).
8. L. Svabova, M. Durica, A closer view of the statistical methods globally used in bankruptcy prediction of companies. *Proceedings of the 16th International Scientific Conference on Globalization and its Socio-Economic Consequences*, pp. 2174-2181 (2016).
9. M. Kovacova, T. Kliestik, K. Valaskova, P. Durana, Z. Juhaszova, Systematic review of variables applied in bankruptcy prediction models of Visegrad group countries. *Oeconomia Copernicana*, **10**(4), 743-772 (2019).
10. P. Durana, K. Valaskova, D. Chlebikova, V. Krastev, I. Atanasova, Heads and tails of earnings management: quantitative analysis in emerging countries. *Risks*, **8**(2), 57 (2020).

11. P. Kral, H. Musa, G. Lazaroiu, M. Misankova, J. Vrbka, Comprehensive assessment of the selected indicators of financial analysis in the context of failing companies. *Journal of International Studies*, **11**(4), 282-294 (2018).
12. P. Kral, V. Valjaskova, K. Janoskova, Quantitative approach to project portfolio management: proposal for Slovak companies. *Oeconomia Copernicana,* **10**(4), 797-814 (2019).
13. P. Jorion, Value at Risk: The New Benchmark for Managing Financial Risk, 2nd ed., McGraw-Hill Trade, New York, (2001).
14. S. Kot, I. Rajiani, Testing and identifying variable dependency through the fisher exact test in central Europe enterprises. *Ekonomicko-manazerske spektrum,* **14**(1), 10-18 (2020).
15. T. Kliestik, M. Misankova, K. Valaskova, L. Svabova, Bankruptcy prevention: new effort to reflect on legal and social changes. *Science and Engineering Ethics*, **24**(2), 791-803 (2018).
16. T. Kliestik, K. Valaskova, G. Lazaroiu, M. Kovacova, J. Vrbka, Remaining financially healthy and competitive: the role of financial predictors. *Journal of Competitiveness*, **12**(1), 74 (2020).
17. B. Y. Lemeshko, Chi-Square-Type Tests for Verification of Normality, Measurment Techniques, **58**(6), 581-591 (2015).
18. Z.J Zhou, Z.Y Ho, J.H. Liu, T.J. Tong, Hypothesis testing for normal distributions: a unified framework and new developments, Statistics and Its Interference, **13**(2), 167-179 (2020).
19. A.K. Bera, A.F. Galvao, L. Wang, Z.J. Xiao, A new characterization of the normal distribution and test for normality, Econometric Theory, **32**(5), 1216-1252 (2016). Garcia-Perez, Chi-square tests under models close to the normal distribution, Metrika, **63**(3), 343-354 (2006).

Promising Areas for Developing HR Managers' Competencies in the Digital Age

V.A. Kapitanov[1], *O.S.* Osipova[2,*], and *L.S.* Chikileva[2]

[1]Polyus Research Institute of M.F. Stelmakh, Moscow, Russia
[2]Financial University under the Government of the Russian Federation, Moscow, Russia

Abstract. The rapid development of technologies is creating a new digital reality in human resource management. In the context of the transition to the digital economy and transformation of traditional classical organizations to digital ones, digital management decentralization will take place on the basis of digital technologies. The most important characteristic feature of a digital organization is interaction of the HR service with employees using mobile applications. Predictive HR analytics, employee-centered professional training, feedback not only at the vertical level but also at the horizontal level requires the development of new competencies of HR specialists. Based on the authors' research, implemented by a temporary creative team, the practice of developing managerial competencies in the process of digital transformation is analyzed using the example of HR management services in organizations at different levels of digital maturity. The study has shown understanding on the part of HR specialists that automation has a significant potential and will be in demand in all the companies, regardless of their size. The transition from a traditional organization to a digital one will require changes in the qualification requirements for HR specialists, readiness of HR specialists to acquire new digital competencies, to develop their professional skills and personal abilities.
Keywords: competencies, skills, digital transformation, HR department, HR management service, organizational development

1 Introduction

In recent years many articles have appeared in the scientific literature on the specifics of the functioning of HR systems in the context of digital transformation [1-5], highlighting the changing role of HR services in the era of digitalization [6]. Human resource management in the digital economy is considered in them as one of the key business processes that directly affects the increase in labor productivity [7-9]. Many scholars have turned directly to the analysis of the digitalization of the HRM system [10-12, 4, 13], to the growing role of HR analytics [14].

The transition to digitalization is a new stage in the development of the HR industry, which imposes new requirements on the competence model of HR specialists, who must

*Corresponding author: ososipova@fa.ru

increasingly use modern digital tools. Scientists develop the typology of personnel competencies for Industry 4.0. For example, E. Flores, X. Xu, and Y. Lu [15] have introduced a new term "workforce architecture", covering the human capital of the future, and suggested a new typology of personnel competencies for Industry 4.0 . This article draws attention to important changes in human capital in the future of Industry 4.0 and offers a human-centered perspective for companies in the new industrial revolution. The interdisciplinary research is based on literature review. However, there are few studies devoted to the new holistic model of competencies of HR specialists. The most advanced area of research in relation to HR services is the study of the evolution of digital competencies of HR employees in the field of recruitment. Analytical works that summarize the first results of the digital transformation of organizations and HR departments are of particular scientific value. Of note is the study of A. DiRomualdo, D. El-Khoury and F. Girimonte [16], in which the authors, using the Hackett Group's extensive database with paid HR benchmarks with Global 1000 companies, show how digital technologies will change HR processes and organizational structure. They emphasize the need to create new roles in HR. The authors pay attention to the fact that HR organizations should plan the changes needed for meeting the requirements of the digital transformation of the enterprise. The article presents some recommendations for the transformation of HR departments at the initial stage of digital transformation [17, 18].

In general, we can talk about a high degree of readiness of Russian HR specialists to automate their work functions, and a significantly higher level of digital literacy of employees of HR departments compared to the staff of organizations as a whole. As recent research shows, Russian HR specialists are able to master digital skills quickly enough [19]. However, there is an imbalance in terms of the training of HR specialists in the field of digital literacy and understanding of the functioning of new business models, especially among the older generation of HR. This is what makes us more closely examine the dynamics of the competence model of HR managers in the process of increasing digital maturity of organizations. The relevance of this article is due to the lack of knowledge of the processes that allow us to come close to developing a model of competence of HR specialists at various levels of digital maturity of organizations.

Research Questions

1. How will the transition from a traditional to a digital organization affect the work of HR managers, the demand for individual skills and abilities?
2. What professional HR-technological and technical competencies do HR specialists plan to acquire/improve?
3. What personal and practical skills do HR-managers plan to get/improve to enhance their professional competence?

Purpose of the Study - to summarize the directions and forms of competence development demanded by HR specialists in the conditions of digitalization of Russian companies.

2 Methodology

The theoretical and methodological basis of this research is the modern understanding of the theory of human capital in the context of the concept of sustainable development, the concept of human development and the concept of decent work. The following methods are used in the research: comparative, logical and methods of statistics. The results of analytical and sociological research, the results of expert surveys are used in the paper. To determine the key drivers of changes, opportunities and limitations of organizations ' human resource management systems in the context of digitalization of the economy, the temporary creative team conducted an original HR survey among specialists responsible for personnel management in organizations operating in the Russian Federation from June 4 to June 26, 2019. The questionnaire was developed on the basis of a review of information from domestic and foreign sources, summarizing the results of expert surveys of IT and HR specialists, and was highly appreciated by experts. The questionnaire consisted of 5 blocks, one of which was devoted to the prospects for the development of the HR service in the context of digital transformation. The survey was conducted remotely using the GoogleDocs form editor. Subsequently, the received data were processed using the Excel program. 152 people who represent different sectors of the economy were interviewed. Respondents represented more than 20 regions of Russia. One in four respondents represented the service sector, including such highly digitalized industries as banking and Finance; media and e-commerce, IT and telecommunications; and one in three represented transport, logistics and distribution. 29% of the respondents represented organizations with more than 10,000 employees, and one in four represented - between 1,000 and 10,000 employees. Such large organizations are expected to be the most automated, according to experts from SAP CIS and Deloitte. Almost one in three respondents noted that their organizations have achieved mature automation or laid the foundations for intelligent HR. Thus, we had a unique opportunity to trace the emergence of new trends in Russian HR management, and for the first time systematically summarize the Russian experience in the field of digitalization of not only organizational HR processes, but also the HR departments. This article analyzes the results of one of the questionnaire blocks (Section 5. The Future of the digital organization's HR service). 64.05% of respondents work in departments called "Human Resources Management", 11.76% – "Human Resources Department", 8.5% –"Human Resources Management".

In accordance with the official affiliation, 68.42% of the respondents were HR managers, Heads of HR Departments - 24.34%, HR partners, HR Directors - 7.24%. The gender distribution of respondents was as follows: females – 74.34%, males -25.66%.

The average age of the respondents was 35 years. It corresponds to the all-Russian indicators of the age of HR specialists. Every second male surveyed belongs to the older age group, this is a "trace" of the Soviet tradition, when the personnel departments preferred to hire retired military personnel. The vast majority of respondents 92.76%, have higher education, 1.97% and 3.95% have secondary specialized and academic degrees respectively. 36.18% of respondents have work experience in the service that provides personnel management processes ranging from 3 to 10 years, 26.32% of respondents have more than 10 years of experience, and 23.03% – from 1 to 3 years. The group of HR employees with experience up to 1 year was 14.47%.

3 Results

The results of our research have shown that there is an uneven involvement of Russian organizations in the process of digital transformation. Almost every second respondent noted that automation of HR processes is mostly relevant for the vast majority of large and medium-sized Russian companies seeking optimization, and only one in four respondents believes that automation is the best way for the vast majority of large companies. About 40% of respondents chose the answer "all the companies are equal in the process of digitalization, no matter what their size is - large, medium or small". Nearly every third person believes that automation of HR processes is also relevant for virtual organizations, as the emergence of new software products opens up opportunities for managing remote employees.

Respondents note that digitalization in HR is necessary primarily to get rid of routine (71%); to improve access to the best labor resources using digital communication channels - 46.7%, to develop remote forms of work (attracting staff from everywhere, at any time - 38.8%), to increase flexibility and individualization of people management - 37.5%.

The transition from a traditional organization to a digital organization, according to respondents, will affect the work of HR managers to change the qualification requirements for HR specialists (according to 56.6% of respondents). Every fourth respondent believes that the role of the HR service will increase, the institution of HR business partnership will not lose its importance, and the main competence of internal HR will be strategic thinking (24.3%).

To understand in what direction HR-specialists competency model HR-specialists should develop, first of all it is necessary to answer a crucially important question: in which direction the development of functionality of the HR specialists will go. 34.9% of respondents believe that HR managers will be experts with a broad range of competences. Three times less respondents assume a different development of events - there will be more highly specialized HR experts (13.8%). The answers to questions from 64 to 66 of the Questionnaire prove the orientation to a broad profile of the development of HR competencies. It is of great importance for the respondents to master a diverse range of new professional competencies that are necessary for the organization's transition to new business models and management of organizational development processes.

Among the professional HR-technological competencies that modern HR specialists would like to improve, the position of strategic HR has undoubtedly taken the leading place. It absolutely correlates with the answers to the previous question. Every second respondent has noted that in order to improve the efficiency of the HR manager's work, it is necessary for him / her to know in which direction the organization is going to develop, and how important is to switch to a new business model. Without analyzing characteristic features of strategic management, HR specialists can neither perform such a professional function as organizational development, nor plan their career growth. As noted above, it has become clear that HR managers can actually claim the position of partner, one of the top positions in digital organizations. 27.6% of respondents would like to acquire or improve their competence in developing organizational culture. Respondents identified change management, employee training and development, the ability to organize project and team work, and electronic document management as the second most promising group of competencies. Almost every third respondent would like to improve such competence as HR branding development.

For the first time, the desire of HR specialists to develop such competencies as managing employees ' emotions and impressions (20.4%) came as close as possible to the desire to develop their competencies in managing candidate / employee evaluation processes (22.4%). In previous years, the latter type of competence clearly led the way, as well as the desire to improve their skills in remuneration management. Almost one in seven respondents would like to learn how to create employees positive experience (14.5%). As early as 2019, HR specialists were clearly aware of the need to improve their skills in organizing fundamentally new forms of work with employees. 17.8% of respondents were in favor of acquiring or improving their skills in organizing flexible forms of employees' employment. One out of four respondents would like to study the management of remote employees and freelancers - 23%, to understand the specifics of working with virtual organizations - 7.2%.

It is important to note the demand for the development of such competencies as building the organizational architecture of the company - 15.1% and the ecosystem of the organization - 19.1%. The study has proved that HR specialists respond quickly to the challenges of the time and are ready to master new methods of work that allow to increase the level of digitalization of HR departments. Answering the question № 60 "How will the transition from a traditional organization to a digital organization affect the work of HR managers?" almost every tenth respondent noted that the main competence of an HR manager will be the ability to make an employee happy - 9.2%.

Question №64 of the questionnaire allowed us to specify the statement "what professional technical skills you would like to acquire / improve". The responses were distributed as follows. Half of the respondents would like to improve their skills in working with electronic document management. Such demand for the development of these skills is explained, in our opinion, by the fact that HR specialists from organizations at various levels of digital transformation took part in the survey.

At the same time, a significant proportion of respondents have demand for the acquisition / development of fundamentally new competencies. The demand for developing such skills as digital design in HR is leading (45%), while working with corporate IT platforms is in the second place (43.4%). This is followed by the demand for such a new competence as data collection, processing and analysis in real time, in other words, HR analytics – 41.5%. Almost the same number of HR managers would like to improve their competencies by using specialized software to implement the HR function - 41%. The control questions confirmed the high reliability of the data obtained: from 41 to 44, 7% of the respondents were in favor of getting these skills.

The second most popular group of technical skills included: working with cloud technologies -33%, working in social media / networks - 32.2%, digital security -29%. One in four respondents wants to improve their skills in working with mobile apps. Every sixth respondent who has taken part in our survey considers it appropriate to involve HR Department employees in the development of the Intranet design and interface. In the age group from 20 to 25 years, one in four respondents has chosen this answer. 22.8% of the HR specialists surveyed would like to improve their skills in advanced data visualization and processes.

In the course of the study, the demand for HR specialists to develop their professional personal and user skills was analyzed. Answering the question № 66 "What professional personal and user skills would you like to acquire " 43.4% of respondents expressed a desire to have digital creativity, 29.6% - the formation, development and promotion of a digital personality (personal digital profile), as well as the development of digital empathy - 13.8%. One in three respondents would like to improve their stress tolerance. Respondents strive to improve user's skills such as working in Photo Shop and similar image editors (39.5%), learn more about personal security when using smart devices (28.9%), and master the production of multimedia content for the Internet (27%). Along with the skills mentioned above, basic skills such as typing speed and the use of mobile communication tools remain quite popular - 23% for both answers.

For the convenience of simultaneous perception of some important new areas of development of HR managers ' competencies, the frequency of responses is shown in Figure 1.

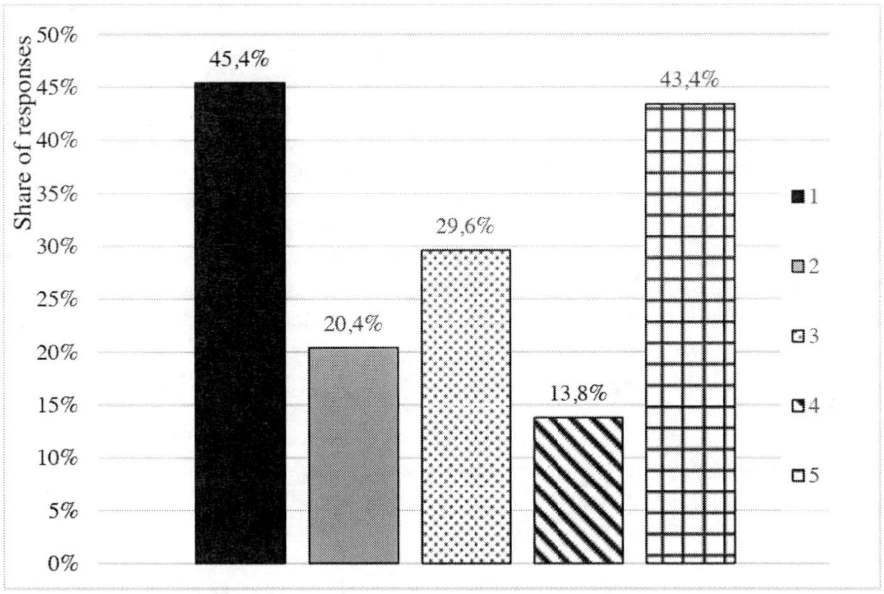

Fig. 1. Promising areas of development of HR managers' competencies identified in the survey of HR specialists. (*All the questions involved choosing one or more answers from several options*).

1. Question № 64: "What professional technical skills would you like to acquire/improve?" Answer option 1: "digital design in HR"
2. Question № 65:" What professional HR-technological competencies would you like to acquire/improve?» Answer option 21: "managing employee emotions and impressions".
3. Question № 66: "What professional, personal and user's skills would you like to acquire/improve?" Answer option 2: "formation, development and promotion of a digital personality (personal digital profile)".
4. The same question. Answer option 3: "digital empathy".
5. The same question. Answer option 4: "digital creativity".

Source: Own processing.

Four out of five consider that competencies are directly related to digital technologies. The respondents ' responses show that they are not only very interested in developing professional HR competencies for the implementation of their work functions, but are also ready to continuously improve their technical skills. All the respondents expressed their willingness to improve their digital literacy if needed, with 27.5% - paying their own money. Experts do not consider the system of professional retraining as the weakest link that hinders the development of software and digital competencies. However, almost every second respondent believes that the system of retraining / advanced training should be improved. Moreover, every fifth respondent believes that the process of retraining should become continuous (19.7%). 36.2% of respondents noted the need to develop mentoring by more IT-advanced employees, 22.4% support the idea of getting acquainted with the experience of software implemented in other companies. Every fourth believes that digitalization will allow to simulate a virtual reality for the training of workers - 24.3 %.

4 Discussion

The increase in the speed of information transfer in the era of computerization at the turn of the Millennium has led to the centralization of organizational management systems, the transition to international quality standards, and the unification of management principles and algorithms. This process includes functional structuring of HR services, specialization of HR managers, development of competence approach and unification of their generalized labor functions within the framework of accepted professional standards. During this period, organizations with varying degrees of intensity have been automating operational HR processes, developing digital competencies for software development, as well as professional specialization of specialists in the main areas of HR departments.

The transition to a new stage of digitalization associated with a further increase in the speed of information and the need for organizations to switch to new business models requires a qualitative revision of the existing model of competence of HR managers.

The conducted sociological research allows us to declare that it is necessary to develop a new model of competencies for HR specialists. This model should be adapted to different levels of digital maturity of HR services. The formation of digital competencies for HR managers is reaching a fundamentally new level of development.

The transition to a new stage of digitalization associated with further increase in the speed of information and the need for organizations to switch to new business models requires a qualitative revision of the existing model of competence of HR managers.

The conducted sociological research makes it possible to declare that it is necessary to develop a new model of competencies for HR specialists. This model should be adapted to different levels of digital maturity of HR services. The formation of digital competencies for HR managers is reaching a fundamentally new level of development.

Modern HR technologies allow you to personalize your interaction with the staff. Large-scale hi-tech exhibitions and technology showrooms held in Russia in recent years show a geometric growth in the introduction of digital technologies in the work of HR services. Every year the scale of digital technologies introduced into the practice of HR work increases. At the fifth HR&Technology Expo 2020, a system platform that gathers the most advanced and effective IT tools for personnel management, more than 70 HR products will be presented this year, and more than 40 applied presentations by representatives of leading Russian and international companies actively implementing digital technologies will be announced. More than 50 HR-Tech experts will take part in the discussions. It is planned to discuss issues related to HR technologies and the evolution of HR functions. Within the framework of HR exhibitions, managers working with people in organizations will not only exchange useful solutions in the field of digitalization of HR functions, but also get the opportunity to test new products from leading IT developers of HR products. Traditionally, the greatest professional interest of the HR community is aroused by the CR digital technology Laboratory, a platform that directly unites HR specialists and developers of HR technologies.

All this creates favourable conditions for rethinking the directions of development of the competence model of HR specialists. Currently, issues of HR Analytics and data management, user's experience, working with vendors and IT teams come to the fore. We are actively implementing marketplaces, services and CRM to automate recruitment, and introducing virtual assistants to implement a number of HR functions-chatbots with artificial intelligence for HR processes in the company. Optimization of recruitment budgets and qualitative evaluation of candidates is based on the introduction of technologies with machine learning, cognitive computing and other data analysis solutions. Automation and e-learning are actively implemented: training, evaluation and staff development. When building a model of HR specialists' competencies in the context of continuous training, an optimal balance of digital, professional and "soft" skills is required, taking into account the multi-level model of digitalization of HR services.

5 Conclusion

In conclusion it should be mentioned that every second respondent believes that it is necessary to improve the system of retraining/advanced training of employees of HR departments. This work should be carried out taking into account the perspective directions of development of the model of competence of HR specialists. During the research, special attention was paid to finding out the subjective vision of HR managers of promising areas for the development of their professional and personal competencies. In recent years, you can often hear such a question: "What will happen to HR?" The answer is: "HR is being transformed". The results of the study have shown the results of this transformation within the competence approach. HR specialists quickly respond to the challenges of the time and are ready to master new methods of work to upgrade the level of digitalization of HR departments. The conducted research has made it possible to determine what competencies of HR specialists are the most important for them today, what is the degree of their development in organizations that are at various stages of digital transformation. More than that, we can identify the degree of demand for fundamentally new skills and abilities necessary for HR specialists to perform their work functions in the context of organizational digital transformation. A significant proportion of respondents are in favor of the acquisition/development of fundamentally new competencies, including the development of skills, such as digital design in HR and management of employees' emotions. The interviewed HR specialists have expressed a desire to develop digital creativity, a personal digital profile, and the development of digital empathy.

Acknowledgements

The reported study was funded by RFBR (Russian Foundation for Basic Research), project number 19-010-01042

References

1. M. Al-Mekhlal, L. Abu Romman, B. A. Al-Saqer, A comprehensive usability evaluation of an online HR system. In Blashki, K. & Xiao Y. (Eds.), *MCCSIS 2018 – Multi Conference on Computer Science and Information Systems. Proceedings of the International Conferences on Interfaces and Human Computer Interaction 2018, Game and Entertainment Technologies and Computer Graphics, Visualization, Computer Vision and Image Processing*. Red Hook, NY: International Association for Development of the Information Society, 93-100(2018).
2. D. Bögenhold, R. Klinglmair, F. Kandutsch, Solo Self-Employment, Human Capital and Hybrid Labour in the Gig Economy. *Foresight and STI Governance,* **11**(4), 23-32(2017). DOI: 10.17323/2500-2597.2017.4.23.32.
3. E.A. Khalimon, M.N. Guseva, I.Z. Kogotkova, I.S. Brikoshina, Digital economy, innovation, information infrastructure, research and development, security, priority projects and programs. In Mantulenko, V. (Ed.), *Proceedings of the International Scientific Conference "Global Challenges and Prospects of the Modern Economic Development". The European Proceedings of Social & Behavioural Sciences*

EpSBS, London: Future Academy (57) 199-213(2018). DOI: 10.15405/epsbs.2019.03.21.

4. E.A. Saveleva, I.V. Bogatyreva? Opportunities of digital economy: Flexible organization of labor. In V. Mantulenko (Ed.), *Proceedings of the International Scientific Conference "Global Challenges and Prospects of the Modern Economic Development". The European Proceedings of Social & Behavioral Sciences EpSBS*, London: Future Academy. (57) 407-415(2019). DOI: 10.15405/epsbs.2019.03.41.
5. N.S. Zagrebel'naya, E.R. Bostoganashvili, Human resources management in the digital economy. *Economics: Yesterday, Today, Tomorrow*, **9**(1A), 374-384(2019). [in Rus.].
6. R. Fenech, P. Baguant, D. Ivanov, The Changing Role of Human Resource Management in an Era of Digital Transformation. Journal of Management Information and Decision Sciences. Weaverville, **22** (2)1-10(2019).
7. D. Bersin, B. Pelster, J. Shwartz, B. van der Vyver, New game rules in digital epoch. *"Deloitte" research "International trends in the sphere of personnel management".* . (2017). Retrieved from: https://www2.deloitte.com/content/dam/Deloitte/ru/Documents/human-capital/russian/hc-2017-global-human-capital-trends-gx-ru.pdf. Accessed: 01.10.19. [in Rus.].
8. O. Osipova, A. Artamonov, Development of the workforce potential of the state company employees in the new industrialization of Russia. In Y. Silin, Y. Animitsa, E. Dvoryadkina, V. Blaginin (Eds.), Proceedings of the 2nd International Scientific conference on New Industrialization: Global, national, regional dimension (SICNI 2018). Advances in Social Science, Education and Humanities Research, Paris: Atlantis Press. 240 (2019). DOI: 10.2991/sicni-18.2019.77.
9. D.K. Zakharov Cloud technologies in the personnel management system. *Vestnik Universiteta* (5) 190-196(2016). [in Rus.].
10. N.F. Altukhova, E.V. Vasileva, M.V. Mirzoyan, Competence-based approach to managing staff in public administration on the basis of ontologies. *Business Informatics*, **1** (43), 17-27(2018). DOI: 10.17323/1998-0663.2018.1.17.27. [in Rus.].
11. Z.F. Mingazov, M.I. Prygunova, Feature of application of an integrated approach to automation of the solution of tasks of human resource management in corporations. *Scientific Works of the Center of Advanced Economic Research* (12) 61-65(2017). [in Rus.].
12. O. Osipova, Digital transformation of personnel management services *International Conference on Digital Transformation in Logistics and Infrastructure (ICDTLI 2019). Atlantis Highlights in Computer Sciences*. **1**. Published by Atlantis Press. 327-329(2019). DOI: https://doi.org/10.2991/icdtli-19.2019.56
13. E.Yates, Cr. Murray, The Future of HR: can HR develop the skills needed to survive?»(2018)Retrieved from: https://pwc.blogs.com/the_people_agenda/2018/10/the-future-of-hr-can-hr-develop-the-skills-needed-to-survive.html Accessed: 23.04.20. [in Rus.].
14. B. A.Vineeta, Human resource analytics. *South Asian Journal of Marketing & Management Research*. **7** (5) 68-77(2017). DOI: 10.5958/2249-877X.2017.00031.5
15. E. Flores, X. Xu, Y. Lu, "Human Capital 4.0: a workforce competence typology for Industry 4.0", *Journal of Manufacturing Technology Management*, **31** (4) 687-703(2020)/ https://doi.org/10.1108/JMTM-08-2019-0309
16. A. DiRomualdo, D. El-Khoury, F. Girimonte, HR in the digital age: how digital technology will change HR's organization structure, processes and roles //

Strategic HR Review, **17**(5)234-242(2018). https://doi.org/10.1108/SHR-08-2018-0074
17. S. Lam, B. Hawkes, From analytics to action: how Shell digitized recruitment, *Strategic HR Review*, **16**(2) 76-80(2017). https://doi.org/10.1108/SHR-01-2017-0005
18. A.V. Vasilchikov, G.P. Gagarinskaya, T.N. Obushchenko, The recruiter's competence model in the era of technology and digital economy // *Labour economics*. **5** (4) 959-974 (2018). DOI: 10.18334/et. 5. 4. 39652
19. I. Ivanova, O. Osipova, V. Pulyaeva, Evolution of process of automation of HR-departments of russian organizations. *The European Proceedings of Social and Behavioural Sciences EpSBS. GCPMED- 2019 II International Scientific Conference "Global Challenges and Prospects of the Modern Economic Development"* 998-1005(2019). ISSN: 2357-1330. DOI: 10.15405/epsbs.2020.03.144

Resource Potential as a Key Factor of Ukrainian Mining Enterprises Competitive Position

Andriy Pochtovyuk[1], *Kostyantyn* Latyshev[1], and *Kateryna* Zaika[1,*]

[1]Kremenchuk Mykhailo Ostrohradskyi National University, Kremenchuk, Ukraine

Abstract. The paper is devoted to formulation of theoretical, methodological and practical recommendations for ensuring enterprise sustainable competitiveness based on an integrated assessment of key resource determinants. The issue of ensuring enterprise competitiveness require increased attention in modern conditions of Ukrainian economy development. It was determined that implementation in production practice of these tasks are possible due to the efficient enterprise resource potential management, which will ensure maximum using of production capacity, cost reduction, improving product quality and competitive position stability. An object of the study is decision-making processes that appear using the enterprise competitiveness resource potential. Subject of the study is expansion of high-quality products' production in conditions of world demand. The authors were solved the following set of tasks: was clarified the term's essence «competitiveness» according scientific approach; was determined the enterprise position based on competitiveness integral index calculation; was identified enterprise competitiveness main components in accordance with the resource concept; an assessment of the enterprise technical and economic condition was carried out by calculating the wear and sustainable coefficients, also products qualitative parameters were established based on economic and mathematical modelling methods. The reasons for the decrease in the rate of metallurgical enterprise development were determined: outdated technologies, the need to comply with international quality standards, increased competition in the market, wear of fixed assets. In the final part, to solve these problems, a set of measures has been developed to improve the state of the enterprise by expanding the production of high-quality products through an investment strategy.
Keywords: competitiveness, mining enterprises, competitiveness assessment, resource potential, product quality, project office.

1 Introduction

Metallurgical complex of Ukraine acts as one of the basic elements of the country's economy [1]. Therefore, one of the most important conditions for ensuring sustainable

*Corresponding author: katya.orlova.977@gmail.com

Ukrainian economy development is increasing of competitiveness of mining enterprises. Sustainable positions in a competitive environment are determined by the potential of the enterprise [2]. It acquires significant importance for enterprises that operating in a transitional economy, which competitive positions remain insufficiently high on modern world markets. The competitiveness of enterprises is a set of interrelated elements that aim to support and develop both existing and create new competitive advantages [3]. Also, the competitiveness of enterprises understood as the ability to produce competitive products due to its effective using of available resources.

Significant scientific contribution to the study of theoretical problems of competitiveness and competitive advantages of enterprises made by the following scientists: Azoiev H.L., Ansoff I., Afanasiiev A.A.; Dolzhanskyi I.Z. and Zahorna T.O., Ivanova Yu.B., Kotler F., Lamben Jean-Jacques; Mescona M., O'Shaughnessi, Porter M., Fathutdinov R.A.; Hayek F.A., Yudanov A.Yu., Evans D. and Berman B. and others [4]. However, a number of problems associated with efficient use of enterprises resource potential remain insufficiently studied. The theoretical provisions on the management quality of the resource potential components are not fully described and systematized, methods for its assessment are poorly developed.

Continuous rivalry market companies will look for sources to obtain a competitive advantage, in order to maintain or improve the existing market position [5]. The competition level of mining enterprises in foreign markets within the European Union requires strengthening their positions through implementation of new approaches to assessing competitiveness [6-11]. But resource approach remains the most important in the process of evaluating long-term advantages formation possibilities.

In the process of approaches evolution to the «competitiveness» concept formation, its basis was considered the availability of enterprise access to resources (production, financial, human). The resource potential was a weighted average supply of the balance reserves of resources main types. The drawback of this approach was the lack of proper attention to using resources process, because the key aspect of competitiveness was considered not the effectiveness of use, but the resources availability. The task of providing material resources is not decisive, since there are no restrictions on access to modern equipment or high-quality materials, raw materials and components. However, there are significant problems with the financial possibilities of ensuring the processes of forming benefits based on traditional resource provision. Also, stable competitive advantages, as evidenced by foreign experience, are based primarily on intellectual and human resources. Thus, to formation and maintenance a competitive advantage enterprise needs not only to respond to the external environment challenges, which are constantly changing, but also to ensure resource potential development. The need to improve the enterprises' competitiveness and ensure their economic growth poses fundamentally new challenges in the development of new models for rationalizing the consumption of all resource's types. The combination of effective structures and forms, as well as methods of managing the resource potential of an enterprise can provide a positive impact on its activities and is the basis of economic growth.

2 Data and methods

The aim of the study is providing practical recommendations for assessing enterprise competitiveness through resource potential key components in today's market conditions.

The research methodological basis is theoretical foundations of domestic and foreign scientists works in the field of enterprise competitiveness, resource potential, enterprise economics, strategic management, the study of which made it possible to formulate and clarify the conceptual apparatus and essence of resource potential. During the research, the following methods were used: *comparative* – to identify the impact of individual factors on the enterprise competitiveness; *economic and mathematical modeling* – to study enterprise qualitative parameters; *graph analytic* – in determining the assessment of the enterprises competitive position; *design and construction* – while substantiating the directions of increasing the enterprises competitiveness.

It is characteristic that the efficiency and the level of material consumption of production are formed under the influence of numerous factors of the external and internal environment. From the intra-production factors, one can single out: products material consumption; the quality of the materials used; introduction of new technological equipment; improvement of material-saving technologies for manufacturing products, etc. It is also important to increase the analysis reliability, so to assess the efficiency of using the resource potential, it is necessary to apply the optimal number of indicators and classification groups. Assessing enterprises' resource potential was proposed by L. Berdnikova methodology with using of the following indicators: human resources; fixed assets; material resources; financial resources. The advantages of the proposed methodology are that it presents a fairly wide range of indicators characterizing the use of fixed assets, and a quantitative assessment of human resources, their application will allow a more complete assessment of the effectiveness of planning the use of fixed assets and human resources. The disadvantages of the method include a limited number of indicators characterizing the efficiency of the use of material resources and human resources of the enterprise, which complicates the development and adoption of optimal management decisions to ensure the rational use of resource potential.

Assessment of the enterprise resource potential competitiveness level consisted of three stages. The first stage: collection of financial and economic information about enterprise state; determination of enterprise main goals, the initial and final characteristics of its economic activities. The second stage: calculation of enterprise actual competitiveness level through the components of the resource potential. The following main tasks were solved: selection of indicators set for each component's category of enterprise's potential; measuring the individual level of enterprises resource potential for each group, accepted for comparison. The third stage: analysis of resource potential use level of enterprise and assessment of existing management system effectiveness for enterprise resource potential. Main data sources for the study was information from materials of official publications of the State Statistics Committee and enterprise annual report for period 2017-2019.

3 Results

3.1 The current state of mining enterprises competitiveness level in Ukraine

In modern economic conditions, the main criterion for the enterprises economic development is the competitiveness indicator. An integral index of enterprise competitiveness should include indicators that directly describe the production process, the financial status, the use of staff, the level of innovations, etc. [12]. Applying of the integral assessment methods involves the use of significance coefficients (Tab. 1).

A visual representation of effective use of resource potential for PJSC «Ferrexpo Poltava Mining» was performed. The main product of enterprise is iron ore fluxed pellets. For the period January – May 2019, the enterprise produced 830.3 Kt of iron ore pellets, while the production of Mikhailovsky GOK (RF) amounted to 1227.0 Kt, Karelsky

Okatysh (RF) – 910.2 Kt of Northern GOK – 849.6 Kt and Stoilensky GOK (RF) – 629.6 Kt (Fig. 1).

Table 1. Integral index assessment of Ukrainian mining enterprises competitiveness

Enterprise name	Product quality	Share in total volume	Margin of financial strength	Capital investment	Return on resources	An integral index	Place
Northern GOK	0.92	1.0	1.0	0.33	1.0	4.25	I
Central GOK	1.0	0.13	0.82	0.22	0.18	2.35	IV
Southern GOK	0.71	0.15	0.70	0.15	0.14	1.85	V
Inguletsky GOK	0.94	0.16	0.89	0.42	0.1	2.52	II
Ferrexpo Poltava Mining	0.9	0.14	0.32	1.0	0.03	2.40	III

Source: Own processing based on [13].

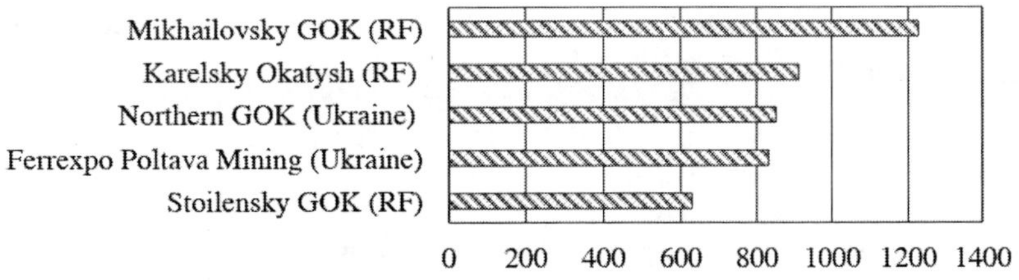

	Stoilensky GOK (RF)	Ferrexpo Poltava Mining (Ukraine)	Northern GOK (Ukraine)	Karelsky Okatysh (RF)	Mikhailovsky GOK (RF)
Kt	629,6	830,3	849,6	910,2	1227

Fig. 1. Enterprises pellets production among the former USSR countries
Source: Own processing based on [14].

One of the main company principles is the desire to increase iron ore production in the future, while at the same time demonstrating sound financial management, creating sustainable value for all interested parties. Today PJSC «Ferrexpo Poltava Mining» is one of the world's largest pellet producers.

3.2 Comprehensive assessment of the resource potential components: personnel, fixed production assets, finances and raw materials quality

A hypothesis is pulled out about the existence on mining enterprises the system of forming competitiveness that consists of 4 subsystems – personal, equipment, raw materials and products [15]. The resource potential is understood as a set of various resources, the availability of which ensures the effective functioning of the technological chain for the production of metallurgical products from ore to finished rolled products. Such resources include production capacities, availability and quality raw materials, basic auxiliary materials, fuel, energy and other objects of labor, human potential in the form of the numerical and professional qualifications of workers, financial resources, their size and the possibility of attracting. An analysis of the human potential assessment of PJSC «Ferrexpo Poltava Mining» is presented in Table 2.

Table 2. The human resources analyses of PJSC «Ferrexpo Poltava Mining»

Categories \ Year	2017		2018		2019		Deviation	
	persons	%	persons	%	persons	%	persons	%
Administrative management	580	7.5	500	6.6	460	6.1	–120	43.2
Industrial	6240	80.2	6100	80.9	6147	81.9	–93	33.5
Auxiliary	965	12.3	942	12.5	900	12.0	–65	23.3
Total:	7785	100	7542	100	7507	100	–278	100

Source: Own processing based on [14].

The pragmatism of the consumers of professional educational services has proved that professional knowledge, and skills are not fully required in the in-house labor market of mining enterprises [16]. The development of a socially oriented market economy in Ukraine, the effective implementation of reforms in this direction and the improvement of the well-being of the population is not possible without qualitative financial support for the competitiveness of workers, above all, in the mining industry [17]. The level of use of fixed assets effects on costs size, product prices and enterprise competitiveness. Based on available data of the enterprise consolidated statement the depreciation coefficient was calculated according to formula No. 1:

$$K_D = D \times 100 \div IS \tag{1}$$

Where the following applies:
D = is the amount of fixed assets depreciation,
IS = is the initial cost of fixed assets.

From the obtained calculations, K_{D2017} = 39%, K_{D2018} = 41%, K_{D2019} = 43%. The following fixed assets suitability was calculated according to formula No. 2:

$$K_S = (IS - D) \times 100 \div IS \tag{2}$$

Results of calculations are K_{S2017} = 61%, K_{S2018} = 59%, K_{S2019} = 57%, reflect the need to update fixed assets and flow of constant financial investments. So, the dynamics of advanced capital is presented in Table 3.

Table 3. Dynamics of PJSC «Ferrexpo Poltava Mining» financial resources

Advanced capital components	Year			2017 in % to 2019
	2017, UAH	2018, UAH	2019, UAH	
Non-current assets				
Unfinished construction	1246317	1559957	2378809	190.9
Fixed assets	7058534	6775459	7139777	101.2
Long-term financial investments	146510	74782	166822	113.9
Total:	104844559	9964357	10450595	9.97
Current assets				
Stocks	3848394	4973315	6822645	177.3
Accounts receivable for goods	2360085	4207966	4606450	195.2
Other Current Accounts Receivable	69315	133108	46300	66,8

Money and their equivalents	320413	1087038	218364	68.2
Total:	8910140	11325661	21290018	238.9
Balance:	19394599	21290018	22829827	117.7

Source: Own processing based on [13].

The quality of pellets is determined by the list of indicators reflected in Table 4. The chemical and mineralogical content (%) and physical properties of unoxidized ferruginous quartzite of the Kremenchuk region are given in Table 5. Ukrainian iron ore products are inferior to foreign ones in the iron content and harmful impurities $Si0_2$, Al_2O_3 (Fig. 2).

Table 4. Quality indicators of pellets production

Chemical content	Fe	FeO	Fe_2O_3	SiO_2	Al_2O_3	CaO	MgO	P_2O_5	SO_2	Others	Weight
Mass fraction, %	62.5	2.0	87.06	8.91	0.48	0.51	0.75	0.02	0.08	0.05	99.8

Source: Own processing based on [18].

Table 5. Unoxidized ferruginous quartzites of Kremenchuk district

Elements and Oxides	Deposit		Minerals and physical properties	Deposit	
	Horishni Plavni	Yeristovo		Horishni Plavni	Yeristovo
Fe FeO Fe_2O_3 SiO_2 Al_2O_3 CaO MgO P_2O_5 SO_2 CO_2	36.16 34.13 16.3 17.33 30.98 41.66 1.51 1.71 2.57 0.075 0.001 3.9	34.2 34.2 16.0 16.2 30.9 42.3 2.83 1.91 3.09 0.11 0.62 1.66	Magnetite Hematite Iron Carbonates Silicate iron Quartz Apatite Carbonates Others Volumetric weight g/cm^3 Porosity (%) by Proto'akonov	41.6 1.4 1.9 16.2 32.2 0.2 5.2 0.9 3.3 4.2-6 15.20	46.5 0.2 0.5 5.2 38.39 0.25 8.0 0.45 3.3 3.5 16.20

Source: Own processing based on [19].

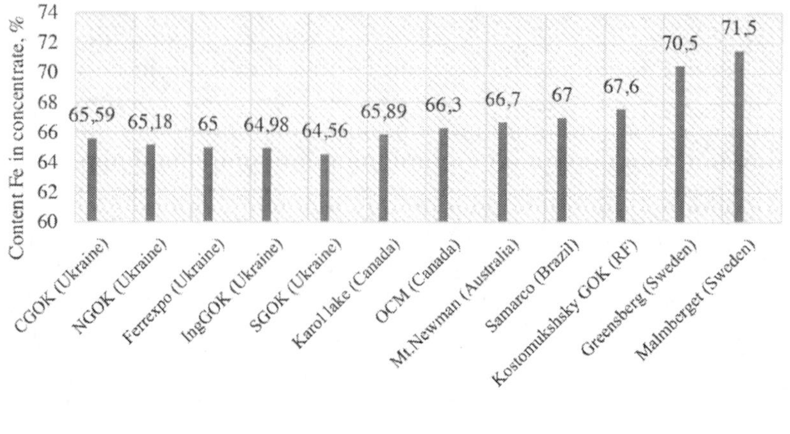

Fig. 2. World quality iron ore concentrate manufacturers
Source: Own processing.

Based on economic and mathematical modeling, the dynamics of the iron content in enterprise pellets over the past decade was studied (Fig. 3). The general demand trend for iron ore pellets in the foreign market shows Figure 4.

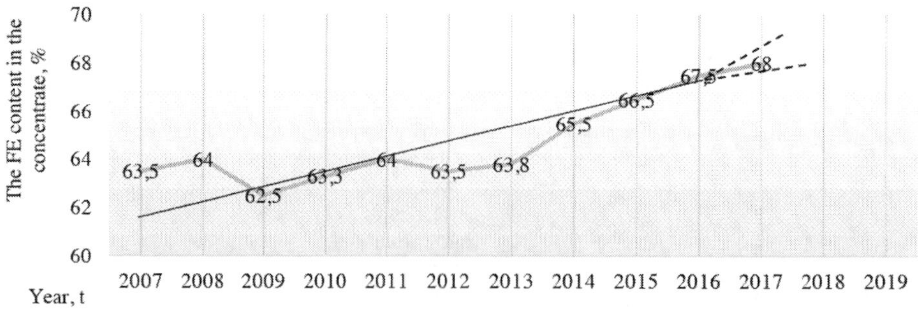

Fig. 3. The dynamics of the iron content in concentrate of PJSC «Ferrexpo Poltava Mining»
Source: Own processing.

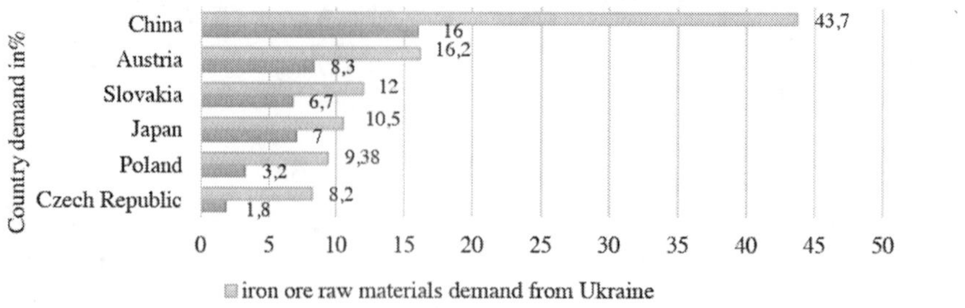

Fig. 4. Iron ore demand trend
Source: Own processing.

So, the strategic objective for PJSC «Ferrexpo Poltava Mining» should be to stimulate an increasing of high-quality iron ore pellets export with a moderate increasing in their total production. Suppliers from Australia and Brazil are crowding out the plant from the Chinese market. Therefore, the enterprise is trying to replace China with more profitable sales markets, to reorient supplies to Japan. If earlier the share of deliveries pellets to China was at the level of 25%, today it has decreased to 13%. Among global oversupply of iron ore, there is a shortage of pellet supply. According to experts, by 2020 the global demand for pellets grew by 4%. Global iron ore production is projected to grow at 2-2.3% annually as a result of substitution of China's domestic production of iron ore with imports, particularly from Australia and Brazil [20]. Thus, the low quality of iron ore raw materials significantly effects on enterprises' competitiveness. The quality improvement of any product is compared with the additional volume of its output.

4 Discussions and Conclusion

Considering the growing demand for concentrate in the world iron ore market, it is proposed to increase the concentrate volumes with a high iron content of 65%-67%, which requires the purchase of new equipment and attraction of additional investments. According to the National Bank of Ukraine, in the first quarter of 2019, the extractive industry was one of the leaders in attracting foreign direct investment [21]. The metallurgical industry is not among the priority, as it requires large investments now. The equipment of world industry leaders is recommended. In addition to investing in basic equipment, other costs should be considered (Tab. 6).

Table 6. Related costs for the project to expand the iron ore production

Costs type	Price, $
Main equipment expenses	97726483.44
Tools and materials	9772648.34
Building an additional workshop construction	18385184.84
Equipment installation services	24431620.86
Equipment installation supervision	2931794.50
Accompanied equipment	14658972.52
Operating expenses	9772648.34
Other expenses	4886324.17
Total:	1825677.01

Source: Own processing.

Provided an economic justification of project costs: main equipment – 97726483.44$ (Metso equipment complex); tools and materials – 10% of equipment cost; building an additional workshop construction – 18385184,84$ (equipping with necessary lighting, two overhead cranes – 50 and 25 tons with lifting capacity, three hoists of 5 tons with lifting capacity, four gates, ventilation and fire extinguishing system); equipment installation services – 25% of equipment cost; equipment installation supervision – 3% of goods cost; accompanied equipment – 15% of main equipment cost (communication systems, additional ventilation, metal structures); operating expenses – 10%; other expenses – 5% (energy, water, sanitation, unpredictable insignificant expenses).

The average income from the sale of 1 ton of iron ore at the company PJSC «Ferrexpo Poltava Mining». amounted to 507.98UAH or 18.814$ in 2019. With the amount of investment costs – 182565677.01$, it will pay off after the production of 9703644 tons of concentrate with an iron content of 65%-67%. With price stability for the next 7 years, if the price of the concentrate is equal to the price of pellets with an iron content of 65%, it is possible to calculate the project payback period according to formula No. 3:

$$Inv/\frac{Inc}{V\,prod}/V\,incr\,prod = \frac{182\,565\,677{,}01}{((5\,923\,879\,000)/(11\,661\,600))\,/\,27}/2915400 \quad (3)$$

Where the following applies:
Inv = investment size,
Inc = net profit for the year, which is calculated taking into account the fixed and variable production costs,
$V\,prod$ = output per year,
$Vincr\,prod$ = production volume increase (by 25% from 11661600),
27 – base hryvnia rate to $.

The project is considered expedient for application through a quick recoupment – 3.3 years from the moment of commissioning, or 7.3 years from the moment of the first investments in the amount of 30% of the basic equipment cost. According to the calculation, the project will pay off after three years and four months (3.3284 years) after commissioning. The investment fund for the supply of equipment and its installation in the amount of 182565677.01$ must be attracted for 4 years. It should be noted that the project will create 1467 work places an additional. To ensure the organization function of project management, it is expedient to create a new structural subdivision – project office, that provides for implementation of appropriate changes in the combine organizational structure. Many project-oriented organizations have implemented a project coordination mechanism, such as a project management office (PMO), to align projects with the organization's strategy, to ensure the success of projects, and to create value for the organization from projects [22]. PMO is a specific infrastructure that ensures the effective implementation of a project within the framework of a computer, communication and information technology systems and established standards for the implementation of activities and communications implementation. The urgency of the proposal to create a project team is that a successful project implementation requires competent project management. Project management education can make a significant contribution to increasing the project implementation success rate and reducing project failures [23].

Thus, it can be argued that the proposed measures are appropriate and effective for the PJSC «Ferrexpo Poltava Mining» and can be recommended for implementation in the enterprise activities. To ensure the stability of other enterprise types it is necessary to systematically assess the effectiveness of resource potential management. Taking into account such an assessment will make it possible to effectively influence the current state and development trends of the enterprise, establish the size and direction of changes, identify more significant factors of such growth, make forecasts and plans for further improving the reproduction processes, using and management of resource potential.

References

1. O. Bugrim, E. Kolesnik, Analysis of competitiveness of enterprises in the scale of the metallurgical industry of ukraine. *Economic Herald of SHEI USUCT*, **1** (11), 120-127 (2020)
2. L. Varava, A. Varava, V. Marchenko, Research of the interconnections of the components of strategic potential of the enterprise and functional strategies. *Proceedings of the 2nd International Scientific Conference*, pp. 73-80 (2020)
3. Yu. Kovtunenko, O. Grabovenko, Competitiveness enterprise in modern conditions: problems and prospects. *Economics: time realities*, **2** (24), 41-46 (2016)

4. A. Khanenko, The assessment of competitiveness and the formation of competitive advantages of an economic entity. *Economic Herald of SHEI USUCT,* **2** (10), 103-111 (2019)
5. S. Grabowska, Business model metallurgical company built on the competitive advantage. *Proceedings of 25th Anniversary International Conference on Metallurgy and Materials,* pp. 1800-1807 (2016)
6. S. Faizova, M. Ivanova, O. Faizova, V. Smiesova, O. Parshyna, O. Zavhorodnia, Use of Balanced Scorecard for Enterprise Competitiveness Assessment. *Journal of Advanced Research in Law and Economics,* **11** (2), 349-361 (2020)
7. M.-H. An, G.-Y. Ri, G.-N. Rim, Intellectual Product and Method of Assessing the Competitiveness of an Enterprise with It. *Journal of the Knowledge Economy,* **11** (3), 1059-1085 (2020)
8. M. Khomenko, K. Pryakhina, K. Latyshev, Prospects for development of Ukraine and EU in the field of renewable energy sources. *SHS Web of Conferences: Innovative Economic Symposium 2018 – Milestones and Trends of World Economy,* **61** (2019)
9. K. Teplicka, S. Hurna, J. Kadarova, K. Čulkova, Competitiveness increasing in mining companies through application of operation research methods. *TEM Journal,* **9** (1), 393-401 (2020)
10. D. Kisel'akova, Hairul, P. Gallo, P. jr. Gallo, V. Čabinova, E. Onuferova Total quality management as managerial tool of competitiveness in enterprises worlwide. *Polish journal of management studies*, **21** (2), 195-209 (2020)
11. T. Pozhuieva O., N. Novik, Competitiveness of the products as a category of modern development of Ukraine. *Economic Herald of SHEI USUCT,* **1** (11), 8-15 (2020)
12. M. Ivanova, O. Varyanichenko, S. Sannikova, S. Faizova, Assessment of the competitiveness of enterprises. *Economic Annals-XXI,* **173** (9-10), 26-31 (2018)
13. M. Abuzyarova, Methodological approaches to ensure the competitiveness of organizations. *Journal of Business and Retail Management Research (JBRMR),* **11** (2), 125-132 (2017)
14. Annual financial statements of enterprises, *Stock market infrastructure development agency of Ukraine (SMIDA)* [online], Available at: https://www.smida.gov.ua/ (2019)
15. O. Temchenko, Estimation of mining enterprises' activity when forming their competiton positions and considering risks. *Economic Annals-XXI,* **3-4** (1), 80-83 (2015)
16. D.Zagirniak, O.Kratt, M.Zagirnyak, Rationalization of the choice of professional education in the context of the needs of business environment. *Naukovyi Visnyk Natsionalnoho Hirnychoho Universytetu,* **4**, 158-163 (2020)
17. I. Perevozoval, N. Daliak, V. Babenko, Modeling of Financial Support for the Competitiveness of Employees in the Mining Industry. *Proceedings of 8th International Conference om Monitoring, Modeling and Management of Emergent Economy: Experimental Economics and Machine Learning for Prediction of Emergent Economy Dynamics,* pp. 444-454 (2019)
18. S. Dwarapudia, T. Ghosha, A. Shankara, V. Tathavadkara, D. Bhattacharjeea, R. Venugopal, Effect of pellet basicity and MgO content on the quality and microstructure of hematite pellets. *International Journal of Mineral Processing,* **99** (1), 43-53 (2011)

19. O. Kratt, A. Pochtovyuk, I. Trunina, K. Pryakhina, Marketing Positions of Kremenchuk Industrial Center in International Markets. *SHS Web of Conference: International Transport Infrastructure, Industrial Centers and Corporate Logistics (NTI-UkrSURT 2019)*, **67** (2019)
20. Iron Ore: 2020 World Market Review and Forecast to 2029 [online], Available at: *https://mcgroup.co.uk/researches/iron-ore* (2020)
21. N. Vyhovska, A. Polchanov, T. Ostapchuk, V. Dovgaliuk, The impact of solvency and business activity on profitability of mining companies in Ukraine. *Naukovyi Visnyk Natsionalnoho Hirnychoho Universytetu*, **4**, 164-170.
22. V. Otra-Aho, J. Iden, J. Hallikas, The Impact of the Project Management Office Roles to Organizational Value Contribution. *International Journal of Information Technology Project Management*, **10** (4), 79-99 (2019)
23. J. Kostalova, M. Bednarikova, M. Patak Project management education in metallurgical companies in the Czech Republic. *Business, Management and Education*, **16** (1), 54-64 (2018)

Sustainable Growth of Airline Companies: Challenges and Prospects Assessment

T.A. Slepneva[1,*] and *N.V.* Svishcheva[1]

[1]Financial University under the Government of the Russian Federation, Moscow, Russia

Abstract. Due to the sufficient openness of the Russian air transportation market, such factors as demand and investment activity, exchange rate fluctuations, economic sanctions against large companies and other unfavorable conditions lead to a slowdown (sometimes a decrease) in sales growth of Russian companies, weakening their economic position and financial stability. In this regard, it is necessary to study the conditions and opportunities for achieving sustainable growth, based on the example of Russian airlines using econometric modeling methods The solution of this problem requires the study of a system of economic and non-economic factors to achieve sustainable growth. It also requires an analysis of the industry specifics of air passenger transportation market to determine the main indicators of economic growth.
Keywords: Airlines growth, Revenue modeling, Sustainable growth, Financial and non-financial growth factors.

1 Introduction

Achieving sustainable growth for many companies in any industry, is an important requirement for economic success, especially in the face of falling demand, declining investment activity, currency fluctuations and other unstable economic conditions.

During general economic downturn the lack of sustainable growth has a negative impact on the development of companies, which leads to a decrease in shareholder income and can also lead to bankruptcy of companies. Despite the annual growth in passenger traffic in recent years, this growth cannot be defined as sustainable in the current economic environment, and airlines are experiencing financial difficulties. Since 2014, due to the increasing losses caused by the decline in passenger traffic and the general decline of the Russian economy, the air transport sector has experienced a difficult economic situation. This has had a negative impact on the sustainability and prospects for further development of companies in the industry.

Financial and economic stability and achieving sustainable growth are important conditions for existence and development of many companies in any industry. Sustainability factors are particularly relevant in the context of falling demand, declining investment activity, currency exchange fluctuations and other negative factors (Ponomarenko, Fedoseev, Korotkiy & Belitskaya, 2016).

[*]Corresponding author: taslepneva@fa.ru

The aviation industry is one of the most developing and promising industries, and the profit from transportation is the main source of income for any airline. In 2019 Russian carriers served 128.1 million passengers, which is 10.2% more than in 2018. The market leader is PJSC Aeroflot, which transported 59.1 million passengers in 2019. The closest competitor of the Aeroflot Group is S7 Group, which transported 17.9 million passengers, 12% more than in 2018. Ural Airlines ranks third in terms of passenger traffic, which in 2019 increased passenger traffic by 7% to 9.6 million people (Aeroflot, 2020).

The Russian market of passenger air transportation is highly consolidated: according to the results of 2019, the share of the listed airlines accounts for more than 60% of the total passenger traffic. Flights of foreign airlines in the Russian market in 2019 carried 19.7 million passengers, which is approximately 13% of the Russian passenger air transportation market.

Airline activity is specific and carries many different risks. One of the main risks is the socio-economic situation in the country where the company operates. Poor social and economic situation relates to instability of political system, presence of armed conflicts, prevalence of various mass diseases. Another important factor in performing business abroad is a high level of insurance coverage of technical resources and passengers. If a region has underdeveloped infrastructure, the airline is obliged to compensate it with its own means of service and maintenance. Instability of the economic situation has a great influence on the demand for air carriers' services, which, first, depends on the mobility of the population. Mobility of the population is characterized by the standard of living, solvency, business activity, intensity of labor migration and others.

Another important feature of the air transportation market is seasonality, which is especially evident during the summer vacation period when the capacity of airlines is fully utilized. There is also a slight increase in demand during the winter vacation period. During periods of decline in service consumption, commercial load factor decreases, and destabilization of cash flow occurs during the year.

The airline industry is highly sensitive to fluctuations of jet fuel prices, since jet fuel costs account for a significant share of airlines' cost of sales. Jet fuel prices are determined by the dynamics of oil and petroleum products prices. Aviation companies also face high operational risks associated with the operation of aircraft as a high-risk vehicle and represent compensations for the onset of adverse events and the severity of their consequences. Safety of flights should be one of the main priorities of any airline, since any incident negatively affects not only the reputation of the air carrier, but also its further functioning. For example, companies "Kogalymavia" and "Tatarstan" went bankruptcy proceedings immediately after the incident of the crash (Bogdanova, 2018).

The main sources of financial risks for airlines are the presence of interest rate and currency risks. Any airline is more likely to finance its operations with borrowed funds (including loans and borrowings, accounts payable, etc. for operational leasing). Changes in interest rates lead to an increase in debt servicing costs and a decrease in the company's profitability.

Currency risk arises due to the dependence of a significant part of airline revenues and expenses on the dynamics of the ruble exchange rate against the Euro and the US dollar. Revenue from the sale of tickets for international flights is either collected or charged in foreign currency (transfer passengers, foreign markets), jet fuel costs, maintenance and leasing payments, payments at foreign airports are denominated and made in foreign currency (Federal air transport Agency, 2020).

The airline's liquidity risk is associated with difficulties in fulfilling its obligations to third parties regarding debt obligations. To reduce the probability of a negative impact of this type of risk, airlines should open current accounts with the most reliable banks with a high reputation. They should also maintain an operational system of control over the use of working capital, which allows to identify cash gaps in advance, create a sufficient liquidity reserve, quickly raise funds in the monetary market and make the most of available funds.

The financial analysis of the leading Russian airlines for 2015-2019 showed a high dependence of their operations on borrowed funds: over the past 5 years, the share of borrowed capital in the aggregate averaged 63%. The expansion of activity was mainly financed by increasing liabilities, which indicates that growth is not stable. At the same time, short-term accounts payable prevail as a relatively cheap source with a high level of default risk. As a result, the balance sheet of the analyzed airlines is mainly illiquid, which also confirms the growth instability.

According to the financial statements, revenue growth of the leading Russian airlines has been observed over the past five years. However, the company's operating expenses are growing even faster, which have a significant impact on profit volumes. As a result of reduced profits, companies annually receive less free cash that can be used to finance current and investment activities (Sabaityte, Davidavičienė & Van Kleef, 2020).

At the same time, the long-term strategy of airline companies is aimed at increasing business activity and shareholders value. However, an increase in business activity does not always lead to the creation of additional value and an increase in the welfare of owners. Practice shows that not every quality of corporate growth contributes to the prosperity of the company in the medium- and long-term time horizons. Unbalanced growth contributes to the accumulation of risks (Khotinskaya & Chernikova, 2017). The high growth rates of transportation that are common for Russian airlines combined with insufficient own resources may lead to depletion of the company's resources and loss of financial stability as a result of increased debt burden and reduced liquidity. Therefore, the actual growth of the company should correspond to its financial capabilities.

2 Methodology

This research is aimed to assess the near-term prospects and quantify the opportunities for Russian airlines to achieve sustainable growth. The aim of the article is to investigate methodological aspects of achieving sustainable growth and to analyze real potential for Russian airlines to achieve sustainable growth.

The research methodology includes descriptive and comparative methods of general scientific knowledge, methods of financial analysis and econometric modeling: regression analysis using the statistical package R. The method of correlation and regression analysis was used to assess the impact of non-financial factors, along with financial ones, on achieving sustainable growth. Based on regression analysis, an econometric model that represents the dependence of the company's sales volume on the values of such factors of sustainable growth as passenger traffic volume, passenger load level, seasonality of operations, profitability of sales, current ratio, assets turnover ratio, debt ratio and cost of non-current assets was created. The method allows to estimate the influence of various company's sustainable growth factors and identify factors that have the greatest impact on the revenue. Understanding these factors allows the company to develop the most rational strategy for its future activities.

The significance of regression coefficient in regression model will be verified using the Student's t-test, determinacy coefficient, Fisher test, and Darbin-Watson test (Dougherty, 2009).

Then the key factor of revenue growth will be determined, and with the help of econometric methods and Monte Carlo simulations the probability of achieving sustainable growth of PJSC "Aeroflot" will be calculated. At the end of the research, some methods of increasing passenger traffic of airlines contributing to the growth of passenger traffic and revenue in these crisis conditions are proposed.

The information base of the research is provided by the data of annual reports and financial statements of the largest Russian airlines (PJSC "Aeroflot", S7 Group, JSC "Ural Airlines") for the period 2014-2019.

3 Results

To assess whether the growth of airlines corresponds to their potential capabilities the models of sustainable growth of Higgins (Higgins, 1977), BCG (Zakon, 1986), Van Horn, James (Van Horn & James, 1998), Ulrich, and Arlow (Ulrich & Arlow, 1980) were used. The purpose of applying the models is to determine the growth rate of the company's sales, which is achievable, on the one hand, with the unchanged operating policy, on the other hand, with a stable financial structure of the company. Sustained growth is the maximum speed at which a company can increase sales without running out of financial resources (Lan, Yang & Tseng, 2019). This determines the possible growth rate of revenue or profit, which is provided with resources, provided that the key indicators of the company's operating, financial (including dividend) policy are constant (Voloshina, Zaharova, Kartashev, & Ulanova, 2019).

Calculations of stable growth rates using these models have shown results indicating that the growth of airline revenue in the analyzed period is not stable. Growth is not supported by its own sources of financing, and the company's activity has been characterized by extremely low profitability over the past two years (table 1).

Table 1. Estimation of the actual and sustainable growth of Russian airlines, %

Sustainable growth models	Aeroflot Group			S7 Airlines		
	2017	2018	2019	2017	2018	2019
Revenue growth	**4.4**	**13.0**	**9.3**	**31.5**	**8.9**	**23.3**
BCG model	18.6	4.5	8.1	26.4	54.9	31.6
Higgins model	18.2	4.6	7.6	26.1	42.5	29.8
Van Horne's model	22.3	4.9	8.2	35.3	73.8	42.5
Ulrich-Arlow's model	34.4	6.7	10.8	30.7	44.2	22.0

Source: Own processing.

Based on the conducted calculations it is possible to identify financial and non-financial indicators of the airlines' stable growth. Financial indicators of sustainable growth are:

- profitability and cost intensity indicators;
- indicators of business activity that assess the intensity of growth. Growth rates analysis based on asset utilization efficiency is carried out by comparing the company's commercial revenue with various operating assets (Makarov, Ryabova & Hvostova, 2019);
- liquidity and financial stability indicators reflecting the company's financial independence;
- the level of interest rates on liabilities that affect the value and structure of companies' capital;
- the level of inflation and economic growth in the country. In a period of high inflation, companies will have a lower debt ratio, which may lead to underfunding of operating activities (Ramli, Latan & Solovida, 2019).

The main non-financial factors for the sustainable growth of airlines are:

- operating results that depend on the industry specifics of the company, such as aircraft load levels, passenger traffic volume, fleet composition, number of flight destinations, etc.;
- the structure of the Board of Directors and the number of independent directors;
- seasonality that affects the company's passenger traffic.

In order to rank and identify key indicators of sustainable growth of Russian airlines, we will conduct a regression analysis of financial and non-financial growth factors.

To identify significant revenue growth factors and assess their impact, we will construct a multiple linear regression model of revenue dependence based on the example of PJSC "Aeroflot". The dependent variable y_t - the company's quarterly revenue from 2015 to 2019, is modeled. Revenue y_t is assumed to be affected by non-financial factors, which represent data from the company's operational statistics, as well as artificially modeled fictitious variables.

The main non-financial factors that affect the company's revenue are the volume of passenger traffic (x_1) and the level of passenger loading of the aircraft (x_2). In addition, the company's revenue is seasonal; therefore, three fictitious variables should be entered for quarterly data:

$$x_3 = \begin{cases} 0, in\ all\ other\ cases \\ 1, second\ quarter\ data \end{cases}$$

$$x_4 = \begin{cases} 0, in\ all\ other\ cases \\ 1, third\ quarter\ data \end{cases}$$

$$x_5 = \begin{cases} 0, in\ all\ other\ cases \\ 1, fourth\ quarter\ data \end{cases}$$

Financial indicators of sustainable growth of companies are considered as financial factors in the model: profit before tax return on sales (x_6). current liquidity ratio (x_7). asset turnover ratio (x_8). financial dependency ratio (x_9). volume of non-current assets (x_{10}).

The parameters of the linear econometric model are estimated using the least squares method. The following regression equation is obtained:

$$\hat{y} = -0.140 + 0.766x_1 + 0.140x_2 - 0.059x_3 + 0.436x_4 + 0.185x_5 + 0.091x_6 - 0.022x_7 + 0.090x_8 + 0.175x_9 + 0.035x_{10}$$

The results of verification the significance of the factors of the obtained model according to the student's criterion are presented in table 2.

Table 2. Analysis of the significance of multiple linear regression coefficients of PJSC "Aeroflot" revenue from sustainable growth factors

	Coefficient estimates, a_i	Standard Error	t-statistics τ_i	p-value
a_0	-0.140	0.291	-0.483	0.641
a_1	0.766	0.153	5.018	0.001
a_2	0.140	0.094	1.491	0.170
a_3	-0.059	0.178	-0.334	0.746
a_4	0.436	0.351	1.244	0.245
a_5	0.185	0.683	0.270	0.793
a_6	0.091	0.096	0.950	0.367
a_7	-0.022	0.063	-0.346	0.737
a_8	0.090	0.372	0.242	0.814
a_9	0.175	0.071	2.455	0.036
a_{10}	0.035	0.051	0.693	to 0.506

Source: Own processing based on student's t-test used.

The lower is the p-value for the coefficient a_i, the higher is the absolute value of the t-statistics for the coefficient and the lower is the probability that this coefficient is equal to

0. That is, the hypothesis that the coefficient is equal to 0 and the factor is not significant in the model is rejected. Therefore, it is necessary to exclude factors with p-value greater than 0.05 in turn.

As a result of alternately excluding insignificant factors, the following regression equation is obtained:

$$\hat{y} = 0.865x_1 + 0.163x_4 + 0.191x_6 + 0.179x_9$$

Let us check the significance of the final model coefficient estimates using the Student's criterion. The data is presented in table 3.

Table 3. Analysis of the significance of multiple linear regression coefficients of PJSC Aeroflot's revenue from sustainable growth factors

	Coefficient estimates. a_i	Standard error	t-statistics. τ_i	p-value
a_1	0.865	0.036	24.168	< 0.0001
a_4	0.163	0.043	3.782	< 0.0005
a_6	0.191	0.039	4.835	< 0.0005
a_9	0.179	0.041	4.417	0.002

Source: Own processing based on student's t-test used.

The calculated values of t-statistics for each coefficient a_i are greater than the critical one. Therefore, all factors are significant in this model. Also, all coefficients a_i for factors are positive, consequently, there is a direct relationship between factors and company revenue. Thus, significant factors for sustainable growth of revenue of PJSC "Aeroflot" are:

- passenger traffic volume (x_1) is the main factor with the largest contribution to revenue;
- the seasonal factor that characterizes revenue in July, August and September (x_4) indicates that it is during these months that there is a high demand for air tickets, and the company can implement various measures to increase revenue with great economic success;
- return on sales by profit before tax (x_6) – a factor that indirectly characterizes the pricing policy or strategy for optimizing operating expenses;
- financial dependency ratio (x_9) – the higher the financial dependency ratio and therefore the amount of borrowed capital, the higher the revenue volume should be for the company to function sustainably.

The determination coefficient for this model is 0.9882. It means that 98.82% of the initial process variability is considered by the created model. The calculated value of Fischer statistics is 313.1. The P-value of this statistic is less than 0.001, therefore, the model is significant and explains the initial process well, and the role of factors in explaining the variability of the analyzed process is essential. Verification of the presence of first-order autocorrelation in the *DW* error row using the Darbin-Watson criterion has shown that the values of the *DW* criterion are ≈ 1.7924. Consequently, the 1st-order autocorrelation coefficient $\rho_1 = 0$. We can conclude that first-order autocorrelation in the error series is rather not observed.

Analyzing the quality criteria of the model created for PJSC "Aeroflot" quarterly revenue volume, it can be concluded that the model generally describes the initial process well. The main factors of sustainable revenue growth are the airline's passenger traffic volume, seasonal factor and the return on sales. The impact on these factors can help airlines achieve sustainable growth.

4 Discussion

The modeling of revenue dependency by the example of PJSC "Aeroflot" has revealed an increase in passenger traffic as a key driver of revenue growth. Calculation of the probability of achieving the range of revenue growth values in the forecast year, calculated at the estimated cost of sales and given levels of profitability, is performed using econometric methods of creating PJSC "Aeroflot" revenue model with seasonal component and Monte Carlo simulation method based on quarterly revenue data for 2015-2019 (table 4). According to the results of the study, it is unlikely that steady revenue growth rates will be achieved in the forecast year.

Table 4. Probability of achieving different revenue growth rates for PJSC Aeroflot in 2020 by 2019

Return on sales, %	Revenue, thousand rubles	Growth rate by 2019, %	Probability, %
0	613 856 737	11.3	39.9
0.5	616 941 444	11.8	37.1
1	620 057 310	12.4	34.2
2	626 384 425	13.5	28.2
3	632 841 996	14.7	21.1
4	639 434 101	15.9	12.4
5	646 164 986	17.1	4.7
6	653 039 081	18.4	0.0

Source: Own processing based on the Monte Carlo method.

In addition, it was found that COVID-19 pandemic development in 2020 may be the reason of unprofitable operating process of airlines. Since March 27, 2020 Russian airlines have stopped almost all scheduled and charter flights abroad (Moscow: RBC, 2020). This situation will lead to significant losses for Russian airlines. At full stop of international flights, airlines certainly will not bear variable costs at the same level (kerosene, airport taxes, passenger service, air navigation services). At the same time, income from international transportation cannot cover losses that arise on domestic flights, so the probability of losses in 2020 is exceedingly high. Moreover, the cancellation of international flights leads to a reduction in transfer passenger traffic within the country.

The International Air Transport Association (IATA) estimates that global passenger airline revenues will fall by 55%, or $314 billion, in 2020 as a result of severe restrictions on air travel (Moscow: Interfax, 2020). Overall, the demand for passenger transportation will drop by 48% in 2020 due to bans and the economic consequences of the crisis, as a result of increased unemployment and a significant decline in income.

According to experts' forecasts, Russian airlines may lose 550 billion rubles of revenues in 2020, if the current situation with air services will remain until the end of June and demand at the level of 2019 will recover only by December 2020, and the ban on international flights will last until the end of July and demand for them will recover only to 60% of last year's figures (Moscow: Vedomosti, 2020).

In order to increase airline sales in the current environment, the following possible measures can be proposed to influence the factors of sustainable revenue growth: changes in airline pricing policy and increase passenger traffic.

However, the low profitability of companies in 2018-2019 and the reduction of passenger traffic in 2020 lead to the impossible pricing policy aimed at reducing the cost of tickets. In 2018, rising kerosene prices cost air carriers more than 50 billion rubles of additional costs. Currency risks are another reason of impossibility to reduce air ticket prices Lease payments and the cost of servicing the company's aircraft fleet are increasing due to the weakening of the ruble. The reason of this situation is that costs are either denominated in foreign currency or calculated in rubles, but at the dollar rate. Under the conditions of instability of the ruble, this leads to instability of operating expenses. The third reason is weather conditions, which imply an increase in airline costs during the five months from November to March of each year, which prevents airlines from reducing the cost of tickets currently. Fourthly, with the spread of coronavirus infection in 2020 and a dramatic decline in the number of flights, airlines are now selling tickets at prices that only cover the cost of fuel, airport services and navigation (Moscow: Kommersant, 2020).

Consequently, it is not possible to reduce air ticket prices. Moreover, it is difficult to maintain prices at current levels as operating costs are decreasing slightly and slightly depending on the airlines' operational and financial policies.

The following measures may be offered to companies as the main ways to increase passenger traffic:

- increasing the number of air transportation destinations by expanding the list of new destinations and increasing flights to existing locations;
- expansion of the air fleet and modification of the aircraft interior;
- cooperation with entertainment companies in different countries in terms of obtaining the opportunity to visit a cultural and entertainment center when purchasing an air ticket of the company;
- changes in luggage fare policy in terms of introducing a progressive luggage fare scale (Augusto, Godinho & Torres, 2019).

5 Conclusion

In order to achieve sustainable growth and increase in value, companies need to balance the key factors of sustainable growth with the key indicators of operational, investment and financial performance. The study identified key financial and non-financial factors for sustainable growth of airlines and proposed several measures to manage these factors, which is practically important for both for individual companies and for the entire airline sector.

Based on the research, it can be concluded that despite the positive dynamics, the growth of Russian airlines in recent years is not sustainable and will not become sustainable in the near future, which is due to the specifics of airline operations and the consequences of the pandemic. In 2020, the situation in the airline industry is characterized as a very difficult one, in which airlines cannot operate without financial support from the government. Companies are also forced to implement additional measures to increase revenues by influencing sustainable growth factors in all possible ways in the face of a significant reduction in demand. To achieve sustainable growth, airlines need to increase operating revenues, primarily by influencing growth factors such as passenger traffic. Implementing proposed ways to attract new customers can help airlines increase sales.

References

1. Augusto, M., Godinho, P. & Torres P. (2019). Building customers' resilience to negative information in the airline industry. *Journal of retailing and consumer services,* 50, 235-248.
2. Bogdanova, M. P. (2018). Specifica razvitiya kompanii "Aeroflot" v usloviyah krizisa otrasli (Specifics of development of the company "Aeroflot" in the context of the industry crisis). *Business strategy,* 10(54), 13-17.

3. Dougherty, C. (2009). Vvedenie v ekonometriku (Introduction to econometrics). Moscow: INFRA-M. 465.
4. Federal'noe agentstvo vozdushnogo transporta (Federal air transport Agency). (2020). Available at: https://www.favt.ru/novosti-novosti/?id=6026. (accessed: 27 August 2020).
5. Godovoy otchet PAO "Aeroflot" za 2018 god (Aeroflot's annual report for 2018). (2020). Available at: https://ir.aeroflot.ru/fileadmin/user_upload/files/rus/reports/annual_reports/aeroflot_ar18_rus.pdf. (accessed: 27 August 2020).
6. Higgins, R.C. (1977). How much growth can a firm afford? *Financial Management*, 6(3), 7-16.
7. Khotinskaya, G.I. & Chernikova, L.I. (2017). Empirical characteristics of Russian business growth. *Academy of Strategic Management Journal*, 16(2), 19.
8. Lan, S., Yang, C. & Tseng, M.-L. (2019). Corporate sustainability on causal financial efficiency model in a hierarchical structure under uncertainties. *Journal of Cleaner Production,* 237, 117769.
9. Makarov, A.S., Ryabova, E.V. & Hvostova, I.E. (2019). Problemy sovershenstvovaniya finansovyh metodov i modeley obespecheniya ustoychivogo razvitiya kompanii (Problems of improving financial methods and models for sustainable development of the company). Moscow: INFRA-M. 175.
10. Ponomarenko, T.V., Fedoseev, S.V., Korotkiy, S.V. & Belitskaya, N.A. (2016)/ Managing the implementation of strategic projects in the industrial holding. *Indian Journal of Science and Technology,* 9, 14, 91517.
11. Profil'naya associaciya sprognozirovala padenie vyruchki aviakompaniy na 55 % (The profile Association predicted a 55 drop in airline revenue %). (2020). Interfax. Available at: https://www.interfax.ru/world/704154. (accessed: 27 August 2020).
12. Ramli, N.A., Latan, H. & Solovida, G.T. (2019). Determinants of capital structure and firm financial performance—A PLS-SEM approach: Evidence from Malaysia and Indonesia. *The Quarterly Review of Economics and Finance*, 71, 148-160.
13. Rossiya ostanovit regulyarnye i charternye polyoty zarubezh (Russia will stop regular and Charter flights abroad). (2020). RBC. Available at: https://www.rbc.ru/society/26/03/2020/5e7c3c2e9a79474c9ad04dc7 (accessed: 27 August 2020).
14. Rossiyskie aviakompanii riskuyut poteryat' poltrilliona rubley iz-za epidemii (Russian airlines risk losing half a trillion rubles due to epidemic). (2020). Vedomosti. Available at: https://www.vedomosti.ru/business/articles/2020/04/30/829368-rossiiskie-aviakompanii (accessed: 27 August 2020).
15. Sabaityte, J., Davidavičienė, V. & Van Kleef, G.F. (2020). The peculiarities of low-cost carrier development in Europe. *Energies,* 13, 3, 639.
16. Ulrich, T. & Arlow. P. (1980). The Financial Implications of Growth. *Journal of Small Business Management,* 27(2), 28–33.
17. Van Horn, J.C. & James. C. (1998). Sustainable growth modeling. *Journal of Corporate Finance*, 1, 19–25.
18. Vladelec S7 Vladislav Filev o pravilah igry vo vremya epidemii (S7 owner Vladislav Filev about the rules of the game during the epidemic). (2020). Kommersant. Available at: https://www.kommersant.ru/doc/4328432?from=doc_vrez. (accessed: 27 August 2020).
19. Voloshina, A.I., Zaharova, E.D., Kartashev, M.A. & Ulanova, N.K. (2019). Upravlenie determinantami ustoychivogo rosta (Managing the determinants of sustainable growth). *Economics and Business: Theory and Practice*, 5(1), 109-112.
20. Zakon, A. (1986). Growth and financial strategies. Boston: Boston Consulting Group.

Scoring Method of Assessment: Relevance and Perspective

Andrey Gusev[1,*], *Alina* Invdeeva[2], and *Polina* Motuzenko[2]

[1]Financial University under the Government of the Russian Federation, Moscow, Russia
[2]Institute of Distance and Open Education, Financial University under the Government of the Russian Federation, Moscow, Russia

Abstract. The article reveals the essence of banking scoring and the importance of compiling high-quality scoring models for assessing risks when working with borrowers. The complexity of processing surface information about the borrower and the need for clarification with additional questions are indicated. The history and causes of the appearance of scoring modeling in the middle of the 20th century are described. The main criteria for scoring used to create algorithms for assessing the financial reliability of the borrower are listed. The contribution of post-Soviet scientists to the improvement of scoring models is indicated. Further prospects of banking scoring and its relevance in the light of the emergence of more progressive evaluation programs.
Keywords: underwriter, scoring, scoring models, application-scoring.

1 Introduction

The creditworthiness of the borrower is one of the most significant factors that creditors need to be aware of in the context of building and improving market relations. Financial institutions create and develop their own algorithms in order to determine the creditworthiness of business entities. However, for this purpose it is necessary to establish precisely what aspects the concept of creditworthiness incorporates in a market economy. It will be useful to pay attention to the fact that in the circumstances of the planned economy, when there was a distortion of lending principles and there was an acute lack of a foundation for the effective creation of credit relations, the idea of creditworthiness as an ability of an economic entity to fulfill its obligations was absent.

Since the planning policy framework of financial resources separation did not involve the use of free market relations, as a result there was no need for an analysis of the economic entity creditworthiness. However, during the transition to market relations, the order of economic relations between borrowers and creditors has significantly changed for the better.

In the new economy, the main role was played by the possibilities of such interaction, in which both the creditor and the borrower would benefit. It should also be noted that at the beginning of the market relations development the general interest in the deal between the borrower and the creditor was directly related to the assessment of creditworthiness. During that time opportunities and tools to assess this aspect of the financial and economic activity of the

*Corresponding author: gusevaa@yandex.ru

economic actor appeared. In this context, different definitions of "creditworthiness" should be considered in order to eventually identify the definition that best reflects the creditworthiness of the borrower.

2 Literary research

Creditworthiness means the ability of an enterprise to meet its credit obligations, while solvency means the ability to meet all obligations, not only credit. It should also be noted that creditworthiness is a forecast of the ability to fulfill obligations to creditors at the current moment, while solvency is a direct opportunity to fulfill credit obligations at the current moment. Therefore, solvency is an opportunity to fulfill credit obligations, and creditworthiness is a forecast of such an opportunity.

One approach to defining creditworthiness assumes that it is a set of different conditions that determine the capacity of the borrower to raise borrowed resources and return them in full and on time - the definition of M.N. Kreinina [1]. In this approach, creditworthiness is considered in a broader sense. In particular, due attention is paid to the prerequisites for obtaining borrowed capital, such as the structure of the working capital used, the financial stability of the enterprise, the valuation of products within the framework of the demand for these products, and the liquidity of the assets of the borrower enterprise.

Thus, when determining creditworthiness, it is necessary to take into account all aspects coming from the three basic principles of the credit - payment, return and urgency. That is, you need to pay attention not only to the fact that the borrowed funds need to be returned within the period established by the loan agreement, but also to the fact that in addition to the principal debt, interest on the loan must be paid [2]. Moreover, it is necessary to take into account all the most important aspects in determining the probability of the borrower to fulfil his credit obligations, notably the factors of business risk, financial risk and credit history risk.

Based on all of the above, it is possible to define the creditworthiness of a borrower enterprise as its complex financial and legal characteristic, represented both by financial risk indicators and business risk indicators. Creditworthiness in this sense shows the capacity of an enterprise to fully meet its credit obligations in the future in accordance with a loan agreement and determines the degree of risk a bank bears while lending to a particular borrower.

It is particularly difficult to directly obtain fair, verified and objective results of assessing the creditworthiness of the business subject. This problem is solved by implementing systematic and comprehensive methods for analyzing a specific enterprise or in other words - the borrower. The use of integrated economic analysis takes into account the most important aspects of the business entity's activities, such as a study of all aspects of the enterprise's activities and a detailed assessment of the relations between the sections of the analysis. Also, in the process of integrated economic analysis, indicators and factors of production are studied.

As part of the integrated economic analysis, creditworthiness is determined by the results of the economic activity of the company and indicates how effectively the management of it allocates financial resources, efficiently uses its own funds, proportionally combines equity and borrowed capital. Also, in this context, creditworthiness shows what are the results of the company's economic activities and what are the relations with counterparties represented by suppliers, buyers, other creditors, the state and others. As a result, the creditworthiness of the economic entity shows the competitiveness of the enterprise, its economic potential. At the same time, for a creditor of an enterprise, information about the creditworthiness of a particular enterprise can reduce the risks of loss due to the likelihood of financial difficulties resulting in the eventual impairment of a loan.

Consequently, information about the financial condition of the company is a serious advantage of the borrower to receive borrowed capital, as it shows the structure of equity and borrowed capital. Also, which is important, according to the structure of the capital, it is possible to assess how capital is distributed among various types of property and how efficiently capital is used. This is followed by an analysis of solvency, financial sustainability, an analysis of the economic potential of the borrower and its attractiveness to creditors. In the context of the borrower's financial status analysis, between financial analysis and creditworthiness arises an interconnection [20].

It should be noted that the fundamental principle of borrower's creditworthiness analysis is the systematic approach as part of a comprehensive economic analysis. This principle means that creditworthiness analysis is an element of a high-level system, to be more precise of integrated economic analysis system, and it has relationships with other subsystems. Also, creditworthiness analysis is an element of a lower level subsystem, that includes some subsystems or in other words stages of analysis. Figure 1 shows the system of complex analysis of business entity's creditworthiness. This scheme is based on the theory of economic analysis by scientists M.I. Bakanov and A.D. Sheremet [2]. The scheme reflects parts that are interconnected and ultimately lead to a single result - a comprehensive assessment of the borrower's creditworthiness.

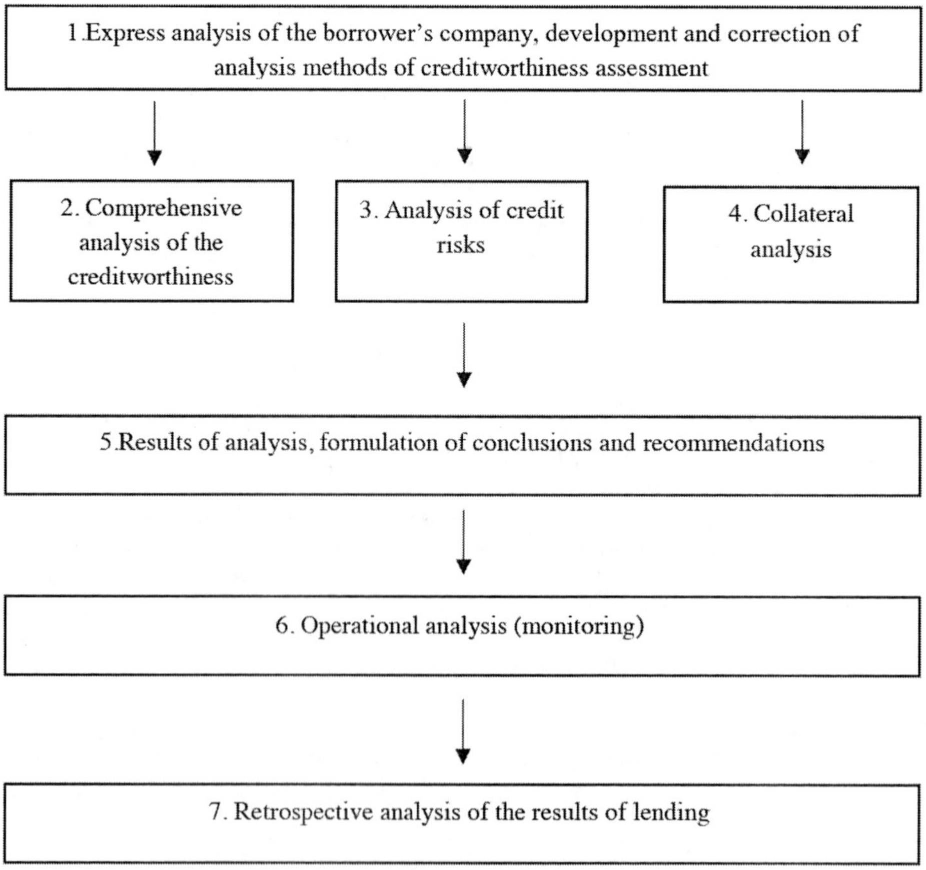

Fig. 1. System of comprehensive analysis of the borrower's creditworthiness
Source: Own processing.

In this figure stages of conducting a comprehensive analysis of the creditworthiness are given, starting from an express analysis of the borrower, or initial analysis till a retrospective analysis of the results of lending or an analysis of previous credit transactions specifically with this borrower. In addition, the scheme reflects the important aspect that after a comprehensive assessment of the borrower's creditworthiness, the financial position of the borrower should be monitored.

Please refer to paragraph 1 of Figure 1.2. - Express analysis of the borrower organization. The concept of express analysis includes clarification of the purpose for issuing credit resources: the purpose of the loan, the regulatory and legal assessment of the documents of the company, an analysis of the company's belonging to a certain industry, a quick assessment of the provided loan security that is guarantors and collateral property. Also, additions to the analysis method can be included in the express analysis. Thus, analysts determine the list of tasks for analysis of creditworthiness at the stage of express analysis. It should be noted that the credit policy of each bank sets the main development directions for the bank, the formation of a loan portfolio. This means that the bank's credit policy provides information on the forms of the loan provided, on the rules for working with collateral on the loan and other important aspects [17].

At the stage of express analysis, loan resources requested by the borrower are checked for its correspondence with the priority directions for the formation of the bank's loan portfolio. Previous experience of the bank interaction specifically with the given borrower is also checked, the availability of all the required documents of the borrower is verified with clarification of its authenticity.

After the express analysis stage, analytical studies are carried out. Results of these studies are of great importance to bank analysts, since the conclusions obtained from a comprehensive analysis of the borrower's creditworthiness, assessments of credit risks and security on the loan are the information on which the decision to grant a loan to the borrower depends. Obviously, it is not possible to draw conclusions about the level of risk when granting a loan to a borrower if you use only an assessment of the creditworthiness of the business entity [21]. This thesis is especially relevant in modern market conditions, since at the moment there is a high probability of such credit risks that can not only destabilize the financial condition of the borrower, but also negatively affect the entire banking industry and the country's economy as a whole for example the financial crisis of 2008 serves as a vivid illustration of this.

The next aspect to be noted is the variability of the types of analysis. This means that operational and retrospective analyses are carried out in separate stages. In Figure 1, these are paragraphs 6 and 7 of the analysis. The special role of operational analysis, or monitoring, is determined by the fact that it is the basis of monitoring for the current financial situation of the borrower.[3] At the same time, the indicators obtained during monitoring are the initial data for the next stage of analysis that is retrospective analysis. The tasks of retrospective analysis include objective analysis's results of interaction between the borrower and the lender in terms of the loan agreement, discovery of negative and positive aspects of such interaction.

Currently, there are many different approaches to assessing creditworthiness. According to professors Bocharova I.V. and Endovitsky D.A., the classification of creditworthiness models proposed by professor Vishnyakov I.V. is successful [4]. This classification is shown in Figure 2. Each of the models presented in the classification will be described in detail below. The main thing to note is that approaches to assessing creditworthiness are divided into two types - classification and complex models [16].

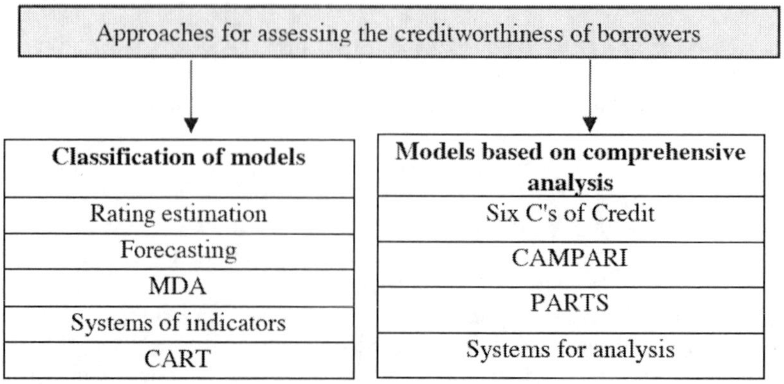

Fig. 2. Classification of models for assessing the creditworthiness of borrowers by Professor Vishnyakova I.V.

The rating estimation is an overall sum of the points. For each measure in the sum of the points, the points are calculated as follows: the value of the measure is multiplied by its weight, or significance factor. As a rule, when assessing creditworthiness based on a system of financial coefficients, groups of indicators are used, such as profitability ratios, liquidity ratios, financial sustainability ratios - among them, special attention is paid to the financial leverage ratio, turnover ratios and debt service quality ratios [18]. A slightly different system of indicators, which include ratios of profitability, liquidity, turnover, and fundraising that are used for defining various aspects of the creditworthiness of the company, was proposed by the American scientist Reed E. [1]. The set of these indicators allows you to determine the liquidity of current assets, analyze the complex financial situation of the borrower's company and its stability, evaluate the possibility of timely payments on borrowed money.

Also, on the basis of E. Reed's integrated system of indicators, it is possible to forecast the boundaries of lowering the volume of profit, in which the constant amount of loan payments is covered. With regard to banking practice, each commercial bank develops its own system of integrated indicators that are relevant to its credit policy. Rating models make it possible to assess the financial position of a borrower based on an indicator - a rating value, which is calculated in points assigned to each indicator value from a group of coefficients [15]. In accordance with the points scored for each borrower, the class of financial position is determined. Each commercial bank has a methodology that defines the most important financial ratios, their weights and scores, and a gradation of values that shows the quality of the borrower's financial position.

When issuing a bank loan, the in-house risk assessors (underwriters) carefully study the borrower's capabilities and determine his reliability. In his work, the underwriter uses the client databases available to him, as well as information received from the loan applicant. To make the information provided by a potential borrower most useful to the bank, underwriters use such a method as credit scoring.

The difference between credit scoring and a rating score is that in the rating score formula, its private score is used instead of the indicator value. For each indicator, several intervals of values are determined, each interval is assigned a certain number of points or a class is determined [2].

If the rating received by the borrower is lower than the value previously established by the bank's employees, then such a borrower will be denied a loan, and if it complies with the standards, the loan application will be satisfied. In this context, it should be noted that initially, the credit scoring model was focused on individual borrowers. Only later, when the model was worked out and its effectiveness was proven in consumer lending, the model was used to assess the creditworthiness of legal entities.

For this work, 33 different enterprises were selected from different fields. All enterprises are different in scale of activity, so in order to bring the sample to a more homogeneous one, semilogarithmic regression was used, which is described below. First, a regression model was built that includes all available regressors in order to draw a conclusion about the most significant regressors based on the residual probability.

The use of regression analysis in the framework of a scoring model involves the construction of a significant regression, a model of the dependence of the score in the sum for financial risk and business risk from various risk factors [22].

In order to analyse the range of enterprises, you need to provide a description of the data with hypotheses.

As an example [23]:

Table 1. Description of the data with hypotheses

№	Variable name	Variable description	Hypothesis
1	Company_name	Company name, text variable	
2	Scoring	Scoring model scores (business risk + financial risk)	Dependent variable
3	Reserve	Reserve for possible losses, in%	The higher the reserve, the lower the scoring points

Source: Own processing.

All characteristics should be given in detail in order to show why exactly they were taken for the regression analysis.

Further, from the set of characteristics of business risk and financial risk, the characteristics that best describe the sample should be taken, usually they describe indicators of business risk, indicators of financial risk, indicators of the structure of assets, the indicator of the structure of liabilities, indicators of the reserve and quality of the loan.

The selection of characteristics was made considering that:

1) most enterprises have this characteristic;

2) it is more important for the Bank;

3) according to opinion about the influence of this characteristic on the creditworthiness of the enterprise.

The next step is to build a regression on the characteristics of the creditworthiness risk, which can reflect both the real and the expected impact on the creditworthiness of the enterprise.

Table 2. Semi-log regression.

Dependent variable: scoring				
Method: least squares				
Sample: 1 33 IF SCORING>=20 AND RESERVE<0.4				
Included observations: 27				
Variable	Coefficient	Std. Error	t-Statistic	Prob.
Market_share	4.408732	6.127710	5.719475	0.0302
Supplier_dependence	4.037292	4.244107	6.951270	0.0228
Log(current_liquidity)	7.994769	7.023251	3.138329	0.0184
Log(sales_profitability)	0.540752	1.742451	3.310340	0.0295
Log(equity_to_debt_relation)	3.479767	1.588918	2.190022	0.0405
Log(shortterm_borrowings)	-0.214523	1.038432	4.206584	0.0384
C	62.45646	9.793678	6.377222	0.0000
R-squared	0.864072	Mean dependent var		49.77222
Adjusted R-squared	0.783293	S.D. dependent var		10.96788
S.E. of regression	10.72783	Akaike info criterion		0.801974
Sum squared resid	2301.727	Schwarz criterion		0.137932
Log likelihood	-98.32665	Hannan-Quinn criter.		0.901872
F-statistic	34.19609	Durbin-Watson stat		1.933245
Prob(F-statistic)	0.348360			

Source: Own processing.

This model has a high explanation - the coefficient of determination is 0.86. The F statistic is also high at 34, which suggests that the model is significant overall. The coefficients are significant individually, since the t-statistics are high, and the residual probabilities are close to unity. Thus, the resulting model of the Scoring dependency was obtained - the sum of scoring points for financial risk and business risk, and individual indicators of financial and business risk.

3 Results

So, the following tasks were set and completed:

1. a basic model of the influence of 6 regressors on scoring was built.
2. it was analysed based on various tests.

3. additional models are built.

Estimation Equation:

Scoring = c(1)*market_share + c(2)*supplier_dependence + c(3)*log(current_liquidity) +c(4)*log(sales_profitability)+c(5)*log(equity_to_debt_relation)+c(6)*log(shortterm_borrowings)+c(7).

During regression analysis, the following conclusions were obtained: firstly, the linear model is not suitable for reflecting the dependence of scoring points on risk indicators. In this case, the reason for this is the high heterogeneity of indicators, high coefficients of variation of explanatory variables.

Thus, a model of the dependence of scoring points on the most important indicators of financial risk and business risk has been built. It should be noted that the model results were obtained based on calculations for 33 enterprises. The model shows changes in scoring points with an increase or decrease in the most important indicators of risk. Using the model, you can find how much the sum of scoring points will increase or decrease when the most important indicators decrease or increase. A possible drawback of the model is the small number of observations, based on which the calculated indicators were obtained.

The essence of the scoring method consists in questioning the client, the result of which affects the bank's decision to issue a loan. Each question in such a scoring card (questionnaire) is assigned a certain number of positive or negative points. In many processing models, the scores do not take negative values and carry a strictly regulated quantitative and evaluative component [5]. The point total gives a general understanding of the client's creditworthiness, and an in-depth analysis of the combination of factors gives the optimal solution that excludes the subjective perception of the underwriter [6].

Bank risk assessors in their work use mathematical algorithms for questioning - scoring models. The scoring algorithm is developed in accordance with the general credit policy of the bank and loaded into the bank's computer system connected to various databases. Scoring models determine the quantity and quality of questions, their scoring and positional placement. In young banks, underwriters are armed with only two or three models of scoring questionnaires. The lending policy of experienced banks forces underwriters to keep a variety of scoring models close at hand, often complicated by querying questionnaires. For example, a positive answer to a question about a criminal record does not give a complete picture of the client's reliability. An extinguished or conditional conviction for fraud carries a significant number of negative points, while a conviction for a petty domestic crime may not have any point weight at all within the scoring model.

The early world experience of bank lending included the mandatory principle of letters of credit and recommendations for new borrowers [8]. The authoritative guarantor was personally responsible with business reputation for the credit good faith of his protégé. With the advent of affordable lending in the United States at the beginning of the 20th century, the banking system was swept by a wave of fraud and loans from clients with overvalued financial capabilities. If before that, US banks dealt with a narrow circle of the financial and industrial establishment, then the new system of "loans for everyone" provoked a massive influx of borrowers. The Great Depression (1929-1939) exposed the depth of the problem. There was a need for a new system for quick and high-quality assessment of the client's potential. In 1941, researcher D. Durand for the first time created an algorithm for recognizing the reliability of customers based on a point calculation. And in 1956, mathematician Earl Isik, in collaboration with engineer Bill Fire, developed the first full-fledged scoring model for determining the creditworthiness of citizens. Simplified scoring models of questionnaires in those years easily identified unreliable loan applicants. This was facilitated by the general illiteracy of citizens, including those who deliberately took out a loan in order to not return it. When the fraudulent environment shifted from audacious adventurism to cunning schemes, credit policy was forced to complicate all its assessment tools. Now, competent scoring models not only analyze the information provided by the client, but also immediately double-check it. Sustainable banks always could check customer data and determine his sincerity when questioning [7].

4 Discussion

Modern algorithms of scoring systems are capable of automatically assessing a client's creditworthiness. The type of such screening models was called Application-scoring. Fraud-scoring algorithms are responsible for identifying suspicious factors indicating an overestimation of personal data. Also, the banking system monitors changes in the client's financial capabilities (new loans, collection by court, irrecoverable loss of property, market reduction in the price of collateral assets, etc.). Changes in the score of the client allows you to predict the risks of non-repayment of the loan taken by him. This turns on the algorithm of the Collection-scoring models (English - a collection of methods for a set of points), which, based on all current parameters and data about the client, give the optimal strategy of actions [8].

These unique tools help the bank to respond flexibly to potential lost profits.

Thus, the problems that the credit scoring model can solve are as follows:

1. subjectivity - often decisions made by loan officers are based only on intuition and personal experience;
2. inflexibility and instability - the quality of the assessment is a random variable that cannot be improved or worsened, and depends on the emotional state and preferences of the expert;
3. lack of a system of training, knowledge transfer and advanced training - before becoming a highly qualified specialist, it is necessary to accumulate a certain level of knowledge based on the acquisition of sufficient experience in this area;
4. limitation of the number of considered applications, which is due to limited physical resources of a person, as a result of this - lost profit from a small number of applications under consideration.

You can try to solve these shortcomings using a scoring model, since its advantages are simplicity (since it is enough to calculate financial ratios and, taking into account the coefficients of their significance, to determine the class of the borrower), the ability to calculate optimal values by particular indicators, the ability to rank organizations by results , an integrated approach to assessing creditworthiness (since indicators are used that reflect various aspects of the organization's activities) [10].

For the first time, Russian researchers used logistic regression models to compile scoring maps [11]. Dependent and independent variables were included in the processing algorithm of the scoring program, and methods for their analysis and correction were determined. This made it possible to significantly improve the predictability of the borrower's creditworthiness based on his verified social and demographic data. Belarusian scientists are not lagging behind their Russian colleagues who have applied statistical analysis of multivariate data of complex structure such as data mining to the scoring model [12]. And although Western scientists continue to hold the palm in the development of models for assessing banking risks, new geopolitical challenges have given extraordinary dynamism to domestic scientific thought in relation to the banking sector of the Russian Federation.

5 Conclusion

The scoring model is one example of the successful application of statistical and mathematical methods. However, the scoring model has not become widespread in Russia.

According to G. Andreeva, the widespread use of scoring is hindered not so much by objective as by subjective reasons, of which the main reason is the mistrustful attitude of bank managers towards mathematical and statistical methods [13]. If we talk about domestic banks, then the use of scoring models in banks is combined with expert judgment.

In banks, for each large enterprise - the borrower, a responsible loan officer is appointed who oversees the financial analysis of this borrower. Due to the low development of scoring models, it will be interesting to consider this topic on the example of a specific bank.

Most experts agree that scoring models were originally a temporary solution for assessing risks. However, the indisputable achievement of scoring modeling has become a quick filtering from borrowers who overestimate their financial capabilities. Namely, this category caused the greatest damage to the banking sector at the time of the emergence of the system of affordable lending [19].

The emergence of new technologies and the merging of the banking sector with state structures predetermines new forms of accounting for the reliability of citizens. In developed countries, banking customer questionnaires, instead of comprehensive testing, involve answering just a few questions. Such decorative formalization is due to the fact that in a highly informative world it is impossible to keep personal data secret [14]. In a modern digitalized state, all information about a loan applicant can be extracted without the participation of the applicant himself. And the data obtained in this way are processed by computer programs with more efficient algorithms than the scoring model with a rough scoring method.

References

1. M. Krejnina, *Financial management*. Moscow: Business and service (2002)
2. Klyachkin, V.N., Shunina, Yu. S., *System for Borrowers' Creditworthiness Assessment and Repayment of Loans Forecasting*, Vestnik komp'yuternykh i informatsionnykh tekhnologii, **11**, p. 45-51 (2015), Available at: http://ezpro.fa.ru:2070/full_record.do?product=RSCI&search_mode=GeneralSearch&qid=1&SID=E1pr5h7xBiIjA2egM5z&page=1&doc=10&cacheurlFromRightClick=no
3. Francisco Louzada, Paulo H. Ferreira-Silva, Carlos A.R. Deniz (2012) *On the impact of disproportional samples in credit scoring models: An application to a Brazilian bank data* // Expert Systems with Applications, **39**(9,3). Available at: http://www.sciencedirect.com/science/article/pii/S0957417412001522
4. M. Bakanov, A.Sheremet, Theory of economic analysis. 5th ed. Moscow: Finances and statistics (2002)
5. E. Solozhencev, N. Stepanova, V. Karasev. Transparency of credit risks and ratings evaluation methods. St. Petersburg: St. Petersburg University Publishing House (2006)
6. I.Vishnyakov. Methods and models of assessing the creditworthiness of borrowers. St. Petersburg: SPbSUEE (1998)
7. V.Starshinov, System realization of credit scoring as a professional system to assess clients' solvency. Information technologies in science, management, social sphere and medicine. -V Intenational science conference (2018)
8. Vernikov, A. Mamonov, M., *Modelling technical efficiency of firms under one-step and two-step approaches (the case of commercial banks)*, Applied Econometrics, **1**, p. 67-90 (2018), Available at: http://ezpro.fa.ru:2070/full_record.do?product=RSCI&search_mode=GeneralSearch&qid=25&SID=E1pr5h7xBiIjA2egM5z&page=1&doc=8&cacheurlFromRightClick=no

9. O. Vorobyeva, Scoring model of creditworthiness evaluation. Moscow: Aktion Management and Finance (2020)
10. I. Vorobyeva, *Improving the Methodology of Credit Scoring of Borrowers in Car Lending*, Den'gi i kredit, **8**, p. 34-39 (2013), Available at: http://ezpro.fa.ru:2070/full_record.do?product=RSCI&search_mode=GeneralSearch&qid=25&SID=E1pr5h7xBiIjA2egM5z&page=2&doc=16
11. Panyagometh, K., Credit scoring by incorporating dynamic networked information, Economics & sociology, **1**, p. 262-269 (2019), Available at: http://ezpro.fa.ru:2070/InterService.do?fromPID=RSCI&product=UA&toPID=UA&returnLink=http%3a%2f%2fapps.webofknowledge.com%2ffull_record.do%3fhighlighted_tab%3dRSCI%26last_prod%3dRSCI%26excludeEventConfig%3dExcludeIfFromFullRecPage
12. V. Selyukov, *Optimization of Credit Risk Management in Banking Scoring Systems*, Herald of the Bauman Moscow State Technical University. Series: Natural sciences, **1**, p. 106-118 (2012), Available at: http://ezpro.fa.ru:2070/full_record.do?product=RSCI&search_mode=GeneralSearch&qid=31&SID=E1pr5h7xBiIjA2egM5z&page=4&doc=36
13. A. Masyutin, Credit scoring based on social network data, National Research University Higher School of Economics, Business Informatics (2015), **3**, p. 15-23, Available at: http://ezpro.fa.ru:2070/full_record.do?product=RSCI&search_mode=GeneralSearch&qid=1&SID=E1pr5h7xBiIjA2egM5z&page=1&doc=9&cacheurlFromRightClick=no
14. Y. Zhosan, Methods of selective data combination in terms of credit scoring limitations. Moscow: Moscow State University Publishing House (2017)
15. A. Sorokin, Creation of scoring schemes with the use of logisctic regression models. Vestnik evrazijskoj nauki, **2** (21) (2014)
16. N.V. Grin', Methodological aspects of scoring models. Available at: https://docplayer.ru/35162944-Metodologicheskie-aspekty-postroeniya-skoringovyh-modeley-n-v-grin-grodnenskiy-gosudarstvennyy-universitet-imeni-yanki-kupaly.html (2012)
17. G. Andreeva, Scoring as a method of credit risk evaluation. Bank technologies, № 6 (2000)
18. Li Yibei, Wang Ximei, Credit scoring by incorporating dynamic networked information, European journal of operational research, **3**, p. 1103-1112 (2020), Available at: http://ezpro.fa.ru:2070/InterService.do?fromPID=RSCI&product=UA&toPID=UA&returnLink=http%3a%2f%2fapps.webofknowledge.com
19. O. Zakharova, *Development of a model of an information system for managing a lending process and an algorithm for cross-selling to bank borrowers*, Software systems and computational methods, **1**, p. 51-58 (2019), Available at: http://ezpro.fa.ru:2070/full_record.do?product=RSCI&search_mode=GeneralSearch&qid=31&SID=E1pr5h7xBiIjA2egM5z&page=1&doc=4&cacheurlFromRightClick=no
20. V. Salin, *Statistical analysis of a banking system of Russia*, Vestnik Finansovogo universiteta, **6**, p. 107-115 (2015), Available at: http://ezpro.fa.ru:2070/full_record.do?product=RSCI&search_mode=GeneralSearch&qid=31&SID=E1pr5h7xBiIjA2egM5z&page=2&doc=13&cacheurlFromRightClick=no
21. V. Aleshin, *Credit scoring is a tool to improve the quality of bank risk management in contemporary conditions*, Terra Economicus, **10** (2-3), p. 27-30 (2012), Available at:

http://ezpro.fa.ru:2070/full_record.do?product=RSCI&search_mode=GeneralSearch&qid=31&SID=E1pr5h7xBiIjA2egM5z&page=3&doc=28&cacheurlFromRightClick=no
22. Christian Bluhm, Ludger Overbeck, Christoph Wagner. «Introduction to Credit Risk Modeling», second edition, USA: Chapman and Hall/ CRC Financial Mathematics Series, 2010, 383 p.

Decent Work During the Pandemic: Indication and Profiling Matters

M.V. Simonova[1,*], *L.V.* Sankova[2], and *F.I.* Mirzabalaeva[3]

[1]Samara State University of Economics, Samara, Russia
[2]Yuri Gagarin State Technical University of Saratov, Saratov, Russia
[3]Plekhanov Russian University of Economics, Moscow, Russia

Abstract. The article is devoted to the study of theoretical and applied aspects of decent work during the period of unstable economy development associated with the pandemic. The aim of research is to identify key labor trends, determine valid indicators and analyze decent work profiles at the level of the world community, Russia and its regions. The authors analyzed the decent work profiles at the beginning and during the downturn of the pandemic; studied the components of decent work in the economy as a whole and at the regional level. Special attention is paid to the criteria of profiling, which determine the necessary investment directions and formats in ensuring decent work conditions in the country context. Indicators and an integral indicator of decent work, taking into account the conditions of the pandemic, including unemployment; expenditure in GDP and payments against the subsistence minimum; loss of working time due to restrictive measures; the proportion of workers transferred to remote employment; Wage change rate; occupational safety index and others are proposed. The materials can be of practical value for the employment policy directions development at the regional and Federal levels of Russia, taking into account the current epidemiological situation and employment digitalization strategic priorities.
Keywords: Decent work, Pandemic, Unemployment; Employment policy; Social welfare.

1 Introduction

The economy instability, as an immanent feature of its development, is projected on the labor and employment sphere, generating shocks of demand for labor, restructuring of employment by sector, remuneration monetary characteristics volatility, active employment policy's new programs updating, the population social welfare formats. The uniqueness of the instability initial period is largely determined by the trigger-coronavirus pandemic and the economic activity compression due to supply restrictions on many components of personal consumption, which in the past were not always affected by the recession [11].

The modern socio-economic space transformation in all countries of the world in the context of the COVID-19 pandemic has a direct impact on the subject field of decent work. It is conditioned with some factors. Firstly there are some changes in the labor market subjects' behavior. Secondly, the cycle of macroeconomic policy measures' impact on

*Corresponding author: m.simonova@mail.ru

labour market indicators is accelerated. Thirdly, new social and occupational risks requiring protection are appeared. Fourth, all the key parameters of decent work that have been defined by ITO are transformed. The request for the definition of indicators of decent work and the analysis of developments in various countries, regions and sectors of the economy during the pandemic is formed.

Let's imagine the modern process of the economy functioning within the framework of the decent work discourse. The International Labour Organization treats decent work as productive work, which at the same time is free, safe, fairly paid, developing and not diminishing human dignity. In summary, the following blocks are proposed within the framework of the Concept of decent work, reflecting the main characteristics of decent work: employment; social welfare; respect for the rights of employees; social dialogue development.

Principles and standards of decent work are developed by foreign and domestic researchers (R. Anker, G. Fields, D. Guy, F. Egger, F. Bonnet, A. Chatenier, G. Standing, V. N. Bobkov, R. P. Kolosova, L. A. Kostin, etc.). Decent work programmes are being implemented in more than 40 countries around the world.

In the context of the pandemic, it is necessary to clarify approaches to the criteria and indicators of decent work, as well as to develop mechanisms for ensuring it for workers. It is necessary to take into account the following aspects: the limited statistical, methodological basis and tools for assessing and monitoring decent work in a pandemic due to the novelty of the problem, the lack of indicators and standards of decent work in a broader context - in relation to the digital transformation of the field of work and employment.

Digitalization of employment, which accelerated significantly during the pandemic, is the leading trend in the labor sector on the one hand, [1,8,16,18], and the only possible mechanism for many companies operating under quarantine measures, on the other hand.

The matters of the coronavirus pandemic's impact on the work sphere have been highlighted in reports by the ILO and other international organizations. The ILO's preliminary assessment (March 2020) of the impact of COVID-19 on the global world of work indicates negative consequences. Unemployment, underemployment, and work poverty, which will require a coordinated and rapid response from government and interstate agencies, are among them [5]. Studies show different effects of the pandemic on workers in different sectors, levels of education, and gender characteristics in different workplaces [15]. Experts focus on the labor market and employment trends evolution during the COVID19 pandemic [4,6,7,9], estimates of supply shocks from labor supply [14].

We propose an approach to the study of decent work as a mechanism for ensuring stability in the period before the pandemic and during the pandemic in the context of an unstable economy. The article will consider methodological issues of indication and profiling of decent work taking into account the influence of the coronacrisis, the limited traditional approaches to its assessment and proposed an integral indicator of decent work. The section "Results" compares a number of decent work indicators in pre-pandemic and pandemic situations at the international, national and regional levels. The grouping of regions of the Russian Federation by unemployment indicator is presented. The section "Discussion" discusses issues of profiling decent work, issues of remote and informal employment during the pandemic, international and Russian experience in implementing anti-crisis measures in the labor market. The Conclusion contains the main conclusions of the study and defines the directions of employment policy in modern conditions, taking into account the emerging regional profiles of decent work.

2 Methodology

The issues of indication and determination of decent work's progress as a key condition for economic stability was at the centre of the ILO's attention even before the pandemic.

The approach to measuring decent work is used by the United Nations development Programme as one of the indicators of human development. To assess the implementation of the principles of decent work, 7 indicators were proposed: the share of employed in the total population; employment in the formal sector of the economy and the ratio of this indicator for women than men; the level of vulnerable employment and increased the ratio of women to men increased; the share of working poor in total employment; the unemployment rate depending on education level; level of child labour use; the duration of paid maternity leave. To rank OECD countries on indicators of decent work in industry, the following components were identified: gender inequality; employment; social dialogue and social protection [10]. To measure decent work, Federal State Statistics Service of the Russian Federation publishes the following indicators: the share of employed in the total population (%), the unemployment rate (%), the share of the informal sector in total employment (%); the share of NEET youth (%); the working poor (%), and others.

We suggest grouping the criteria for describing decent work profiles reflecting pandemic conditions as follows:

- criteria determined by the influence of external factors. They include measures of fiscal and monetary policy that affect employment and employee health; business response to measures to support jobs and business activity, compensation for losses; effectiveness of regional management in keeping employment and supporting the income of the population;
- criteria determined by the institutional aspects for the labor market and employment functioning (labor legislation, employment policy, work of the Employment Services in pandemic conditions);
- criteria determined by the possibility of using the digital format of employment by organizations and employees;
- criteria reflecting the specifics of population social welfare during the pandemic (coverage, amount of payments, share of expenditures in GDP).
- criteria that reflect the interaction of the parties to the social dialogue (government, business, trade unions, NGOs) in protection of employment and safety in workplaces, income support.

To assess changes in work sphere, we suggest calculating decent work profiles and integrated decent work indices. After selecting the components for each specified criterion, the profile and index of decent work in the context of a pandemic are calculated. It seems that the choice of indicators of decent work should be based on the following principles: comparability of indicators; availability and completeness of statistical and empirical base; flexibility, the possibility of structuring by elements of decent work.

Taking into account both quantitative and qualitative parameters, we can propose the following formula for calculation of the integral indicator of decent work - Idw:

$$Idw= 0{,}2(1\text{-}Iur) + 0{,}3Isw + 0{,}1(1\text{-}Iwtl) + 0{,}1Isre ++ 0{,}1Isc + 0{,}1Ils + 0{,}05Ige + 0{,}05Isd \quad (1)$$

Iul–unemployment rate; Isw – social welfare indicator (complex indicator, including expenses in DGP share and size of the compensation in relation to living wage); Iwtl – working time loss indicator due to restrictive measures (in percent); Isdw –share of employees transferred to remote employment format; Isc – salary changes indicator (share of employees getting the same wages as before pandemic); Ils – labour safety indicator (it is calculated as a share of employees working in conditions that meet standards and regulations, the proportion of employees covered by COVID protection measures is also taken into account); Ige – gender equality indicator; Isd – indicator, reflecting social dialogue efficiency.

The integral indicator of decent work during the pandemic will be defined as the weighted average of these groups of indicators. Each of the indicators takes values in the range from 0 to 1, the sum of the weights must be equal to 1. To calculate these indicators, you can use official statistics, as well as data from surveys and surveys (taking into account the quality parameters of decent work).

3 Results

Taking into account the limitations of statistics on a number of the proposed index components, consider the decent work profile based on available data. The analysis of regional problems in the decent work principles implementing revealed the main limitations and potential risk zones even in the period before the pandemic [17]. Compare the decent work indicators of the Russian Federation in the period before the pandemic and the situation with a number of decent work indicators in pandemic conditions. Analysis of the data of the Federal State Statistics Service of the RF shows that despite the positive trends in the change in a number of indicators of decent work, even before the pandemic, alarming dynamics remain. This also applies informal employment, inequality in income distribution. The proportion of NEET youth is also declining. It should be noted that it is the representatives of the informal sector who have found themselves in the most difficult situation in the context of the pandemic.

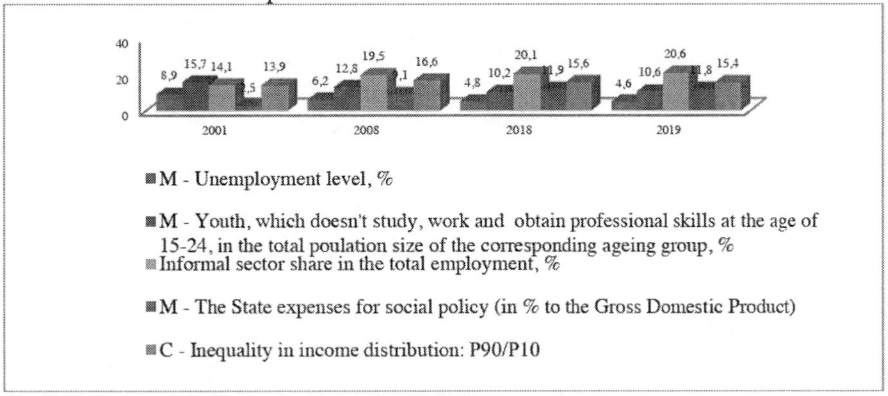

Fig. 1. Decent work indicators in Russia in 2001-2019
Source: Own processing

In the context of a pandemic, the instability rate is the most sensitive to changes in the socio-economic situation (Fig.2). In the period of April and May 2020, The Federal State Statistics Service of the Russian Federation fixes an increase in the indicators growth both of overall unemployment rate (from 4.6 to 6%), and registered unemployment (from 1 to 2.8%).

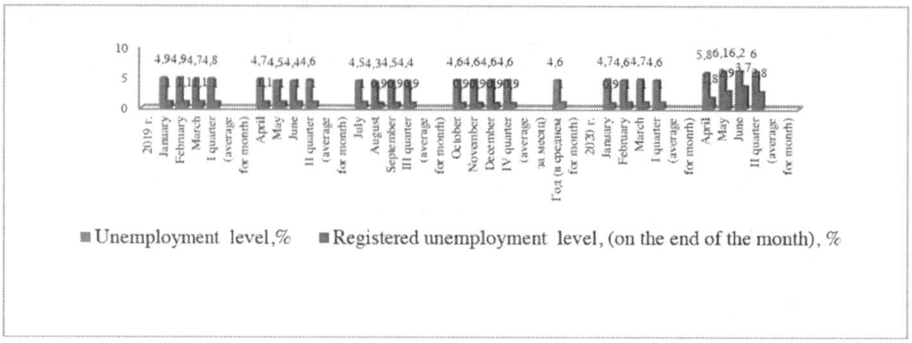

Fig. 2. Unemployment level dynamic in 2019- 2020
Source: Own processing.

In all previous periods of instability and crises, price and time adjustment in the Russian labor market prevailed over quantitative adjustment. This reduced the incentives of companies to release, contributed to the preservation of employment. During the pandemic period, these company strategies were also widely used.

Researchers identify the following changes in the registered labor market in the 2nd quarter of 2020[3]:

- a significant increase in the share of citizens who have not previously worked and are looking for work for the first time, among the unemployed registered with the employment service. Thus, the total number of unemployed (aged 15 years and older) in June 2020 was 4.6 million people or 112.6% compared to the first half of 2019, and the number of officially registered unemployed in the 1st quarter of 2020 was 179.1 % compared to the first quarter of 2019; registered unemployment grew at a faster rate in relation to total unemployment (according to the ILO methodology), which increased by 31% over the same period (according to preliminary data from Rosstat) [3].
- increase in the share of young people aged 20 to 29, as well as the share of unemployed citizens aged 30 to 49 in the structure of the unemployed;
- significant increase in the share of unemployed people who do not have basic General education, as well as unemployed people with basic General and secondary General education;
- exceeding the number of resumes over the number of vacancies.

In regional view decent work profiles are primarily formed by the employment block, which is reduced. In this context we have identified four groups of regions based on the unemployment rate (see table.1). The table shows regions in descending order of the unemployment rate. Before the pandemic period (January-March 2020), the minimum unemployment rate (1.5%) was observed in Moscow, and the maximum was 26.6% (Republic of Ingushetia). In the 2nd quarter of 2020, there is an increase in unemployment in all regions. The maximum value is still observed in Ingushetia (30.1%), the minimum value-in Moscow (2%).

Table 1. Unemployment rate of the population aged 15 years and older in the regions of the Russian Federation, April-June 2020 (in %)

Unemploy mentrate,%	Regions
2,0-5,4	Magadan region, Smolensk region, Ryazan region, Kaliningrad region, Rostov region, Orenburg region, Saratov region, Belgorod region, Leningrad region, Sevastopol, Tyumen region without auth. districts, Samara region, Tambov region, Kaluga region, Nizhny Novgorod region, Chukotka Autonomous district, Voronezh region, Lipetsk region, Tula region, Tver region, Bryansk region, Khabarovsk territory, Kamchatka territory, Republic of Tatarstan, Moscow region, Tyumen region, Saint Petersburg, Yamalo-Nenets Autonomous district, Moscow
5,5-8,7	The Republic of Adygea, Yaroslavl region, Komi Republic, Kurgan region, Murmansk region, Irkutsk region, Volgograd region,Astrakhan region, Arkhangelsk region, Arkhangelsk region without auth. districts, Republic of Sakha (Yakutia), Chuvash Republic, Jewish Autonomous region, Udmurt Republic, Oryol region, the Republic of Mari El, Krasnoyarsk territory, Pskov region, the Republic of Crimea, Sverdlovsk region, Chelyabinsk region, Perm region, Kemerovo region, Vologda region, Novosibirsk region,Vladimir region, Krasnodar territory, the Republic of Bashkortostan, Altai territory, Amur region, Stavropol territory, Kirov region, Kostroma region, Ivanovo region, Primorye territory,Kursk region, Penza region, Novgorod region, the Republic of Mordovia, Ulyanovsk region, Sakhalin region
9,4-11,0	the Republic of Buryatia, the Republic of Kalmykia, the Republic of Khakassia, Omsk region, TRANS-Baikal territory, Tomsk region, the Republic of Karelia
12,3-30,1	The Republic of Ingushetia, the Republic of Tyva, the Republic of North Ossetia-Alania, the Republic of Dagestan, the Republic of Karachay-Cherkessia, the Republic of Kabardino-Balkaria, Chechen Republic, Altai Republic

Source: Rosstat.

Pandemic conditions are changing the portrait of an unemployed person. In particular, his comparison in the 1st and 2nd quarters of 2020 shows significant changes in age, professional qualification, and educational aspects [3]. This perspective allows us to judge the shortage of decent work for different population groups.

Trade, construction, services have been more affected by the significant deterioration of decent work profiles during the pandemic. To a greater extent, the problem of reducing employment and declining income due to the liquidation of organizations affected trade; repair of motor vehicles and motorcycles (86.3 liquidated organizations per 1000 organizations in the 1st half of 2020), construction (79.1), transportation and storage (62.9). Trade, construction, and services were more likely to experience a significant deterioration in the "profiles" of decent work during the pandemic.

Wages have always been one of the key indicators of decent work. During the pandemic (for the three spring months of this year, which included a period of non-working days and severe economic restrictions, the total wage Fund for employees decreased by 4.7% compared to last year's spring – to RUB 6.8 trillion, (according to FinExpertiza). The peak collapse was recorded in April (more than 20%).This process involves a direct reduction in salaries or allowances, a reduction in the number of employees, and the transfer of part-time employees.

Let us turn to the international aspect. Consider the dynamics of the unemployment rate in a number of OECD countries before and during the pandemic (figure 3).

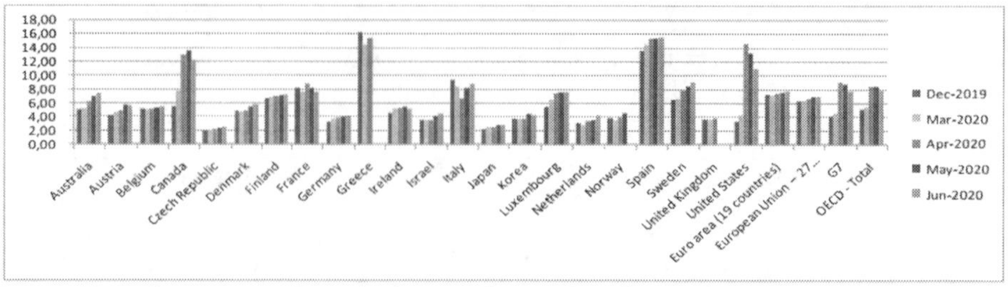

Fig. 3. Unemployment rate dynamics in several countries of OECDin December, 2019 - June 2020
Source: Own processing based on https://stats.oecd.org/index.aspx?queryid=36324.

Analysis of the data shows, first, an increase in the unemployment rate in all the countries represented; second, the presence of common patterns in the dynamics of unemployment in recent years in many countries; and third, different rates of the spread of the pandemic, restrictive measures and the effectiveness of policies to preserve and protect employment in countries. For example, the highest unemployment rate in June 2020 was observed in Spain (15.6%), the lowest - in the Czech Republic (2.6%). At the same time, according to forecasts, the average employment rate in these countries in 2020 will be 4.1–5% lower than in 2019. The share of working people is expected to remain below pre-crisis levels even at the end of 2021.

According to the ILO, by the end of May 2020, more than 90 countries had introduced or announced fiscal measures totaling more than $10 trillion, while the same number of countries reduced interest rates after the outbreak. The approved international financing for health and social protection amounted to $ 46.9 billion, of which 51% is a soft loan, 46% is a loan. According to ILO estimates, from the point of view of institutions, the IMF represents 47.3%, 15.4% - the World Bank, 14.2% - the Asian Development Bank.

Considering the problem of financing anti-crisis measures in Russia, we note that assistance packages worth more than 2 trillion rubles have been developed and are being implemented, which primarily affects the field of employment and income. To restore the economy, a three-tier program has been developed until 2021, which includes the "adaptation stage", the "recovery stage" and the "active growth" stage. At each of these stages, the objectives of employment policy and the measures applied may vary.

4 Discussion

In the context of a pandemic, traditional criteria and parameters for decent work are becoming more informative. In particular, the total indicator of the impact of the COVID-19 crisis on the labor market is the loss of working time estimated by the ILO using the forecasting model [13]. The amount of losses is determined by the institutional structure of national labor markets.

In this context, we are faced with differences in the assessment of the components of "losses" in the labor market and in employment, which allows us to form an idea of the profile of decent work in a particular country or region. For example, the traditional indicator of the decent work deficit is unemployment, which has increased to some extent in all countries. However, the ILO focuses on the loss of working time. At the same time, the structure of working time losses can include the following components: reduction of the average working time per week; the presence of workers formally present at work, but not really working; unemployment; leaving the labor market.

In terms of changing decent work profiles, there is a significant cross-country differentiation in the pattern of time losses. For example, in Great Britain and Northern Ireland, the loss of working hours was mainly due to reduced working hours and the number of workers who were employed but not employed. At the same time, the share of the unemployment component in the structure of labor losses was very small. In countries with stringent measures to contain unemployment, there has been little growth. There has been a significant loss of jobs in Canada, leading to both unemployment and "unnecessary" employment. Finally, in the US, job losses amounted to about 2/3 of the reduction in working hours, and unemployment accounted for almost half of this reduction.

As shown by international practice and the ILO's activities in the context of economic shocks, it is the issues of decent work that are becoming a priority for the fight against COVID-19 on the basis of international labor standards. In particular, the following measures are key ones in various countries of the world:

First, protecting workers in the workplace; strengthening occupational safety measures; various job adaptations formats (such as telework); preventing discrimination and exclusion; ensuring access to health for all groups of workers; and increasing access to paid leave.

Second, stimulating labor demand: active fiscal policy; enabling and adaptive monetary policies; lending and financial support to specific sectors, including health care. Fiscal policy measures in advanced economies account for 88 per cent of global financial stimulus, accounting for an average of 5 per cent of GDP.

Third, support for employment and income through expansion of social welfare for all; various formats of employment retention (short-term work, paid leave, etc.), subsidies, financial/tax benefits for small and medium-sized businesses.

In terms of overcoming the economic consequences of the new coronavirus infection, prepared By the government of the Russian Federation, the following measures to support income and employment should be highlighted: an interest-free salary loan (coverage-805 thousand people); a 2% salary loan with the possibility of write-off (coverage-3.5 million employees); financial assistance to businesses (in April and may - 3.3 million people will keep their salaries); deferred and reduced tax and insurance payments for small and medium-sized businesses (in the affected industries, 3 million people will be affected); for all industries, halving insurance premiums for salaries above the minimum wage (coverage of more than 20 million people); preferential loans to system-forming enterprises to maintain working capital and preserve jobs (coverage -911 thousand employees); increasing the minimum amount of sick leave payments per month-possible coverage-2.3 million people; increasing the maximum amount of unemployment benefits; additional benefits for families where parents are unemployed;additional benefits for families with children (up to 1.5 years, up to 3 years, unemployed with children under 18 years; families in need for children 3-7 years old, etc.); participation in temporary employment programs, assistance to self-employment, distance learning, advanced vocational training, compensation to employers for partial remuneration of employees dismissed from other organizations due to liquidation or reduction - 67 thousand people.

In the context of the pandemic, employment in the informal sector is most severely affected. However we have to take into account the fact that, during the recovery period, informality is rapidly recovering.

The next aspect to pay attention to is the possibility of remote employment in the context of a pandemic. According to an analytical report by the Russian Public Opinion Research Center and Social Business Group, the share of Russians working remotely during the self-isolation regime due to coronavirus has increased eight times. The main part of them are specialists with higher education working in the public sector (60%, of which 47% have switched to remote work completely), and 13% - partially. Among specialists with higher education employed in the commercial sector, 51% of respondents switched to "remote". At the same time, there is a high differentiation of remote employment by region (in Moscow and St. Petersburg-29%, in rural areas - 10%). According to RBC, only 34% of companies switched to remote operation in the affected regions of Russia. Amendments to the law are being developed to legalize this format of work in the future, which allows us to speak about the institutionalization of key changes in decent work standards. The possibility of prolongation of this format for the post-pandemic period forms a request to assess an individual's satisfaction with their work. According to a study by Colliers International, conducted in Europe and the Middle East, 45% of office employees want to return from a remote location: they are not comfortable working at home. About 65% of respondents said that they are more used to working in the office.

5 Conclusion

The authors analyzed indicators and profiles of decent work and proposed an approach to calculating its integral index during the pandemic. Based on the index proposed by the authors, it is possible to determine the profiles of decent work at the national and regional levels, which will allow developing a set of measures to ensure employment and protect the population in conditions of economic instability, taking into account existing opportunities. It was noted that the differentiation of regions and sectors of the Russian economy according to the profiles of decent work has deepened. There is a need for flexibility or change of approaches at different stages in joint efforts to overcome the pandemic [11]. The study emphasizes that decent work is an institutional, economic and organizational issue.

Changes in decent work profiles will be determined largely by the effectiveness of measures to support unemployed and small and medium-sized businesses, as well as opportunities for rapid economic recovery (saving and generating new jobs). Indicators of decent work will be determined by accurate termination of support facilities at all stages of the recession.

As unemployment becomes one of the key indicators of the state of the labor sector during the pandemic, the transformation of institutions in the labor market takes on a special role. Interregional cooperation of employment services can help: to smooth out the shortage of vacancies in the region based on expanding remote operation formats in other regions; to allocate the resources of the territorial mobility assistance program; to create temporary and seasonal jobs as a buffer to unemployment, etc.

Decent work in the post-pandemic period should go in two ways and take into account short-term and long-term effects. New jobs need to meet decent work standards in terms of designated criteria and national labour standards. At the same time, it is necessary to implement the principles of decent work in the field of job security, wages, security, social protection, gender equality and access to social dialogue in existing jobs.

It seems that one of the productive areas of further study of the problem is the close connection of indicators of decent work with poverty. The decline in income, the increase in unemployment during the pandemic generates an increase in poverty in all countries; require the development of urgent measures to support the income of the population, investment in human capital. There is a need to explore the experiences of different countries in generating and sustaining income in pandemic conditions [12] and scale up good practices.

Declining incomes and rising unemployment during the pandemic generate an increase in poverty in all countries, and require the development of urgent measures to support the population's income and investment in human capital [12].

When analyzing indicators of decent work, it should be taken into account that the situation with the coronavirus pandemic (which is certainly a force majeure factor) cannot cancel out (although it can partially correct) the long-term trends and problems that have developed in the Russian labor market, and will require greater flexibility and adaptability from the state employment service.

References

1. Agrawal, A., Horton, J., Lacetera N., Lyons, E. Digitization and the Contract Labor Market: A Research Agenda. Economic Analysis of the Digital Economy /A. Goldfarb, Sh. M. Greenstein, and C. E. Tucker, ed. http://www.nber.org/chapters/c12988(2015)
2. Alfaro,L., Becerra,O., Eslava, M. EMES and COVID-19: Shutting Down In A World Of Informal And Tiny Firms, Working Paper 27360 http://www.nber.org/papers/w27360(2020)
3. Antonova G.V., Mirzabalaeva F.I., Bondarchuk A.G. Unemployed portrait transformation at the registered market under impact of institutional and economic factors. *Labour Economics*. **7**(8). doi: 10.18334/et.7.8.110781.
4. Baker, S., Bloom, N., Davis S. J. & Terry St.J. COVID-Induced Economic Uncertainty. Working Paper 26983http://www.nber.org/papers/w26983(2020)
5. Cajner, T., L. D. Crane, R. A. Decker, A. Hamins-Puertolas, and C. Kurz (2020). Tracking Labor Market Developments During theCOVID-19 Pandemic: A Preliminary Assessment. Finance and Economics Discussion Series (2020)
6. Coibion, O., Gorodnichenko, Y., and Weber, M. Labor Markets During the COVID-19 Crisis: A preliminary view. NBER w27017. (2020)
7. COVID-9 and World of Work: Impacts and Responses, 18 March. URL: ilo.org/global/
8. Digital Labour Platforms and the Future of Work: Towards Decent Work in the Online World and the Future of Work: Towards Decent Work in the Online World URL: https://www.ilo.org/wcmsp5/groups/public/---dgreports/---dcomm/---publ/documents/publication/wcms_645337.pdf
9. Furceri, D, P Loungani, J D Ostry, & Pizzuto, P. COVID-19 will Raise Inequality if Past Pandemics are a Guide. (2020)
10. Gay, D. Decentwork: concepts and indicators. International labour survey. T. 142. №1–2. 2003. M., C. 3–38. (2004)
11. Grigoriev, L.A., Pavlushina, V.A. Musychenko, E.Ye. Drop in the World Recession 2020. Economics Issues, 5,5-24. (2020)
12. Han, J., Meyer B. D. ,Sullivan, J. X. Income And Poverty In The COVID-19 Pandemic.Working Paper 27729http://www.nber.org/papers/w27729 (2020)
13. ILO Monitor: COVID-19 and the World of Work. Fifth edition. Updated Estimates And Analysis. https://www.ilo.org/wcmsp5/groups/public/@dgreports/@dcomm/documents/briefingnote/wcms_749399.pdf
14. Leibovici, F., Santacreu, A.M., &Famiglietti, M. Social Distancing And Contact-Intensive Occupations. On the Economy, St. Louis FED. (2020)
15. Montenovo, L., Xuan, J., Felipe, L, Schmutte, R., Simon, K., Weinberg, B., Coady, W. Determinants Of Disparities In COVID-19 Job Losses //Working Paper 27132. http://www.nber.org/papers/w27132 (2020)
16. Non-Standard Forms Of Employment: Resent Trends And Future Prospect European Foundation for the Improvement of Living and Working Conditions / Background paper for Estonian Presidency Conference "Future of Work: Making It e-Easy", 13-14 September. (2017)
17. Simonova, M. V., Sankova, L. V., Mirzabalaeva, F. I., Shchipanova, D. Y., & Dorozhkin, V. E. Assessment problems and ensuring of decent work in the Russian regions. International Journal of Environmental and Science Education, **11**(15), 7608–7626. (2016).

18. Ziemann, V. Inclusive Labour Markets In The Digital Era: The case of Austria, OECD Economics Department Working Papers,WP-17-143.OECD Publishing, Paris. URL: http://dx.doi.org/10.1787/c2331c20-en. (2017)

Innovation and its importance in the context of a company's marketing strategy

Veronika Olexova[1,*] and *Martina* Gogolova[1]

[1]University of Zilina, Faculty of Operation and Economics of Transport and Communications, Department of Economics, Zilina, Slovak Republic

Abstract. The current business environment is characterized by high competition between companies as well as global challenges and growing technological progress. Businesses must constantly adapt to changes in the business environment, changes in the market environment and the constantly changing requirements of customers. There are currently several tools and ways to respond to these changes. One of the most important areas can be considered corporate strategy, because it determines the basic direction of the company. An important area is also marketing, which affects the overall business activity. Determining the right marketing strategy has an important impact on the overall direction of the company. Another important element of the current time in the field of business is innovation, which can be considered a basic prerequisite for the success of the company. The issue of innovation management is currently one of the basic conditions for maintaining a market position and maintaining competitiveness. The aim of the article is to point out the theoretical basis of the corporate strategy, marketing strategy and the importance of innovation in business. Subsequently, based on secondary findings, we analyze available selected data related to innovation in Slovakia.
Keywords: company, innovation, marketing, strategy

1 Introduction

The ever-changing conditions in today's global world create a complex environment for companies, to which the company must constantly respond. In recent years, the business environment has undergone significant changes. That is why one of the key factors for success is a quick response to market changes. The company must be able to respond quickly to market and customer requirements. The aim of every company is to achieve success and maintain and improve its market position. It is essential that the company is constantly looking for new customers, but also to maintain and strengthen the relationship with the company's existing customers [1]. The economy is constantly undergoing significant changes, the role of the company is to anticipate these changes if it wants to move ahead of the competition. The means of this are innovation strategies and the rapid introduction of innovation into all business activities of the production process. Every business that wants to be successful in the long term needs an innovation management

*Corresponding author: veronika.paurova@fpedas.uniza.sk

system [2]. It is therefore essential for businesses to have a well-defined long-term strategy that takes innovation into account. Company can suggest an appropriate marketing strategy that can bring it new customers [3]. At present, innovation is considered an important condition for competitiveness, which is a key strategic factor in economic performance. The importance of corporate strategy and marketing strategy should be a priority for companies. Marketing strategy is concept of a business venture in the field of marketing, that task is the allocation of resources enterprise with the aim of meeting the fundamental objectives of the company and to make a profit and competitiveness and achieve customer satisfaction [4]. The established strategy is then transformed into basic marketing elements, such as product policy, communication with target customers, pricing and distribution. Strategy setting is a continuous process that must adapt to changing market conditions. Innovation is now considered a prerequisite for a company's competitiveness [5]. Innovation is an important tool for business growth [6]. Implementing and creating innovation is a way in which a company can respond to the requirements of customers and the market itself, and thus ensure change in the company. Innovation is a process where the result is something new. With legal response to the market, an innovative approach and adaptation to market conditions lead to satisfied consumers, which are the foundation of a company's success. Innovation is the key to the success of many sectors of the economy [7].

1.1 Marketing strategy of the company

Strategy is the way in which a company achieves its goals [8]. It is the long-term intention of the organization that affects the current position [9]. The strategy expresses basic ideas about how long-term corporate goals should be achieved. The use of the company's resources and the course of individual strategic operations, which have the task of achieving the fulfillment of individual corporate goals, are part of the company's strategy. One of the specific areas of business strategy is marketing strategy. A marketing strategy is a process or plan that is used to achieve a company's marketing goals. Strategic direction is important for marketing managers [10]. Marketing strategy is the way a company achieves its marketing goals. It consists of decisions on marketing costs, the marketing mix and the allocation of marketing resources in relation to expected competitive conditions. [11]. The marketing strategy is also deciding on the program for the development of individual products and services, on the method of their sale, on the production and sales program and on the development of all components of the marketing mix [12]. Marketing strategy can be considered as the most complex tool of marketing strategic management, because its role is long-term operation and taking into account internal and external conditions in order to achieve the expected market position [13]. The marketing strategy is elaborated in detail in sub-strategies, which include product strategy, pricing strategy, distribution strategy and marketing communication strategy. The process of building and managing brand value should be part of the strategic marketing of every business [14]. The marketing strategy forms the basis for the development of the company's plans and the process of creating and maintaining a strategic balance between its goals and capabilities. The basis of the marketing strategy is the definition of the company's mission, determination of the company's goals, creation of a business portfolio and coordination of marketing and other functional strategies. Marketing strategy is a tool by which a company achieves its set marketing goals. The marketing strategy should also address the market segments that the company intends to focus on and focus on. The strategy that the company adopts is reflected in the design and development of the product, in the product policy, in the choice of the target market, in the subsequent communication with the target group, in the pricing policy and in the choice of distribution channels. Creating an optimal marketing strategy is not a one-time activity, but a continuous process. A strategy that has been successful in its time, or is currently successful, must constantly adapt to changing market conditions. The company must be able to understand consumer behavior and motivation to buy a product [15]. A strategic analysis that shows the situation of the company is necessary for the development of a quality company strategy. In general, strategic analyzes can be divided into internal analysis and external environment analysis [5]. The basic strategic analysis includes, for example SWOT analysis, EFE and IFE matrix, Porter's analysis of the five forces, PEST analysis, BCG matrix.

1.2 Innovation

Innovation is new developmental changes which the company apply in practical activities [5]. Innovation is the foundation of sustainable growth in the business and important source of competitiveness, economic growth, as well as the image of each country [16]. Innovation is a process from idea, through development to implementation. The term innovation can be understood as the introduction of a new or possibly significantly improved product or service on the market. The same applies to the process, i.e. the introduction of a new or significantly improved process within the company. Innovation is obtained through new or existing company technologies, or also using company knowledge. Innovation can be developed by a company or other companies and should represent a new change for the company. In the past, non-technological innovation, which has been based on experience, has predominated, but at present technological innovation is being applied [5. The result of an innovation is an improvement in a product, service or process, or anything else. Innovation is one of the most important factors influencing consumer attitudes [17]. Innovation can be technological or non-technological in nature. Technological innovation represents product innovation, ie. new or significantly improved product (product, service) introduced to the market or process innovation, a new or significantly improved process introduced within the company. Non-technological innovation includes organizational innovation and marketing innovation [18]. A product innovation is the launch of a new product (product or service) whose characteristics, or intended use, differ significantly from previous products. This includes new or significant changes to technical specifications, components and materials, embedded software, user acceptability (accessibility) or other functional characteristics. Innovation help business to extend the development of new products [19]. Process innovation is the introduction of a new or improved production method, logistics, supply and distribution systems. This includes significant changes to specific techniques, equipment or software that are designed to improve the quality, efficiency or flexibility of production or supply activities, or to reduce environmental or safety risks. Organizational innovation is the introduction of significant changes in business practice, the organization of work responsibilities and decision-making, which includes training or education to increase qualifications and responsibilities; and in the organization of external relations with other undertakings or public institutions. Their goal is to improve the innovative capacity of the company or performance characteristics, such as. quality and efficiency of workflows. Organizational innovation usually involves changes in more than one part of a company's supply chain and are less technology dependent than process innovation. Marketing innovation represents significant changes in the way products and services are traded, including changes in design and packaging. Many of them are included only in the first use by companies, for example the first use of the Internet for product placement is an innovation, but the second use of the Internet for the placement of other products, or. in various geographic markets it is no longer an innovation [18]. An important marketing tool is the brand, which is a key factor in changing the perception of competing brands [20]. In relation to customers, CRM (Customer Relationship Management) has an important position, which is defined as a marketing strategy that integrates technology, process and all business activities around the customer [21]. Innovation also plays an important role in this area. In connection with the innovation have an important role employee. Correcting executive stocks can stimulate innovation as well as meet employee expectations and needs [22]. Among the tools to support innovation, primarily by the state, we include state support. Support for innovation takes the form of a targeted subsidy from public sources. Funds are provided from the budget chapter of the Ministry of Economy. Other sources of support for innovation include the European Union Structural Funds for co-financing innovative projects under the operational program and measures to support innovation under a special regulation and the Innovation Fund established under a special regulation [23]. Innovative enterprise is an enterprise which in its activities is based primarily on an innovation strategy. It is characterized by a high share of innovative products in the sales structure, the use of advanced technology and innovative management. The structural features of an innovative enterprise are high share of investment in research, know-how, product preparation and technology. In addition to the positive benefits of innovation, companies may also face the impact of innovation barriers that may limit innovation activity [24]. In the context of innovation, the basic task of the strategy is to determine the future image of the company through goals and innovative practices. The basic groups of strategic goals of innovation include minimization of fixed costs (investments), increasing the added value of products and services, reducing production costs. Marketing innovation means the introduction of a new marketing method or methods. These methods may include changes related to the design of the product or its packaging, or the promotion of the product and its location, but also include new pricing and service methods. The aim of marketing innovation can be to increase the attractiveness of products and services, or to enter new markets. What we no longer consider a marketing innovation are, for example, seasonal changes, advertising or publicity, unless it is an advertisement or advertisement that is first introduced and based on the use of a new media medium.

2 Methods

Innovation is important not only from the point of view of individual specific companies, but also from a global point of view. The level of innovation in individual companies creates an overall position in the field of innovation at the national, transnational and global level. One of the tools that provides a comparative assessment of the research and innovation performance in EU countries is the European Innovation Scoreboard [25]. The research is based on secondary findings from the European Innovation Scoreboard and from the Statistical Office of the Slovak Republic [26]. In the following part of the paper we analyze the development of Innovation index in Slovakia in the years 2012-2019 (Table 1) and Share of enterprises with innovation activity from the total number of enterprises in industry and selected services available from the Statistical Office of the Slovak Republic in 2012-2018 (Table 2). In the Results and Discussion section, we analyze the data from Table 1 and Table 2 using descriptive statistics in aim to make these data clear by graphical means.

We set 2 research questions:

Research question 1: What is the development of the Innovation Index in Slovakia?

Research question 2: What is the development of the structure of companies with innovation activity by size in Slovakia?

https://doi.org/10.1051/shsconf/20219101021

Table 1. Innovation index of Slovakia

Year	Score
2019	72,48
2018	70,55
2017	*68,21*
2016	71,3
2015	68,89
2014	67,24
2013	71,36
2012	70,42

Source: European Innovation Scoreboard 2020.

Table 2. Share of enterprises with innovation activity from the total number of enterprises in industry and selected services

	2012	2014	2016	2018
small businesses	29,80	28,50	24,60	25,80
medium businesses	40,00	37,90	42,70	39,10
large businesses	62,10	54,70	60,10	61,10

Source: The Statistical Office of the Slovak republic.

3 Results and Discussion

Research question 1: European Innovation Scoreboard shows individual data of innovation, one of indicator is Innovation Index, based on which individual countries are classified into 4 categories: Innovation Leader, Strong innovator, Moderate innovator and Modest innovator. Slovakia is currently based on data and various indicators identified as Moderate innovator with a score of 72.48. In comparison, at the top of this ranking and thus the top Innovation leader is Switzerland with a score of 179.73. Slovakia is in 26th place out of 37 European countries. Figure 1 shows the development of the score of the Innovation index of the Slovak republic in the years 2012-2019.

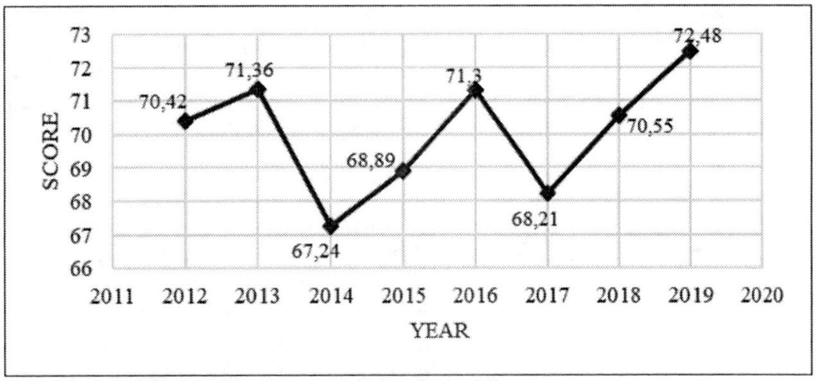

Fig. 1. Innovation index development of Slovak republic
Source: Own processing based on European Innovation Scoreboard.

In Figure 1 we see the development of the Innovation index of the Slovak republic. In the picture we can see that for the last year, ie 2019, the index had the highest value so far. The highest decline can be seen from 2013 to 2014, followed by an increase in the index until 2016. Subsequently, there was a decline again. Since 2017, the score has grown to its current value.

Research question 2: Figure 2 shows secondary data from the Statistical Office of the Slovak republic. This is the development of the percentage of enterprises with innovation activity in Slovakia according to the size of enterprises, showing 3 basic groups of enterprises: small enterprises (10-49 employees), medium enterprises (50-249 employees) and large enterprises (250 and more employees) in years 2012-2018.

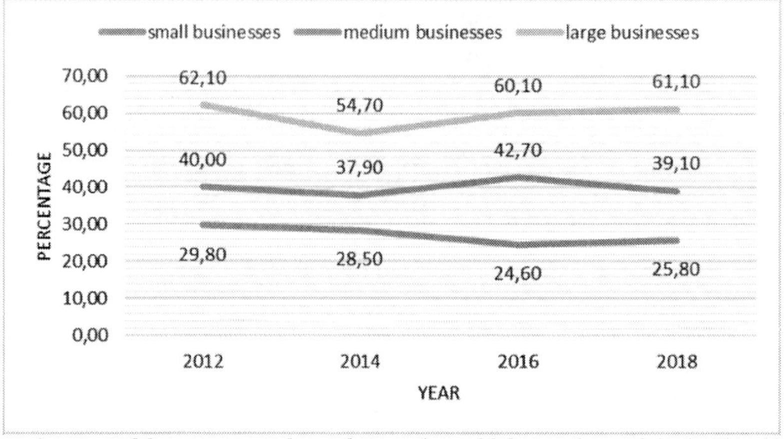

Fig. 2. Development of the percentage share of enterprises with innovation activity in Slovakia according to the size of enterprises
Source: Own processing based on the Statistical Office of the Slovak republic.

In Figure 2 we see the development of the percentage of enterprises with innovation activity in Slovakia according to the size of enterprises. Based on the data, we can see that in the years 2012-2014, the share of companies with innovation activity decreased in all types of companies. Subsequently, in the period 2014-2016, in the case of medium and large businesses, the share of enterprises with innovative activity increased and small businesses recorded a decrease in this period. In the period 2016-2018 large businesses and as well as small businesses recorded a slight increase. At medium businesses we see in this period decrease. Based on Figure 1, we can see that the innovation index did not increase continuously. In individual years, there were increases as well as decreases in this index. It is important that the innovation index of the Slovak Republic as well as other countries increases, therefore companies should include innovation activities in their business activities. Based on Figure 2, we can conclude that, despite scientific and technical progress and greater information, the last figure from 2018 is lower for all 3 groups of companies compared to the share of companies with innovation activity in 2012. The European Commission published the results of the European Innovation Scoreboard 2020 in June 2020, which shows that innovation performance continues to increase across the EU and is surpassed by the US for the second year in a row. However, more needs to be done to catch up with global innovation leaders such as South Korea, Australia and Japan. Results from the 2019 data point to opportunities for better coordination of EU innovation policies, helping Europe to improve its global competitiveness and strengthen the key role that innovation plays in overcoming the coronavirus pandemic [25]. Today's globalized and technology-influenced times provide a lot of room for innovation. This makes them among the most important sources of gaining a competitive advantage. The priority position in the market is given to those companies that can flexibly adapt to changing conditions and those that are interested in the field of innovation. Thanks to innovation, work is becoming more efficient, products are being developed and new products are being produced. The quality management requires effective innovation in the enterprise. A quick and flexible response to changing customer needs is also important. In recent years, the business environment has undergone significant changes. In addition to the traditional excess of supply over demand, the degree of globalization of the competitive environment is constantly increasing. Therefore, the efforts of many companies at present are not only focused on economy, but also on the development of their revenues by offering the customer extra added value. One way that may represent the full potential of this is the expansion of customer relationships. Limit our research is to use on the secondary dats of 2018, which are relatively active, but only one currently that are available. Another limit to research is the use of descriptive statistics. The main recommendation for further research is to examine the causes of declines in individual values between years, as well as suggestions and ways to improve the innovative position of companies.

4 Conclusion

If a company wants to be successful today, it is essential that it monitors the competition and adapts to new trends. Innovations are important in the context of the current situation in the market environment. Creating the right innovation strategy means ensuring that innovation in the company is long-term and effective. The innovation strategy needs to be created and integrated into various parts of the company and used to manage and develop the company's innovation performance. The aim of paper was to point out the theoretical basis and to analyze the available data on innovation in Slovakia. We analyzed the development of Innovation index of Slovakia and we could see that the innovation index did not have a continuous increase, because in individual years there were increases as well as decreases of this index. Slovakia is currently identified as Moderate innovator with a score of 72.48 and Slovakia is in 26th place out of 37 European countries. We then analyzed Development of the percentage share of enterprises with innovation activity in Slovakia according to the size of enterprises. Data show that the last figure from 2018 is lower for all 3 groups of enterprises compared to the share of enterprises with innovation activity in 2012. Businesses as well as whole countries should focus on increasing innovation activities by incorporating innovation activities into their business activities. In order for a company to be successful and keep pace with the market and provide its customers with the best products and services, it must constantly innovate. It's not just about innovating the products and services it offers businesses, but also about innovating processes, strategies, plans, and many other business-related things.

Acknowledgment

This paper is an output of scientific project VEGA no. 1/0718/18: The impact of psychographic aspects of pricing on the marketing strategy of companies across products and markets.

References

1. M. Nadanyiova, The Customer Satisfaction with Services Railway Company Cargo Slovakia as a Factor of Competitiveness. Proceedings of the 18th International Conference on Transport Means Location, pp.120-124 (2014)
2. L. Michalkova, K. Frajtova Michalikova, A. Tănase, Factor analysis and its application in innovation management within manufacturing enterprises in Romania. *Ekonomicko-manazerske spektrum*, **13**(2), 37-45 (2019)
3. M. Durica, L. Svabova, Improvement of Company Marketing Strategy Based on Google Search Results Analysis. *Procedia Economics and Finance*, 26, 454-460 (2015)
4. M. Jedlicka, Marketingove strategie. Trnava: Univerzita sv. Cyrila a Metoda v Trnave (2006)
5. V. Trommsdorff, F. Steinhoff, Marketing inovaci. Praha: CH Beck (2009)
6. S. Adisak, Logistics service innovation for business growth: a case study of logistics service entrepreneurs. *Polish Journal of Management Studies*, **21**(1), 394-407 (2020)
7. G. A. Pazilov, L. A. Bimendiyeva, N. P. Ivashchenko, A. N. Aitymbetova, Textile industry: issues of managing the growth of innovative activity in enterprises. *Polish Journal of Management Studies*, **21**(1), 297-315 (2020)
8. I. Chambers, D. Gray, D. Hall, Business studies. Harlow: Pearson Education (2010)
9. A. Krizanova, The current possition and perspecives of the integrated transport systems in Slovak republic. *Eksploatacja i Niezawodnosc – Maintenance and Reliability*, 4, 25-27 (2008)
10. J. Kliestikova, A. Krizanova, T. Corejova, P. Kral, E. Spuchlakova, Subsidies to increase remote pollution?. *Science and Engineering Ethics*, **24**(2), 755-767 (2018)
11. P. Kotler, Marketing Management. Praha: Grada (2013)
12. D. Lesakova, Strategicky marketing. Bratislava: Sprint (2014)
13. N. Kumar, Marketing jako strategie vedouci k uspechu. Praha: Grada (2008)
14. J. Majerova, T. Kliestik, Brand valuation as an immanent component of brand value building and managing. Procedia Economics and Finance, 26, 546-552 (2015)
15. M. Nadanyiova, Neuromarketing - An opportunity or a threat?. *Communications - Scientific letters of the University of Zilina*, **19**(4), 90-94 (2017)
16. P. Durana, A. Zauskova, L. Vagner, S. Zadnanova, Earnings drivers of Slovak manufacturers: Efficiency assessment of innovation management. *Applied Sciences*, **10**(12), 4251 (2020)
17. K. Valaskova, J. Kliestikova, A. Krizanova, Consumer perception of private label products: an empirical research. *Journal of Competitiveness*, **10**(3), 149-163 (2018)
18. OECD/Eurostat (2019), Oslo Manual 2018: Guidelines for Collecting, Reporting and Using Data on Innovation, 4th Edition, The Measurement of Scientific, Technological and Innovation Activities, OECD Publishing, Paris/Eurostat, Luxembourg.
19. P. Durana, K. Valaskova, L. Vagner, S. Zadnanova, I. Podhorska, A. Siekelova, Disclosure of strategic managers' factotum: behavioral incentives of innovative business. *International Journal of Financial Studies*, 8, 17 (2020)
20. A. Krizanova, J. Majerova, The proposal of activities of pricing policy in the process of building and managing brand value in Slovak Republic. Proceedings of the 2013 international conference on information, business and education technology (ICIBET 2013), pp.416-419 (2013)
21. A. Krizanova, L. Gajanova, M. Nadanyiova, Design of a CRM Level and Performance Measurement Model. *Sustainability*,**10**(7) 2567 (2018)

22. M. Zabolotniaia, Z. Cheng, Z. Dacko-Pikiewicz, Influence of leadership style on employees' innovative activity. *Polish Journal of Management Studies*, **20**(1), 478-496 (2019)
23. M. Kovac, Inovacie a technická tvorivosť. Kosice: SjF TUKE (2003)
24. A. Jakimowicz, D. Rzeczkowski, Do barriers to innovation impact changes in innovation activities of firms during business cycle? The effect of the Polish Green Island. *Equilibrium. Quarterly Journal of Economics and Economic Policy*, **14**(4), 631-676 (2019)
25. European Innovation Scoreboard 2020 [online] https://www.interactivetool.eu/ (2020)
26. The Statistical Office of the SR [online] http://www.statdat.statistics.sk (2020)

Impact of Credit Market Development on Project Financing

Monika Poradova[1,*] and *Katarina* Janoskova[1]

[1]University of Žilina in Žilina, Faculty of Operation and Economics of Transport and Communications, Department of Economics, Žilina, Slovakia

Abstract. The purpose of our report is to assess the prospects and sentiment for bank financing in the real estate sector in Europe, based on the views of bank representatives from 15 European countries. The presented paper is devoted to the issue of project financing. The aim is to provide an analytical overview of the issue of project financing as a specific credit financing instrument with the current access to finance for banks in Europe. The paper consists of four parts. The first part deals with the issue of project financing, the different dates of project financing and classifies project financing in terms of penalties. The second part of the paper aims to clarify the alternative method of project financing and analysis of banks in Europe. The third part of the article addressed the concrete results of the analysis and carried out discussion. Describes the volumes of credit transactions. The fourth part deals with the conclusion of the issue of project financing. A descriptive method, analysis, mathematical-statistical methods, graphic methods, comparison and synthesis was used in the processing of the paper submitted. The result of this article is that, in the future, the general financing conditions are expected to remain relatively favourable across Europe. This may be due in particular to competitive pressure contributing to milder credit conditions. However, trade tensions, declining investments, as well as policy concerns on a global scale, may curb such optimism for the next few years.
Keywords: project financing, banks, credit financing, investment project

1 Introduction

The first mentions of project financing have been recorded since the thirteenth century. Of course, this method of financing has undergone considerable development over several centuries and has been transformed into the form we know it today. In the twentieth century, project financing has seen significant expansion, particularly in America and Europe. Today, we can also meet with him in developing countries, where he has his justification for building infrastructure and development as a whole. Project financing is a form or also a technique by which we can finance a project. In practice, there are many types of projects where the form of project financing is applied. From the implementation

*Corresponding author: monika.poradova@fpedas.uniza.sk

of industrial projects, infrastructure development, energy, water management, real estate construction to internationally significant projects. [1]

Project financing is a specific form of financing which has some differences from other loans. Obtaining a loan is preceded by the creation of a separate economic unit specifically for a specific project. An interesting feature is the form of repayment of the credit, which is based on the flows that the project produces after its completion in the operation itself. [2] The implementation of large-scale projects requires close cooperation between several entities. The success of the whole project is conditional on the preparation detailed project, technical documentation, selection of right entities to participate in the project.

The growing interest in this form of funding has prompted authors, economists to dedicate and explore project financing. In publications, we may encounter several definitions of the concept of project financing, but by its very nature, we class it as part of investment banking. [3]

Project financing as individual financing of an economic unit, which first takes into account the future earnings and cash flow of the project from which the credit will be repaid. [4] Another definition refers to a legally independent project. It emphasises that the repayment of the credit is directly conditional on the fund's real institutions of the economic unit. Project financing involves the creation of a legally and economically independent project company, which is financed in a non-recourse manner, usually with a fixed maturity. Lenders and sponsors will view the economic unit as a single cash flow generator and a source of security and credit repayment. Creditors have no or limited penalty against the sponsor. [5]

Project financing is characterised by the separation of project financing from other business activities of the investment firm. The granting and repayment of the credit are closely linked to the cash flows of the project both at the construction stage and in its subsequent operation. The subject of financing is a large-scale long-term investment project, where the loan repayment scheme is derived from the project's planned revenues. At the forefront of the interest is the elimination of the risk of the creditor in the phase of construction of the project and implementation of the investment plan. [6]

Project financing is mainly linked to forecasting the effect of the selected project and relies only to a small extent on the company's own resources. The starting point is the existence of a large-scale investment project with a detailed project and financial documentation. With a special technical and financial-economic evaluation of the project by an independent expert. Separating project financing from other activities of the company will ensure the establishment of a special project company, which will focus all resources on the project envisaged. [7]

Bartosova, Kral talks in its publication about the common features of project financing and syndicated credit. Although project financing is essentially characterised by certain specificities, it advises it into a sub-group of syndicated loans. There is no exact line between project financing and other types of funding where a large amount of funding is needed to finance a project. Ultimately, we can categorize project financing as a form of structured financing falling within the subject of investment banking. [8]

1.1 Types of project financing

Project financing is individually tailored to the specific needs of the project. Efficiency is linked to the cash flow of the project and the total assets of the project. Risks are divided between shareholders and investors. Another criterion for a successful project is its ability to function independently as a legal and economic entity. [9] In project financing, we distinguish two different types, which differ from each other in the different application of recourse to investors in a given project. The essence of project financing lies in the fact that it is not possible to apply a full sanction to project investors, but only a limited sanction. Project funding without non-recourse and limited recourse are therefore one of the key factors underlying project funding.

1.1.1 Non-limited recourse project financing

Non-limited recourse project financing is a relationship between entities where project creditors and investors have no sanction on the project sponsor's assets. Non-retrospective financing, therefore, depends purely on the project's ability to generate cash flow as well as on the creditworthiness of the project sponsor. [10] Repayment of credit and liabilities, therefore, places the greatest emphasis on the project's projected cash flow, which serves as a source of repayment. Repayment of the credit is independent of the creditworthiness of the project sponsors. In this case, the project sponsor has no direct legal obligation to repay the project debt or interest payments. This type of project financing is relatively unique because it carries potential risks. [11]

1.1.2 Limited-recourse project financing

In most cases of project financing, the obligations and responsibilities of the project sponsor are limited, meaning that it is project financing with limited-recourse finance. [12] It provides creditors and investors with a certain recourse against project sponsors. In this case, the creditors are therefore entitled to demand repayment from the sponsor only in part or only at certain stages of the project. It can be said that the project sponsors undertake to end the project in the event of failure. The sponsors of the project are therefore persons who invest the share capital in a special entity. An important measure for 16 successful completion is to design the project documentation so that it can respond flexibly to unwanted changes that may occur. Nevertheless, investors are primarily focused on the success of the project. This second type of project financing is generally more widely used, especially in market developing countries.

The number of recourses needed to support funding depends on the unique characteristics of the project and its complexity. The large impact on the risk and the overall risk profile of a particular project depends on the extent of the aid to the investor. [13] The forms and methods of recourse can, of course, be the subject of negotiations between the bank, the creditor and the investor.

2 Methodology

Project financing is a progressive specific technique of credit financing of investment-intensive projects, which began to be applied especially in the 70s of the last century. Using this technique, it is possible to accumulate financial resources in a sufficient amount to finance infrastructure projects and at the same time redistribute investment and project risk among the participants. In developed economies, it is currently a common way of financing investment-intensive infrastructure and other projects.

2.1 Bank analysis

The banking sector has undergone many structural changes since the 1990s, which were necessary for its stability and efficient performance. In the mid-1990s, at the time of privatization, the banking sector was in a bad state in most countries and threatened to collapse the entire system and capital markets. Undercapitalization, low liquidity led to remedial measures and restructuring of the entire banking system. Project financing, which belongs to the investment banking portfolio, represents a dynamically developing form of financing. [14]

The aim is to provide an analytical overview of the issue of project financing as a specific credit financing instrument with the current access to finance for banks in Europe. As part of specific globalization processes, capital is not only focused on local markets but is moving to different parts of the world. Project financing is mainly associated with the financing of construction projects, but the possibilities of using this product are much wider. All the largest banks were involved in project financing. The analysed banks are mainly involved in financing projects in the area of:

- Real Estate: construction of apartment houses, administrative projects, industrial halls, logistics centres, hospitality, shopping and entertainment centres, sports parks,
- Energy projects: renewable energy sources, photovoltaic power plants, wind power plants, biomass heating plants,
- Municipal projects,
- Transport projects,
- Projects supported by EU funds. [15]

Depending on the survey participants' organisational structure, interviewees were the heads of real estate, project financing or risk management departments. Banks were selected from among the leading financial institutions operating in each individual country. The majority the survey participant banks were local, i.e. those operating predominantly within one European country, whilst the rest were mainly regional banks, as well some multinational banks participated.

The paper includes input from close to 70 banks, collected primarily via in-depth interviews and online questionnaires. Representatives from leading financial institutions have provided their views on the key issues influencing property lending. The questionnaire survey was conducted for the whole year 2019. The first half of 2020 the questionnaire survey was evaluated. The following countries are represented in the 2019 survey: Bulgaria, Croatia, the Czech Republic, Cyprus, Finland, Germany, Greece, Hungary, Ireland, the Netherlands, Poland, Romania, Serbia, Slovakia and Slovenia. The banks were selected from the leading financial companies of the institution operating in each individual country.

Selected countries were compared on the basis of geographical locations and two categories of analysis were created. The first is the "Economies of Central and Eastern Europe". These countries are located in Central and Eastern Europe, including most of the Balkans. Due to their geographical proximity and the relatively comparable stage of economic and development of these countries, they are assumed to be similar mainly due to their advantages and challenges. Therefore, the created group of countries consists of Bulgaria, the Czech Republic, Hungary, Poland, Slovakia, Romania.

Unlike Central and Eastern Europe, other European countries form the second category: "Other European Economies". Precisely because of their different economic capabilities, they provide a benchmark for evaluating data from Central and Eastern Europe. This category includes: Cyprus, Finland, the Netherlands, Ireland and Greece. In figure 1 is a geographical representation of the banks of the countries included in the sample.

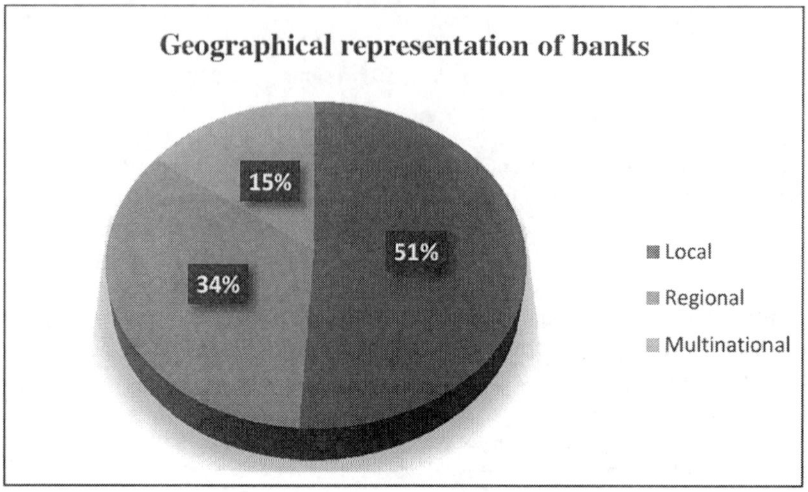

Fig. 1. Geographical representation of banks by countries included in the sample
Source: Own processing according to Lending Barometer 2019.

3 Results and Discussion

When financing investment plans through project financing, the greatest emphasis is placed on finding the optimal sources of project financing. When obtaining resources, it is most important to make the best use of the resources that are available and to obtain security from external sources as advantageously as possible. Traditionally, in this case, credit from one or more banks are used. Project financing involves obtaining a large amount of funds, so it is most often a combination of several sources. As credit as an external source of project financing is the most important and most frequent alternative for the implementation of such projects, the analysis will be based primarily on credit in selected countries.

First, the breakdown of the volume of credit transactions by individual countries of Central and Eastern Europe in 2018 and 2019 was analysed, which are shown in figure 2.

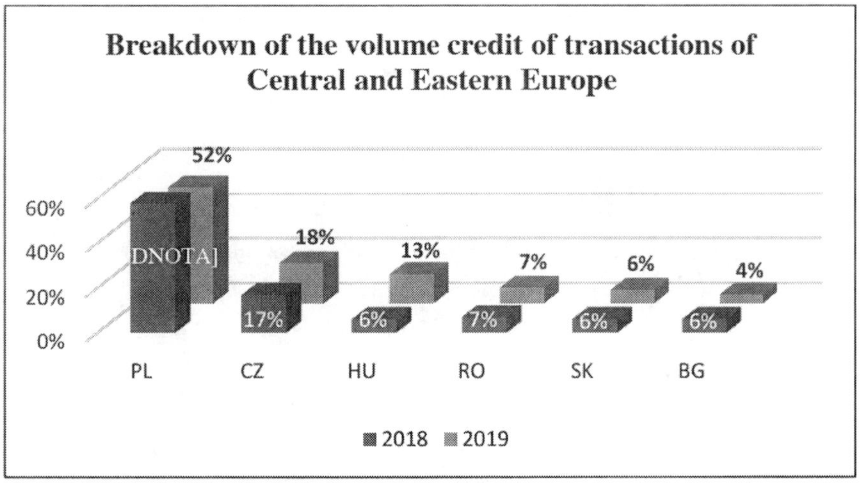

Fig. 2. Credit transaction volume by country of Central and Eastern Europe
Source: Own processing according to Lending Barometer 2019.

In the medium-term trend since 2014, the continuing growth of credit activity in Central and Eastern Europe is in almost all credit categories. In 2019, this trend was expected to continue, which proved to be correct assumption. This is mainly due to competitive pressure, which contributes to simpler credit conditions, especially in the larger countries of Eastern and Central Europe. Sustained quantitative easing by the European Central Bank has had a positive effect on many borrowing opportunities in Europe for many years. This was a major factor in improving the forecast for bank credit.

The increased level of activity in the credit portfolio market has maintained its dynamics over the last year. However, pressure on banks to consolidate their balance sheets and sell more of their outstanding credit portfolios remains a key driver of the market. As a result, traders have been closed in the European portfolio market amounting to approximately 50 billion in the first half of 2019.

Based on the analysis of the volume of credit transactions, an analysis was made of the investments that are most often provided through credit in Central and Eastern Europe, which are shown in figure 3. In this analysis, five main categories of credit were identified. These include Office, Retail, Industrial, Hotel and Residential.

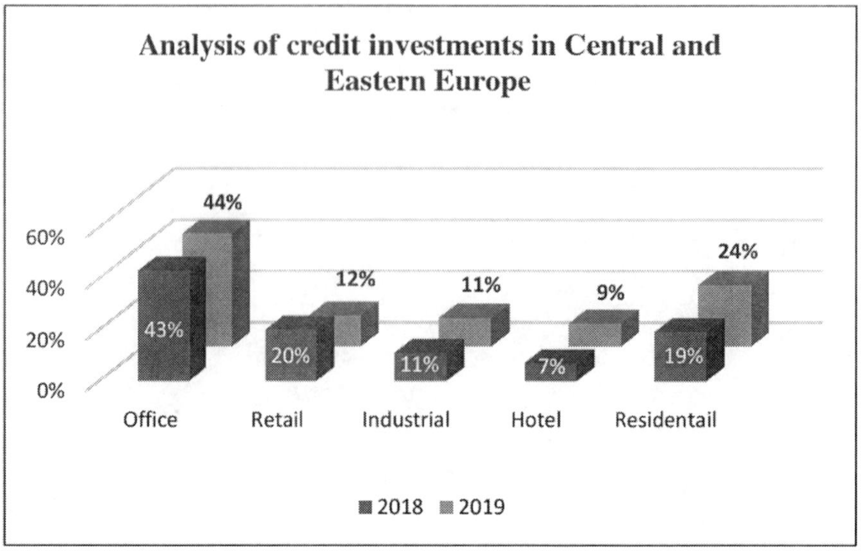

Fig. 3. Analysis of credit investments, which are most often provided through credit in Central and Eastern Europe
Source: Own processing according to Lending Barometer 2019.

In the breakdown of investment in Europe in the Residential sector, there was a remarkable increase in its share last year. Data for 2019 represent that the office sector is still the most important, reaching up to 44%. Subsequently, the retail sector reaches 20%, the industrial sector reaches 11% and the hotel sector reaches 9%. The retail sector suffered a decline of 51% compared to the decline in 2018. On the other hand, the hotel sector grew by 3%.

Sustained quantitative easing by the European Central Bank has had a positive effect on many borrowing opportunities in Europe for many years. This was a major factor in improving the forecast for bank credit.

3.1 Managing impaired credits

Banks apply a variety of credit measures to address the global economic crisis, including the restructuring, closing or sale of non-performing credit portfolios. Therefore, in this article, we have assumed that Western European markets are recovering faster and more widely in terms of disadvantaged loans than Eastern and Central Europe. The results of the analysis of the established assumption are shown in Figure no. 4 and 5. Western European markets are recovering faster and more widely in terms of bad credit.

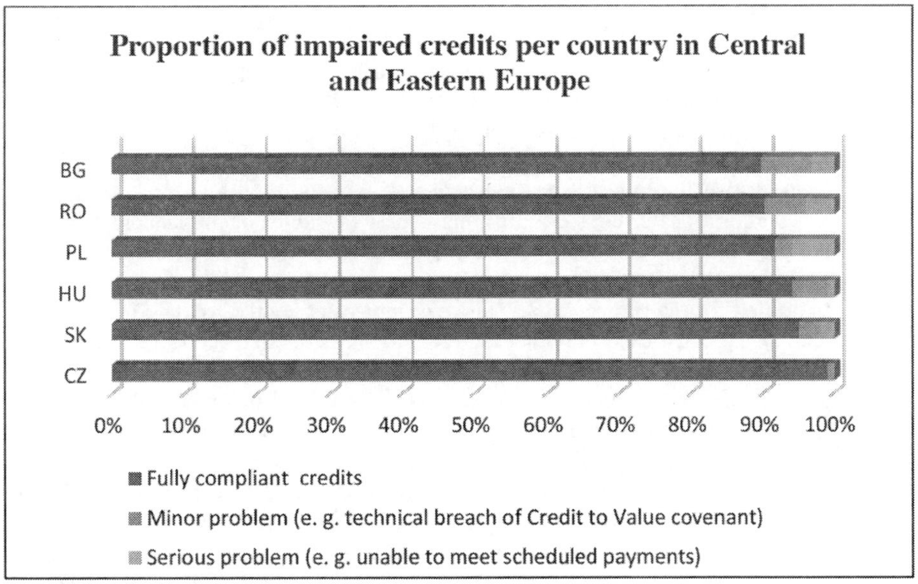

Fig. 4. Proportion of impaired credits per country in Central and Eastern Europe
Source: Own processing according to Lending Barometer 2019.

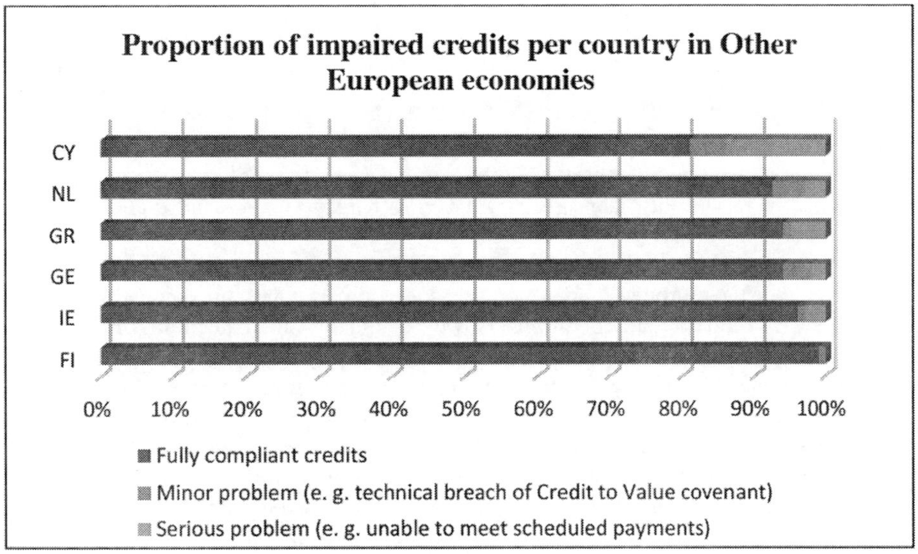

Fig. 5. Proportion of impaired credits per country in Other European economies
Source: Own processing according to Lending Barometer 2019.

The share of fully compliant credits is particularly high in Finland and the Czech Republic. With the current rate of fully compliant credits, exceed 99%. While only some countries, such as in our survey with a ratio of less than 90%, represent the country of Bulgaria and Cyprus. Other countries in the survey range on average between 90-98%.

The analysis of the set assumption confirmed the successful management of impaired loans in all countries, not only in Western European countries.

Project financing offers the creation of a separate economic unit for a specific project. In the case of the concept of non-recourse, the credit is repaid from the cash flow generated by a separate economic unit. In this way, the sponsor gets a significant advantage by not guaranteeing the repayment of the credit. In this case, the credit is covered by the project assets. [16] The advantage of project financing is also that the sponsor does not show any liabilities to the project in its balance sheet statement. The form of off-balance-sheet financing does not affect the overall indebtedness of the sponsor. This is especially important when the project is not developing favourably and could have a negative impact on the sponsors' financial profile.

In the event that the ratio of the sponsor's equity to the total resources needed for the project is disproportionate, we speak of a high share of debt. For the lender, this option is acceptable only after a thorough analysis of the entire project, when the project is high quality and viable. Lenders generally require a ratio of 25% to 55% of equity to the total resources needed for the project. [17] This requirement is related to the complexity of the project and the expected cash flow that the project will generate. The level of indebtedness in project financing is conditioned by the distribution and limitation of risks. When selecting, the sponsors take into account the risks, the size of the equity, the length of the credit repayment, the capital intensity of the entire project and its profitability. [18] When financing projects, the classic form of credit financing is often used. It is therefore up to the project sponsors what form of funding they will decide on with careful consideration of their options and expectations. [19] One of the main differences between project financing and traditional financing in the form of credit is in the process of the flow of funds that flow into a specific project. In credit financing, the resources provided by the creditor are transferred directly to the sponsor, who finances the project together with its own capital. [20] In the form of project financing, the creditor transfers external resources directly to the project, not through the sponsor.

4 Conclusion

This paper provides an overview of the European market as a whole, by focusing on key issues such as the strategic importance of real estate financing for banks, and the proportion of impaired loans. We also consider areas such as banks' average and preferred loan/deal size, as well as the length of loan contract term.

In the future, the general financing conditions are expected to remain relatively favourable across Europe. This may be due in particular to competitive pressure contributing to milder credit conditions. However, trade tensions, declining investments, as well as policy concerns on a global scale, may curb such optimism for the next few years.

When completing the questionnaire, bank representatives also answered the question of what other sectors they considered attractive in terms of financing. Overall, the food and manufacturing industries were selected from the answers, which banks prefer over financing development projects. Banks in general continue to offer more favourable terms when financing income-generating projects than for new developments. Among alternative lenders, respondents say that non-local commercial banks are banks' strongest competitors in most European countries, while private equity/debt funds provide an equally significant competitive threat in the other European economies' country group.

The preparation of project financing is an extremely complex and often costly process that requires a thorough examination of the future development of the project environment and individual solutions. Compared to traditional operating financing, it represents a more expensive way of financing projects, as financial institutions are aware of the increased risk, which is offset by a higher interest margin. Properly functioning project finance can help break down the long-standing barrier between business and financial institutions. Both parties are existentially dependent on each other. This creates a suitable space for closer cooperation and the joint realization of the goals of both parties. Financing through credits -

it means mostly by foreign sources, is more advantageous in the area of project financing than in the case of financing with own capital. Therefore, project financing has become an important tool in business practice, which is associated with the creation of new opportunities and the implementation of new investments.

The contribution is an output of the project VEGA 1/0619/20 Fundamental research of quantitative and qualitative determinants of enterprise innovation potential and innovation performance in relation to increasing its competitiveness.

References

1. E. Nica, A. M. Potcovaru, R. E. Hurdubei (Ionescu), Resilient cyber-physical systems and big data architectures in Industry 4.0: smart digital factories, automated production systems, and innovative sustainable business models. *Economics, Management, and Financial Markets*, **14**(2), 46–51 (2019)
2. W. Thierie, L. De Moor, Loan tenor in project finance. *International journal of managing projects in business*, **12**(3), 825-842 (2019)
3. D. Satish, EXIM Bank's Rampal Project Financing Dilemma. *Asian case research journal*, **23**(2), 237 (2019)
4. B. Aimurzina, M. Kamenova, A. Omarova, G. Bodaubayeva, A. Dzhunusova, G. Kabdullina, Major sources of financing investment projects. *Entrepreneurship and sustainability issues*, **7**(2), 1555-1566 (2019)
5. E. Alani, S. Kamarudin, L. Alrubaiee, R. Tavakoli, A model of the relationship between strategic orientation and product innovation under the mediating effect of customer knowledge management. *Journal of International Studies*, **12**(3), 232-242 (2019)
6. V. Oliinyk, O. Kozmenko, I. Wiebe, S. Kozmenko, Optimal control over the process of innovative product diffusion: the case of Sony corporation. *Economics and Sociology*, **11**(3), 265-285 (2018)
7. A. Jakimowicz, D. Rzeczkowski, Do barriers to innovation impact changes in innovation activities of firms during business cycle? The effect of the Polish Green Island. *Equilibrium. Quarterly Journal of Economics and Economic Policy*, **14**(4), 631-676 (2019)
8. V. Bartosova, P. Kral, A methodological framework of financial analysis results objectification in the Slovak Republic, *Proceedings of the 3rd International Conference on Business and Economics*, pp. 189-197 (2016)
9. D. Vrtana, A. Krizanova, Brand Strategy of Successful Companies. *Vision 2020: sustainable economic development and application of innovation management*, pp. 8007-8014 (2018)
10. P. Bednar, M. Halaskova, Innovation performance and R&D expenditures in Western European regions: divergence or convergence?. *Journal of International Studies*, **11**(1), 210-224 (2018)
11. E. Geroe, Non-recourse project financing for concentrated solar thermal power. *Utilities policy*, (2019)
12. P. Durana, K. Valaskova, L. Vagner, S. Zadnanova, I. Podhorska, A. Siekelova, Disclosure of strategic managers' factotum: behavioral incentives of innovative business. *International Journal of Financial Studies*, **8**, 17 (2020)

13. Y. N. Wang, X. H. Jin, Structural risk of diversified project financing of city investment company in China based on the best worst method. *Engineering construction and architectural management*, (2019)
14. P. Kral, V. Valjaskova, K. Janoskova, Quantitative approach to project portfolio management: proposal for Slovak companies. Oeconomia Copernicana, **10**(4), 797-814 (2019)
15. R. M. Visconti, D. Morea, Healthcare Digitalization and Pay-For-Performance Incentives in Smart Hospital Project Financing. *International journal of environmental research and public health*, **17**(7), (2020)
16. K. Ajaz Khan, G. Çera, V. Nétek, Perception of the selected business environment aspects by service firms. *Journal of Tourism and Services*, **10**(19), 111-127 (2019)
17. G. A. Pazilov, L. A. Bimendiyeva, N. P. Ivashchenko, A. N. Aitymbetova, Textile industry: issues of managing the growth of innovative activity in enterprises. *Polish Journal of Management Studies,* **21**(1), 297-315 (2020)
18. K. Valaskova, V. Bartosova, P. Kubala, Behavioural aspects of the financial decision-making. *Organizacija*, **52**(1), 22-32 (2019)
19. K. Valaskova, P. Durana, P. Adamko, J. Jaros, Financial compass for Slovak enterprises: modeling economic stability of agricultural entities. *Journal of Risk and Financial Management*, **13**(5), 92 (2020)
20. J. C. Hollowell, B. Kollar, J. Vrbka, E. Kovalova, Cognitive decision-making algorithms for sustainable manufacturing processes in Industry 4.0: networked, smart, and responsive devices. *Economics, Management, and Financial Markets*, **14**(4), 9–15 (2019)
21. KPMG [online], Available at: https://home.kpmg/xx/en/home.html (2019)

Comparison of Selected Subjects of the Non-profit Sector in the Conditions of the Slovak Republic

Zuzana Rosnerova[1,*] and *Dagmar* Hraskova[1]

[1]University of Zilina, Faculty of Operation and Economics of Transport and Communications, Department of Economics, Zilina, Slovak Republic

Abstract. Non-profit organizations are a specific type of entities in the national economy that significantly helps to improve the living standards of the country's population. The article focuses on the financial performance of entities in the non-profit sector of the Slovak Republic depending on the financial instrument of tax allocation, taking into account the time factor during which the entities operate in the non-profit sector. The aim of the article is to analyse the financial performance of non-profit organizations providing services of general interest and foundations with the subsequent specification of the results of the issue. We work with the hypothesis that in terms of financial performance, foundations will perform better compared to non-profit organizations. Segmentation, comparison and descriptive methods were used in the solved issue. Only in one of the analysed groups did we come to the result that the non-profit organization was able to obtain more funds than the foundation through the tax allocation tool in 2019. In the case of the other groups analysed, foundations were more financially efficient, whose performance significantly exceeded that of non-profit organizations. In terms of market life, based on the analysis, the most efficient organizations were organizations with a market life of 1-5 years and with a longer market life, their financial performance decreased. In the case of foundations, we found that the highest performance is achieved by foundations with a market existence of 15 - 20 years. We can therefore state that the established hypothesis was only partially confirmed, although in most cases foundations were more financially efficient.
Keywords: non-profit organization, performance of non-profit subject, tax allocation, non-profit sector

1 Introduction

In the national economies of countries, there is a certain area of organizations or institutions that deal with specific activities, and ultimately these organizations are very beneficial to certain groups of end users, whom they help in many areas. These are organizations or institutions that belong to the non-profit sector and their existence is linked

*Corresponding author: zuzana.rosnerova@fpedas.uniza.sk

to their mission. Their basic goal is to cover a certain segment through the direct production of benefits, which takes the form of a service that is provided publicly [1]. Within their scope, non-profit organizations provide a different type of services compared to business entities, the primary difference being that the services are provided without the right to payment [2]. According to Novotny and Lukes, a non-profit organization fills the space that arises between the state, the market and the family and at the same time serves to capture a wide range of civic activities through which it contributes to positive changes in people's way of life [3]. Incentives that precede the emergence of non-profit entities stem from the situation and needs of a particular state, as well as their emergence depends on the motivation of their founders and the purpose for which they will serve. A specific feature that significantly distinguishes these organizations from the group of business entities is the fact that the primary goal of these organizations is not to make a profit, but to provide a service that is characterized by its public benefit characteristics. This means that such a service is provided to the user free of charge [4]. According to the author Tetrevova [5] a non-profit organization performs several functions, which include functions: social, participatory, political, and at the same time economic, informational and educational functions.

The non-profit sector includes several types of non-profit entities, the basic selection of which is offered by the author Rektorik. According to him, the typology of non-profit entities is as follows:

1. private non-profit organizations mutually beneficial,
2. private non-profit organizations of public benefit,
3. non-profit public organizations such as organizational units and contributory organizations of the state, or organizations of self-governing territorial units,
4. other public non-profit organizations,
5. private non-profit organizations such as companies [6].

The Slovak Republic distinguishes the following types of non-profit organizations in terms of legal form: associations, foundations, public benefit organizations and non-investment funds.

A frequently discussed topic is the issue of financing non-profit organizations, which is very specific in nature. A non-profit organization can make a profit during its operation, but it must never be divided among its founders and must be spent solely on ensuring the mission and purpose of the non-profit subject. The financial factor is also very closely connected with the performance of a non-profit organization and can also fulfill a certain descriptive function in terms of its performance. When dealing with the performance of a non-profit entity, it is possible to take into account to a large extent the amount of funds that the non-profit entity was able to obtain for a specific period for the needs of financing its subsequent operation and maintenance and survival in the market [1]. In connection with the issue of financing, in the environment of non-profit entities we encounter the concept of fundraising. Fundraising is essentially an opportunity for non-profit entities to obtain financial and other resources to ensure their activities using various methods and procedures [7]. In the case of fundraising, the question is addressed as to how it is possible to obtain external support and thus secure resources for self-financing. In their publication, the authors Vrtana and Krizanova describe the importance of spreading enlightenment about his existence and the possibility to convince others of the importance of his activities and how important the organization is for society as such [8]. The goal of fundraising is basically to gain people's interest, their time and their trust. In practice, not only one source is sufficient for the needs of financing a non-profit entity, but the principle of multi-source financing is used. In this case, it is true that the more supporters - donors there is for a non-profit entity, the better and that is why non-profit entities try to develop multi-source funding to a large extent [9]. In her publication, the author Bartosova mentions as the most common sources of financing forms such as monetary donation, which is one of direct support, endowment funds or non-monetary donation [10]. In practice, however, there are often cases where non-profit entities are very burdened by a lack of funding and in general face problems with their funding [11]. The author Valaskova considers behavioural elements of decision-making to be a very important factor in solving financial issues, which can serve as a means to justify why certain market areas are created as inefficient for a non-profit organization [12].

If we compare the system of financing non-profit organizations and companies, entities of the non-profit sector are at a significant disadvantage in terms of monitoring financial performance. In the case of companies, this performance is easier to monitor and analyse through various financial indicators [13]. The risk of financial distress has a negative impact on their operations both in the case of companies and in the case of non-profit entities [14]. The author Shemyatikhina considers the influence of marketing tools through which non-profit entities can spread awareness about their mission to be an important element in addressing their performance, and in general their performance can be supported [15].

A non-profit entity can be seen as a kind of dynamic structure. This dynamism stems from the fact that the non-profit entity is subject to certain changes that occur in the national economy in which the entity operates and thus constantly evolves and develops, eventually declines and thus responds to stimuli from the environment. Basically, every non-profit entity is characterized by its life cycle. The author Carvalho emphasizes in the life cycle of organizations to focus as much as possible on their responsibility as a key factor in securing their mission, through which they can influence the time of their existence in the market [16].

In his publication, Alani emphasizes the strategic solution to the problematic issues facing the organization, and according to him, through such an approach, the organization can innovate. These are an integral part of the advancement and, in general, the survival of the subject on the market [17]. In connection with the implementation and evaluation of the success of the strategic approach to the issues addressed, the use of artificial intelligence is justified [18].

Based on the specified criteria, it is possible to determine at what stage of development the non-profit entity is and what development trends await it in the future. The life cycle of an organization can be represented by a curve that is compared to the silhouette of an "elephant's back", as shown in the following figure. The horizontal axis shows time and the vertical axis captures qualitative changes - that is, the progress or decline of the organization.

Fig. 1. The elephant's back as the life cycle of a nonprofit organization
Source: [19]

The aim of the article is to analyse the performance of selected entities of the non-profit sector through the most striking factor for ensuring the operation - funding through the collection of funds, i.e. through tax allocation as the most important means of financing to ensure the mission of non-profit entities.

2 Methods

In the article we deal with the analysis of the performance of selected subjects of the non-profit sector. As part of the analysis, we decided to compare the performance of selected entities in the non-profit sector, focusing primarily on the financial side of these entities in 2019, which operate in the Slovak Republic. From among the entities of the non-profit sector, we have chosen two groups of entities of the non-profit sector, namely foundations and non-profit organizations that provide services of general interest. The hypothesis was determined as an assumption that foundations will be stronger in terms of their financial performance. Comparative and descriptive methods were used in this analysis. To carry out the conclusions of the analysis, we selected entities in terms of their time in the non-profit sector in the conditions of the Slovak Republic. In the analytical part, we worked with the available data of these subjects, taking into account the subjects placed in the first three positions from each selection group. The data come from the financial statements of non-profit organizations. The selection groups for performing the analysis are as follows:

- non-profit organizations providing services of general interest and foundations with a duration of 1-5 years,
- non-profit organizations providing services of general interest and foundations with a period of operation of 5-10 years,
- non-profit organizations providing services of general interest and foundations with a period of operation of 10-15 years,
- non-profit organizations providing services of general interest and foundations with a duration of 15-20 years,
- non-profit organizations providing services of general interest and foundations with a duration of 20-25 years.

1. **Selection Group: Non-profit organizations providing services of general interest and foundations with a period of operation 1 – 5 years**

Table 1. Volume of funds raised by non-profit organizations providing services of general interest through tax allocation in 2019

Non-profit organizations providing services of general interest	
Name of non-profit organization	Assignment of 2% income tax, year 2019
To children with cancer	117 289 €
Man in danger	20 354 €
KUBKO	20 008 €
Foundations	
Name of foundation	Assignment of 2% income tax, year 2019
X-BIONIC®SPHERE Foundation	64 892 €
F.P. Mitterbauer Foundation	33 445 €
Gabor	17 106 €

Source: Own processing.

2. **Selection Group: Non-profit organizations providing services of general interest and foundations with a period of operation 5 - 10 years**

Table 2. Volume of funds raised by non-profit organizations providing services of general interest through tax allocation in 2019

Non-profit organizations providing services of general interest	
Name of non-profit organization	Assignment of 2% income tax, year 2019
Liberta	233 698 €
The light of help	97 390 €
ViaVitae	34 043 €
Foundations	
Name of foundation	Assignment of 2% income tax, year 2019
Kia Motors Slovakia	1 423 798 €
EPH Foundation	1 300 497 €
Adeli Foundation	1 205 344 €

Source: Own processing.

3. **Selection Group: Non-profit organizations providing services of general interest and foundations with a period of operation 10 - 15 years**

Table 3. Volume of funds raised by non-profit organizations providing services of general interest through tax allocation in 2019

Non-profit organizations providing services of general interest	
Name of non-profit organization	Assignment of 2% income tax, year 2019
The good angel	363 606 €
DEPAUL SLOVAKIA	62 334 €
People to people	36 155 €
Foundations	
Name of foundation	Assignment of 2% income tax, year 2019
VOLKSVAGEN SLOVAKIA Founation	1 597 497 €
Orange Foundation	662 984 €
Allianz Foundation	496 037 €

Source: Own processing.

4. **Selection Group: Non-profit organizations providing services of general interest and foundations with a period of operation 15 - 20 years**

Table 4. Volume of funds raised by non-profit organizations providing services of general interest through tax allocation in 2019

Non-profit organizations providing services of general interest	
Name of non-profit organization	Assignment of 2% income tax, year 2019
Flame	872 023 €
For Lumen	51 265 €
Ronald McDonald House Charity	34 515 €
Foundations	
Name of foundation	Assignment of 2% income tax, year 2019
SPS Foundation	1 143 912 €
VUB Foundation	1 039 912 €
Slovak Savings Bank Foundation	1 007 046 €

Source: Own professing.

5. Selection Group: Non-profit organizations providing services of general interest and foundations with a period of operation 20 – 25 years

Table 5. Volume of funds raised by non-profit organizations providing services of general interest through tax allocation in 2019

Non-profit organizations providing services of general interest	
Name of non-profit organization	Assignment of 2% income tax, year 2019
The sun center	13 282 €
New house	11 600 €
The rehabilitation center - HARMONY	9 522 €
Foundations	
Name of foundation	Assignment of 2% income tax, year 2019
SPS Foundation	2 661 939 €
VUB Foundation	374 134 €
Slovak Savings Bank Foundation	211 955 €

Source: Own processing.

3 Results

Based on the analysis of selected subjects, we came to the results, which are described below according to the individual selection groups.

1. Selection Group: Non-profit organizations providing services of general interest and foundations with a period of operation 1 – 5 years

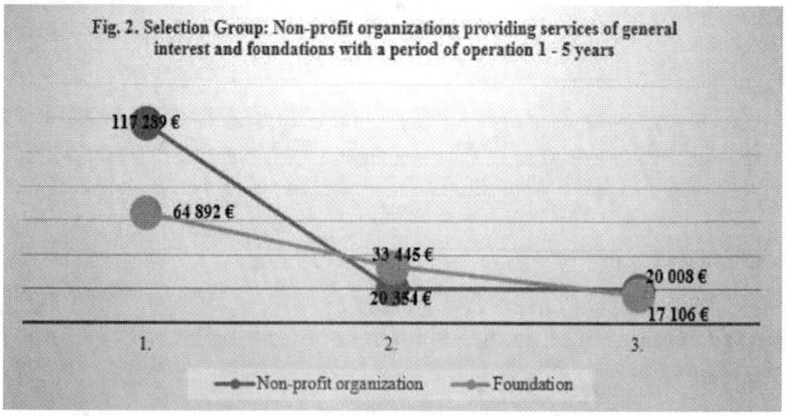

Fig. 2. Obtained amounts of funds of non-profit organizations providing services of general interest and foundations during their existence on the market in the range of 1-5 years.
Source: Own processing.

The difference in performance in terms of contributions collected through the form of tax allocation is the strongest in the case of non-profit organizations in the first - leading position in the national economy compared to the foundation. The non-profit organization To children with cancer was able to collect € 117,289 through this form of funding, which in relative terms is 44,67 % more than the foundation, which in 2019 occupied the leading position in this category in the non-profit sector in Slovakia. Non-profit organizations providing services of general interest in the second and third place noted the following differences. In the second place, the situation was more favourable in terms of performance in the case of the foundation, which in 2019 raised more funds through tax allocation compared to a non-profit organization providing services of general interest, namely by 39,14 %. In the third position, there was a difference in the funds obtained in favour of a non-profit organization providing services of general interest, which raised 14,50 % more funds compared to the foundation.

2. Selection Group: Non-profit organizations providing services of general interest and foundations with a period of operation 5 - 10 years

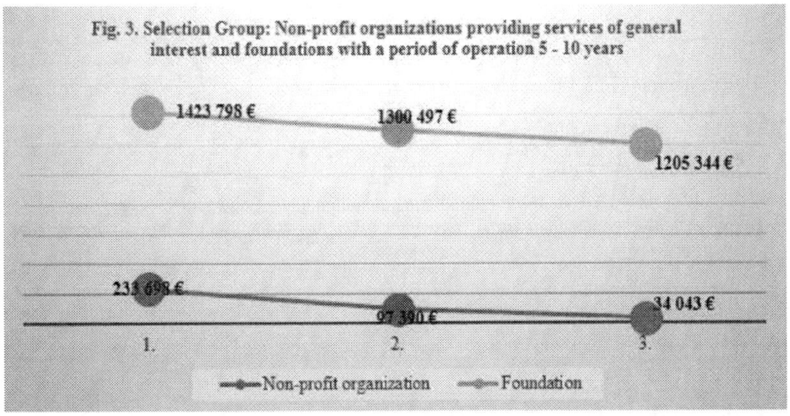

Fig. 3. Obtained amounts of funds of non-profit organizations providing services of general interest and foundations during their existence on the market in the range of 5 - 10 years.
Source: Own processing.

According to the graph, it is clear that foundations were able to raise more funds in all three positions through tax allocation compared to non-profit organizations. In the first place in 2019, the foundation led, which received 83,59 % more funding than a non-profit organization. In the second place, the foundation led by 92,51 % in the lead in the collected funds. In the third position, the foundation collected 97,18 % more than a non-profit organization.

3. Selection Group: Non-profit organizations providing services of general interest and foundations with a period of operation 10 - 15 years

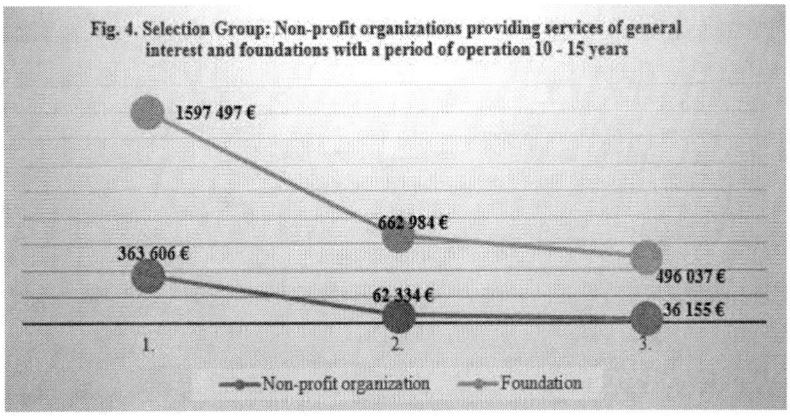

Fig. 4. Obtained amounts of funds of non-profit organizations providing services of general interest and foundations during their existence on the market in the range of 10 – 15 years.
Source: Own processing.

In the case of non-profit sector entities with a period of 10-15 years, foundations were able to obtain more funds in comparison with non-profit organizations providing services of general interest. In the first position, the foundation was able to collect up to 77,24 % more compared to a non-profit organization. In the second place, the difference in percentage expression was 90,60 % higher in favour of the foundation. In the third position, the foundation raised 92,71 % more funds than a non-profit organization.

4. Selection Group: Non-profit organizations providing services of general interest and foundations with a period of operation 15 - 20 years

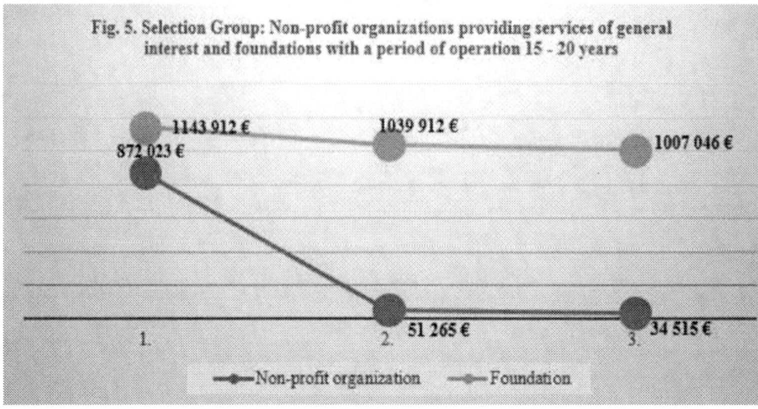

Fig. 5. Obtained amounts of funds of non-profit organizations providing services of general interest and foundations during their existence on the market in the range of 15 – 20 years.
Source: Own processing.

Within the fourth selection group, they again led the foundations in the volume of funds raised. In the comparison of both entities in the first position was won by the foundation, which collected 23,77 % more than the non-profit organization. In the second place, a deep drop in the non-profit organization was recorded, as the foundation managed to raise 95,07 % more money. The third position brought a difference, which represented up to 96,57 % greater success in the contributions received for foundations.

5. Selection Group: Non-profit organizations providing services of general interest and foundations with a period of operation 20 - 25 years

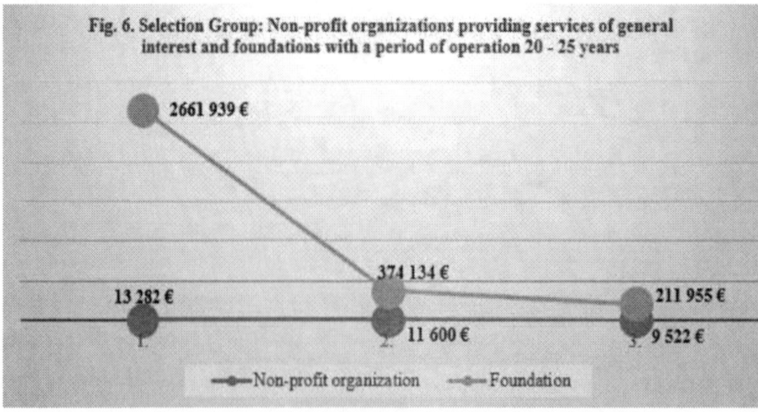

Fig. 6. Obtained amounts of funds of non-profit organizations providing services of general interest and foundations during their existence on the market in the range of 20 – 25 years.
Source: Own processing.

The last analysed group shows a deep drop in non-profit organizations compared to foundations. The foundation, with a period of operation on the market of 20-25 years, achieved up to 99,50 % higher inflows of funds than a non-profit organization. In the second position, the foundation received 96,90 % more funds, in the third position it was 95,51 % more compared to a non-profit organization providing services of general interest.

4 Discussion

Only in one of the analysed groups did we reach a result where the non-profit organization was able to raise more funds through the tax allocation tool for 2019 than the foundation. Such a result occurred in the first analysed group with the existence of the subjects 1 - 5 years in positions 2 and 3. In other cases, the first place was always won by the foundation, which significantly exceeded the volume of funds collected by the foundation in individual cases. Regarding the evaluation of the financial performance of non-profit entities, we can state that in view of the extent of contributions collected, the vast majority of foundations were more successful and thus their performance is clearly stronger than in the conditions of non-profit organizations providing services of general interest. If we wanted to evaluate the performance of non-profit entities in terms of market life, we can say that the strongest non-profit organizations are in the category of life on the market 1 - 5 years. Foundations with a period of operation of 15 to 20 years were able to raise the most funds, and a significant amount of funds was also obtained by the foundation in the first position with a period of existence on the market of 20 to 25 years. In the last analysed group - non-profit entities with a period of operation of 20-25 years, we can see a significant decline in non-profit organizations. In each of the above positions, there is a significant drop in performance of these entities compared to the foundation.

Based on the above analysis, we can state that in terms of market life, the most productive non-profit organizations in the group with a market life of 1 - 5 years. On the contrary, their productivity is weakened with increasing market life. In the case of foundations, we could state that in the existence of 5 - 10 years and in the period of 10 - 15 years there is a significantly higher performance and as can be seen from Fig. 6. and in the case of the Pontis Foundation, which has a market life of 20-25 years, the performance of that foundation is very strong. In terms of building a lasting position in the market, the author Truffnell emphasizes the need to constantly monitoring the processes that affect the performance of each entity [20].

5 Conclusion

The issue of the article focuses on the performance of selected entities of the non-profit sector of the Slovak Republic in terms of the ability to secure funding to ensure its operation and existence in the market through the most important tool of these entities - tax assignment. Based on the analysis and subsequent comparison of the results of the analysed entities, we came to the conclusion that in the initial period of existence, i.e. 1-5 years were financially more efficient non-profit organizations, but in other periods of the foundation achieved better results compared to non-profit organizations providing services of general interest. In the last group of entities with a period of existence on the market of 20 - 25 years, we came to the result that non-profit organizations lagged significantly behind the foundations and lose their performance. Based on the findings, we can state that the established hypothesis was not confirmed in all cases in terms of the existence of the entity on the market.

The paper is an output of the science project VEGA 1/0544/19 Formation of the methodological platform to measure and assess the effectiveness and financial status of non-profit organizations in the Slovak Republic.

References

1. J. Crotty and S. Ljubownikow, Restrictive Institutions and Critical Resources: Non-Profit Organisations and Volunteer Resources in the Russian Federation. *Europe-Asia Studies,* **72**(1), 99-116 (2019)
2. K. Ajaz Khan, G. Çera, V. Netek, Perception of the selected business environment aspects by service firms. *Journal of Tourism and Services,* **10**(19), 111-127 (2019)
3. J. Novotny and M. Lukes et al., *Success factors of non-governmental non-profit organizations*. Prague: Oeconomica (2008)
4. W. Zheng, N. Ni and D. Crilly, Non-profit organizations as a nexus between government and business: Evidence from Chinese charities. *Strategic Management Journal,* 10(4), 658-648 (2019)
5. L. Tetrevova, *Verejna ekonomie*. Prague: Professional Publishing (2008)
6. J. Rektorik et al., *Non-profit sector organization*. Prague: Ekopress (2007)
7. H. Majduchova, *Non- profit organizations*. Bratislava: Sprint dva (2009)
8. D. Vrtana and A. Krizanova, Brand Strategy of Successful Companies. *Sustainable Economic Development And Application Of Innovation Management,* 8007-8014 (2018)
9. R. Rajnoha and S. Lorincova, Strategic Management of Business Performance Based on Innovations and Information Support in Specific Conditions of Slovakia. *Journal of Competitiveness*. **7**, 3–21 (2015)
10. Z. Bartosova, *Guide to corporate philanthropy*. Bratislava: Donors forum (2007)
11. T. Andjarwati, K. E. Susilo, A. K. Audah, Predictors of job satisfaction in non-profit organization. *Polish Journal of Management Studies,* **20**(1), 19-28 (2019)
12. K. Valaskova, V. Bartosova, P. Kubala, Behavioural Aspects of the Financial Decision-Making. *Organizacija,* **52**(1), 22-32 (2019)
13. J. Salaga, V. Bartosova, E. Kicova, Economic Value Added as a measurement tool of financial performance. *Procedia Economics and Finance,* **26,** 484-489 (2015)
14. K. Valaskova, P. Durana, P. Adamko, J. Jaros, Financial compass for Slovak enterprises: modeling economic stability of agricultural entities. *Journal of Risk and Financial Management,* **13**(5), 92 (2020)
15. L. Shemyatikhina, K. Shipitsyna, M. Usheva, Marketing management of a non-profit organization. *Ekonomicko-manazerske spektrum,* **14**(1), 19-29 (2020)
16. A. O. Carvalho, M. R. Ferreira, P. A. Silva, Partners in a caring society – a nonprofit organization case study. *Economics and Sociology,* **12**(2), 129-146 (2019)
17. E. Alani, S. Kamarudin, L. Alrubaiee and R. Tavakoli, A model of the relationship between strategic orientation and product innovation under the mediating effect of customer knowledge management. *Journal of International Studies,* 12 (3), 232-242 (2019)
18. E. Nica, R. Miklencicova, E. Kicova, Intelligence-supported Workplace Decisions: Big Data Algorithmic Analytics, Sensory and Tracking Technologies, and Metabolism

Monitors. *Psychosociological Issues in Human Resource Management,* **7**(2), 31–36 (2019)

19. D. Ondrusek et al., *Reader for non-profit organizations*. Bratislava: Center for Conflict Prevention and Resolution (1998)
20. C. Tuffnell, P. Kral, P. Durana, T. Krulicky, Industry 4.0-based manufacturing systems: smart production, sustainable supply chain networks, and real-time process monitoring. *Journal of Self-Governance and Management Economics,* **7**(2), 7–12 (2019)

The usage of the CSR concept and its communication in Slovak companies in the process of globalization

Margareta Nadanyiova[1,*], *Jana* Majerova[1], and *Lubica* Gajanova[1]

[1]University of Zilina, Faculty of Operation and Economics of Transport and Communications, Department of Economics, Univerzitna 8215/1, 010 26 Zilina, Slovak Republic

Abstract. The socially responsible communication is focused on the visibility of the social, economic and environmental responsible activities of the company through a variety of communication tools. In the reason of the gradual globalization that brings many changes whit the impact on the business and the environment including the growing importance of socially responsible corporate behaviour, the paper is aimed on the usage of the CSR concept in Slovak companies and the determination of their communication in the process of globalization. This includes providing the theoretical background and analysis of the corporate social responsibility and its communication from the viewpoint of Slovak and foreign authors. The important source for secondary data was scientific researches, statistical databases, published professional publications. In order to determine the level of usage, the CSR concept and its communication in companies in Slovak Republic, a questionnaire survey was conducted among Slovak consumers. Based on the analysis and results of the questionnaire survey, benefits of the efficient socially responsible company's communication are highlighted such as gaining competitive advantage, increasing the brand value, improving customers loyalty and relationships with stakeholders.
Keywords: corporate social responsibility (CSR), socially responsible company, socially responsible communication, stakeholders, sustainability, globalization, company

1 Introduction

Globalization and internationalization have brought about changes that have affected the business and the environment. At present, the process of globalization and internationalization continues and affects all areas of social life. In these changed conditions the importance of socially responsible corporate behaviour is growing [1].

Thus, it is obvious, that the issue of corporate social responsibility is currently very topical and is becoming the focus of attention of companies, consumers, media and the whole society as well. Companies are expected to use their power and financial resources to influence the surrounding environment [2]. Companies that are referred to as socially responsible therefore

* Corresponding author: margareta.nadanyiova@fpedas.uniza.sk

strive not only to meet traditional economic goals, but also to meet the social and environmental aspects of their activities, which in practice manifests itself by voluntarily setting high ethical standards, avoiding corruption, minimizing the negative impacts of their business to the environment, effort to be a good employer, support the region in which they operate [3]. They strive to stand on the three basic pillars in the long run and as firmly as possible – economic efficiency, environmental responsibility and social commitment. CSR is thus becoming not only an important means for companies to be competitive in the long run, but also for communities and society to develop successfully on a lasting basis [4, 5].

The concept of CSR was developed mainly in countries with stable market economies such as USA, Canada and Western Europe, where it became part of the management of many companies [6, 7]. In the beginning, companies focused mainly on corporate philanthropy, social security for their employees, as well as assistance to local governments and sponsorship.

The situation in Slovak Republic as a country of post-communist Eastern Europe was different due to political and economic influences. It was paid less attention to the corporate social responsibility in contrary to developed countries. Despite this fact, the concept of social responsibility has gradually begun to spread to companies, especially on their own initiative with minimal support from civil society, media and public authorities. Later, after the reform in Slovak economy, the country opened up to the world and thus the inflow of foreign investment was increased from large investors who have brought CSR principles from their home countries into business strategies in Slovak conditions.

Currently in Slovak Republic, CSR principles are applied not only in global companies but increasingly in the home businesses. However, when implementing CSR principles in Slovak Republic, it is necessary to take into account its specific conditions that are different than in developed countries. The Slovak Republic is unique, it has the distinctive culture and history. Thus, the implementation of CSR principles is different than in other countries. The reason is especially the level of national economy as well as overall social system of the Slovak Republic.

According to Lewicka-Strzalecka, this concerns the phenomenon of rationality asymmetry when rational procedures are developed by the institution for a market with different characteristics and historical experience introduced in post-communist countries. It means, that when some of the mentioned countries try to implement effective solutions transferred from developed economies, this will not be successful, but on the contrary, poverty as well as social differences will increase. Thus, the promotion and implementation of CSR activities should be adapted to the country's specific conditions [8].

For many companies, CSR is an important part of their corporate strategy and represents a tool to improve image, competitive advantage or the way to differentiation. CSR contributes to the creation of the company value as well as the brand and builds customer loyalty and post-purchase behaviour. According to Hansen, the reward for the company is to influence the customer's positive awareness of the company and their shopping behaviour. The customer is the central determinant of who will be "rewarded" and who will be "punished". Through its purchasing decisions, it influences the profits of companies competing in the market and, indirectly, the direction in which the economy is heading. Not only companies benefit from CSR. Other stakeholders also become profiteers [9].

In particular, several authors highlight the following most important practical benefits for companies implementing the principles of the CSR concept in their corporate strategy [3, 10]:

- CSR contributes to creating a good business environment
- CSR increases the attractiveness for investors
- CSR builds the trust and reputation of the company as well as the brand in the eyes of the public

- CSR allows to gain a competitive advantage and increases customer loyalty
- CSR reduces costs
- CSR increases profits in the long run
- CSR improves risk management and reduces risk management costs
- CSR can attract new partners
- CSR enables the achievement of the status of "preferred supplier"
- CSR makes it possible to increase the efficiency of supplier-customer relationships.
- CSR increases the value of intangible assets, such as company reputation, brand value, human capital or relationships of trust and partnership

As mentioned before, nowadays many companies in Slovak Republic deal with the topic of social responsibility. They try to invest part of their profits in meaningful projects that will help a good cause. However, in order for a company's CSR activities to be valuable to customers, they must first be informed about them. For companies, therefore, the communication of socially responsible activities is as important as its own implementation. In this context, the paper is focused on the usage of the CSR concept in Slovak companies and the determination of their communication in the process of globalization.

Already when creating a CSR strategy, it is necessary to deal with communication of the company's socially responsible activities, because this can significantly contribute to the streamlining of CSR corporate policy. The appropriate communication with target groups represents the key factor of the socially responsible company success [11]. Without efficient and properly targeted communication of socially responsible activities, the company cannot achieve its goals. The socially responsible communication is also necessary when creating a corporate identity, reputation and image improvement. Thus, the CSR philosophy needs to be incorporated directly into the company's mission, vision and strategy. By communicating its CSR efforts, the company provides people with information about the company's values as well as its product portfolio. In addition, thus, it offers a positive example for other companies. Thus, socially responsible communication means for the company to clearly define its values and formulate a statement that will concisely summarize its approach to socially responsible business. This step gives the company an opportunity to think about what is important and beneficial for it and for the whole society. Currently, there is enormous influence of the digital marketing on the way customers communicate and receive information about companies [12, 13]. In this reason, appropriate and well targeted socially responsible communication focused on digital media represents a great opportunity to revive and specialize traditional communication.

Recently, many domestic and foreign authors have dealt with the issue of corporate social responsibility and its communication.

According to several authors, socially responsible communication is an important part of company success [14-15] with many benefits such as creation the desired reputation effects [17]. Also, it is necessary to mention opinions of scholars which emphasize the impact of stakeholders in long-term value creation of the socially responsible company [18-20].

Luo and Bhattacharya highlighted the right timing of CSR communication and the way it affects the customer loyalty, brand value and its reputation, increase the sales performance and stakeholder identification [21].

A survey by Focus Agency shows that awareness of the concept of corporate social responsibility by companies in Slovakia was relatively low recently. However, companies have been interested in this information, especially through the Internet. The main motives for which the companies implemented the concept are, in particular, keeping up with the competition, attracting and retaining quality employees, as well as improving the reputation [22].

According to a survey by NMS Market Research, currently, most of the companies (85%) use their websites to promote their socially responsible activities, even though they are mostly convinced that it will bring only little benefit for the resulting marketing communication. On the other hand, the social media represents the greatest potential in the way of socially responsible communication activities for the majority of companies (56%). Their share is expected to gradually increase and thus overtake the dominant position of the website. However, the traditional PR remains to be actual and an important tool of company's CSR communication, especially the media relations. It is realized by up to 70% companies and supported by most of the business managers who see the considerable potential in this communication tool [23].

Based on the mentioned above, we can claim, that the aim of socially responsible communication is to make the social, economic and environmental responsible activities of the company visible through a variety of communication tools.

2 Methodology

The paper is aimed on the on the usage of the CSR concept in Slovak companies and the determination of their communication in the process of globalization. This includes providing the theoretical background and analysis of the corporate social responsibility and its communication from the viewpoint of Slovak and foreign authors. Based on the analysis and results of the questionnaire survey, benefits of the efficient socially responsible company's communication are highlighted such as gaining competitive advantage, increasing the brand value, improving customer's loyalty and relationships with stakeholders.

The important source for secondary data was scientific researches, statistical databases, published professional publications. In order to determine the level of usage of the CSR concept and its communication in companies in Slovak Republic, a questionnaire survey was conducted among Slovak consumers.

The questionnaire survey was aimed at finding out the level of usage the CSR concept and its communication in companies in Slovak Republic, and was divided into three parts. The first part was devoted on a general profile of the company. The second part was focused on finding out the perception of the CSR concept. In the third part we determined the communication of company's CSR activities.

In the survey, the size of the population was determined based on data obtained from the organizational statistics of the Statistical Office of the Slovak Republic. The population consists of economic units, and the size of the base population was 596.042 companies (December 2019). As 532 companies participated in the marketing survey, the minimum sample size 384 was met. The confidence interval 5% was determined, and confidence level 95%, thus the error rate 5 % was admitted. The survey took the form of questionnaire method through e-mail communication during January - March 2020.

General scientific methods were used for data processing, as well as mathematical and statistical methods in the evaluation of data from the survey.

In order to achieve the main goal of the paper, based on the theoretical background and survey results, following research questions and hypotheses are formulated:

- *Research question 1:* Is the majority of companies in the Slovak Republic socially responsible?
  Hypothesis 1: More than 50% companies in the Slovak Republic are socially responsible.
- *Research question 2:* Are the socially responsible companies in the Slovak Republic aware of the importance of its communication?
  Hypothesis 2: More than 50% of companies in the Slovak Republic communicate information about their CSR activities towards their stakeholders.
- *Research question 3:* Do the socially responsible companies in the Slovak Republic prefer its communication through digital media?

Hypothesis 3: More than 50% of companies in the Slovak Republic prefer socially responsible communication focused on company websites.

Hypothesis 4: More than 50% of companies in the Slovak Republic prefer socially responsible communication focused on social media.

The marginal rate 50% used in hypotheses was set as the expression of the simple majority [24].

Statistical hypotheses testing is defined as a statement of the assumption of unknown parameters in the basic set, which is formulated as a statistical hypothesis and its validity is verified by statistical procedures based on selection characteristics. The role of statistical hypotheses testing is to decide whether to accept or reject the hypothesis regarding the basic set in accordance to the information from the available choice. In verification of the hypotheses, the methodology of statistical hypothesis testing consisting of the following steps was met [25]:

1. Formulation of the null hypothesis (H_0).
2. Formulation of the alternative hypothesis (H_1).
3. Determining the level of significance (α).
4. Calculation of test statistics and probability.
5. Decision.

The test statistic for hypotheses 1 – 4 we calculated by using the method testing a single proportion by one-tailed testing because it is commonly used [26, 27]. Significance level α was determined at 0.05.

3 Results

3.1 Results of questionnaire survey focused on the level of usage the CSR concept and its communication in companies in Slovak Republic

Of the 532 respondents in the sample set, 63.2% were micro-enterprises, 25% small enterprise, 7.3% medium-sized enterprises and 4.5% large enterprises. Due to the fact that in Slovak Republic predominate micro enterprises, these numbers are sufficient. The results of the questionnaire survey show that 79.5% respondents belong to the commercial companies. Most of the respondents operate in retail and wholesale (21.7%), communication (15.3%) and industrial production (11.5%).

Based on the survey results, 83.4% respondents know the CSR concept, and 75.2% state that they implement the principles of CSR concept in their corporate strategy and 71.5% respondents consider that the involvement in social responsibility should be as a necessary part of their corporate strategy. According respondents, the most important benefits of implementation CSR in their corporate strategy includes good reputation and image of the company (72.4%), customer's loyalty (68.1%), improving relationships with stakeholders (55.2%) and increasing competitive advantage (46.8%). From the survey results, it is clear, that 40.4% respondent's identity the most important area of CSR as achieving profits in an honest and ethical manner, next it is caring for employees (21.7%) and compliance with legal standards and the payment of taxes (12%).

53.9% respondents inform about company's CSR activities with the external environment, mainly stakeholders. They mostly communicate due company websites (74.5%), social media (62.9%), annual company reports (44.1%), CSR reports (40.3%) and newsletters (38%).

3.2 Answering research questions and verification of statistical hypotheses

To verify the statistical hypotheses, we used the method testing a single proportion. The hypothesis testing for a single proportion includes the following steps:

1. Determination of the null hypothesis:

$$H_0: \pi = \pi_0 \tag{1}$$

2. Determination of the alternative hypothesis:

$$H_1: \pi > \pi_0 \tag{2}$$

3. Selection of the significance level

$$\alpha = 0{,}05 \tag{3}$$

4. Calculation of the sample proportion:

$$p = \frac{m}{n} \tag{4}$$

where m is the number of respondents which indicate the particular option and n is the total number of respondents.

5. Satisfaction of the condition

$$n * \pi_0 * (1 - \pi_0) > 5 \tag{5}$$

Application of the test criteria:

$$T = \frac{p - \pi_0}{\sqrt{\frac{\pi_0 * (1 - \pi_0)}{n}}} \tag{6}$$

6. Critical field of the test:

Use the tables of the normalized normal distribution to find the critical value for the right-tailed test $z_{2\alpha}$.

$$T > z_{2\alpha} \tag{7}$$

7. Decision on the test result:

If the inequality does not apply, do not reject the hypothesis H_0.
If the inequality applies, reject hypothesis H_0, i.e. accept the alternative hypothesis H_1.
Results of verification of statistical hypotheses are shown in Table 1.

Table 1. Verification of statistical hypotheses

Calculation of the sample proportion: $p = \frac{m}{n}$	**Satisfaction of the condition $n * \pi_0 * (1 - \pi_0) > 5$**	**Test criteria**	**Critical field**	**Inequality**	**Acceptance or rejection of the hypothesis**
Hypothesis 1: H_0: 50% companies in the Slovak Republic are socially responsible. H_1: More than 50% companies in the Slovak Republic are socially responsible.					
p = 0.752	133 > 5	11.625	1.645	11.625 > 1,645	H_0 rejected
Hypothesis 2: H_0: 50% of companies in the Slovak Republic communicate information about their CSR activities towards their stakeholders. H_1: More than 50% of companies in the Slovak Republic communicate information about their CSR activities towards their stakeholders.					
p = 0,539	133 > 5	1.799	1.645	1.799 > 1.645	H_0 rejected
Hypothesis 3: H_0: 50% of companies in the Slovak Republic prefer socially responsible communication focused on company websites. H_1: More than 50% of companies in the Slovak Republic prefer socially responsible communication focused on company websites.					
p = 0.745	133 > 5	11.302	1.645	11.302 > 1.645	H_0 rejected
Hypothesis 4:					

H_0: 50% of companies in the Slovak Republic prefer socially responsible communication focused on social media. H_1: More than 50% of companies in the Slovak Republic prefer socially responsible communication focused on social media.					
p = 0.629	133 > 5	5.951	1.645	5.951 > 1.645	H_0 rejected

Source: Own processing.

As is obvious from Table 1, in all cases, the inequality applies, so we reject the hypothesis H_0, we accept the alternative hypothesis H_1.

To answer the first research question, we asked the respondents whether they are socially responsible. The survey results show, that 75.2% are socially responsible (implement the principles of CSR concept in their corporate strategy). Subsequently, we used the primary data from this question to test the statistical hypothesis 1. The results are shown in Table 1. Therefore, we can say that more than 50% companies in the Slovak Republic are socially responsible.

To answer the second research question, we asked the respondents whether they are aware of the importance of its CSR communication. The survey results show, that 53.9% respondents inform about company's CSR activities with the external environment, mainly stakeholders. Subsequently, we used the primary data from this question to test the statistical hypothesis 2 (Table 1). Therefore, we can say that more than 50% of companies in the Slovak Republic communicate information about their CSR activities towards their stakeholders.

To answer the third research question, we asked the respondents whether they prefer the CSR communication through digital media. The survey results show, that 74.5% respondents inform about company's CSR activities through company website and 62.9% respondents through social media. Subsequently, we used the primary data from this question to test the statistical hypotheses 3 and 4 (Table 1). Therefore, we can say that more than 50% of companies in the Slovak Republic prefer socially responsible communication focused on company websites as well as more than 50% of companies in the Slovak Republic prefer socially responsible communication focused on social media.

4 Discussion

Based on the mentioned above, it is obvious, that the majority of companies in Slovak Republic want to be socially responsible, thus they implement the CSR concept in their corporate strategy. They see the most important benefits of implementation CSR in their corporate strategy in the good reputation and image of the company, customer's loyalty, improvement of relationships with stakeholders and increase of competitive advantage. These results are quite similar that advantages highlighted in surveys by several scholars mentioned before [17-21].

More than half of companies in Slovak Republic are aware of the importance of CSR communications, especially with stakeholders who represent their target groups. Same as results realized by the NMS Market Research [23], they prefer company website for CSR communication as well as social media. In contrary, PR activities are not so important for Slovak companies as on a global scale.

We can claim, that creating an effective CSR communication within CSR strategy is a very crucial step, the implementation of which should be addressed by the company's management. The creation of the CSR strategy is based on the analysis of corporate stakeholders and their comparison with the development intentions and company's goals. Thus, the CSR communication may be focus on the dialogue with stakeholders, and it means to find out their expectations from the socially responsible company, improve the interrelationship and gain their loyalty. Individual stakeholders may have very specific demands and requirements. If the company wants to stay in the market for a long time and function smoothly, it should respect and adequately satisfy these interests of stakeholder groups. Its CSR communication has to be appropriate and well targeted on stakeholders, focused on modern forms of communication such as digital media. Next, it needs to address which stakeholders' expectations are not being met and try to find a possible solution. After identifying and analysing stakeholders, it is necessary to identify key areas of CSR and create structures for the long-term integration of CSR values across all levels of the company, including CSR communication.

5 Conclusion

In the gradual globalization, competition in the market is constantly deepening, which means that the company's reputation and its image have become vulnerable, and therefore the protection of companies is becoming a priority. The growing impact of CSR on corporate reputation has been reflected in the increasing importance of social communication effort. CSR communication can even help a company build a reputation that protects its image from negative publicity, as well as increase public interest. If a company wants to succeed in today's competitive market, it is no longer enough to offer high-quality products at the desired price, but it should respond to the emerging sensitivity of stakeholders to environmental and social issues and communicate success in these areas.

As is obvious from the outcomes of the research, the implementation of CSR does not in itself lead to successful results, but the communication of these activities with stakeholders is essential for success. The most visible benefits of implementation CSR concept and its efficient communication in Slovak companies includes gaining competitive advantage, increasing the brand value, improving customer's loyalty and relationships with stakeholders. The survey results can have a significant application in practice, especially in appropriate managing and targeting of socially responsible communication on stakeholders in Slovak companies. However, there are some barriers in CSR implementation, including the increase of the number in CSR scandals over the years, which has caused public scepticism. Growing mistrust and scepticism have put pressure on businesses to show that they are doing good and to communicate their CSR commitments. CSR communication thus has a significant effect on reducing scepticism.

Based on the above mentioned, the main managerial recommendation is to focus on the appropriate and well targeted communication of the socially responsible companies in the Slovak Republic aimed on the digital marketing such as company website, social media and so on. This includes customizing the communication according to preferences of the individual stakeholders to increase their attention and loyalty.

It follows from the above, that the issue of CSR and its communication in Slovak companies has the great potential for deeper research in the future. There is also a space for further discussions on socially responsible companies and their communication in relation to individual stakeholders and so on.

This paper is a partial output of the project VEGA no. 1/0064/20: Behaviorism in a socially responsible communication strategy of enterprises

References

1. H. Nemeth Mazarova, Social responsibility in response to globalization and social responsibility programs in practice. *Better Responsibility Initiative*, 14 p. (2018)

2. W. Sroka, R. Szanto, Corporate Social Responsibility and Business Ethics in Controversial Sectors: Analysis of Research Results. *Journal of Entrepreneurship Management and Innovation*, **14**(3), 111-126 (2018)
3. V. Kunz, Corporate social responsibility. Prague: Grada Publishing (2012)
4. E. Stonkute, J. Vveinhardt, W. Sroka, Training the CSR Sensitive Mind-Set: The Integration of CSR into the Training of Business Administration Professionals. *Sustainability*, **10**(3), art. no.754 (2018)
5. M. Hitka, G. Pajtinkova-Bartakova, S. Lorincova, H. Palus, A. Pinak, M. Lipoldova, M. Krahulcova, N. Slastanova, K. Gubiniova, K. Klaric, Sustainability in marketing through customer relationship management in a telecommunication company. *Marketing and Management of Innovations*, **4**, 194-215 (2019)
6. A. Buriak, I. Voznakova, J. Sulkowska, Y. Kryvych, Social trust and institutional (bank) trust: Empirical evidence of interaction. *Economics and Sociology*, **12**(4),116-129 (2019)
7. R. Myskova, P. Hajek, Sustainability and Corporate Social Responsibility in the Text of Annual Reports-The Case of the IT Services Industry. *Sustainability*, **10**(11), art. no. 4119 (2018)
8. A. Lewicka-Strzalecka, Opportunities and Limitations of CSR in the Postcommunist Countries: Polish Case, *Corporate Governance*, **6**, 440-448 (2006)
9. A. Pomering, S. Dolnicar, Assessing the Prerequisite of Successful CSR Implementation: Are Consumers Aware of CSR Initiatives? *Journal of Business Ethics*, **85**, 285-301 (2009)
10. M. Pavlik, M. Belcik. Corporate Social Responsibility – CSR in practise and how to proceed with it. Praha: Grada Publishing (2010)
11. E. Kicova, A. Krizanova, Building a sustainable brand. *Springer Proceedings in Business and Economics: Advances in Applied Economic Research*, pp. 367-379 (2017)
12. L. Lizbetinova, P. Starchon, S. Lorincova, D. Weberova, P. Prusa, Application of Cluster Analysis in Marketing Communications in Small and Medium-Sized Enterprises: An Empirical Study in the Slovak Republic. *Sustainability*, **11**, art. no. 2302 (2019)
13. M. Vetrakova, M. Hitka, M. Potkany, S. Lorincova, L. Smerek, Corporate Sustainability in the Process of Employee Recruitment through Social Networks in Conditions of Slovak Small and Medium Enterprises. *Sustainability*, **10**(5), art. no. 1670 (2018)
14. A. McWilliams, D. Siegel, Corporate Social Responsibility: A Theory of the Firm Perspective. *The Academy of Management Review*, **26**, 117-127 (2001)
15. A. Diacikova, A. Dankova, Green Marketing Communication. *Proceedings of the 2nd International Scientific Conference of the Poprad Economic and Management Forum*, pp. 83-89 (2019)
16. D. Bratic, M. Palic, M. T. Lisanin, D. Gajdek, Green marketing communications in the function of sustainable development. *Proceedings of the 11th Annual Conference of the EuroMed Academy of Business – Research Advancements in National and Global Business Theory and Practice*, pp. 185-198 (2018)
17. E. P. Scholder, D. J. Webb, L. A. Mohr, Building Corporate Associations: Consumer Attributions for Corporate Socially Responsible Programs. *Journal of the Academy of Marketing Science*, **34**(2), 147-157 (2006)
18. M. Morsing, M. Schultz, Corporate social responsibility communication: Stakeholder information, response and involvement strategies. *Business Ethics: A European Review*, **15**, 324-338 (2008)

19. A. A. Bailey, A. S. Mishra, M. F. Tiamiyu, Application of GREEN scale to understanding US consumer response to green marketing communications. *Psychology & Marketing*, **35**, 863-875 (2018)
20. T. Wagner, R. J. Lutz, B. A. Weitz, Corporate Hypocrisy: Overcoming the Threat of Inconsistent Corporate Social Responsibility Perceptions. *Journal of Marketing*, **73**, 77-91 (2009)
21. X. Luo, C. B. Bhattacharya, The Debate over Doing Good: Corporate Social Performance, Strategic Marketing Levers, and Firm-Idiosyncratic Risk. *Journal of Marketing*, **73**, 198-213 (2009)
22. Pontis – Focus. *Perception of the concept and practices of socially responsible business by small and medium enterprises in Slovakia*. [online], Available at: http://www.sbagency.sk/sites/default/files/sprava_csr_sme.pdf (2010)
23. Mediaguru. *Social media will be central for CSR*. [online], Available at: https://www.mediaguru.cz/clanky/2015/06/pro-csr-budou-stezejni-socialni-media/ (2015)
24. B. P. Macfie, P. M. Nufrio, Applied Statistics for Public Policy. New York: M. E. Sharpe (2006)
25. M. Rimarcik, Statistics for practice. Kosice: Marian Rimarcik (2007)
26. G. D. Ruxton, M. Neuhauser, When should we use one-tailed hypothesis testing? *Methods in Ecology and Evolution*, **1**, 114-117 (2010)
27. C. M. Lombardi, S. H. Hurlbert, Misprescription and misuse of one-tailed tests. *Australian Ecology*, **34**, 447–468 (2009)

The Agile Content Marketing Roadmap Based on the B2B Buyer's Journey – The Case Study of the Slovak republic

Lubica Gajanova[1,*]

[1]University of Zilina, Faculty of Operation and Economics of Transport and Communications, Department of Economics, Zilina, Slovakia

Abstract. The rise of new technologies has created opportunities for the emergence of many trends today. This fact is a chance for the visibility of companies and in recent years internet marketing has become an integral part of marketing strategies of companies across all markets. Today's marketing world give new opportunities to engage and gain the attention of relevant users. But customers including B2B buyers are overwhelmed by advertising on the Internet. Emphasis on the creation of quality content therefore becomes necessary and creates another of the right ways to address customers. This paper deals with the use of content marketing strategy in the B2B market in the conditions of the Slovak Republic. As there are a large number of tools usable in the content marketing strategy, the aim of the paper is to use factor analysis to obtain an alternative description of the variability of the observed variables (B2B content marketing tools) with a potentially lower number of latent variables.
Keywords: content marketing, B2B, factor analysis, buyer's journey, Slovakia

1 Introduction

The awareness of B2B buyers is currently growing thanks to better connectivity and the use of modern technologies. The communication resources that the current environment offers influence them, but it also offers a choice between an inexhaustible number of messages and appeals coming from the market. [1]. B2B buyers therefore choose the information they want to receive. As a result, they become more resilient to marketing incentives [2]. The spiral of advertising chaos or advertising blindness sends a clear signal to companies about the unsustainability of this phenomenon, which forces them to make significant efforts to communicate the message [3]. The problem outlined therefore represents a space for finding effective marketing ways to help break down these barriers. In this connection, it is necessary to realize that the customer is basically not so interested in the goods or services of a particular company, they are primarily interested in themselves and their needs, which they want to satisfy through these products [4, 5]. This finding is not new, but it took some time for such marketing forms to develop that were truly and

*Corresponding author: lubica.gajanova@uniza.sk

effectively able to absorb this fact into its essence. Therefore, they do not primarily prioritize the product itself, but first solve the customer's problems, needs and interests, offer him a solution and sensitively incorporate advertising messages into the communication [6, 7]. The notion of repeatedly targeting the customer with the idea that someone will be eventually attract by the product to buy it is no longer enough. The technique, which is defined as content marketing, tries to solve this problem [8, 9]. Content marketing plays an important role for business-to-business (B2B) companies due to the higher (than for business-to-customers (B2C) companies) dependence on customers [10, 11]. Nearly all of the most successful B2B content marketers (90%) prioritize the audience's informational needs over their sales/promotional message, compared with 56% of the least successful [12].

Despite the assumptions, content marketing is currently used by more than 90% of companies worldwide in the B2B sector [13]. In fact, up to 93% of B2B companies say that content marketing can attract more potential customers than traditional marketing strategies [14]. The data confirm that in the B2B sector, more than half of all buyers will encounter several digital communication tools during the purchasing process. And more than 80% of buyers view various publications about the company, its successes and references [15]. Almost 75% of buyers say that the content they find significantly affects their purchase decision. And 62% say they can complete their selection criteria based only on digital content when making a purchase [16]. The research carried out by B-inside in 2013 focused directly on the use of content marketing in the B2B segment in the Czech Republic. The results were very favourable. 95% of B2B marketers used at least one of the content marketing tools. Most of them were articles in magazines, newspapers. Furthermore, verbal references of clients, articles on the Internet, seminars, etc. [17]. According to further research by this company in 2017, companies are deviated away from using content. Companies started to pay less attention to content marketing, PR, e-mail marketing, etc. On the contrary, they started to pay more attention to sales support, sponsorship, etc. [18]. However, the survey in 2019 again showed an increase in interest in content marketing in the B2B segment. While in 2017 only 16% of B2B in the Czech Republic used it, in 2019 it was already 36%. However, most (77%) B2B companies engage in content marketing unsystematically, but at least create and distribute some content. The systematic approach existed only in a minority of cases. That is the situation which has not changed much compared to 2016. Even then, only about a quarter of B2B companies regularly created and communicated interesting content [19].

How it is in Slovakia is not known. Likewise, academic research on this topic is in its infancy. We encounter a situation in which technology and society are evolving faster than the ability of companies to adapt to new conditions. This phenomenon is collectively referred to digital Darwinism [20]. The vast majority of the approach to content marketing discussed so far has not been differentiated with regard to a specific area of targeting, and therefore we do not know what quality content means specifically for the customer in the B2B segment in the Slovak republic. That represents the research gap for this paper, which therefore focuses on communication with corporate clients, which has its own specifics playing an important role in marketing communication, decision-making and purchasing process. There is a fundamental difference between an average content marketing strategy and an excellent content marketing strategy. This is the ability of the marketer to reveal what kind of content is appropriate in each given situation. According to Content Marketing Institute's 2020 report, 69% of successful B2B marketers have a documented content marketing strategy [16].

2 Methodology

As there are a large number of tools and techniques that can be used in content marketing, the aim of the paper is to use factor analysis to obtain an alternative description of the variability of observed variables with a potentially lower number of hidden compact latent variables (factors). Factor analysis assumes that the factor construction of the model may be simpler than the data on which it is based [21]. This construction then represents an aggregated abstraction of a large number of observed phenomena based on the presence of their common characteristics, which allows to significantly simplify and summarize large data matrices into smaller ones, without significant loss of information [22]. The information obtained in this way can lead to the creation of a content marketing roadmap of the B2B buyer's journey in conditions of the Slovak republic.

Factor analysis simplifies the description of the original configuration to the smallest possible number of dimensions. This simplification is based on the magnitude of the linear interdependence between the input variables - if the variables are linearly dependent on other variables, these variables can be described through common linearly independent hidden variables. By preserving only a subset of linearly independent hidden variables, other variables can be explicitly described by expressing their relationship to this subset [23, 24]. The assumption is that the size of the set of latent variables will be smaller than the original set of data, i.e. the configuration of the data matrix Y of size $K \times I$ can be reduced to r dimension $(r \leq I)$. These r-dimensions then form a factor space into which the original I vectors can be projected [24].

As input variables for factor analysis were selected content marketing tools usable in the B2B market [9, 10, 11, 25, 26], which are shown in Figure 1.

Newsletter / Email Marketing	Social Media Content	Reviews and Testimonials
Blogs	Podcasts	LinkedIn Lead Generation
Articles	Live In-Person Events	Sales Conversation
E-books	Webinars	Demos
White Papers	Social Media Engagement	Free Assessment
Tip Sheets	Case Studies	Quotes/Proposals
Checklists	Videos	Free Consultations
Infographics/Data Viz	Live Streaming	Free Trial

Fig. 1. Content marketing tools
Source: Own processing.

Respondents assigned data to these content marketing techniques within the research according to a liqueur scale, which expressed the degree of influence of respondents in the purchasing process within the B2B market. The respondents were customers who had been part of the B2B purchasing team at least once during the past year.

Factor analysis is based on correlations and these are stabilized only on relatively large samples. The literature in this case states that the sample of 500 respondents is very good and some authors state that the number of respondents (cases) must be at least 5 times more than the variables for factor analysis [27]. The size of the examined sample was 500 respondents, which fulfilled both cases of conditions for determining the sufficient sample size.

3 Results

The initial conditions of factor analysis include: a) high correlations of a large number of variables and b) low partial correlations. Whether the correlation and partial correlation matrices satisfy the assumptions is therefore tested by various coefficients and indices, usually KMO and Bartlett's sphericity test [21, 22, 23], see Table 1.

Table 1. KMO and Bartlett's Test

Kaiser-Meyer-Olkin Measure of Sampling Adequacy.		0,927
Bartlett's Test of Sphericity	Approx. Chi-Square	11064,325
	df	276
	Sig.	0,000

Source: Own processing according SPSS.

The value of the Kaiser-Meyer-Olkin measure (KMO), i.e. the measure of the adequacy of the selection at the level of 0.951 and the significance of Bartlett's Test of Sphericity at the level of 0,000 show that the mentioned initial assumptions of factor analysis are correct, so the condition of the sample adequacy is met [22].

Table 2 shows the output of the Principal Component Analysis procedure. The input variables contain an evaluation of 24 questions, which were obtained in the survey concerning the degree of influence of respondents in the purchasing process within the B2B market.

Table 2. Total Variance Explained

Comp.	Initial Eigenvalues			Extraction Sums of Squared Loadings			Rotation Sums of Squared Loadings		
	Total	% of Variance	Cumulative %	Total	% of Variance	Cumulative %	Total	% of Variance	Cumulative %
1	11,476	47,815	47,815	11,476	47,815	47,815	5,810	24,209	24,209
2	3,247	13,530	61,346	3,247	13,530	61,346	4,149	17,288	41,497
3	1,613	6,719	68,065	1,613	6,719	68,065	4,047	16,861	58,359
4	1,466	6,110	74,175	1,466	6,110	74,175	3,796	15,816	74,175
5	0,881	3,670	77,846						
6	0,728	3,034	80,880						
7	0,564	2,351	83,231						
8	0,431	1,796	85,027						
9	0,385	1,604	86,631						
10	0,349	1,454	88,085						
11	0,331	1,379	89,464						
12	0,317	1,322	90,786						
13	0,281	1,171	91,958						
14	0,277	1,155	93,113						
15	0,245	1,022	94,134						
16	0,223	0,931	95,065						
17	0,196	0,818	95,882						

18	0,187	0,779	96,661						
19	0,173	0,719	97,380						
20	0,148	0,615	97,995						
21	0,137	0,573	98,568						
22	0,129	0,539	99,107						
23	0,114	0,474	99,581						
24	0,101	0,419	100,000						

Source: Own processing according SPSS.

The procedure with the extraction Principal Component Analysis revealed four significant factors in the input data, which explain 74,175 % of the variability of the observed variables.

Table 3 shows the Rotation matrix by Varimax with Kaiser Normalization and contains the structure of the individual components and their loads expressing the degree of correlation of the variable with the given component [21]. In this case, all component loads came out high enough for unambiguous interpretation.

Table 3. Rotated Component Matrix

	Component			
Attributes (Variables)	1	2	3	4
Newsletter / Email Marketing	0,765	0,146	0,089	0,178
Blogs	0,754	0,036	0,166	0,253
Articles	0,677	0,036	0,192	0,224
E-books	0,845	0,141	0,182	0,110
White Papers	0,834	0,209	0,129	0,100
Tip Sheets	0,852	0,180	0,089	0,133
Checklists	0,821	0,159	0,130	0,168
Infographics / Charts / Data Viz	0,879	0,168	0,158	0,094
Social Media Content	0,211	0,216	0,304	0,755
Podcasts	0,215	0,271	0,230	0,787
Live In-Person Events	0,273	0,231	0,236	0,771
Webinars	0,215	0,215	0,288	0,787
Social Media Engagement	0,178	0,265	0,159	0,723
Case Studies	0,203	0,783	0,250	0,303
Videos	0,144	0,773	0,289	0,181
Live Streaming	0,168	0,762	0,317	0,243
Reviews and Testimonials	0,158	0,822	0,204	0,252
LinkedIn Lead Generation	0,169	0,826	0,161	0,223
Sales Conversation	0,166	0,170	0,837	0,249
Demos	0,165	0,236	0,818	0,217
Free Assessment	0,254	0,440	0,654	0,167
Quotes/Proposals	0,131	0,194	0,774	0,313
Free Consultations	0,215	0,222	0,723	0,334
Free Trial	0,209	0,412	0,576	0,070

Source: Own processing according SPSS.

The first component strongly correlates with the eight variables (Newsletter / Email Marketing, Blogs, Articles, E-books, White Papers, Tip Sheets, Checklists, Infographics / Charts / Data Viz) and together form one dimension of content marketing tools and techniques. Similarly, the second component brings together five other variables (Case Studies, Videos, Live Streaming, Reviews and Testimonials, LinkedIn Lead Generation). The third component consists of an exclusive representation of six variables (Sales Conversation, Demos, Free Assessment, Quotes / Proposals, Free Consultations, Free Trial). The last fourth component contains five variables (Social Media Content, Podcasts, Live In-Person Events, Webinars, Social Media Engagement).

4 Discussion

The results of the factor analysis of content marketing tools and techniques showed that the respondents divided the given tools into four dimensions within the examination of their degree of influence of the respondents during the purchasing process within the B2B market. By examining their structure, we can assign them properties derived from contributing attributes that have clearly demonstrated the creation of dimensions in accordance with the each stages of the B2B buyer's journey based on the See-Think-Do-Care framework for the digital age [28]. It is a method that enables clear content planning and generally helps in creating a marketing strategy for the company's products and services, so it is suitable for content marketing [10]. The model works on the assumption that one customer is in four different phases during the purchasing process [29].

For a content marketing strategy, it is essential to reveal what type of content is appropriate in each given situation. Figure 2 shows a closely related link between content marketing tools and techniques with a specific phase of the B2B buyer's journey.

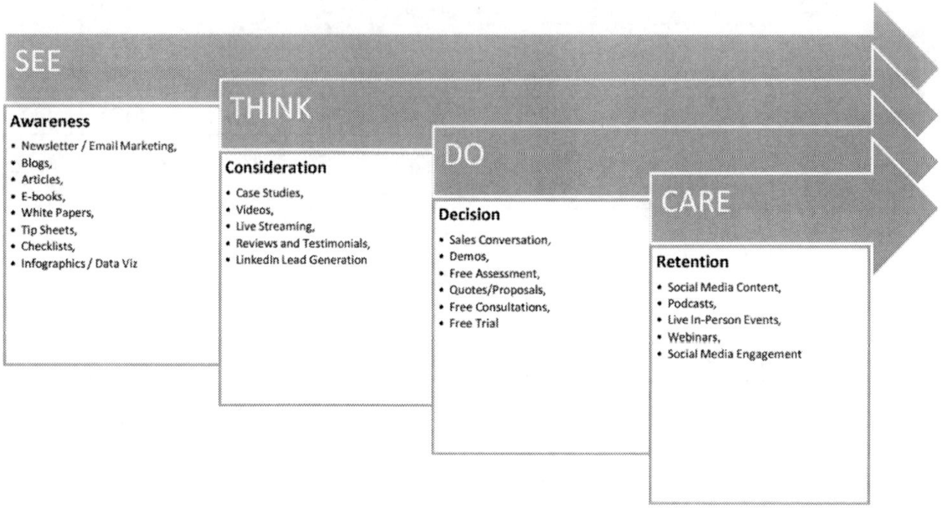

Fig. 2. The agile content marketing roadmap based on the B2B buyer's journey
Source: Own processing.

The B2B buyer's journey is one of the most important elements of a marketing strategy. For a content marketing strategy to be successful, the seller must be able to target it to the right audience and address it with the right message at the right time [10]. In addition, B2B marketers will understand what their buyers are doing and help them move from one stage of decision making to another.

5 Conclusion

The paper dealt with the issue of content marketing and its connection to the creation of a content marketing roadmap on the B2B buyer's journey in the conditions of the Slovak republic. The main idea of the work was to show how the tools of content marketing can be optimally used in the B2B market. Along with a full understanding of buyer's journey, it should be possible to build a targeted marketing strategy. The construction of factor analysis abstracted a large number of content marketing tools and techniques based on the statistical existence of their common features and characteristics, and thus significantly reduced data matrices to four dimensions in accordance with the each stages of the B2B buyer's journey based on the See-Think-Do- Care framework. The information thus obtained led to the creation of a content marketing roadmap of the B2B buyer's journey in conditions of the Slovak republic.

Acknowledgement

This paper is an output of scientific project VEGA no. 1/0718/18: The impact of psychographic aspects of pricing on the marketing strategy of companies across products and markets.

References

1. A. Domańska, Cooperation between knowledge-based institutions and business: empirical studies and network theories. *Forum Scientiae Oeconomia*, **6**(4), 81-94 (2018)
2. J. Majerova, Analysis of Slovak Consumer's Perception of the Green Marketing Activities. *Proceedings of the 4th World Conference on Business, Economics and Management (WCBEM)*, pp. 553-560 (2015)
3. A. Aribarg, E. M. Schwartz, Native Advertising in Online News: Trade-Offs Among Clicks, Brand Recognition, and Website Trustworthiness. *Journal of Marketing Research*, **57**(1), 20-34 (2020)
4. M. Nadanyiova, The Possibility of Introducing the CRM System in Railway Company Cargo Slovakia. *Proceedings of the 17th International Conference on Transport Means*, pp 87-92 (2013)
5. A. Krizanova, J. Majerova, K. Zvarikova, Green Marketing as a Tool of Achieving Competitive Advantage in Automotive Transport. *Proceedings of the 17th International Conference on Transport Means*, pp. 45-48 (2013)
6. J. Řezníček, T. Procházka, *Obsahový marketing: nakrmte Internet svým obsahem.* 1st ed. Brno: Computer Press (2014)
7. M. Valaskova, A. Krizanova, The passenger satisfaction survey in the regional integrated public transport system. *Promet-Traffic & Transportation*, **20**(6), 401-404 (2008)
8. S. Yaghtin, H. Safarzadeh, M. Karimi Zand, Planning a goal-oriented B2B content marketing strategy. *Marketing Intelligence & Planning*, Early Access (2020)
9. J. Jarvinen, H. Taiminen, Harnessing marketing automation for B2B content marketing. *Industrial Marketing Management*, **54**, 164-175 (2016)
10. E. Bakhtieva, The role of customer journey in B2B content marketing. *Proceedings of the 12th Annual International Bata Conference for Ph.D. Students and Young Researchers (DOKBAT)*, pp. 31-42 (2016)
11. J. Horvath, M. Olearova, Importance and effectiveness of content marketing in conditions of e-commerce B2B model. *Proceedings of the* International Scientific Conference on Marketing Identity: Online Rules, pp. 93-103 (2017)
12. Content Marketing Institute. *B2B Content Marketing 2019 Benchmarks, Budgets, and Trends* [online], Available at: https://contentmarketinginstitute.com/wp-content/uploads/2018/10/2019_B2B_Research_Final.pdf (2019)
13. Content Marketing Institute. *B2B Content Marketing 2018 Benchmarks, Budgets, and Trends* [online], Available at: https://contentmarketinginstitute.com/wp-content/uploads/2017/09/2018-b2b-research-final.pdf (2018)
14. Forbes. *Obsahový marketing a hodnotný obsah. Tu sú 3 stratégie, ktoré v roku 2019 fungujú* [online], Available at: https://www.forbes.sk/obsahovy-marketing-hodnotny-obsah-tu-su-3-strategie-ktore-v-roku-2019-funguju/ (2019)

15. P. Andersen, R. Archacki, N. de Bellefonds, D. Ratajczak, *How Digital Leaders Are Transforming B2B Marketing* [online], Available at: https://www.bcg.com/en-us/publications/2017/marketing-sales-how-digital-leaders-transforming-b2b (2017)
16. Content Marketing Institute. *B2B Content Marketing 2020 Benchmarks, Budgets, and Trends* [online], Available at: https://contentmarketinginstitute.com/wp-content/uploads/2019/10/2020_B2B_Research_Final.pdf (2020)
17. B-inside. *Ministudie: Content marketing v B2B v ČR a USA, 2013* [online], Available at: http://www.b2bmonitor.cz/wp-content/uploads/2013/11/B2B-monitor-VI-Content-marketing-bez-sektor%C5%AF.pdf (2013)
18. B-inside. *B2B marketing v ČR v roce 2017* [online], Available at: http://www.b2bmonitor.cz/wp-content/uploads/2017/10/B2B-marketing-v-%C4%8CR-v-roce-2017-ve%C5%99ejn%C3%A1-studie-FINAL.pdf (2017)
19. B-inside. *Content marketing v B2B v roce 2019* [online], Available at: http://www.b2bmonitor.cz/wp-content/uploads/2019/03/B2B-monitor-Content-marketing-v-B2B-v-roce-2019.pdf (2019)
20. B. Solis, *WTF?: What's the Future of Business?: Changing the Way Businesses Create Experiences*. 1st ed. Wiley (2013)
21. P. Durana, J. Kliestikova, M. Kovacova, The Quality of Brand Products: Expected Attributes vs. Perceived Reality. *Economics and Culture*, **16**(1), 98-107 (2019)
22. J. Kliestikova, M. Kovacova, P. Durana, Quo vadis brand loyalty? Comparative study of perceived brand value sources. *Polish Journal of Management Studies*, **19**(1), 190-203 (2019)
23. J. Kliestikova, P. Durana, M. Kovacova, Naked consumer's mind under branded dress: Case study of Slovak Republic. *Central European Business Review*, **8**(1), 15-32 (2019)
24. D. J. Tigert, J. N. Sheth, *Factor Analysis in Marketing*. Marketing Classics Press (2011).
25. W. L. Wang, E. C. Malthouse, B. Calder, B2B content marketing for professional services: In-person versus digital contacts. *Industrial Marketing Management*, **81**, 16-168 (2019)
26. F. Stroch, M. Stoklasa, Methods for creating personalized marketing content in B2B online communication. *Proceedings of the International Scientific Conference on Marketing Identity: Online Rules*, pp. 434-439 (2017)
27. A. B. Costello, J. W. Osborne, Best Practices in Exploratory Factor Analysis: Four Recommendations for Getting the Most From Your Analysis. *Practical Assessment, Research & Evaluation*, **10**, 7 (2005)
28. A. Kaushik, *See, Think, Do, Care Winning Combo: Content +Marketing +Measurement!* [online], Available at: https://www.kaushik.net/avinash/see-think-do-care-win-content-marketing-measurement/ (2015)
29. D. Rowles, *Digital Branding: A Complete Step-by-Step Guide to Strategy, Tactics and Measurement*. Kogan Page (2014)

Factors of Entrepreneurial Success: A qualitative multiple case study of new small businesses

Jan Urban[1,*]
[1]Institute of Technology and Business in České Budějovice

Abstract. In most countries, small businesses represent a majority of firms providing close to fifty percent of the total employment. However, on average, more than half of new small-business start-ups usually fail within the first 5 years of operation. Therefore, learning from the success of small-business owners who have sustained in business for at least 5 years can provide helpful insights not only for individual firms, but also for the continued growth of local economies which usually rely on small business to a great extent. The purpose of this case study based article was to explore common approaches, strategies and skills of successful business founders/small business owner´s that contribute to their sustained development. The article is based on a qualitative survey of sixteen, locally operating, small, retail and service-industry firms which, having successfully survived at least 5 years in business, continue to grow. The method of the survey covered semi structured interviews led with individual entrepreneurs and focused on identifying the crucial business as well as personal approaches and skills important for their market survival.
Keywords: small business, business owner characteristics, entrepreneurial skills, leadership skills, business planning, networking, market differentiation

1 Introduction

Even though the definition of an entrepreneur/small-business owner varies in the literature, the prevailing understanding is that he/she is the (sole) proprietor of a firm with fewer than 500 employees (SBA, 2016). The SBA (2012) has also defined a small business as one that is independently owned and operated, organized for profit, and is not dominant in its field.

The importance of small-businesses is based on the fact that they account for a significant portion of new job creation, total employment, private sector payroll as well as innovations that fuel the economy. Furthermore, they tend to be highly important for the continued growth of local economies.

Most new businesses, however, are fragile and often fail: worldwide, in many industries, as many as 80% of new start-ups are out of business within five years (Artinger & Powell, 2015; Cader & Leatherman, 2011), and even less start-ups survive 10 years (Small Business

* Corresponding author: 21537@mail.vstecb.cz

and Entrepreneurial Council, 2016). Thus, in spite of being an important vehicle of economic growth and innovation, they are also „responsible" for a tremendous amount of job destruction.

2 Factors of Entrepreneurial Success: Literary Research

The higher vulnerability of small businesses to economic slow-downs, bigger firms' competition, as well as their lack of funding are usually quoted as the main factors of their "early mortality" (Osei-Assibey, Bokpin, and Twerefou, 2012). In this regard, Schiff, Hammer, and Das (2010) noted that an important factor of their undercapitalization is the failure of the small-business owners to estimate the amount of pre-tax income that the business must generate to sustain operations and profitability. Furthermore, collecting sufficient capital for expanding their operations is another crucial factor since growing to a level necessary for economic efficiency (Salazar, Soto, & Mosqueda, 2012) is quite often is a prerequisite for small-businesses market survival.

Cronin-Gilmore (2012) suggested that while small-business owners have usually good knowledge of their product or service, they typically lack skills to develop a strategic and operational business plans. Eddleston, Kellermanns, Floyd, Crittenden, and Crittenden (2013) noted business owners who developed and updated the strategic business plans experienced fewer failures than other small-businesses. Even for initially successful firms, the challenges may include how to develop innovation strategies to sustain long-term growth (Adams, Kauffman, Khoja, & Coy, 2016).

Similarly, creating effective marketing plans, determining what to offer and to whom, can be challenge for a small-business owner (Perry, 2014). As Box (2011) concluded, the new businesses owners may have to adopt aggressive „guerrilla" marketing techniques and/or create a niche market. Communication strategies are also important in this regard; and small-business owners are increasingly using cost effective social media as a platform for their marketing, advertising, and recruitment activities (Schaupp & Bélanger, 2014).

Last but not least, as Fahed-Sreih and Morin-Delerm (2012) stressed, a lack of leadership skills contributes to small-business owners' failures in producing efficient organizations (Sakiru, D'Silva, Othman, DaudSilong, & Busayo, 2013) with motivated and loyal employees.

In this regard, Hansen, Shrader, and Monllor (2011) suggested that personal characteristics of small-business owners often include a strong product focus as well as focus of internal control and the need for achievement. Successful small-business owners, however, typically have to have also a leadership charisma and passion, contagious to their employees (Kao, Pai, Lin, & Zhong, 2015). As for their people management skills, they claim, their goal alignment and delegation skill is crucial.

Simiralry, Byrd (2010) identified that the leadership practices of successful and unsuccessful small-business owners differed significantly. More precisely, according to Surdez-Pérez, Aguilar-Morales, Sandoval-Caraveo, López-Parra, & Corral-Coronado, 2014), personal traits contributing to small business sustainability and growth include (a) strong negotiation skills, (b) creativity, (c) hard work, and (d) self-discipline. To achieve higher employee productivity and creativity, entrepreneurs should create an atmosphere that incorporates shared values, feeling of uniqueness and even play (Kauanui, Thomas, Sherman, Waters, and Gilea, 2010).

Furthermore, Insead (2011) identified social networking, interpersonal ties, and the ability to create new connections as essential characteristics of sustainability of a business.

O'Donnell (2014) noted that networking is critical to knowledge exchange, and an increase in small-business owners' knowledge led to a higher sustainability rate in their business operations. Fifty percent of small-business owners identified their business venture came from ideas from networking contacts and/or innovations, slight variations, or creativity from existing business (Martinez & Aldrich, 2011). Most small-business owners lack experience, do not have a history of previous success, and need funding information so networking can be of value (Martinez & Aldrich, 2011).

3 Methods and Data

The purpose of this qualitatively oriented, multiple-case study was to follow the above mentioned line of research and explore the skills, knowledge, and strategies small-business owners use to succeed in business. The primary research question, guiding the study, was "*What skills, knowledge, and strategies characterize small-business owners who succeed in business beyond 5 years?*". In identifying these skills, stress was put on common personal features and views of entrepreneurs operating in a particular industry.

3.1 Selection of Survey Participants

For the purpose of the study, the definition of an entrepreneur/small-business owner was a person who began a privately owned small business, took on financial risks in seeking profits, and employs not more than 50 employees (Lawrence, 2014).

The sample comprised 16 small business owners residing in Central Bohemia and operating small retail shops and personal services outlets who sustained their business beyond 5 years, employed 10 to 50 employees and led a growing and viable/profitable business entity.

3.2 Methods Used

Face-to-face semi structured interviews as well as the basic financial business records provided the data for answering the research question. Data saturation occurred when no new or relevant themes or supporting information emerged during the data collection process (methodology suggested by Denzin & Lincoln, 2011; Walker, 2012). An audio recording of the interviews was to ensure accuracy. Scheduling of the face-to-face semi structured interviews occurred at times the participants selected.

The appropriateness of the qualitative methodology used to conduct the study leans on Denzin and Lincoln conclusions (2011), that qualitative method is appropriate when (a) the problem requires more than a yes or no hypothesis, (b) a large sample may not be available, and (c) common themes are expected to result from exploring the data to address the research question.

In a similar context, this methodology was used Box (2011) who conducted a qualitative case study included small business owners and entrepreneurs in which business owners were divided into three categories - differentiators, cost leaders, or stuck in the middle - when sustaining a business.

Furthermore, a case study of this kind is an adaptable approach for analysing societal and developmental disciplines (Yin, 2014).

Yin noted that the case study can be a single case or multiple cases bounded by time and location. For this study, a multiple-case study approach was used, as obtaining information from more than a single participant provided additional insights, understanding, and assurance of validity in identifying strategies small-business owners use to increase success rates (Denzin & Lincoln, 2011).

4 Results

The success of a newly established company is always, at least in part, a matter of luck. As the interviewed entrepreneurs admitted, it depends on factors that entrepreneurs do not have complete control over.

Nevertheless, there are certain strategies, traits and ways of acting, by which successful founders, according to interviews results differ. They can be summarized in the following themes that emerged from the collected data. These themes enrich, to a great extent, earlier studies in this field, while not contradicting them, and can thus be regarded as an original study of factors contributing to an entrepreneurial success, especially in early stages of business development.

Successful entrepreneurs

- **don't necessarily adhere to what competitors are doing,** and are not worried about seemingly "crazy" ideas or thoughts.

  While many new entrepreneurs are strongly influenced in their decisions by what their competitors do (sometimes even tend to imitate them and believe that if they follow their example, they will succeed), the members of the survey were convinced a simple recipe that companies should take over or imitate does not exist, i.e. that successful new business owners learn from the experiences of others, but „not too much.“ They understand that if all firms in a given industry took over the best and successful experiences of others, these experiences – as well as the companies that imitated them - would only become mediocre: a company that only adheres to or even takes over the procedures of others can never overrun its competitors, even if the methods themselves were successful.

  The tendency not to refrain from strange-looking or even seemingly "strange" business ideas can be regarded as one of the main features of successful business founders. What seems like an almost crazy idea to others is, for them and often in fact, more of a sign that they are on the right track.

- **are not afraid of markets regarded as "mature" or already occupied.** Even though conventional thinking often claims that a newly founded company should locate a place in the market that is not yet fully occupied, and this strategic approach is often successful "occupied" markets can still offer opportunities, and thus do not necessarily have always have to be avoided.

  The reason why even apparently mature and saturated markets often offer for innovative companies interesting opportunities. is that a new company can almost always find a field in which it can do better, such as building friendlier, more personal or otherwise better relations with its clients.

- **in forming their marketing strategy, they rely on their personal experience.** They even quite often trust their own first-hand experience more than various would-be professional marketing researches and market surveys Also, they are interested in what both their current or potential clients desire or complain about, and use information they gain from everyday personal contacts with people.

- **understand the importance of leadership skills and surrounding themselves with a motivated team.** They understand that even a promising and interesting idea as such is, in most cases, not sufficient to develop a new business. It needs a successful implementation, and for this, the company's founders need a team.

  The prerequisite is to find people who are interested in the business idea and able to contribute or help with its implementation, but also people who are "team players". Simultaneously, successful entrepreneurs do not fear to delegate gradually at least some decision-making competencies to their team. They do not mind if others decide on certain aspects of their idea or company. They trust the people around them enough and even though they sometimes cultivate a certain "cult of personality" within their company, they understand that unwillingness to delegate leads usually to stopping growth of the firm and inability to motivate or even retain important employees.

- **they tackle the problems of their clients and understand the pitfalls of certain innovations.** In the age of digitization, many products are becoming more customer friendly and/or easier to use. Sometimes, however, or for some customers, new products may become more complicated. And often unnecessarily.

  Successful entrepreneurs or company founders are aware of these pitfalls and try to solve their problems for their customers, not to assign new ones. They allow them to orient themselves better in the "digitization jungle".

- **understand they have to invest not only money but also time.** Many advisors give guidance on how to better reconcile the requirements of work and personal life. This is important if we want to avoid burnout. At the same time, however, it must be understood that this is not the case for the founders of successful companies, at least in the first years.

  People who start thinking at the end of the weekend about when to take their next vacation are not the most suitable for business. Even those who feel guilty about their time at work and not with their family are not entirely suitable for them. It may not and should not always be the case, but the initial stages of business require time from the founder.

- **do not always try to keep their new business ideas secret.** Sometimes a newly established company may not be able to complete a new product or concept completely. In that case, it will usually do better if it publishes it or even introduces it to the market as unfinished, and only tries to complete it afterwards, even in cooperation with other companies.

  Thus, an interesting business idea does not always have to be kept secret, contrary to ordinary views. Sometimes it may be more appropriate to share it with others and work it out or finalize it with them. If a newly established company tries to present its product only in a definitive and perfect form, it will often find out at the moment of its launch that other companies have already entered it with a similar idea. By negotiating with potential partners, this danger can be avoided.

- **use various forms of professional external support.** Successful company founders are no longer mostly "lone fighters". They use all forms of support that are offered to them and do not resist them. It can be the support of the state and economic or business associations, but also help from schools and their professional or research background.

The goal of this support can be not only the development of a new product, but also the acquisition of managerial know-how, which the founders of companies do not always have at their disposal. It can also include help in solving problems with the management and leadership of people or organizations of the company, which the founders sometimes underestimate.

5 Discussion

Most business founders surveyed admitted that while they trusted in their business idea, they began to operate without adequate professional background. Specifically, they were lacking at least some business skills, knowledge, and strategies to succeed. Mainly, they admitted to have a very basic knowledge of accounting procedures and financial planning in the initial stages of business development, and often sought help from family and friends to learn how to become more adept in these areas.

Most participants, however, asserted that while business background is very helpful, friendliness, hospitality, and a welcoming spirit were key catalysts for achieving customer satisfaction. On top of that, most of them were convinced that not just the quality of customer service but also an interest in the community belonged to the most important ingredients in their capability to survive successfully for 5 or more years.

Some of them, in this regard, described customers as family, knowing most, if not all, regular customers by name as well as their usual order. All owners appeared to have customer relationships that perhaps a larger chain store could not achieve.

As far as people management and leadership skills, most participants mentioned that initial challenges of their business development involved a high rate of employee turnover and/or hiring the right people.

All participants noted the importance of the challenge of keeping staff motivated and happy, one of them shared that "open dialog and clear communications of expectations are critical." Another asserted that being involved on a personal level and being supportive of employees reduces turnover as the employees feel personally invested in the business. Most offered motivational rewards, some celebrated holidays with the employees. Most, however, expressed also a need to improve their interview process to make better hiring decisions and get the right fit of employees for the job as well as the culture of the business.

Several owners believed that while running an efficient business is important, it should also be fun.

When asked what the best advice they could offer to future small-business owners, the participants responded as follows: "*Have a good attitude and radiate your passion and dedication to your business and the customers", "Do not offer what everyone else if offering. Set yourself apart from the crowd. Keep it local with great products and fresh quality."* One should "*learn from your mistakes. Just because something did not work out the first time, do not give up. Give the customer what they want and a great experience and they will return with friends.*"

In terms of business development, word of mouth was one of the primary sources of initial advertising for all participants. Each of the participants noted overcoming the challenge of how to attract customers after they established credibility as a legitimate business.

5.1 Recommendations for Further Study

Recommendations for further research include focusing on different sizes of small business in different industries and varying geographical locations. Future researchers should also consider studying small-business owners who have increased profitability and examine the relevance of gender and age variances of the small-business owners through quantitative designs. Moreover, focusing on each of the specific themes identified within this study would provide more detailed areas for additional research.

6 Conclusion

The purpose of this descriptive, qualitative multiple case study was to explore what skills, knowledge, and strategies small-business owners use to succeed in business beyond 5 years. The study covered business from retail and personal service industries and the data sources included participant interview data, information concerning their financial results and observations from site visits of the businesses concerned.

The main and, as we believe, original findings of the study can be summarized in the following conclusions.

Successful entrepreneurs (a) feel the need for marketing differentiation, and don't necessarily adhere to what their competition is doing. Nor are they worried about apparently strange business concepts or ideas, (b) are not afraid of "mature" or occupied markets, (c) rely on personal information they get from daily contacts with people, both in the area of their business and outside it, (d) surround themselves with a motivated and empowered team, (e) tackle the problems of their clients, and specifically try not to come up with new ones for their clientele, (f) are not afraid to invest not only money but also time (g) do not always try to keep their new business ideas secret if the business development asks for co-operation with other subjects, (h) they do not mind using various forms of external professional support including the support of local business owners networking.

References

1. ADAMS, J., R. G. KAUFFMAN, F. M. KHOJA and S. COY, 2016. Looking at purchasing development through the lens of small business. *Journal of Managerial Issues*, **28**(3/4), 145-170. ISSN 1045-3695.
2. ARTINGER, S. and T. C. POWELL, 2015. Entrepreneurial failure: Statistical and psychological explanations. *Strategic Management Journal*, **37**(6), 1047–1064. ISSN 1097-0266.
3. BOX, T. M., 2011. Small firm strategy in turbulent times. *Academy of Strategic Management Journal*, **10**(1), 115–122. ISSN 1939-6104.
4. BOX, T. M. and W. D. MILLER, 2011. Small-firm competitive strategy. *Academy of Strategic Management Journal*, **10**(1), 55–59. ISSN 1939-6104.
5. BYRD, W., 2010. *The personal leadership practices of successful entrepreneurs* (Doctoral dissertation). Available from ProQuest Dissertations and Theses database. (UMI No. 3412018)
6. CADER, H. A. and J. C. LEATHERMAN, 2011. Small business survival and sample selection bias. *Small Business Economics*, **37**(2), 155–165. ISSN 0921-898X.
7. CRONIN-GILMORE, J. 2012. Exploring marketing strategies in small businesses. *Journal of Marketing Development and Competitiveness*, **6**(1), 96–107. ISSN 2155-2843.

8. DENZIN, N. K., and Y. S. LINCOLN, 2011. *The Sage handbook of qualitative research* (4th ed.). Thousand Oaks, CA: Sage Publications. ISBN-13 978-1412974172.
9. EDDLESTON, K. A. et al., (2013). Planning for growth: Life stage differences in family firms. *Entrepreneurship: Theory and Practice,* **37**(5), 1177–1202. ISSN 1042258.
10. FAHED-SREIH, J. AND S. MORIN-DELERM, 2012. A perspective on leadership in small businesses: Is the need for achievement a motive in predicting success? *International Journal of Entrepreneurship*, **16**(1), 1–23. ISSN: 1939-4675.
11. HANSEN, D. J., R. SHRADER and J. MONLLOR, 2011. Defragmenting definitions of entrepreneurial opportunity. *Journal of Small Business Management*, **49**(2), 283–304. ISSN 2515-1274.
12. HUNTER, M. G., 2011. Understanding the common causes of small business failures: A qualitative study. *Journal of Applied Management and Entrepreneurship,* **16**(1), 87–104. ISSN 1077-1158.
13. INSEAD, L. C., 2011. Competitors' resource-oriented strategies: Acting on competitors' resources through interventions in factor markets and political markets. *Academy of Management*, **33**(1), 197–121. ISSN 0001-4273.
14. JACKSON, E. A. and H. F. JACKSON, 2016. The role of corporate social responsibility in improving firms' business in directions of sustainable development, accountability and transparency. *Economic Insights – Trends and Challenges*, **68**(1), 39–50. ISSN 2284-8576.
15. KAUANUI, S. K. et al., 2010. An exploration of entrepreneurship and play. Journal of Organizational Change Management, **23**(1), 51–70. ISSN 0953-4814.
16. KAO, P. et al., 2015. How transformational leadership fuels employees' service innovation behaviour. *Service Industries Journal*, **35**(7/8), 448–466. ISSN 0264-2069.
17. LUSSIER, R. N. and C. E. HALABI, 2010. A three-country comparison of business success versus failure prediction model. *Journal of Small Business Management*, **48**(3), 360–377. ISSN 2515-1274.
18. MARTINEZ, M. A. and H. E. ALDRICH, 2011. Networking strategies for entrepreneurs: Balancing cohesion and diversity. *International Journal of Entrepreneurial Behaviour and Research*, **17**(1), 7–38. ISSN 1355-2554.
19. MCLARTY, R., M. PICHANIC and J. SRPOVA, 2012. Factors influencing the performance of small to medium-sized enterprises: An empirical study in the Czech Republic. *International Journal of Management*, **29**(3), 36–47. ISSN 2277- 5846.
20. O'DONNELL, A., 2014. The contribution of networking to small firm marketing. *Journal of Small Business Management*, **52**(1), 164–187. ISSN 2515-1274.
21. OSEI-ASSIBEY, E., G. BOKPIN, and D. TWEREFOU, 2012. Microenterprise financing preference: Testing POH within the context of Ghana's rural financial market. *Journal of Economic Studies*, **39**(1), 84–105. ISSN 0144-3585.
22. PERRY, M. (2014). Market orientation in small businesses: Creative or lacking? *Marketing Management Journal,* **24**(1), 96–107. ISSN 2333-6099.
23. SAKIRU, O. K. et al.,2013. Leadership styles and job satisfaction among employees in small and medium enterprises. *International Journal of Business and Management*, **8**(13), 34–41. ISSN 2321-8916.
24. SALAZAR, A. L., R. C. SOTO and R. E. MOSQUEDA, 2012. The impact of financial decisions and strategy on small business competitiveness. *Global Journal of Business Research*, **6**(2), 93–103. ISSN 1931-0277.
25. SCHAUPP, L. and F. BÉLANGER, 2014. The value of social media for small businesses. *Journal of Information Systems,* **28**(1), 187–207. ISSN 0306-4379.
26. SCHIFF, A., S. HAMMER and M. DAS, 2010. A financial feasibility test for aspiring entrepreneurs. *Entrepreneurial Executive*, **15**(3), 33–38. ISSN 1087-8955.

27. SENGE, P. M. et al., 2010. The necessary revolution: How individuals and organizations are working together to create a sustainable world. New York, NY: Doubleday. ISBN 978-0-385-51904-5.
28. Small Business and Entrepreneurial Council. (2016). *Small business facts and data*. Retrieved from http://sbecouncil.org/about-us/facts-and-data/
29. Surdez-Pérez, E. et al., 2014. The relationship between company performance and owner characteristics: Evidence from Mexico. *Global Journal of Business Research*, **8**(2), 13–22. ISSN 1931-0277.
30. U.S. Small Business Administration (SBA). (2012). *Growing your business*. Retrieved from http://www.sba.gov/content/ideas-growing-your-business
31. U.S. Small Business Administration (SBA). (2016). Frequently asked questions. Retrieved from https://www.sba.gov/sites/default/files/advocacy/SB-FAQ-2016_WEB.pdf
32. YIN, R. K., 2014. *Case study research: Design and methods*, (5th ed.). Thousand Oaks, CA: Sage Publications. ISBN 978-1-4522-4256-9.

Non-profit organization's endowment as a source of financing to improve its sustainability

Filip Rebetak[1,*] and *Viera* Bartosova[2]

[1]University of Zilina, Faculty of Operation and Economics of Transport and Communications, Department of Economics, Univerzitna 1, 010 26 Zilina, Slovakia
[2]University of Zilina, Faculty of Operation and Economics of Transport and Communications, Department of Economics, Univerzitna 1, 010 26 Zilina, Slovakia

Abstract. Non-profit organizations are an important part of the economy and are needed to fulfill many beneficiary roles that neither market, nor state would fill. Non-profit organizations in Slovakia became important after the 1989 Velvet revolution, the dissolution of the Czechoslovakia and creation of the free and democratic Slovak Republic. Ever since them, non-profit played a role in everyday life of people in Slovakia. Non-profit organizations exist with the aim of providing generally beneficiary services without the aim of making profit for their founders – which is what makes them distinct from for profit businesses. However, non-profit organizations need funding for their operations in the same way as any other organizations. They need to pay their bills, salaries, purchase materials, services, etc. Because of their non-profit nature, the sources of financing for these costs are different. The aim of this paper was to look at the endowment of non-profit organizations and its investment as a source of non-profit financing and the current situation in Slovakia regarding it use. We analysed financial statements of foundations available through Finstat.sk database. We found that the use of investing the endowment for self-financing of non-profit organizations is almost non-existent in conditions of Slovak non-profit sector, with only 13,7% of organizations having any long-term financial assets and only 3,9% having any substantial gains from them. Further, we proposed a framework for possible use of the endowment as a source of funding for more non-profit organizations in order to improve the financial situation and sustainability of the non-profit sector.
Keywords: non-profit organization, non-profit endowment, foundations, investment, non-profit self-financing

1 Introduction

Non-profit organizations are an important part of the economy and are needed to fulfill many beneficiary roles that neither market, nor state would fill in areas like healthcare, social care, education, culture, sports and culture, community development, etc. The importance of non-profit organizations is rising in the economic activity of contemporary societies. [1] Non-

* Corresponding author: filip.rebetak@fpedas.uniza.sk

profit organizations in Slovakia became important after the 1989 Velvet revolution, the dissolution of the Czechoslovakia and creation of the free and democratic Slovak Republic. Ever since them, non-profit played a role in everyday life of people in Slovakia. They have been only getting more and more important since. Currently, there are 51 784 non-profit organizations as per the latest data available for 2018 by the Statistical Bureau of the Slovak Republic. Out of which, 45 938 are associations, 4874 non-profit organizations, 468 foundations and 504 non-investment funds. [2]

Non-profit organizations exist with the aim of providing generally beneficiary services without the aim of making profit for their founders – which is what makes them distinct from for profit businesses. However, they need funds for their functioning the same way as for-profit businesses do. There are many ways they can get the funds they need. Some of them have already been studied by Slovak authors. [3, 4] Authors usually focus on fundraising from external sources, either public or private, or self-financing through business activities. Non-profit organizations may carry out income-generating activities, if their statutes so provide, only insofar as this serves the purposes for which they were established, and if it meets those objectives [5] and some authors think of it as an insurance [6] However, we identified one source that has been neglected and underdeveloped in the body of literature on non-profit organizations financing and funding. It is the creation and use of endowments to self-fund the non-profit organizations. In this article, we understand an endowment as money given to the organization to provide it with an income. It is a source of financing for non-profit organizations that is rarely mentioned in studies or articles, however, we find in an interesting option to be explored. Endowments, if properly invested and generating return, can help provide stable source of income for non-profit organizations independent on outside fundraising, current profitability of their business activities or grant applications.

The aim of this paper is to explore the endowment and its investment as a source of non-profit financing and the current situation in Slovakia regarding it use. Furthermore, we intend to offer a framework for possible use of the endowment as a source of funding for more non-profit organizations.

This paper is divided into 3 parts. The first part is a literature review of both domestic and foreign authors. The second part describes the methodology used in this paper. The third part represents the results of the analysis and discussion of possible solutions.

1.1 Literature review

Non-profit organizations in general have been and still are an object of study by many authors, both internationally and domestically. Sources of non-profit financing are also among the topics studied although are only marginal. Nationally, we can mention Hraskova [3] who wrote about dimensions of financing non-profit organizations in Slovakia, their different sources of financing which she identified as follows:

- state funding;
- partnership with other sectors;
- obtaining finance from own activities,
- use of EU funds and others.

Valaskova [4] focused on principles of non-profit financing and assess the system and structure of financing of a non-profit organization taking into account conditions of Slovak Republic. They both agree on vulnerability of non-profit organizations to the dependence on a single funding sources and that multiple sources of financing help achieve their long-term sustainability and existence. Valaskova [4] goes even further and mentions that historically, non-profit organizations in Slovakia have been dependent on a main source of financing from large, mainly foreign foundations and their existence is threatened by depletion of this source of financing.

As for the use of endowment for investing and generation of income, it has been mentioned by Marcek [7] who mentioned it as a form of self-financing among others, Svidronova and Vacekova [8] included it in their study on the sources of financing in 2011. They found it being used by 7,69 % of respondents, however she had only 136 of them. They didn't go further and focused only on the use of the sources, not on their principles.

As for internationally, this topic has been much more studied, since it is more common in other countries for non-profit organizations to have endowments. Main body is consisting of the articles from the US. There, the topic research was mainly the endowments of universities, who are non-profit and invest their endowment in search of a revenue stream. Merton [9] studied optimal investment strategies for university endowment funds back in 1993. Davidson [10] researched the same topic even before in 1971 while focusing on the differences in using internal and external management for the funds. Dahiya and Yermack [11] made an exhaustive study of non-profit endowment funds in the US on a sample of 29 762 organization in years 2009-2017. Endowment funds in the US is $0.7 trillion asset class, with majority of them (54%) being the higher education endowments while constituting only 6% of observations in the study. Best known example of a higher education endowment might be the one of Yale University, which is reported to be $29.4 billion in 2018. However, they found that typical endowment underperforms a 60-40 portfolio of equity and bond market indexes by about 4,46 percentage points. Liaw [12] also analysed investment strategies and return of university endowments and found their investment performances underperformed passive benchmarks for the studied years. In general, he found that stellar returns achieved by high profile endowments could not be duplicated by endowments of all sizes. He also underscored the importance of choosing and having access to top performing managers. There are also articles from other countries on the topic of university endowments – some analysed the methods of managing endowments and their applicability in Russian universities [13] others studied university endowments in Pakistan [14]. They both stressed importance of the endowments for the stability of the organizations.

Other type of research focused mainly on the technical aspects of non-profit endowments studying the efficiency of institutional spending and investment rules [15] or asset allocations [16]. Big data is also a new emerging trend. [17, 18] Others focus on strategic orientation [19] and employees [20].

As with all investing, there are financial risks involved for the organization and it is important to know how to assess them. [21, 22]

2 Methodology

The aim of this paper is to explore the endowment and its investment as a source of non-profit financing and the current situation in Slovakia regarding it use. Furthermore, we intend to offer a framework for possible use of the endowment as a source of funding for more non-profit organizations.

We use the data from publicly available sources such as the Statistical Bureau of the Slovak Republic, Database of non-profit organizations of Finstat (available at https://finstat.sk/databaza-neziskoviek), the Registry of financial statements and so on. Accounting legislation in the Slovak Republic gives entities the space to choose, within the regulated financial accounting, some alternative accounting procedures. [23]

This paper is aimed at Slovak non-profit organizations. In Slovak law, there is a broad definition of what a non-profit organization is - a taxable person not founded or incorporated for doing business. This would include churches, universities or state television and radio corporation. A stricter definition would include only organizations that are a part of what is more generally perceived as a non-profit, third, sector in Slovakia. That would be civic associations, foundations, non-investment funds and non-profit organisations providing generally beneficial services. However, due to the lack of availability and accessibility of information, we will focus on the assessment of foundations. It is also the best type legal entity to create an endowment and invest it.

The study focuses on the use of basic scientific methods of analysis, synthesis, induction, deduction and abstraction to accomplish the determined aim.

3 Results and Discussion

To analyze the current state of the use of endowments as a source of financing for non-profit organizations in Slovakia, we chose to analyze 51 foundations in Slovakia. The criterion for their choice was the availability of data from their financial statements available through Finstat for the year 2019. We found data for 51 foundations and analysed all of the available data. This constitutes approximately 10% of all foundations in Slovakia. We analyzed their balance sheets and income statements to look for evidence of any long-term capital and gains from it. We conducted this research on foundations, as they are the best vehicle to create an endowment and invest it and also, they are the most likely to publish their financial statements online. However, we accept this as a limitation for our research.

Only 13,7% of foundations had any kind of long-term financial assets on its book. Moreover, we found that only 3,9% (2 out of the 51 studied) had any substantial gains from their capital. It was invested in bonds held till maturity. Coincidently, both of these organizations were corporate foundations of banks operating in Slovak republic. Corporate foundation can be defined as founded by a corporation and having the corporation's name in its name. That might point to a better understanding of the use of the foundation's capital in foundations tied to banks. However, not every foundation founded by a bank was using its capital this way.

Most of foundations had their assets in form of cash in the bank, some of them even to the tune of millions of euros (7,8%), but usually in hundreds of thousands (58%). That is a lot of capital sitting in the accounts that could maybe be used for more productively – to gain some income that would further help finance organization's mission. Most of these were sitting idle gaining no income in interest, or only a modest interest in tens of euros.

A few of the analyzed foundations had other some other type of assets (buildings, land), but only one had earnings from lease.

What we thus see is that non-profit organizations in Slovakia in general don't use their capital endowment to generate income but usually only redistribute this capital or spend it on their mission.

What we want to propose for them is to create an endowment with principal that would not be touched and will be invested for gains. The central idea is to set aside a certain sum of money which can be invested to earn a sustainable form of income for the non-profit organization. Which would make it less dependent on outside financing.

We see this as a possibility mainly for the bigger foundation with already a large amount of money and an established donor base, that can redirect a part of their cash into investments and progressively create a nest egg that would be sufficient. Exact rates of investments are dependent on the ability to explain this plan to their stakeholders or find a replacement source of income in order not to decrease their current funding for their missions. Some of the biggest foundations in Slovakia are corporate foundations which usually serve as a sort of philanthropic arm of their parent corporations. They usually also depend on their parent company for most of their funding, either through direct donations or through tax assignation, which is a special form of support for non-profit sector whereby companies can direct a percentage of their income tax paid to a non-profit organization or organizations. Creation of an endowment might allow them to flatten the fluctuations in donations and assignations from the parent company stemming from the different stages of the economic cycle and help them create more stable income flows. For smaller non-profit organizations, we see this as an opportunity for creating a community behind a shared goal of creating an endowment that would finance the concrete missions – small local community projects, specific research, etc. for future generations. For example, with a creation of endowments for specific causes, they might persuade a big donor or a group of donors to put the principal in exchange for naming rights and then just support the mission from the realised income from the investments.

As for potential ways to invest the endowment, there are many possibilities. The choice depends on the amount of money invested, the timeframe for the endowment or choice of endowment spending policy. The bigger the amount of the endowment, the more possibilities it opens, since big sums open access to investment previously unavailable. Also diversification can better be attained while keeping in mind the costs of investing. The timeframe of endowment and choice of spending policy are tied together. The timeframe of the endowment depends on how much of income do we drain each year and whether we use its "body". The wisest spending policy would be to use only a part of the gains and reinvest in order to grow. This should however be decided in accordance with the means and goals of an organization.

As for investment vehicles, options are limitless, however many would be available only at higher sums and not for retail. Safest bets would be bonds. That's what some foundations already own. Bonds are generally considered less risky form of investment and pay yearly coupon payments, which can be translated into the spendable income. The safest would be government bonds, however with interest rates as low as 1%, they would not provide enough of an income. Corporate bonds are a riskier possibility; however, they can provide a more attractive interest rate. Stocks are another option with a possibility of more upside but also downside. Stock picking and trading stocks might not be what a typical non-profit organization might have the necessary know-how to do. For example, they might use technical analysis for their trading activities or P/E models. [24, 25] Slovak stock market is underdeveloped and thus they would be trading probably bigger European markets or the US market. [26] However, it is more probable they will use mutual funds managed by some domestic or international manager. Their disadvantage is the management fees they charge. In fact, there was even a mutual fund dedicated to non-profit organizations created in 2003 called CharityFund managed by Tatra Asset management, however, it's not available anymore. [7] A creation of a new such fund might be also a possibility to support creation of endowments in non-profit organizations in Slovakia, as it might provide easy way to invest, if it also kept fees low. However, as already mentioned, Dahiya and Yermack found that typical endowment in the US underperforms a 60-40 portfolio of equity and bond market indexes by about 4,46 percentage points. [11] Therefore, it might be wiser to shun stock picking and passively invest in market indexes, maybe even the 60-40 split suggested in their study. Pursuing this idea, a creation of an index fund aimed at Slovak non-profit organizations with the support of one of the big foundations supporting the non-profit sector giving it the proof of trustworthiness might be a right push to encourage Slovak non-profit organizations to invest. Another asset class non-profit organization might invest in is real estate. Whether commercial or residential, it might provide income through from renting it out and also might appreciate in value. Non-profit organizations functioning locally might have an added insight while investing in this type of assets and can also use these to fulfil their missions by being a sort of social entrepreneurs. Social entrepreneurship, socially responsible investing and impact investing might also be choices for an investment although we don't see many non-profit organizations going that way because of the complexity.

There are also some disadvantages. First and foremost, with almost any investment, there is a risk of a loss – loss of the income generated but also of a part or the whole of principal invested. For non-profit organizations, such a loss might threaten its financial stability and sustainability and also bring about an important reputational damage. Secondly, there are costs involved. Most of the investments mentioned have some costs either for the transaction or there are costs for the management of investments. These have to be taken into account too. They might be restrictive for non-profit organizations with smaller endowments, that's also why we proposed an establishment of a low-cost index fund for non-profit organizations.

Creation of such a fund might help keep costs low while letting the largest number of organizations to invest their capital in order to help sustainable fund their missions.

4 Conclusion

Non-profit organizations need funding for their operations in the same way as any other organizations. They need to pay their bills, salaries, purchase materials, services, etc. Because of their non-profit nature, the sources of financing for these costs are different. The aim of this paper was to look at the endowment of non-profit organizations and its investment as a source of non-profit financing and the current situation in Slovakia regarding it use. We found that the use of investing the endowment for self-financing of non-profit organizations is almost non-existent in conditions of Slovak non-profit sector. Further, we proposed a framework for possible use of the endowment as a source of funding for more non-profit organizations in order to improve the financial situation and sustainability of the non-profit sector. Limitation of our research might be its reliance on data only from foundations and not other types of non-profit organizations. However, we chose foundations due to the availability of data.

Acknowledgement
This paper is an outcome of project VEGA: 1/0544/19 Formation of the methodological platform to measure and assess the effectiveness and financial status of non-profit organizations in the Slovak Republic.

References

1. A. O. Carvalho, M. R. Ferreira, P. A. Silva, Partners in a caring society – a nonprofit organization case study. *Economics and Sociology*, **12**(2), 129-146 (2019)
2. Statistical Bureau of the Slovak Republic *Number of non-profit organizations* [online], Available at http://datacube.statistics.sk/#!/view/sk/VBD_SLOVSTAT/ns2003rs/v_ns2003rs_00_00_00_en
3. D. Hraskova, The dimension of financing of non-profit organizations in Slovak Republic. *Economic and Social Development 55th International Scientific Conference on Economic and Social Development Development*, **1**, 93-99 (2020)
4. K. Valaskova, Financing principles of non-profit organisations: A case of Slovakia. *Global Journal of Business, Economics and Management: Current Issues*. **9**(2), 55-62 (2019)
5. L. Shemyatikhina, K. Shipitsyna, M. Usheva, Marketing management of a non-profit organization. *Ekonomicko-manazerske spektrum*, **14**(1), 19-29 (2020)
6. J. Bennett, E. Iossa, G. Legrenzi, Commercial activity as insurance: the investment behaviour of non-profit organizations. *Annals Of Public And Cooperative Economics*, **81**(3), 445-465 (2010)

7. E. Marcek, Financing of non-profit sector in Slovakia after 1989 (Financovanie neziskového sektora na Slovensku po roku 1989) [online], Available at: http://www.cpf.sk/files/files/VV%20PANET_Financovanie_NS_89.pdf (2004).
8. M. Svidronova, G. Vacekova, Current state of selffinancing of private non-profit organizations in the conditions of the Slovak Republic. *Technological and Economic Development of Economy*, **18** (3), 438-451 (2012)
9. R. Merton, Optimal Investment Strategies for University Endowment Funds. In Clotfelter, C.T. & Rothschild M. (Eds.). *Studies of Supply and Demand in Higher Education* (pp. 211-242), University of Chicago Press (1993)
10. H. A. Davidson, Investing College Endowment Funds: A Comparison of Internal and External Management. *Financial Analysts Journal*, 69-74 (1973)
11. S. Dahiya, D. Yermack, *Investment Returns and Distribution Policies of Non-Profit Endowment Funds NBER Working Paper No. 25323* [online], Available at: https://www.nber.org/papers/w25323
12. K. T. Liaw, Investment Strategies and Returns of University Endowment Funds. *Review of Business*, **40**(1), 51-60 (2020)
13. E. Dyachkova, Managing Endowment Fund Income in Universities, *Voprosy Obrazovaniya-Educational Studies Moscow*, 1, 225-239 (2016)
14. N. Fardows, S. Nayer, S. Jaffar, H. Yousaf, Z. Nayer, Impact of endowment fund in the financial sustainability of an academic institute: a case study of Edwardes college of Khyber Pakhtunkhwa Pakistan. *9th International conference on education and new learning technologies (edulearn17)*, 10297-10305 (2017)
15. JM. Schumacher, Efficiency of institutional spending and investment rules. *Scandinavian Actuarial Journal,* **2020**(5), 454-476 (2019)
16. JL. Zhao, M. Zhou, A Review of the Research on Asset Allocation Theory. *Proceedings Of The 2017 International Conference On Sports, Arts, Education And Management Engineering*, **105**, 430-433 (2017)
17. L. Ashander, J. Kliestikova, P. Durana, J. Vrbka, The decision-making logic of big data algorithmic analytics. *Contemporary Readings in Law and Social Justice*, **11**(1), 57–62 (2019)
18. T. D. Meyers, L. Vagner, K. Janoskova, I. Grecu, G. Grecu, Big Data-driven algorithmic decision-making in selecting and managing employees: advanced predictive analytics, workforce metrics, and digital innovations for enhancing organizational human capital. *Psychosociological Issues in Human Resource Management,* **7**(2), 49–54 (2019)
19. E. Alani, S. Kamarudin, L. Alrubaiee, R. Tavakoli, A model of the relationship between strategic orientation and product innovation under the mediating effect of customer knowledge management, *Journal of International Studies*, **12**(3), 232-242 (2019)
20. T. Andjarwati, K. E. Susilo, A. K. Audah, Predictors of job satisfaction in non-profit organization. *Polish Journal of Management Studies*, **20**(1), 19-28 (2019)
21. K. Valaskova, T. Kliestik, M. Kovacova, Management of financial risks in Slovak enterprises using regression analysis. *Oeconomia Copernicana*, **9**(1), 105–121 (2018)
22. L. Svabova, M. Durica, Being an outlier: a company non-prosperity sign?. *Equilibrium. Quarterly Journal of Economics and Economic Policy*, **14**(2), 359-375 (2019)

23. V. Bartosova, P. Kral, A methodological framework of financial analysis results objectification in the Slovak Republic, *Proceedings of the 3rd International Conference on Business and Economics*, 189-197 (2016)
24. J. Chutka, K. Frajtova-Michalikova, N. Elvira, Technical Analysis as a Basis for Management of Trading Activities in Conditions of Slovak Republic. *Education Excellence And Innovation Management Through Vision 2020,* 7462-7468 (2019)
25. J. Chutka, K. Kramarova, Usage of P/E earning models as a tool for valuation of shares in condition of global market. *19th International Scientific Conference Globalization And Its Socio-Economic Consequences 2019 - Sustainability In The Global-Knowledge Economy,* **74**, (2020)
26. J. Salaga, V. Bartosova, E. Kicova, Economic Value Added as a measurement tool of financial performance. *Procedia Economics and Finance*, **26**, 484-489 (2015)

Impacts of COVID-19 on the Real Estate Market in the Czech Republic

Eduard Hromada[1,*]

[1]Czech Technical University in Prague, Faculty of Civil Engineering, Thakurova 7, Prague 6, Czech Republic

Abstract. The article deals with the description of the impacts of COVID-19 on the real estate market in the Czech Republic. The article focuses on the housing market - sales and rentals of apartments. The article contains graphs that show the development before COVID-19 and during COVID-19. Trends are indicated as the real estate market will develop in the next period. All results published in this article were created using the EVAL software, which the author of the article has been developing since 2007. This software continuously maps real estate advertising within all cities in the Czech Republic.
Keywords: real estate market, software, big data, statistics

1 Introduction

The presence of COVID-19 affected all areas of human activity, including the real estate market. At the beginning of the pandemic, many experts expected a significant drop in prices in the real estate market. These expectations turned out to be wrong. COVID-19 caused the real estate market to freeze in the months of March to June 2020, but from July 2020 the situation has changed. Currently, the average price level of the real estate market (sales of apartments) is higher than before the start of the pandemic COVID-19. Only the rental housing market in Prague is still affected by the pandemic.

2 Literature review

Many authors examine the development of the real estate market. They use various methods and procedures for this, which are further innovated. Various specialized software is used that can collect, analyse and evaluate large data. The problem is the evaluation of results, which due to inappropriate aggregation can provide the public with inaccurate information about trends in the real estate market. For example, a common mistake is to average results for too large a location (such as a state). In a given state, there are usually diverse areas with different socio-economic characteristics and population structure.

These articles describe the development of the real estate market, for example [3], [2], [11]. The authors focus on the market in their home state or home city. The dependences of individual parameters of the real estate market, for example on employment, the extent of

* Corresponding author: eduard.hromada@fsv.cvut.cz

mortgages, the age of the population, the balance of migration, the extent of new construction, etc., are examined. The paper [1] studies the housing market in Florida. Paper [9] discusses about the life-cycle cost analysis in relation to building's value assessment. The article [6] examines the dependencies of how immigration affects real estate prices in the locality. It was found that there is a big difference between whether rich immigrants come to the locality or immigrants without sufficient financial background.

Many authors [8], [4], [10], [7] also search for models that evaluate the real estate market. These models are used, for example, for investors to decide on the most advantageous real estate or to manage the risks associated with the acquisition and operation of real estate. These models are usually based on collecting a large set of data using automated algorithms.

Paper [5] elaborates on methods of automated data acquisition based on web mining, reviews methods of econometric spatial modelling with impact of proximity in order to deduct efficiency parameters within different categories.

3 Methodology of data collection

The results on the development of the real estate market in the Czech Republic were processed using the software tool EVAL, which the author of this paper is developing at the Faculty of Civil Engineering of the Czech Technical University in Prague. The software systematically collects, analyses and evaluates real estate price offers published on the real estate server www.sreality.cz. Currently, the scope of the software application covers all published advertisements in the category of apartments for sale and apartments for rent from this real estate server. Information is collected regularly over a period of one month, starting in 2007. Due to the situation with COVID-19, the data collection interval was shortened to a period of 15 days, always on the first and fifteenth day of the month.

The software recorded the following number of advertisements for individual cities in 2020 (January to September 2020): Prague (56,460 records), Brno (6,627 records), České Budějovice (1,673 records), Ostrava (4,887 records) and Ústí nad Labem (1,299 records).

The data show the offer prices published in the real estate advertisement. These are not actually realized prices. The results show trends in the development of the real estate market, in the current time. The data set was adjusted for advertisements that contain obviously incorrectly entered data by advertisers, as well as for advertisements for which the price was not filled.

4 Results

4.1 Apartments for sale in large cities in the Czech Republic

The effects of the measures on the COVID-19 pandemic have not yet had a significant effect on real estate sales prices in the Czech Republic. The real estate market usually reacts with a certain delay. Immediately after the measures were taken due to the pandemic, the real estate market froze. Since June 2020, there has been an increase in turnover on the real estate market and a further rise in prices. At present, real estate prices in Prague and the surrounding area are already higher than they were before the onset of the pandemic.

In the next period, we can expect further growth in sales prices of apartments. This is mainly due to the following factors:

- Central bank policy - quantitative easing. People are afraid of inflation and the devaluation of their savings. This newly printed money is usually not covered by any labour, value or production. Unfortunately, this money is used mainly for immediate consumption and is thus fully transferred to inflation (especially to the stock market, real estate, bonds). Many people are already realizing this risk and are trying to protect their capital. This is a key factor that will prevent real estate prices from falling even in the event of a recession.

- Decrease in mortgage interest rates, lack of alternative investment for capital appreciation and failure of pension reform.
- In the next period, developers want to keep part of their production (approximately 20-30% of the offer of flats they produce) and rent them themselves. Furthermore, they want to sell a large part of their production (up to 20-50%) directly as whole apartment buildings to funds and corporations. So the offer of new individual flats available to the general public (natural persons) will be significantly reduced with all the impacts. What is lacking is expensive.
- There is a lack of labour in the construction industry from abroad. There is no one to replace this workforce. Czech workers usually do not want to work in manual professions. If, however, some foreign workers are forced to be replaced by Czech workers, the work will become more expensive and thus the results of their work.
- The effects of the pandemic slowed down ongoing construction and delayed approval processes at building authorities. So new construction output will emerge more slowly.

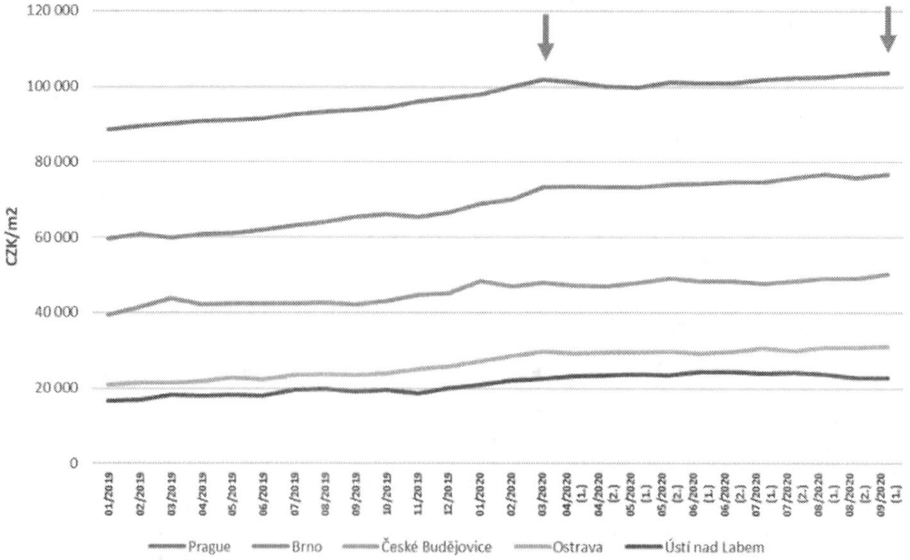

Fig. 1. Development of offer prices of apartments for sale in selected cities of the Czech Republic – medians
Source: Own processing.

The number of apartments offered for sale has a long-term downward trend. Only the AirBnB phenomenon in Prague and Brno and limited travel cause a temporary fluctuation in the number of apartments offered for sale upwards. In the medium term, a further decline in the number of apartments offered for sale throughout the Czech Republic can be expected. Construction companies and developers will limit new construction due to expectations of a global recession and a reduction in the demand side (lower purchasing power of the population, reduced availability of mortgages relative to household income, lower willingness to borrow, lower willingness to take risks, etc.).

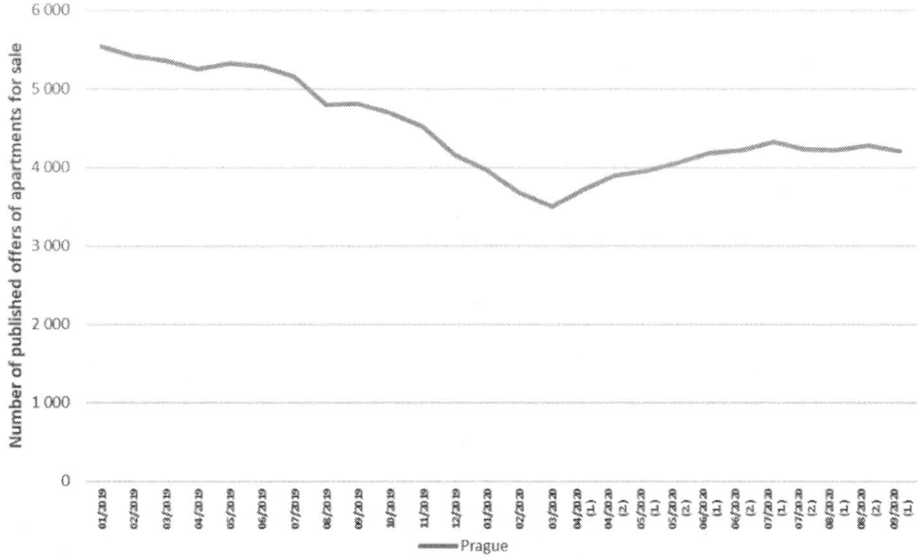

Fig. 2. Development of the number of apartments offered for sale in Prague
Source: Own processing.

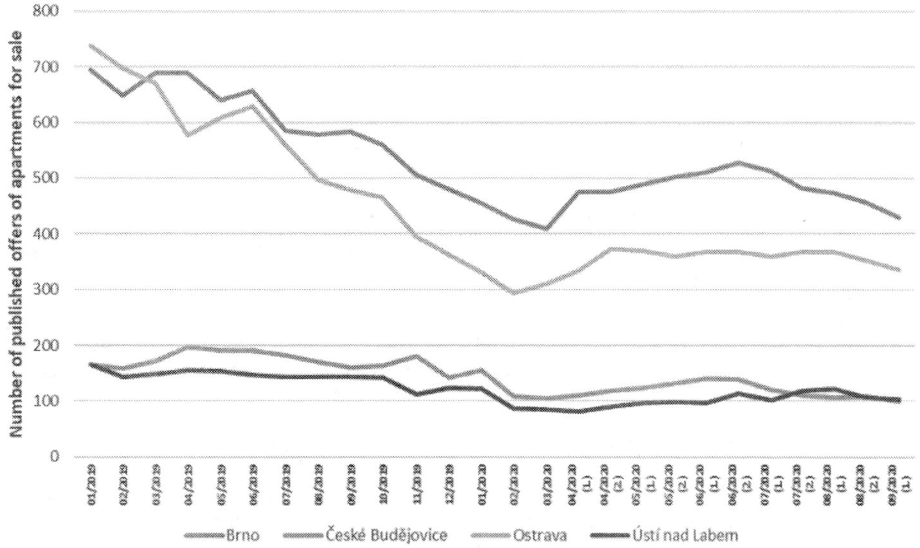

Fig. 3. Development of the number of apartments offered for sale in selected cities of the Czech Republic
Source: Own processing.

4.2 Apartments for rent in large cities in the Czech Republic

The effects of the COVID-19 pandemic were significantly reflected in the price of apartment rentals in Prague. This applies especially to apartments in the centre of Prague. Many AirBnB homeowners have started offering long-term rentals instead of short-term rentals. With this step, the offer of rental flats was expanded and prices fell. There is a paradoxical situation that rental prices in Brno are close to rental prices in Prague, while job opportunities and wages are significantly higher in Prague than in Brno.

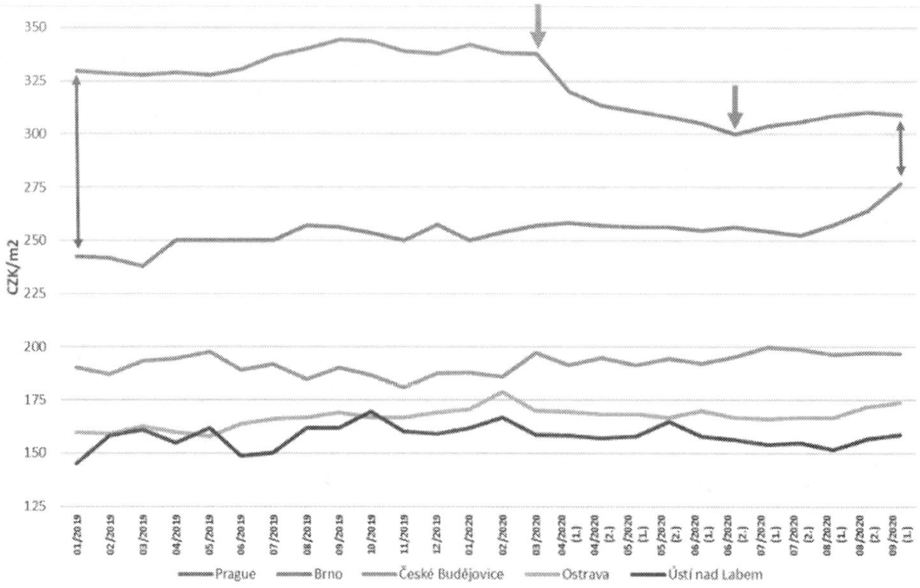

Fig. 4. Development of offer prices of apartments for rent in selected cities of the Czech Republic – medians
Source: Own processing.

The impacts of COVID-19 and the related shortfall in income from short-term leases caused a significant increase in the number of flats offered for rent in Prague and Brno. The situation was accelerated by the fact that at one point (suddenly) thousands of flats in Prague were returned to the rental market. The real estate market was not able to absorb this oversupply quickly. In other cities in the Czech Republic (with the exception of Český Krumlov and other attractive tourist destinations for foreign visitors), the impact of the AirBnB phenomenon is negligible.

In terms of the number of published advertisements offering apartments for rent, in April, May and June 2020 there was a massive increase in the number of offers for Prague and Brno apartments. From July 2020, the beginning of a change in the trend in Prague and Brno is evident. In Prague and Brno, the number of advertisements for rent is already declining. Other cities in the Czech Republic, where the AirBnB phenomenon is not significantly present (Ústí nad Labem, Ostrava, Zlín, etc.), maintain a similar development as before COVID-19. In addition, the related Figure 5 and 6 show a drop in the number of offers in August, September and October 2019 due to increased demand for rental housing from university students.

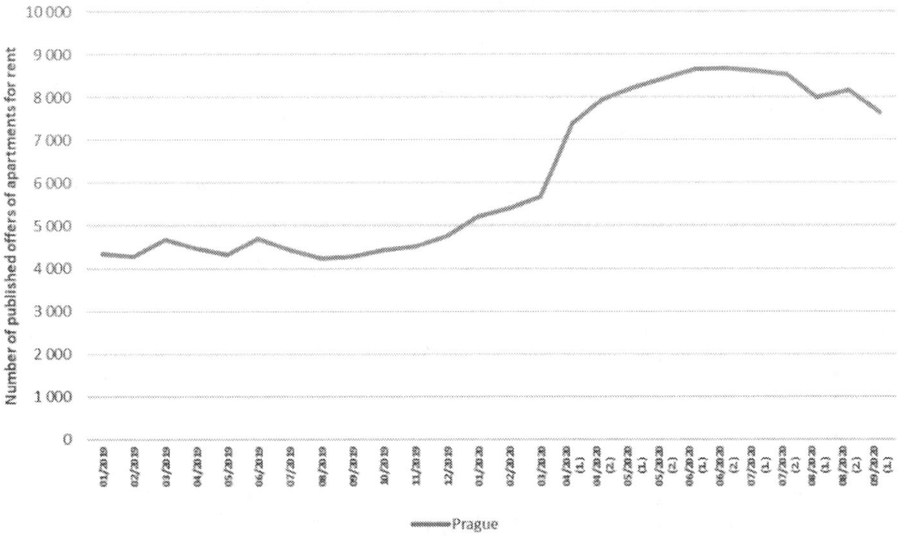

Fig. 5. Development of the number of advertisements for apartments intended for rent in Prague
Source: Own processing.

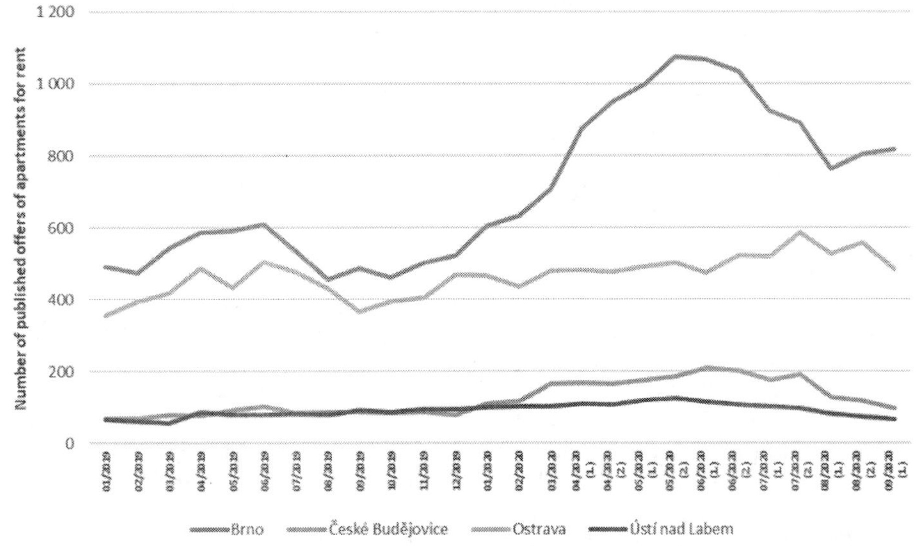

Fig. 6. Development of the number of advertisements for apartments intended for rent in selected cities in the Czech Republic
Source: Own processing.

5 Conclusion

In the next period, a deterioration in the financial affordability of owner-occupied and rented housing can be expected. In the area of apartment sales, AirBnB apartment sales can be expected from financially weaker property owners to stronger market players, including foreign ones. These can be, for example, wealthy individuals and corporate entities. It can be expected that developers will focus primarily on corporate clients in the coming period. The COVID-19 pandemic will accelerate the already started trends in the real estate market, such as accelerating the accumulation of assets towards the richest group of the population and strengthening the rentier capital.

The demand will increase especially for cheap flats and rooms for rent with good public transport access to the centre of Prague. There will be a fluctuation of tenants from more expensive real estate to cheaper real estate. Price will be the deciding factor, not the quality of housing. This trend will affect especially owners of AirBnB apartments in the centre of Prague, who have newly renovated apartments, equipped with quality and expensive furniture, who have invested a large amount of initial capital in the acquisition and reconstruction and cannot afford to rent them at lower prices.

References

1. A. Temilola, H. Chi, Investigation of Florida Housing Prices Using Predictive Time Series Model. *Proceedings of the Practice and Experience on Advanced Research Computing*, pp. 1-4 (2018)
2. G. Massimo, The Predictability of Real Estate Excess Returns: An Out-of-Sample Economic Value Analysis. *The Journal of Real Estate Finance and Economics*, (2020)
3. H. Xin, Search Benefit in Housing Markets: An Inverted U-Shaped Price and TOM Relation: Search Benefit in Housing Markets. *Real Estate Economics*, **48**(3), 772-807 (2020)
4. L. Beatriz, A Computer-Assisted Expert Algorithm for Real Estate Valuation in Spanish Cities. *Environment and Planning B: Urban Analytics and City Science*, (2020)
5. G. Muggenhuber, Geospatial Data Mining and Analytics for Real-Estate Applications. *Service-Oriented Mapping*, *Springer International Publishing*, pp. 225-40 (2019)
6. A. Pavlov, T. Somerville, Immigration, Capital Flows and Housing Prices. *Real Estate Economics*, **48**(3), 915-49 (2020)
7. J. Pike, The Future of Sustainable Real Estate Investments in a Post-COVID-19 World. *Journal of European Real Estate Research*, (2020)
8. H. Rabiei-Dastjerdi, Gap Analysis in Decision Support Systems for Real-Estate in the Era of the Digital Earth. *International Journal of Digital Earth*, 1-18 (2020)
9. R. Schneiderova Heralova, Building's value assessment using the utility and the LCC. *CESB 2007 Prague International Conference - Central Europe Towards Sustainable Building*, (2007)
10. S. R. Kumar, Factors Influencing Dividend Decisions of Indian Construction, Housing and Real Estate Companies: An Empirical Panel Data Analysis. *International Journal of Finance & Economics*, (2020)
11. R. Wisniewski, J. Brzezicka, Glocal Real Estate Market: Evidence from European Countries. *Journal of European Real Estate Research*, (2020)

Long memory in stock returns: Evidence from the Eastern European markets

Rui Dias[1] , *Paula* Heliodoro[2,*] , *Paulo* Alexandre[2], *Hortense* Santos[2] , and *Ana* Farinha[2]

[1]School of Business and Administration, Polytechnic Institute of Setúbal, Portugal and CEFAGE-UE, IIFA, University of Évora, Portugal
[2]School of Business and Administration, Polytechnic Institute of Setúbal, Portugal

Abstract. This essay aims to analyze the impact of the 2020 global pandemic on the memory properties of the Eastern Europe stock markets, from the period between 1 January 2016 to 2 September 2020, the sample was divided in two subperiods: 1 January 2016 to 30 August 2019 (before Covid 19) and 2 September 2019 to 2 September 2020 (after Covid 19). To perform this analysis, different approaches were undertaken to analyze whether if: (i) the global pandemic (Covid-19) accentuated the exponents *Detrended Fluctuation Analysis (DFA)* and the *Detrended Cross-Correlation* Analysis ($pDCCA$) in the Eastern European stock markets?. The daily returns do not have normal distributions, they have negative asymmetries, leptocubtic, and also exhibit conditional heteroscedasticity. The exponents Detrended Fluctuation Analysis (DFA), during the Covid-19 period, range from 0.64 to 0.75, showing significant long memories in all markets, except for the SLOVAKIA market (0.45). When we compared the 2 subperiods, we found that 41 pairs of markets have cross-correlation coefficients without trend (λ_{DCCA}) strong (out of 45 possible), and 4 pairs of markets decreased the $rhoDCCA$ in particular the markets ESTONIA-SLOVAKIA, LITHUANIA-SLOVAKIA, HUNGARY-SLOVAKIA, POLAND-SLOVAKIA. These findings show that the assumption of the market efficiency hypothesis may be in question, since the prediction of market movement can be improved if we consider the out-of-lag movements of the other markets, enabling the occurrence of arbitrage operations and some difficulties in portfolio diversification.
Keywords: Eastern European markets; long memories; arbitration; portfolio diversification.

1 Introduction

International financial markets have seen a succession of major setbacks in recent months, triggered by Covid-19, followed by a series of collapses, the oil war, and currency fluctuations. The economic turbulence associated with the coronavirus pandemic in 2019-

* Corresponding author: paula.heliodoro@esce.ips.pt

2020 has serious repercussions on financial markets, notably on the stock, bond and commodity markets (including crude oil and gold). The main events were an oil price war between Russia and Saudi Arabia after an OPEC agreement was not reached, which led to the collapse of oil prices, and a significant drop in stock markets in March 2020 [1].

Financial instability is a very important factor for society since a financial crisis or a stock market crash can directly or indirectly affect the level of economic well-being of the inhabitants of a country. If a given stock market is strongly linked to the stock market of another country, the financial stability of the former depends in part on the financial stability of the second. Thus, the occurrence of significant causation between markets may question the diversification of portfolios [2].

Statistical physicists began to analyze financial data, creating an interdisciplinary research area called Econophysics [3]. To answer the research question, we will use the *Detrended Fluctuation Analysis (DFA)* methodology, and Zebende's *Detrended Cross-Correlation Analysis (DCCA)* [4]. *DFA* is used to analyze the dependency pattern of a given time series and is typically identified as a way to study the level of efficiency of financial markets. While *DFA* analyzes the behavior of individual time series, $rhoDCCA$ will be applied to analyze the far-reaching correlations between central European stock markets.

This essay aims to analyze the memory properties of eastern European stock markets from 1 January 2016 to 2 September 2020. To perform this analysis, different approaches were undertaken to analyze whether: (i) the global pandemic (Covid-19) accentuated the exponents *Detrended Fluctuation Analysis (DFA)* and the *Detrended Cross-Correlation Analysis*($pDCCA$) in the Eastern European stock markets)?. The results suggest that the exponents *Detrended Fluctuation Analysis (DFA)* exhibit long memories, which may show some predictability in these regional markets. However, despite our findings, the hypotheses of anomic profitability have not been tested without incurring additional risk. The cross-correlation coefficients trendless λ_{DCCA} show 41 pairs of markets with strong $rhoDCCA$ ($\cong$ 0.666 $\rightarrow$ $\cong$ 1.000). The Estonia-Slovakia, Lithuania-Slovakia, Hungary-Slovakia, Poland-Slovakia markets decreased $rhoDCCA$. These findings show that the stock markets analyzed have high $rhoDCCA$ levels, which may question the hypothesis of portfolio diversification.

This essay adds two main contributions to the literature. The first contribution concerns the study of risk diversification in eastern European markets. As far as we know there are studies that have analyzed risk diversification in these regional markets, namely the authors Filipovski and Tevdovski [5], Ferreira [6], Muharam, Mawardi, Arfinto, and Najmudin [7], Moagar-Poladian, Clichici, and Stanciu [8], but the research questions, the sample, and the approach were different from those followed in this paper.

The second contribution is econophysical in nature, due to the comparative models *Detrended Fluctuation Analysis* (DFA) and *Detrended Cross-Correlation Analysis* ($pDCCA$) that have the possibility of evaluating correlations in the context of non-stationarity. DFA is an analysis method that examines temporal dependence on non-stationary data series. This technique by assuming that time series are non-stationary avoids spurious results when the analysis focuses on the relationships of the data series in the long term. *Zebende's* trendless *cross-correlation coefficient* ($pDCCA$) [4] is a method for quantifying the level of cross-correlation between two non-stationary time series.

In terms of structure this test is organized in 5 sections. Section 1 is represented by the current introduction. Section 2 presents a Literature Review regarding the presence of long memories in international stock markets. Section 3 describes the data and methodology. Section 4 contains the results. Finally, section 5 presents the general conclusions of the work.

2 Literature Review

The theme on the efficient market hypothesis (EMH) has motivated other studies to analyze the implications in the market efficiency hypothesis, according to which the current price of assets reflects all available information at a given time, and the price adjusts rapidly as new and unforeseen information comes to market. The hypothesis of reversal of the mean, also called negative series correlation, has been interpreted as an efficient correction mechanism in developed markets and a sign of a speculative bubble in emerging financial markets [9].

Nisar and Hanif [10], Mehla and Goya [11], El Khamlichi, Sarkar, Arouri, and Teulon [12], Li, Nishimura, and Men [13], tested the random walk hypotheses in asian markets. Nisar and Hanif [10] show that South Asian stock markets reject the hypothesis of random walk evidencing (in) market efficiency in its weak form. Mehla and Goya [11] examined india's stock market, showing that the market does not show the characteristics of random walk and, as such, is not efficient in weak form. El Khamlichi, Sarkar, Arouri, and Teulon [12] suggest that Islamic indices have the same level of (in)efficiency as benchmarks but MSCI and FTSE indices are less inefficient. Li, Nishimura, and Men [13] examined the fractal dimensions of 30 stock exchanges from 2006 to 2013, the authors argue that most markets are efficient, under the assumption of non-arbitration.

Ngene, Tah, and Darrat [14], Ali, Shahzad, Raza, and Al-Yahyaee [15], Pernagallo and Torrisi [16] investigated whether international markets follow the random walk hypothesis. Ngene, Tah, and Darrat [14] examined 18 emerging markets in the presence of multiple unscheduled and successive structural breakdowns. When single break tests are used, the random walk hypothesis is rejected. However, when double-level change tests are used in the mean and due tolerance is made to various structural breaks, the results are consistent with the random walk hypothesis in most markets. Ali, Shahzad, Raza, and Al-Yahyaee [15] using MF-DFA analysis, they show that developed markets are relatively more efficient, followed by BRICS stock markets. Comparative analysis of efficiency shows that almost all Islamic stock markets, with the exception of Russia, Jordan and Pakistan, are more efficient than their conventional peers. Pernagallo and Torrisi [16] analyzed whether the behavior of daily profitability of the stock indexes of 12 emerging economies corroborate the hypothesis of "fat tails" and whether these series show long memory. Hurst exponents range from 0.51 to 0.62 showing significant long memories.

Lahmiri, Bekiros, and Bezzina [17], Lahmiri and Bekiros [18] tested the presence of long memories in the stock markets. Lahmiri, Bekiros, and Bezzina [17] show that long-term price fluctuations are persistent, while short-term price fluctuations are anti-persistent. In addition, the level of anti-persistence and the content of information in short-term fluctuations are similar in the four European markets. Lahmiri and Bekiros [18] tested the efficient market hypothesis, in its weak form, on the Casablanca Stock Exchange (CSE),

Dow Jones and S&P500 stock markets. The authors show that prices are potentially predictable in the 3 markets and, in all industrial sectors.

Caporale, Gil-Alana, and Poza [19], Milos, Hatiegan, Milos, Barna, and Botoc [20] analyzed the central European stock markets. Caporale, Gil-Alana, and Poza [19] show that these regional markets have high levels of integration, which could call into question portfolio diversification. Milos, Hatiegan, Milos, Barna, and Botoc [20] verified the presence of long memories and long-term correlations, showing that stock markets are not efficient, in their weak form, and have not yet reached a mature stage of development.

In summary, this work aims to contribute to the provision of information to investors and regulators in the stock markets of Eastern Europe, where individual and institutional investors seek diversification benefits, as well as to help promote the implementation of policies that contribute to the efficiency of these markets. Therefore, the aim of this trial is to analyze the impact of the global 2020 pandemic on the memory properties of Eastern European stock markets.

3 Methodology

3.1 Data

The data used for the preparation of the paper were the prices index (diaries) of the stock markets of CZECH REP, BULGARIA, CROATIA, ESTONIA, LITHUANIA, HUNGARY, POLAND, SLOVAKIA, SLOVENIA, RUSSIA, from 1 January 2016 to 2 September 2020 and the sample was divided in two subperiods: from 1 January 2016 to 30 August 2019 (before Covid) and from 2 September 2019 to 2 September 2020 (Covid-19). The source of the information used was the DataStream platform with local currency quotes to mitigate exchange rate distortions.

3.2 Methodology

The development of the research took place through several stages. The characterization of the sample used was performed through descriptive statistics, the adherence test of Jarque and Bera [21] to verify if the data follow a normal distribution. In order to answer the research question, we will use the *Detrended Fluctuation Analysis (DFA)* methodology, and *Detrended Cross-Correlation Analysis* $(\rho DCCA)$. *DFA* is an analysis method that examines temporal dependence on non-stationary data series. This technique by assuming that time series are non-stationary avoids spurious results when the analysis focuses on the relationships of the data series in the long term. *DFA* has the following interpretation: $0 < \alpha < 0{,}5$: anti-persistent series; $\alpha = 0{,}5$ series features *random walk*; $0{,}5 < \alpha < 1$: persistent series. The function of this technique is to examine the relationship between values x_k and x_{k+t} at different moments [22]. The non-trend cross-correlation coefficient $(\rho DCCA)$ from Zebende [4] is a method to quantify the level of cross-correlation between two non-stationary time series. The coefficient is based on the *DFA* [23]and *DCCA* [23] methods. The cross-correlation coefficient depends on the length of the box s (time scale). One of the advantages of this cross-correlation coefficient is centered on the possibility of measuring the correlations between two non-stationary time series at different time scales. The function of this technique is to examine the relationship between values x_k and x_{k+t} at different moments [24]. Tables 2 and 3 show the interpretation of exponents α_{DFA} and $\rho DCCA$.

Table 1. Detrended Fluctuation Analysis

Exponent	Type of signal
$\alpha_{DFA} < 0.5$	long-range anti-persistent
$\alpha_{DFA} \simeq 0.5$	uncorrelated, white noise
$\alpha_{DFA} > 0.5$	long-range persistent

Source: Own processing.

Table 2. Detrended cross-correlation coefficient, $pDCCA$, levels

Weak	Medium	Strong
$\cong 0.000 \rightarrow \cong 0.333$	$\cong 0.333 \rightarrow \cong 0.666$	$\cong 0.666 \rightarrow \cong 1.000$

Source: Own processing.

4 Results

Figure 1 shows the evolution, in % of the differences, of the 10 stock markets in Eastern Europe. In all series, a relatively high dispersion around the mean is noted, as well as a relatively synchronized behavior between the data series. Through graphical analysis, high volatility is observed, especially in January, February and March 2020.

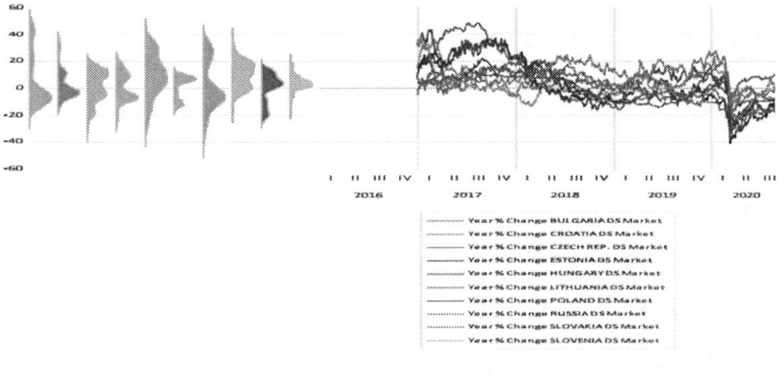

Fig. 1. Evolution, in % of differences, of the 10 stock markets, in the period from 1 January 2016 to 2 September 2020
Source: Own processing.

Table 3 shows the main descriptive statistics of the ten East European stock markets, yields present positive diary means, with the exception of CZECH REP, POLAND and SLOVAKIA. The HUNGARY market has the highest standard deviation (0.011803), while the LITHUANIA market has the highest level of kurtosis (70.51083) and asymmetry (-3.432789). In addition, the Jarque-Bera test confirms that we are faced with time series that do not follow a normal distribution, evidence confirmed by the levels of asymmetry (< 0) and kurtosis (> 3).

Table 3A. Descriptive statistics on profitability of the 5 stock markets from 1 January 2016 to 2 September 2020

	BULGARIA	CROATIA	CZECH REP	ESTONIA	HUNGARY
Mean	6.95E-05	3.26E-05	-9.64E-05	2.32E-05	0.000229
Std. Dev.	0.005066	0.005983	0.008566	0.007335	0.011803
Skewness	-2.145286	-1.497950	-1.551421	-4.697442	-1.507607
Kurtosis	29.18512	16.07891	18.42601	69.25563	18.52451
Jarque-Bera	35760.87***	9144.195***	12575.48***	227448.8***	12703.09***
Sum	0.084769	0.039718	-0.117520	0.028329	0.278575
Sum Sq. Dev.	0.031259	0.043601	0.089366	0.065531	0.169668
Observations	1219	1219	1219	1219	1219

Source: Own processing.

Table 3B. Descriptive statistics on profitability of the 5 stock markets from 1 January 2016 to 2 September 2020

	LITHUANIA	POLAND	SLOVAKIA	RUSSIA	SLOVENIA
Mean	8.69E-05	-6.88E-05	-6.47E-06	0.000373	8.13E-05
Std. Dev.	0.005160	0.011463	0.007668	0.010603	0.007269
Skewness	-3.432789	1.457475	-0.536518	-1.151359	-2.258049
Kurtosis	70.51083	20.33156	8.716673	20.08998	33.40669
Jarque-Bera	233887.9	15688.53	1718.371	15103.92	47996.20
Sum	0.105954	-0.083915	-0.007889	0.454651	0.099117
Sum Sq. Dev.	0.032434	0.160038	0.071624	0.136931	0.064355
Observations	1219	1219	1219	1219	1219

Source: Own processing.

In table 4 we can see the *DFA* exponents referring to the 10 financial markets of Eastern Europe. Period I regarding the Pre-Covid time scale and we can verify the presence of long memories in the markets BULGARIA (0.54), CROATIA (0.59), ESTONIA (0.55), LITHUANIA (0.59), SLOVENIA (0.55). The CZECH REP. (0.41), HUNGARY (0.44), POLAND (0.46), SLOVAKIA (0.36), RUSSIA (0.48) markets show signs of anti-persistence, i.e. the presence of short-term memories. In period II the covid-19-time scale is contained, and we found that the exponents *Detrended Fluctuation Analysis (DFA)* range from 0.64 to 0.75, showing significant long memories in all markets, except for the SLOVAKIA market (0.45). These findings show that prices do not fully reflect the information available and that price changes are not i.i.d. This situation has implications for investors, as some profitability can be expected, creating opportunities for arbitration and abnormal income. These results are corroborated by the authors Lahmiri and Bekiros [18], Milos, Hatiegan, Milos, Barna, and Botoc [20] who evidence long memories in the profitability of the stock markets.

Table 4. DFA exponent for index and return. The values of the linear adjustments for α DFA always had R2 > 0.99

Stock market	DFA exponent (before crisis)	DFA exponent (Covid period)
CZECH REP. DS Market	0.41 ≅ 0.0039	0.68 ≅ 0.0024
BULGARIA DS Market	0.54 ≅ 0.0076	0.74 ≅ 0.0074
CROATIA DS Market	0.59 ≅ 0.0058	0.72 ≅ 0.0012
ESTONIA DS Market	0.55 ≅ 0.0008	0.73 ≅ 0.0049
LITHUANIA DS Market	0.59 ≅ 0.0003	0.64 ≅ 0.0062
HUNGARY DS Market	0.44 ≅ 0.0011	0.65 ≅ 0.0007
POLAND DS Market	0.46 ≅ 0.0097	0.66 ≅ 0.0034
SLOVAKIA DS Market	0.36 ≅ 0.0041	0.45 ≅ 0.0039
SLOVENIA DS Market	0.55 ≅ 0.0056	0.75 ≅ 0.0011
RUSSIA DS Market	0.48 ≅ 0.0041	0.71 ≅ 0.0012

Note: The hypotheses are H_0: α = 0.5 and H_1: α ≠ 0.5.
Source: Own processing.

Table 5 shows the *Detrended cross-correlation coefficient* for $pDCCA$ period from January 1, 2016 to September 2, 2020. Period I comprises the time lapse from January 1, 2016 to August 30, 2019 (Pre-Covid), and we can verify that $rhoDCCA$ shows 28 cross-correlation coefficients trendless (λ_{DCCA}) weak (≅ 0.000 → ≅ 0.333), and 17 medium correlation coefficients (≅ 0.333 → ≅ 0.666), in 45 possible pairs. When we analyzed period II (Covid) we found 36 cross-correlation coefficients trendless (λ_{DCCA}) strong (0.666 → ≅ 1,000), and 9 weak correlation coefficients (≅ 0.000 → ≅ 0.333). When compared the 2 subperiods we found 41 pairs of markets cross-correlation coefficients trendless (λ_{DCCA}) strong (out of 45 possible), and 4 pairs of markets decreased $rhoDCCA$, namely ESTONIA-SLOVAKIA, LITHUANIA-SLOVAKIA, HUNGARY-SLOVAKIA, POLAND-SLOVAKIA. These findings show that the stock markets analyzed have high levels $rhoDCCA$, which may question the hypothesis of efficient portfolio diversification.

Table 5. Summary table of coefficients, pDCCA referring to the 10 stock markets, in the period from 01/01/2016 to 02/09/2020

	ρDCCA (before Covid-19)		ρDCCA (Covid-19)		
Index	*pDCCA*	Time scale (days)	*pDCCA*	Time scale (days)	Tendência
CZECH REP. / BULGARIA	**0.13**	n > 236 days	0.91	n > 64 days	↑
CZECH REP. / CROATIA	**0.19**	n > 102 days	0.88	n > 60 days	↑
CZECH REP. / ESTONIA	**0.26**	n > 66 days	0.86	n > 56 days	↑
CZECH REP. / LITHUANIA	0.34	n > 138 days	0.85	n > 64 days	↑
CZECH REP. / HUNGARY	0.42	n > 22 days	0.88	n > 56 days	↑
CZECH REP. / POLAND	0.41	n > 187 days	0.87	n > 64 days	↑
CZECH REP. / SLOVAKIA	**0.03**	n > 166 days	**0.23**	n > 64 days	↑
CZECH REP. / SLOVENIA	0.36	n > 210 days	0.93	n > 64 days	↑

CZECH REP. / RUSSIA	**0.28**	n > 14 days	0.83	n > 62 days	↑
BULGARIA / CROATIA	**0.08**	n > 210 days	0.89	n > 64 days	↑
BULGARIA / ESTONIA	**0.29**	n > 236 days	0.86	n > 60 days	↑
BULGARIA / LITHUANIA	**0.10**	n > 26 days	0.80	n > 58 days	↑
BULGARIA / HUNGARY	**0.15**	n > 236 days	0.86	n > 60 days	↑
BULGARIA / POLAND	0.42	n > 42 days	0.81	n > 64 days	↑
BULGARIA / SLOVAKIA	**0.06**	n > 156 days	**0.28**	n > 64 days	↑
BULGARIA / SLOVENIA	**0.11**	n > 85 days	0.90	n > 62 days	↑
BULGARIA / RUSSIA	**0.19**	n > 75 days	0.80	n > 60 days	↑
CROATIA / ESTONIA	0.35	n > 166 days	0.84	n > 64 days	↑
CROATIA / LITHUANIA	**0.28**	n > 166 days	0.79	n > 58 days	↑
CROATIA / HUNGARY	**0.17**	n > 176 days	0.83	n > 64 days	↑
CROATIA / POLAND	**0.19**	n > 147 days	0.87	n > 64 days	↑
CROATIA / SLOVAKIA	**0.11**	n > 198 days	**0.20**	n > 64 days	↑
CROATIA / SLOVENIA	0.35	n > 138 days	0.93	n > 62 days	↑
CROATIA / RUSSIA	0.33	n > 223 days	0.80	n > 64 days	↑
ESTONIA / LITHUANIA	0.44	n > 187 days	0.84	n > 56 days	↑
ESTONIA / HUNGARY	0.36	n > 210 days	0.83	n > 64 days	↑
ESTONIA / POLAND	0.35	n > 198 days	0.77	n > 64 days	↑
ESTONIA / SLOVAKIA	**0.40**	**n > 223 days**	**0.14**	**n > 64 days**	↓
ESTONIA / SLOVENIA	0.36	n > 138 days	0.87	n > 64 days	↑
ESTONIA / RUSSIA	**0.31**	n > 198 days	0.72	n > 64 days	↑
LITHUANIA / HUNGARY	**0.11**	n > 108 days	0.69	n > 45 days	↑
LITHUANIA / POLAND	**0.26**	n > 130 days	0.79	n > 60 days	↑
LITHUANIA / SLOVAKIA	**0.32**	**n > 166 days**	**0.12**	**n > 64 days**	↓
LITHUANIA / SLOVENIA	0.53	n > 236 days	0.83	n > 64 days	↑
LITHUANIA / RUSSIA	**0.16**	n > 130 days	0.76	n > 64 days	↑
HUNGARY / POLAND	0.53	n > 96 days	0.80	n > 62 days	↑
HUNGARY / SLOVAKIA	**0.22**	**n > 122 days**	**0.19**	**n > 58 days**	↓
HUNGARY / SLOVENIA	**0.20**	n > 90 days	0.85	n > 62 days	↑
HUNGARY / RUSSIA	**0.29**	n > 236 days	0.81	n > 48 days	↑
POLAND / SLOVAKIA	**0.18**	**n > 236 days**	**0.08**	**n > 62 days**	↓
POLAND / SLOVENIA	0.46	n > 122 days	0.86	n > 64 days	↑
POLAND / RUSSIA	0.49	n > 14 days	0.87	n > 62 days	↑
SLOVAKIA / SLOVENIA	**0.12**	n > 147 days	**0.20**	n > 64 days	↑
SLOVAKIA / RUSSIA	**0.10**	n > 130 days	**0.16**	n > 36 days	↑
SLOVENIA / RUSSIA	**0.15**	n > 20 days	0.80	n > 64 days	↑

Source: Own processing.

5 Conclusion

The overall conclusion to be withheld and sustained in the results obtained, through tests carried out with econophysical models, shows that the global pandemic has a significant impact on the adjustment of the financial markets of Eastern Europe. The results indicate that these regional markets have significant persistence during the Covid period, which motivates some arbitrage opportunities. Additionally, *Detrended cross-correlation coefficient* $\rho DCCA$ increased in the Covid period when compared to the previous subperiod, which may call into question the hypothesis of portfolio diversification. These findings also make room for market regulators to take steps to ensure better information among international financial markets. In conclusion, we consider that investors should diversify their portfolios, and invest in less risky markets, with the purpose of mitigating risk and improving the efficiency of their portfolios.

References

1. M. T. S. G.Sudha, V.Sornaganesh, "IMPACT OF INDIAN STOCK MARKET DUE TO CRISIS IN MARCH 2020," 2020.
2. R. Dias, J. V. da Silva, and A. Dionísio, "Financial markets of the LAC region: Does the crisis influence the financial integration?," *Int. Rev. Financ. Anal.*, **63**, (February), 160–173 (2019).
3. M. Gallegati, "Beyond econophysics (not to mention mainstream economics)," *Eur. Phys. J. Spec. Top.* (2016).
4. G. F. Zebende, "DCCA cross-correlation coefficient: Quantifying level of cross-correlation," *Phys. A Stat. Mech. its Appl.*, **390**, (4), 614–618 (2011).
5. V. Filipovski and D. Tevdovski, "Stock market efficiency in south eastern Europe: Testing return predictability and calendar effects," in *Regaining Global Stability After the Financial Crisis* (2018).
6. P. Ferreira, "Long-range dependencies of Eastern European stock markets: A dynamic detrended analysis," *Phys. A Stat. Mech. its Appl* (2018).
7. H. Muharam, W. Mawardi, E. D. Arfinto, and Najmudin, "Volatility spillovers under difference in the degree of market integration: Evidence from the selected asian and eastern European stock markets," *J. Int. Stud* (2019).
8. S. Moagar-Poladian, D. Clichici, and C. V. Stanciu, "The comovement of exchange rates and stock markets in Central and Eastern Europe," *Sustain.* (2019).
9. Lawrence H. Summers, "Does the stock market rationally reflect fundamental values," *J. Finance* (1986).
10. S. Nisar and M. Hanif, "Testing weak form of efficient market hypothesis: Empirical evidence from South-Asia," *World Appl. Sci. J.* (2012).
11. S. Mehla and S. K. Goyal, "Empirical Evidence on Weak Form of Efficiency in Indian Stock Market," *Asia-Pacific J. Manag. Res. Innov.* (2013).
12. A. El Khamlichi, K. Sarkar, M. Arouri, and F. Teulon, "Are Islamic equity indices more efficient than their conventional counterparts? Evidence from major global index families," *J. Appl. Bus. Res.* (2014).
13. D. Li, Y. Nishimura, and M. Men, "The long memory and the transaction cost in financial markets," *Phys. A Stat. Mech. its Appl.* (2016).
14. G. Ngene, K. A. Tah, and A. F. Darrat, "The random-walk hypothesis revisited: new evidence on multiple structural breaks in emerging markets," *Macroecon. Financ. Emerg. Mark. Econ.* (2017).
15. S. Ali, S. J. H. Shahzad, N. Raza, and K. H. Al-Yahyaee, "Stock market efficiency: A comparative analysis of Islamic and conventional stock markets," *Phys. A Stat. Mech.*

its Appl. (2018).
16. G. Pernagallo and B. Torrisi, “An empirical analysis on the degree of Gaussianity and long memory of financial returns in emerging economies,” *Phys. A Stat. Mech. its Appl.* (2019).
17. S. Lahmiri, S. Bekiros, and F. Bezzina, “Multi-fluctuation nonlinear patterns of European financial markets based on adaptive filtering with application to family business, green, Islamic, common stocks, and comparison with Bitcoin, NASDAQ, and VIX,” *Phys. A Stat. Mech. its Appl.* (2020).
18. S. Lahmiri and S. Bekiros, “Nonlinear analysis of Casablanca Stock Exchange, Dow Jones and S&P500 industrial sectors with a comparison,” *Phys. A Stat. Mech. its Appl.* (2020).
19. G. M. Caporale, L. A. Gil-Alana, and C. Poza, “High and low prices and the range in the European stock markets: A long-memory approach,” *Res. Int. Bus. Financ.* (2020).
20. L. R. Milos, C. Hatiegan, M. C. Milos, F. M. Barna, and C. Botoc, “Multifractal detrended fluctuation analysis (MF-DFA) of stock market indexes. Empirical evidence from seven central and eastern european markets,” *Sustain.* (2020).
21. C. M. Jarque and A. K. Bera, “Efficient tests for normality, homoscedasticity and serial independence of regression residuals,” *Econ. Lett.*, **6** (3), 255–259 (1980).
22. E. F. Guedes *et al.*, “Statistical test for ΔρDCCA: Methods and data,” *Data Br.* (2018).
23. C. K. Peng, S. V. Buldyrev, S. Havlin, M. Simons, H. E. Stanley, and A. L. Goldberger, “Mosaic organization of DNA nucleotides,” *Phys. Rev. E*, **49** (2), 1685–1689 (1994).
24. B. Podobnik and H. E. Stanley, “Detrended cross-correlation analysis: A new method for analyzing two nonstationary time series,” *Phys. Rev. Lett.*, **100**, (8), (2008).
25. P. Ferreira, A. Dionísio, E. F. Guedes, and G. F. Zebende, “A sliding windows approach to analyse the evolution of bank shares in the European Union,” *Phys. A Stat. Mech. its Appl.*, **490**, 1355–1367 (2018).

SHS Web of Conferences **91**, 01030 (2021) https://doi.org/10.1051/shsconf/20219101030
IES2020

Current Needs of Business Owners and Managers, Business Schools' View

V.V. Mantulenko[1,*] and *D.V.* Aleshkova[1]

[1]Samara State University of Economics, Samara, Russia

Abstract. This article investigates actual training programs of leading Russian business schools for entrepreneurs, managers, and business owners. The aim of this research is to analyze current trends in the Russian business education and compare them with experts' forecast and opinions about the orientation of Russian entrepreneurship to general management issues (complex, integrated consideration of business processes, cultural and value based approaches to business management). The main research methods are analysis, systematization, and generalization. Based on the results of this research work, the authors make a conclusion about the presence of a new development stage of Russian entrepreneurship characterized by new tasks and requirements, urgent needs and skills in demand.
Keywords: business schools, business education, entrepreneurial culture, Russian entrepreneurship.

1 Introduction

The transition from the industrial mitre to the cognitive world that we are now witnessing and actively participating in is more like a quantum leap from one reality to another, built on different principles, new values, and a reinterpretation of traditional truths. The experience accumulated by mankind is becoming obsolete faster and faster. Business experience is certainly historically significant, but it is less and less used as a basis for modern directors when making important strategic decisions. An information-based economy requires business owners and managers to master the tools for obtaining, processing, broadcasting and preserving information resources, to synthesize new knowledge, to create unique products and services, and to make decisions in conditions of high uncertainty and instability of the environment. At the same time, according to the observations of scientists, experts, and practitioners, today many Russian entrepreneurs and senior managers have an intellectual hunger, due to the fact that they know "their" business very well and manage it quite competently [19, 22]. Quite young Russian businesses as a class today have less and less pragmatic needs related to a global rethinking of the process of creating business models, intellectual communication on topics that are not directly related to their professional sphere.

* Corresponding author: mantoulenko@mail.ru

Increasingly, scientific discussions use the term "intellectual entrepreneurship" as a new conceptual framework that requires its own unique methodology. The source of the main transformations of modern business education is the thesis that a person should be able to build their intellectual world, including in the business space. Entrepreneurship as a human activity is increasingly associated with creativity. Educational systems of various levels are aimed at forming an entrepreneurial and creative attitude to one's own life [24]. Life itself is considered as a creative project, and entrepreneurship is not just a way to make money ("profit-making machine"), but is defined as the search for opportunities that are beyond the resources that are available today, as an opportunities for development.

Today we can clearly see the drift of Russian entrepreneurship towards society, more and more often business is not about money, but about pleasure, and more and more clearly entrepreneurs realize that they sell not just goods, services and works, but "emotions". The term "earn" is increasingly found not only in combination with the word "money", but also "trust and loyalty of customers", "respect", "reputation". There is a new format of competition - competition for the most accurate satisfaction of the needs of society, designing, foreseeing the needs of the future right now. In these conditions, analytics skills are extremely in demand. It is proved that digital technologies solve analytical problems faster and better than a human. However, an economy based on knowledge and information is not only an economy of algorithms and diagrams, graphs and indicators, it is also an economy of intuition and emotions, impressions and unconventional thinking. In this context, human intelligence today and in the near future is unlikely to be able to give up its leading position to artificial intelligence.

The purpose of this work is to consider the current needs of Russian entrepreneurship, the required competencies in the context of the global transformation of labor markets and the economy as a whole. In accordance with this goal, the main research tasks were to analyze the current requirements of the labor market, trends, development trends of entrepreneurial competencies, as well as the suggestions of the largest business schools in the Russian Federation in this context.

2 Literature review

Modern researchers and experts point out a number of skills as key to the development of entrepreneurial competencies in the near future. These include creativity [20], readiness for change [14, 15], the ability to solve complex problems, and critical thinking in the context of general competencies and skills that are already in demand in the global labor market. Entrepreneurs, business consultants, coaches and business coaches themselves are often talking about the transition from "business models" to network interaction and cooperation, creating business ecosystems (value networks and business ecosystems) [7, 12, 13].

On the one hand, the sphere of educational services today (especially business education) strives to adapt to the requirements of customers and their needs in order to bring theory as close as possible to practice. On the other hand, the problems of the gap between the competencies and knowledge generated by different business schools and the real requirements of the rapidly changing labor market and business context are not new, but they are still relevant [1, 8]. Peter Trkman points out the need for a deeper study of the content value of existing business schools [16]. Ahuja & Purankar consider factors which impact the quality of business school education and help understand employer expectations from different sectors [21].

The effectiveness of business schools' operation is considered in various studies. First of all, the focus is on the composition of experts who are engaged in the training process. In most cases, it is the quality of teaching that helps to master a particular curriculum more successfully. It is necessary to take into account the specifics of the target audience in order to adapt business programs to them [11]. Practical experience shows that the most successful method for the main target groups is discussing and solving various business cases. It is used in business schools of different countries [4] and important for real entrepreneurs and those who are only entering the business and want to implement the gained knowledge as fully and efficiently as possible, form competencies and skills which are relevant in the modern VUCA world. The issue of competencies that are especially significant for the contemporary business world is also highly debatable. The new economic paradigm provides modernization of educational business programs in the direction of their local sensitivity, interpersonal relations, and socialization of the individual [18]. Such trends shape needs of target groups and require the educational processes modernization that is why the analysis of existing business schools is so important for potential consumers at the present time.

3 Methodology

The main research methods were the analysis, synthesis and generalization of scientific literature on the studied issues, official information from business school websites, interviews with experts in the field of labor economics, management and economic sociology. New Professions Atlas, developed by the Agency for Strategic Initiatives and the SKOLKOVO International School of Management in 2014, publications of the World Economic Forum in 2019, PwC in Russia in 2018 were analyzed [17]. Within the framework of the research work, the author relied on the principles of theoretical and practical significance of the research results, its relevance and novelty, consistency. Information brochures, descriptions of training courses and programs of leading business schools in Russia, whose target audience is entrepreneurs and managers, were the actual material.

The research sample consists of 6 leading Russian business schools, according to the National Ranking, and it was composed based on the following criteria:

- the sum of scores by independent experts and students,
- estimation of learning parameters (arithmetic mean): income growth, career growth, usefulness of business communications, personal and professional development,
- cost of training [23].

4 Results

Based on the analysis of educational programs and courses of leading business schools in Russia, we identified the most popular entrepreneurial competencies and the intersection points of the real offer to entrepreneurs and senior and middle executives with the forecasts of expert analysts regarding changes in this area (Table 1).

Table 1. Educational programs for entrepreneurs

Name of the program (Business school)	**Target group**	**Key points of the program**
Startup Academy (SKOLKOVO)	Young entrepreneurs, founders of startups with an innovative business model	- combinations of techniques that work through thinking; - formation of the entrepreneur's *networking* for building investment-attractive projects; - creating and testing business models;

		- interaction with (potential) investors.
SKOLKOVO Workshop (SKOLKOVO)	Entrepreneurs, managing owners, general and functional directors of companies that already own a business and have more than 10 years of management experience.	- «rebuilding of the business model and analysis of new markets»; - creating your own business model; - development of business strategy and *personal development trajectory*; - changing the scale of business and expanding planning horizons.
Practicum Global Shift (SKOLKOVO)	Managing owners, top managers, and functional directors.	- design and test a global *strategy*; - "entrepreneur ontology" - basic concepts and schemes that define the global nature of business activity; - project work on creating a global company; - business modeling is a rebuilding of the business model for global markets.
Global Shift Expedition (SKOLKOVO)	Managing owners, top managers, and functional directors.	- opportunity to explore new territories, markets, consumer culture and entrepreneurial *ecosystems*; - to see new business models and partnerships, to get socio-cultural experience and to exchange insights behind the history of the country; *partnerships*, to get *socio-cultural experience* and exchange insights; - scenario planning and research-based learning.
SKOLKOVO Workshop center (SKOLKOVO)	Entrepreneurs and their teams that build a strategy in situations of market change and business transformation.	- development of strategic development projects and their preparation for implementation; - formation of a common *strategic vision* of the business and organization of strategic communication between the company's key leaders - representation system for business scaling and deployment of new business directions; - business transformation practices for reaching the next scale; - development of competencies necessary for managing the transformation process aimed at changing the scale of the business.
"Entrepreneurship and company management" (Ranepa)	Business owners who participate in management or perform control functions, directors with an entrepreneurial mindset.	- formation of skills and abilities to carry out business activities and manage the company at different stages of its life cycle; - *system representation* of business, practical tools in the field of organization management.
"Managerial competence of a director" (Business school IMISP)	Directors who already have initial experience in creating and developing a business, as well as those who plan to become a successful director.	- improving *personal efficiency, management efficiency* and business interaction; - development of key management competencies; to move to a new quality level.
"Unconscious scenarios in business: how much does your choice cost?" (Business school IMISP)	The target group is not specified.	- the discovery of their own unconscious scenarios in business; - development *of a conscious implementation* scenario and obtaining quality changes; - developing a conscious scenario as a basis for implementing and obtaining quality changes;

		- mastering effective communication *tools* - filters, metaprograms, emotional intelligence.
NASA program: influencing people through a model of the communication process (Process Communication Model©) (Business school IMISP)	Directors and leaders of business teams, business owners.	- constructive communication and business *efficiency* improvement; - choosing the optimal *strategy* for interaction in negotiations, management, partnership, etc.; - formation of effective *business teams*; - motivating and managing your own motivation; - prevention of manipulation.
Business Storytelling (Business school IMISP)	Entrepreneurs-owners of companies, middle and senior managers, directors of functions and directions.	- attracting the audience's attention; - retention of attention during a public speech; - meaningful performance; - designing an advertising/PR campaign.
"Project management in the field of social entrepreneurship» (High School of Management of St. Petersburg State University, St. Petersburg)	Current and potential social entrepreneurs, as well as managers of companies that develop projects in the field of social entrepreneurship.	- studying of the comparative experience of social entrepreneurship in Russia and in the world; - *comprehensive study* of the most important aspects of project management in the field of social entrepreneurship; - development of professional management competencies necessary for the successful implementation of projects in the field of social entrepreneurship.
MBA "Challenges of the digital world" (High School of Management of St. Petersburg State University, St. Petersburg)	Business owners and directors who want to take the company to the next level using innovative and technological methods.	- *strategic management,* digital leadership, *change management*; - *value-based management;*; - effective communication; - flexible team management.

«Effective Director» (Moscow International Higher School of Business MIRBIS)	Middle and senior managers, entrepreneurs interested in increasing their own capitalization, as well as developing managerial competencies and finding new solutions for business.	- *comprehensive view* on the company's work; - the ability to systematize knowledge, create and then put into practice your own business project, taking into account the knowledge and competencies obtained in the program (an interdisciplinary group project).
Personal and professional management skills - soft skills (Higher school of business of the State University of Management)	Directors and managers.	- issues of improving personal and professional effectiveness of directors; - leadership and management communications; - *change control*; - stress management and *emotional intelligence*; - effective presentations and public speaking.
«General Management» (Higher school of business of the State University of Management)	Owners of small and medium-sized businesses, middle managers.	- *a systematic view* on the business processes of companies, enterprises and organizations; - optimization of strategic planning skills and all management activities; - fundamentals *of strategic thinking*;; - ability to analyze processes, identifying factors that determine their effectiveness and productivity; - an idea of how to successfully *adapt* to changing conditions in the business environment.

Source: Own processing based on [2, 3, 5, 6, 9, 10].

Analyzing the data in the table, there are several key areas in which work is conducted in all these business schools, it is possible to allocate 4 basic directions: the creation and transformation of business models; globalization and expansion of business; development of business development strategies; personal and managerial development. The directions were formulated based on the key points of the programs identified by the business schools themselves. To a greater extent, only 3 business programs deal with issues of globalization and business expansion. A quarter of the listed business schools are focused on training in the development of their own business models, and 6 training programs out of 15 are focused on training in strategic development and strategic planning within their own business structures. Issues of personal and managerial development of entrepreneurs have the maximum coverage. These programs are implemented in all the analyzed business schools. Based on this, we can clearly identify relevant areas for our own development, which are of great importance for modern entrepreneurs.

For young Russian entrepreneurship, an interest in business modeling, strategic planning and management, building effective teams, and generally raising awareness in business, focusing on the value aspects of business is an extremely important stage of development that are relevant for today. It is also important that the offer of business schools to Russian entrepreneurs and managers includes aspects that are important today on a global scale: networking, development of soft skills (critical, systems thinking, strategic thinking, emotional intelligence), creation of business ecosystems, and change management. However, the wording in the course descriptions is not always clear and specific. For example, a combination of techniques that work through thinking (which?); development of key management competencies (which?). The SKOLKOVO Practicum program describes itself as "the only program for working with entrepreneurial thinking", in fact, it is not true compared to other business schools. Programs that include the study of the comparative experience of entrepreneurship in Russia and in the world, addressing issues of personal and professional effectiveness are interesting.

Speaking about the target groups that the activity of the considered business schools is aimed at, we can also distinguish 4 general groups, which include managing owners, functional executive, project managers and young entrepreneurs. At the same time, some business schools identify managing owners and functional managers as the target group. Project managers and young entrepreneurs are identified as their target groups for only 2 and 4 business programs. All this indicates that business schools are more focused on active entrepreneurs than on those who are just starting their way in business. This is primarily due to the desire of entrepreneurs to develop their own business, maintain its importance and relevance in a changing world, follow global trends and at the same time do not stop in personal development. Such programs combine the needs of mature entrepreneurs to the fullest extent.

5 Discussion

Program suggestions of various schools are developed based on the needs of the target audience. Business schools are more focused on experienced entrepreneurs and practicing managers. This trend may be due to the fact that the thinking of experienced entrepreneurs does not have a sufficient level of flexibility and requires additional incentives from educational institutions. Some business schools still identify young, aspiring entrepreneurs who have just started their business career or are at the stage of the idea and development of a business plan, as their target group. Considering the difference between these target groups, it makes sense to study their features, which may affect the content of programs presented by business schools (Table 2).

Table 2. Features of the needs of experienced and novice entrepreneurs

Experienced entrepreneurs	Novice entrepreneurs
• The presence of a business that has already existed for a long time and is currently operating; • Conservative views on business management are possible; • An established, not always flexible mindset is possible; • Outdated knowledge in the field of entrepreneurship or only accumulated own entrepreneurial experience is possible.	• Having an idea of your own business or a young business that is already opened; • Availability of fresh knowledge, usually field-specific education. • Flexible thinking, quick adaptation to changing environmental conditions; • Lack of experience in business management and planning.
• Desire to develop and expand your own business (existing or new); • Creating and(or) developing / transforming a business model; • Personal development and self-improvement.	

Source: Own processing.

As can be seen from Table 2, the two target groups under consideration have both differences and common features. It can be assumed that business schools often choose existing entrepreneurs due to the fact that experienced entrepreneurs are already more morally stable and know the specifics of the market functioning. All this suggests that they will be easier to work with. However, experience shows that a lack of flexibility of thinking can significantly complicate the task. For successful development there is not always enough experience, it is important to respond to new trends and constantly evolve. This explains the need to retrain this target group. Experienced entrepreneurs realize that the reorientation of society forces entrepreneurs to rebuild their own business in accordance with the needs of society.

Younger entrepreneurs have sufficient flexibility of thinking, they are ambitious and easily generate ideas for their own business. Most novice entrepreneurs have specialized education, which greatly facilitates their further training. However, their lack of experience does not allow them to fully assess the risks and features of the market functioning. It can be concluded that the general desires for experienced and young entrepreneurs to be engaged in self-development and development of their own business is largely dictated by society, but business schools believe that training is more necessary for experienced entrepreneurs, since they are the ones who need a radical restructuring of thinking. It is important to direct young entrepreneurs in the right direction and help them to build their own business model. This shows the specifics of training specific target groups and the choice of programs for them.

6 Conclusion

The current situation in society interprets new directions for business development, which makes it necessary to reorient business in accordance with existing socio-cultural trends and tendencies. More and more attention is being paid to areas aimed at developing soft-skills, emotional intelligence, and other innovative solutions that are being actively implemented in modern business. There is a reorientation of entrepreneurship towards the formation of cognitive skills designed to revolutionize the field of modern business management. This is primarily due to the reluctance of entrepreneurs to stop at what they have achieved and the need to find new, sometimes not obvious solutions to open up horizons. Constant modernization, moving away from the strictly material to the mental, allows the entrepreneur to quickly adapt their business to changing environmental conditions. At the same time, it gets the opportunity not only to preserve the material and intellectual resources embedded in the business, but also to significantly increase them. These areas are the basis of the programs offered by business schools to modern and intellectually curious entrepreneurs.

Considering the proposals of the most popular business schools, 4 main directions of their activities were identified, the most popular of which were programs related to the personal and managerial development of entrepreneurs. At the same time, the activities of business schools are aimed more at existing entrepreneurs and functional managers. Experts' forecasts regarding the orientation of young Russian entrepreneurship towards a comprehensive view of business that takes into account the socio-cultural features of society's development are confirmed by the range of current programs offered to entrepreneurs by existing business schools.

References

1. A. Blue: 5 things we know about the jobs of the future. *World Economic Forum 2020*. URL: https://www.weforum.org/agenda/2020/01/future-jobs-and-skills-in-demand/. Accessed: 29.04.2020. (2020).
2. Agency for Strategic Initiatives. Atlas of new professions. 2014. URL: https://asi.ru/reports/16344/. Accessed: 29.04.2020.

3. Business School IMISP: Programs. (2020). URL: https://www.imisp.ru/programmes/. Accessed: 26.08.2020.
4. F. Mu, J. Hatch: Accelerating the introduction of the case method into Chinese business schools: An action research approach. *Educational Action Research*, **28**(4), 609-625. DOI:10.1080/09650792.2019.1642774. (2020).
5. F.R. David, M.E. David, F.R. David: What are business schools doing for business today? *Business Horizons*, **54**(1), 51-62. DOI: 10.1016/j.bushor.2010.09.001. (2011).
6. High School of Management of St. Petersburg State University: Programs. (2020). URL: https://gsom.spbu.ru/programmes/exed/open_programs/dop_upravlenie_proektami_v_oblasti_socialnogo_predprinimatelstva/. Accessed: 26.08.2020.
7. Higher School of Business of the State University of Management (GUU): Programs. (2020). URL: https://hbs-guu.ru/program/increase-qualitication/personal-and-professional-management-skills-soft-skills/. Accessed: 26.08.2020.
8. Institute of Business and Business Administration of the Russian Academy of National Economy and Public Administration (Ranepa): Retraining and skills development. (2020). URL: https://ibda.ranepa.ru/programs/povyshenie-kvalifikatsii-profperepodgotovka/povyshenie-kvalifikatsii/moduly-mba/. Accessed: 26.08.2020.
9. L. Broccardo, A. Zicari: Sustainability as a driver for value creation: A business model analysis of small and medium enterprises in the Italian wine sector. *Journal of Cleaner Production*, **259**, 120852. DOI: 10.1016/j.jclepro.2020.120852. (2020).
10. M. Farashahi, M. Tajeddin: Effectiveness of teaching methods in business education: A comparison study on the learning outcomes of lectures, case studies and simulations. *International Journal of Management in Education*, **16**(1), 131-142. (2018).
11. M. Rhaiem, N. Amara: Determinants of research efficiency in Canadian business schools: Evidence from scholar-level data. *Scientometrics*, **125**(1), 53-99. DOI:10.1007/s11192-020-03633-z. (2020).
12. Moscow International Higher School of Business MIRBIS: Programs. (2020). URL: https://mirbis.ru/mba/effective-director/. Accessed: 26.08.2020.
13. Moscow Management School Skolkovo: Educational programs. (2020). URL https://common.skolkovo.ru/ru/educational-programmes/. Accessed: 26.08.2020.
14. P. Gregori, P. Holzmann: Digital sustainable entrepreneurship: A business model perspective on embedding digital technologies for social and environmental value creation. Journal of Cleaner Production, 272, 122817. (2020). DOI: 10.1016/j.jclepro.2020.122817.
15. P. Leviäkangas, R. Öörni: From business models to value networks and business ecosystems – What does it mean for the economics and governance of the transport system? *Utilities Policy*, **64**, 101046. DOI: 10.1016/j.jup.2020.101046. (2020).
16. P. Trkman: Value proposition of business schools: More than meets the eye. *The International Journal of Management Education*, **17**(3), 100310. DOI: 10.1016/j.ijme.2019.100310. (2019).
17. PwC in Russia: Workforce of the future. The competing forces shaping 2030. (2018). URL: https://www.pwc.ru/ru/publications/the-future-of-the-labour-market.html. Accessed: 29.04.2020.
18. S. Waddock: Will businesses and business schools meet the grand challenges of the era? *Sustainability (Switzerland)*, **12**(15), 60-83. DOI:10.3390/su12156083. (2020).
19. S.I. Ashmarina, V.V. Mantulenko, E.P. Troshina: Readiness to changes as one entrepreneurial value of the innovation-oriented economy. In S. Ashmarina S., M. Vochozka (Eds.), *Sustainable Growth and Development of Economic Systems. Contributions to Economics* (pp. 157-166). (Springer, Cham, 2019). DOI: 10.1007/978-3-030-11754-2_12.

20. S.L. de Vasconcellos, I.L. Garrido, R.C. Parente: Organizational creativity as a crucial resource for building international business competence. *International Business Review*, **28**(3), 438-449. DOI: 10.1016/j.ibusrev.2018.11.003. (2019).
21. V. Ahuja, Sh. Purankar: Quality business school education and the expectations of the corporate – A research agenda. *Procedia Computer Science*, **139**, 561-569. DOI: 10.1016/j.procs.2018.10.209. (2018).
22. V.V. Mantulenko, D.V. Aleshkova: Comparison of leading business schools in Russia and in the world. In V.V. Mantulenko (Ed.), *Proceedings of PEDTR 2019 18th International Scientific Conference "Problems of Enterprise Development: Theory and Practice". European Proceedings of Social and Behavioural Sciences,* **82** (pp. 1-6). (European Proceedings, London, 2020). DOI: 10.15405/epsbs.2020.04.1.
23. The rating of Russian business schools MBA.SU (2020). URL: https://www.mba.su/narodnyj_rating_bschools_2019/. Accessed: 23.10.2020.
24. Your Training: Soft skills. (2019). URL: https://utraining.ru/2019/11/13/soft-skills-2/. Accessed: 29.04.2020.

Prospects for the development of the entrepreneurial potential of the regions of the Russian Federation

S.I. Ashmarina[1,*] and *G.M.* Murzagalina[1]

[1]Samara State University of Economics, Samara, Russia

Abstract. This article analyzes the business environment of regions on the example of the Russian Federation through a system of indicators that characterize the level of its development within the 8 selected groups of factors (general economic, social, demographical, innovative, natural and geographical, factors of living standards, financial, infrastructural). The main research methods are correlation analysis, regression analysis, and modeling. Based on the results of the research, the authors come to the conclusion about the key role of socio-demographic, general economic and innovative infrastructure conditions for the development of the business environment of the regions.
Keywords: economy of Russia, regions of Russia, entrepreneurial environment, entrepreneurial potential, regional economy.

1 Introduction

The business environment of the region is a complex category that is influenced by many factors of different nature. Statistical science is designed to systematize this variety of determinants, to make a logical element in their structure. To solve this problem, the authors divided the factor indicators into several enlarged groups and performed a correlation analysis with the indicators of the business environment of the region. The business environment of the region is a complex economic phenomenon that characterizes the degree of business development (primarily small) and its activity. The business environment of the region is determined by the socio-economic, natural-geographical, demographic, legal and other conditions of the region's development. Due to its complexity and versatility, the business environment of the region can only be evaluated by a comprehensive system of indicators [5]. The source of information for our research was data from the All-Russian center for public opinion research of the Russian Federation (VCIOM)[18].

* Corresponding author: asisamara@mail.ru

2 Literature review

Modernization of the Russian economy is a priority task for the country that means first of all strengthening the market infrastructure and creating innovative and investment capacity of companies [1, 2]. The main focus of contemporary publications today is on technological and economic issues in the context of digitalization: efficiency increase, reduction of costs, strengthening the competitiveness, improving the quality of products and services, etc. [3, 7, 8]. However, the issue of creating and transforming business models in the context of the VUCA world is still not sufficiently developed. The integration in the intellectual economy requires more attention to the issues of knowledge-based, innovative activities of regional business.

Prospects and some specific features of entrepreneurship development in Russia under digital conditions are considered by Rybakova & Nazarov [4]. Yakovlev, Streltsov, & Nikulina see consequences of digital economy in effective platforms that are being formed, where production capacities of organizations with high scientific and technical potential play a special role [7]. Issues of reorienting entrepreneurial support infrastructure are analyzed by Muñoz, et al. [12]. Based on the analysis of indicators characterizing the innovative development of the Russian regions, some researchers distinguish between leading and lagging regions depending on the effectiveness of socio-economic and scientific-technical potential implementation [11, 13].

The literature review shows that the entrepreneurial potential of regions is determined by a number of macro and mezo-economic factors that create favorable or unfavorable conditions for the birth, functioning and development of entrepreneurship [6, 9, 10, 14-17]. The study and analysis of these factors, systematization and generalization of theoretical research and experience in the practical implementation of business projects are valuable for the creation of strategic programs of regional development.

3 Methodology

The main research methods used in this study include correlation analysis, regression analysis, and modeling. These methods allowed the authors to consider the main factors influencing the entrepreneurial potential of regions, to measure the closeness of the relationship between variables and quantify factors that have the greatest impact on the performance indicator.

The group "General economic factors» includes the following indicators: gross regional product (GRP) per capita, rub/person $X_{1.1}$; GRP volume index, %, $X_{1.2}$; actual final consumption of households per capita, rub/person, $X_{1.3}$; share of agriculture, hunting and forestry in the gross value added of the region, %, $X_{1.4}$; share of mineral extraction in the gross value added of the region, %, $X_{1.5}$; share of manufacturing in the region's gross value added, %, $X_{1.6}$; share of education in the region's gross value added, %, $X_{1.7}$; industrial production index, %, $X_{1.8}$; consumer prices index, %, $X_{1.9}$; real income index of the population, %, $X_{1.10}$; unemployment level, %, $X_{1.1}$.

These indicators characterize the state of the region's economy as a whole, as well as its most important segments – the industry structure of the economy, the labor market, and the price level. Hypothetically, they can be linked to the business environment of the region both directly and vice versa. On the one hand, the high level of economic development, growth in the physical volume of production of goods and services is an incentive for creating new businesses and active growth of existing ones. On the other hand, in economically developed regions, there is a high probability of high competition in the market, which is an obstacle to entering the market (Table 1).

Table 1. Matrix of paired correlation coefficients for the group "General economic factors"[†]

r	Y_1	Y_2	Y_3	Y_4	Y_5	Y_6	Y_7	Y_8	Y_9	Y_{10}	Y_{11}	Y_{12}
$X_{1.1}$	-0.228	0.835	0.026	0.023	-0.014	0.217	0.021	-0.078	-0.098	0.126	0.044	0.354
$X_{1.2}$	-0.066	0.271	0.005	0.089	0.108	-0.081	-0.019	-0.144	-0.004	0.137	-0.033	0.160
$X_{1.3}$	0.149	0.457	0.462	0.382	0.016	0.400	-0.065	-0.045	-0.001	0.232	0.012	0.660
$X_{1.4}$	-0.319	-0.416	-0.511	-0.114	0.065	-0.092	-0.133	-0.205	-0.227	-0.280	-0.023	-0.498
$X_{1.5}$	-0.124	0.776	-0.076	-0.059	0.003	0.160	0.152	0.097	-0.077	0.041	-0.024	0.219
$X_{1.6}$	0.507	-0.253	0.290	-0.091	0.001	-0.222	-0.043	0.268	0.319	0.296	0.227	0.114
$X_{1.7}$	-0.455	-0.394	-0.424	-0.180	-0.117	-0.079	0.004	-0.213	-0.207	-0.376	-0.301	-0.629
$X_{1.8}$	-0.003	-0.118	-0.085	0.037	-0.035	0.005	0.168	0.111	0.304	0.134	0.183	0.060
$X_{1.9}$	0.017	-0.075	0.217	-0.042	-0.076	-0.093	-0.126	0.042	0.037	-0.001	-0.025	-0.047
$X_{1.10}$	-0.180	0.167	-0.015	0.223	0.133	-0.051	-0.081	-0.309	-0.123	-0.006	-0.067	0.140
$X_{1.11}$	-0.415	-0.203	-0.513	-0.317	-0.264	0.075	0.172	-0.138	-0.153	-0.387	-0.283	-0.605

Source: Own processing.

The data in Table 2 largely confirm the first part of our hypothesis. Thus, the high positive value of the correlation coefficient of the Multidimensional assessment of the business environment of the region (Y_{12}) and the turnover of private organizations (Y_2) with the level of GRP ($X_{1.1}$) and actual final consumption ($X_{1.3}$) is noteworthy. This suggests that the economically developed regions of the Russian Federation are still creating favorable ground for business development. The most successful business environment is formed in industrially developed territories (both in the field of mining and processing - $X_{1.5}$ и $X_{1.6}$), but agricultural regions and large educational centers are not successful in terms of private enterprise development. At the same time, the high share of education in GRP ($X_{1.4}$ и $X_{1.7}$) negatively affects the indicators of entrepreneurial activity of enterprises, that is, the qualitative growth of the business sector. In addition, businesses need workers and personnel, but the unemployed population is not the main reserve of the private sector's labor force: regions with high unemployment ($X_{1.11}$) have low values of the business environment of the region indicators, such as a multidimensional assessment and many private indicators (the

[†] Correlation coefficients that are significant at the level $p < 0.05$ are highlighted in red

number of private and small enterprises, entrepreneurial activity). People who have lost their jobs tend to trust more to the work in the municipal and public sector, seeing it as a stable source of their life resources, and even more so are not always willing to risk starting their own business. We also note that the factor of inflation ($X_{1.9}$) was insignificant for the development of business initiatives in the region.

"Demographic factors" group contains the following indicators: $X_{2.1}$ – proportion of urban population, %; $X_{2.2}$ – gender of the population (number of women per 1000 men); $X_{2.3}$ – birth rate, ‰; $X_{2.4}$ – death rate, ‰; $X_{2.5}$ – infant mortality rate, ‰; $X_{2.6}$ – migration balance rate, ‰; $X_{2.7}$ – the ratio of divorces to marriages (there are divorces per 1000 marriages); $X_{2.8}$ – life expectancy at birth, years; $X_{2.9}$ – share of the working age population, %.

Since business is primarily focused on meeting the needs of various segments of the population, its features of settlement can serve as important factors for the state of the business environment of the region. This influence may be mediated by the volume and specialization of regional production (for example, a high share of manufacturing or agriculture). The main factor in this group, in our opinion, is the urbanization of the territory (Table 2).

Table 2. Matrix of paired correlation coefficients for the "Demographic factors" group

r	Y_1	Y_2	Y_3	Y_4	Y_5	Y_6	Y_7	Y_8	Y_9	Y_{10}	Y_{11}	Y_{12}
$X_{2.1}$	0.481	0.256	0.652	0.067	-0.031	0.060	-0.090	0.090	0.003	0.057	-0.011	0.365
$X_{2.2}$	0.503	-0.395	0.292	-0.156	-0.029	-0.297	-0.151	0.200	0.148	0.088	0.119	-0.070
$X_{2.3}$	-0.364	0.152	-0.259	-0.027	-0.197	0.267	0.143	-0.075	-0.115	-0.195	-0.271	-0.195
$X_{2.4}$	0.315	-0.371	0.270	-0.119	0.200	-0.315	-0.159	0.053	0.123	0.171	0.237	0.058
$X_{2.5}$	-0.434	0.005	-0.344	0.075	-0.151	0.085	0.007	-0.136	-0.116	-0.134	-0.204	-0.266
$X_{2.6}$	0.460	-0.116	0.451	0.085	0.126	-0.067	0.025	-0.027	0.175	0.087	-0.031	0.305
$X_{2.7}$	0.214	0.146	0.171	-0.087	0.201	-0.162	-0.094	-0.026	-0.066	0.121	0.261	0.202
$X_{2.8}$	0.182	-0.006	0.004	0.213	-0.183	0.132	0.007	0.107	0.066	-0.067	-0.063	0.033
$X_{2.9}$	-0.128	0.529	0.002	0.328	-0.104	0.319	0.046	-0.090	-0.122	-0.026	-0.136	0.251

Source: Own processing.

Paired correlation analysis confirmed the validity of the last thesis: it is the urban environment ($X_{2.1}$) that accumulates the main sources of business activity and feeds the business environment. We consider the influence of the gender factor to be very interesting ($X_{2.2}$) – the greater the quantitative predominance of the female population in the region, the more there are private enterprises of different sizes, but the volume of their average turnover is smaller. Conversely, an increase in the working age population ($X_{2.9}$) has a positive impact on the financial results of the business. In terms of correlation coefficients, we note the positive impact of the migration balance ($X_{2.6}$) on the number of private organizations and on the business environment of the region as a whole. It is interesting that the indicators of business demography do not correlate with similar indicators of classical demography.

The "Social factors" group includes $X_{3.1}$ – provision of places for pre-school children in organizations that carry out educational activities under pre-school educational programs, supervision and care of children, places (per 1000 children); $X_{3.2}$ – number of students in general education organizations, people (per 1000 inhabitants); $X_{3.3}$ – number of students enrolled in training programs for qualified workers, employees (per 10,000 inhabitants); $X_{3.4}$ – number of students enrolled in the programs of mid-level specialists, people (per 10,000 inhabitants); $X_{3.5}$ – number of educational institutions of higher education and scientific organizations, units (per 10,000 inhabitants); $X_{3.6}$ – number of students enrolled in bachelor's, specialist's, master's programs, people (per 10,000 inhabitants); $X_{3.7}$ – number of hospital beds, units (per 10,000 inhabitants); $X_{3.8}$ – capacity of outpatient organizations, visits per shift (per 10,000 residents); $X_{3.9}$ – number of doctors of all specialties, people (per 10,000 inhabitants); $X_{3.10}$ – morbidity rate (registered diseases of patients with a diagnosis established for the first time in life), units (per 1000 inhabitants); $X_{3.11}$ – number of theater audience, people (per 1000 inhabitants); $X_{3.12}$ – number of museum visits, units (per 1000 inhabitants); $X_{3.13}$ – number of gyms, units (per 10,000 inhabitants); $X_{3.14}$ - library fund, copies (per 1000 inhabitants); $X_{3.15}$ – number of collective accommodation facilities, units (per 10,000 inhabitants); $X_{3.16}$ – crime rate, units (per 100,000 inhabitants).

The state of the social sphere is characterized by very multifaceted indicators, which can be briefly described as "education", "health", "culture (including physical)" and "criminogenicity". Accordingly, the task is to study the impact of the educational and cultural level of the population, its health and law abidance on the ES of the region (Table 3).

Table 3. Matrix of paired correlation coefficients by "Social factors" group

r	Y_1	Y_2	Y_3	Y_4	Y_5	Y_6	Y_7	Y_8	Y_9	Y_{10}	Y_{11}	Y_{12}
$X_{3.1}$	0.043	0.264	0.193	-0.132	0.177	-0.190	-0.137	-0.075	-0.109	0.088	0.149	0.133
$X_{3.2}$	-0.538	0.169	-0.439	-0.148	-0.090	0.133	0.098	-0.163	-0.171	-0.213	-0.242	-0.320
$X_{3.3}$	-0.179	0.002	-0.217	-0.346	0.196	0.050	0.075	-0.058	-0.057	-0.028	-0.028	-0.133
$X_{3.4}$	-0.103	-0.111	-0.097	-0.177	0.013	0.055	-0.010	0.017	-0.206	-0.236	-0.058	-0.243
$X_{3.5}$	0.286	-0.297	0.544	0.147	-0.068	0.085	-0.040	0.202	0.121	0.027	-0.008	0.157
$X_{3.6}$	0.399	-0.340	0.508	0.108	0.014	-0.075	0.008	0.233	0.105	-0.045	-0.060	0.084
$X_{3.7}$	-0.296	0.024	0.000	-0.079	0.134	0.024	0.007	-0.155	-0.082	0.031	0.052	0.027
$X_{3.8}$	-0.100	0.056	0.192	-0.065	0.128	-0.166	-0.099	-0.101	-0.155	-0.002	0.095	0.030
$X_{3.9}$	-0.061	0.174	0.283	-0.068	-0.001	0.159	0.149	0.103	0.050	0.064	0.025	0.196
$X_{3.10}$	-0.062	0.443	0.136	-0.147	-0.065	-0.020	-0.082	-0.032	-0.234	-0.133	-0.081	-0.030
$X_{3.11}$	0.486	-0.264	0.712	0.020	-0.061	-0.001	0.089	0.222	0.170	0.050	0.042	0.257
$X_{3.12}$	0.222	-0.072	0.511	-0.078	0.193	-0.126	0.129	0.034	0.226	0.157	0.043	0.303
$X_{3.13}$	-0.536	0.338	-0.222	-0.198	0.086	-0.038	-0.133	-0.207	-0.228	-0.081	0.033	-0.155
$X_{3.14}$	-0.383	-0.046	0.055	-0.142	0.168	-0.130	-0.168	-0.215	-0.117	0.000	0.133	-0.061
$X_{3.15}$	-0.115	0.000	0.008	-0.126	0.465	-0.163	0.280	0.020	0.198	0.163	0.050	0.184
$X_{3.16}$	-0.118	0.077	0.084	-0.375	0.060	0.058	0.075	-0.068	-0.092	0.033	-0.020	-0.031

Source: Own processing.

According to table 3, we can see that the values of paired coefficients are mostly small, and these factors should not play a decisive role in the territorial differentiation of the business environment of the region. At the same time, the cultural level of the population matters, factors $X_{3.11}$ и $X_{3.12}$ have a positive relationship with almost all indicators of the business environment of the region. Development of higher education ($X_{3.5}$ и $X_{3.6}$) gives birth to new entrepreneurs, the number of private organizations is growing, but this business is rarely financially successful and active.

Group "Factors of living standards" contains $X_{4.1}$ – per capita monetary income of the population, rubles; $X_{4.2}$ – average consumer spendings per capita average consumer spendings per capita (per month), rubles; $X_{4.3}$ – retail trade turnover per capita, rubles; $X_{4.4}$ – volume of paid services per capita, rubles; $X_{4.5}$ – share of household expenditure on food, %; $X_{4.6}$ – share of household expenditures on housing and utilities, and fuel, %; $X_{4.7}$ – share of the population with monetary incomes below the subsistence minimum, %; $X_{4.8}$ – level of consumption of meat and meat products per capita, kg; $X_{4.9}$ – availability of cars (number of own cars), units (per 1000 inhabitants); $X_{4.10}$ – the total area of residential premises, on average per one inhabitant, m^2/ person; $X_{4.11}$ – the share of dilapidated housing stock in the total area of the entire housing stock, %. This group of factors is also very broad. It takes into account both the indicators of income of the population, and the provision of essential material goods (cars, housing), the consumption of basic food (in particular, meat and meat products). These indicators also correlate with each other. Thus, a high standard of living (in particular, a high income) implies appropriate housing conditions.

We note that this group contains factorial characteristics that are diametrically opposite to the standard of living. Thus, high values of features $X_{4.1}$ … $X_{4.4}$, $X_{4.8}$ … $X_{4.10}$ are indicators of a high standard of living, while other signs are the opposite (Table 4).

Table 4. Matrix of paired correlation coefficients of the group "Factors of living standards"

r	Y_1	Y_2	Y_3	Y_4	Y_5	Y_6	Y_7	Y_8	Y_9	Y_{10}	Y_{11}	Y_{12}
$X_{4.1}$	0.083	0.426	0.435	0.461	-0.022	0.315	-0.151	-0.135	-0.093	0.035	-0.111	0.383
$X_{4.2}$	0.381	0.282	0.590	0.481	-0.041	0.411	-0.152	-0.070	-0.045	0.068	-0.062	0.523
$X_{4.3}$	0.410	0.259	0.528	0.488	-0.110	0.438	-0.210	-0.051	-0.052	0.068	-0.022	0.487
$X_{4.4}$	0.272	0.311	0.570	0.386	0.127	0.289	-0.026	-0.085	-0.020	0.083	-0.095	0.516
$X_{4.5}$	-0.287	-0.300	-0.323	-0.025	-0.052	0.152	0.126	-0.015	0.132	-0.075	-0.058	-0.138
$X_{4.6}$	-0.146	0.080	-0.117	-0.075	-0.135	-0.085	0.045	-0.086	-0.040	0.028	0.019	-0.102
$X_{4.7}$	-0.524	-0.352	-0.453	-0.497	0.010	-0.215	0.214	-0.082	-0.030	-0.086	-0.069	-0.466
$X_{4.8}$	0.233	0.174	0.218	0.047	0.193	-0.141	-0.136	-0.030	-0.089	-0.011	0.137	0.156
$X_{4.9}$	0.509	0.156	0.386	0.068	0.141	-0.130	-0.144	-0.037	-0.083	0.081	0.082	0.264
$X_{4.10}$	0.321	0.248	0.203	0.135	0.147	-0.179	-0.407	-0.107	-0.178	-0.039	0.154	0.107
$X_{4.11}$	-0.167	0.043	-0.067	-0.145	-0.033	0.245	0.015	-0.127	-0.122	-0.062	-0.017	-0.036

Source: Own processing.

The correlation analysis in this group showed the importance and closeness of the relationship between consumer activity ($X_{4.1}$ … $X_{4.4}$) and the business environment of the region.. This is quite logical, since the business is focused on the consumer market and, in turn, is financially dependent on the willingness of buyers to fulfill their demand. This logic also includes negative coefficients for the factor variables "the share of household expenditures on food" ($X_{4.5}$) and "the share of the population with monetary incomes below the subsistence minimum" ($X_{4.7}$). These indicators are a characteristic of the impoverished population, its inability to fully finance its needs. In such regions, the private sector (especially in the service sector) has a low rate of return, and according to the laws of the economy, capital will not stay in it. "Natural and geographical factors" group consists of $X_{5.1}$ – population density, people/km^2; $X_{5.2}$ – woodiness area, %; $X_{5.3}$ – average annual air temperature, degree; $X_{5.4}$ – average annual precipitation, mm. The difference of these factors is that they are almost independent of humans (some exceptions are the factor of forest area). They mainly characterize the geographical area of the subject of the Federation. The population density factor adds an aspect of the nature of residents settlement across the territory, their concentration or deconcentration on a limited area of surface (Table 5).

Table 5. Matrix of paired correlation coefficients of "Natural and geographical factors" group:

r	Y_1	Y_2	Y_3	Y_4	Y_5	Y_6	Y_7	Y_8	Y_9	Y_{10}	Y_{11}	Y_{12}
$X_{5.1}$	0.207	-0.022	0.626	0.216	-0.101	0.030	-0.048	0.021	0.074	0.065	-0.088	0.269
$X_{5.2}$	0.209	-0.090	0.181	-0.228	-0.129	0.140	-0.036	-0.041	-0.135	-0.086	-0.102	-0.069
$X_{5.3}$	0.095	-0.388	-0.068	0.196	0.078	-0.126	-0.037	0.060	0.167	0.034	0.091	-0.024
$X_{5.4}$	0.196	-0.147	0.332	-0.023	0.064	-0.123	-0.206	-0.099	-0.060	0.028	0.040	0.064

Source: Own processing.

Of all the natural and geographical facts, let's highlight the "population density" ($X_{5.1}$). It has a particularly strong impact on the quantitative indicators of small businesses (Y_3,Y_4) and on the business environment of the region as a whole (Y_{12}): the more concentrated the population, the more active the business is in this area. Under these conditions, logistics is simplified, business relationships become closer and simpler, and advertising and delivery work better and faster. "Financial factors" group includes $X_{6.1}$ – revenues of the consolidated budgets of the subjects of the Russian Federation per capita, thousand rubles; $X_{6.2}$ – expenditures of the consolidated budgets of the subjects of the Russian Federation per capita, thousand rubles; $X_{6.3}$ – fixed capital investment per capita, thousand rubles; $X_{6.4}$ – foreign direct investment in fixed assets per capita, dollar; USA; $X_{6.5}$ – balanced financial result (profit minus loss) of organizations' activities per capita, thousand rubles; $X_{6.6}$ – share of unprofitable organizations, %. This group contains factors that show the potential (theoretical) and real financial possibility of investing in the business sector – investment activity and financial condition of organizations and regions (Table 6).

Table 6. Matrix of paired correlation coefficients of "Financial factors" group

r	Y_1	Y_2	Y_3	Y_4	Y_5	Y_6	Y_7	Y_8	Y_9	Y_{10}	Y_{11}	Y_{12}
$X_{6.1}$	-0.240	0.370	0.060	0.237	0.071	0.172	0.014	-0.109	-0.064	0.075	-0.039	0.247
$X_{6.2}$	-0.251	0.340	0.061	0.234	0.071	0.173	0.014	-0.118	-0.067	0.070	-0.044	0.235
$X_{6.3}$	-0.266	0.809	-0.078	-0.028	-0.029	0.201	0.009	-0.086	-0.068	0.051	0.051	0.235
$X_{6.4}$	0.062	0.095	0.235	0.178	-0.183	0.468	-0.046	-0.062	-0.016	0.145	0.069	0.316
$X_{6.5}$	0.009	0.655	0.150	0.160	-0.002	0.151	0.006	0.041	-0.015	0.096	-0.009	0.331
$X_{6.6}$	-0.175	0.156	-0.011	-0.259	0.180	0.021	0.187	0.049	0.195	0.156	0.206	0.161

Source: Own processing.

All paired correlation coefficients with a multidimensional assessment of the business environment of the region have positive values, that is, there is an influence of financial factors on the degree of regional business development. The coefficient $r_{y2x6.3}$ = 0,809: should be especially noted: the level of investment significantly determines the volume of private enterprises turnover. This is a positive feature, as it highlights the importance and necessity of investment in the economy. At the same time, among the financial factors, there was not a single one that had a statistically significant relationship with the indicators of blocks 4 and 5.

The group of "Innovative factors" contains $X_{7.1}$ – innovative activity of organizations, %; $X_{7.2}$ – share of innovative goods, works, services, % of the total volume of shipped goods, performed works, services; $X_{7.3}$ – internal costs on research activities and developments per capita, thousand rubles; $X_{7.4}$ – expenditures on ICT (information and communication technologies) per capita, thousand rubles; $X_{7.5}$ – share of organizations that used special software tools, % of the total number of surveyed organizations; $X_{7.6}$ – number of active subscribers of fixed and mobile broadband Internet access, units (per 100 inhabitants); $X_{7.7}$ – digital TV coverage, %. This group includes indicators of the economy's readiness to transition to the era of digitalization and to an innovative development path. At the same time, statistical indicators consider this phenomenon both at the level of enterprises and at the level of households (Table 7).

Table 7. Matrix of paired correlation coefficients of the "Innovative factors" group

r	Y_1	Y_2	Y_3	Y_4	Y_5	Y_6	Y_7	Y_8	Y_9	Y_{10}	Y_{11}	Y_{12}
$X_{7.1}$	0.314	0.000	0.333	0.161	0.058	-0.098	-0.151	0.019	-0.042	0.005	0.030	0.162
$X_{7.2}$	0.321	-0.145	0.203	0.140	-0.010	-0.024	-0.125	-0.048	-0.081	-0.101	-0.042	0.048
$X_{7.3}$	0.435	-0.037	0.672	0.203	-0.192	-0.017	-0.062	0.114	0.053	0.065	-0.101	0.241
$X_{7.4}$	0.013	0.498	0.421	0.204	-0.122	0.218	-0.076	-0.006	0.007	0.137	-0.026	0.379
$X_{7.5}$	0.212	0.063	0.270	-0.099	0.079	-0.159	-0.020	0.228	0.181	0.281	0.316	0.226
$X_{7.6}$	0.593	0.067	0.689	0.177	-0.036	-0.107	-0.172	0.140	-0.047	0.002	-0.009	0.270
$X_{7.7}$	0.128	-0.069	0.043	-0.003	-0.204	-0.161	-0.203	-0.021	-0.099	-0.085	-0.037	-0.182

Source: Own processing.

Scientific research, innovation and the active use of information and communication technologies are reliable factors for activating the business environment of the region. This is especially evident in factor $X_{7.6}$ – "The number of active subscribers of fixed and mobile broadband Internet access", which has a high positive pair correlation coefficients with the number of private sector enterprises and with a multidimensional assessment of the business environment of the region. Out of the entire list of factor variables in this block, only the digitalization factor of TV broadcasting (X7.7) has no statistically significant relationships ($X_{7.7}$). Perhaps, at present, this advanced technology of telematics services is more social in nature and is not related to the development of business, but to the activities of the state.

The group "Infrastructural factors» is presented through $X_{8.1}$ – number of personal computers, units (per 100 employees); $X_{8.2}$ – depreciation rate of fixed assets, %; $X_{8.3}$ – share of paved roads in the total length of public roads, %; $X_{8.4}$ – passenger turnover of public buses per capita, thousand pass-km; $X_{8.5}$ – automobile transport cargo turnover per capita, thousand t-km; $X_{8.6}$ – commissioning of residential buildings, m^2 of total residential area (per 1000 inhabitants); $X_{8.7}$ – emissions of pollutants into the atmosphere from stationary sources, thousand tons (per 1000 inhabitants). This group covers a wide range of factors that take into account the environmental situation, the state of the basic segment of the economy – fixed assets, as well as the availability and use of elements of information, transport and social infrastructure (Table 8).

Table 8. Matrix of paired correlation coefficients for a group "Infrastructure factors»

r	Y_1	Y_2	Y_3	Y_4	Y_5	Y_6	Y_7	Y_8	Y_9	Y_{10}	Y_{11}	Y_{12}
$X_{8.1}$	0.262	-0.160	0.707	0.014	0.192	-0.085	-0.021	0.016	0.017	0.038	- 0.082	0.243
$X_{8.2}$	0.079	0.151	-0.148	-0.027	-0.147	-0.122	0.227	0.391	0.267	0.195	0.225	0.041
$X_{8.3}$	0.289	0.182	0.242	0.164	0.219	-0.071	0.162	-0.005	0.092	0.114	- 0.055	0.345
$X_{8.4}$	0.101	-0.069	-0.012	-0.234	0.029	-0.027	0.370	0.411	0.408	0.274	0.388	0.172
$X_{8.5}$	-0.103	0.293	0.022	0.113	0.013	-0.091	-0.189	-0.147	-0.122	0.057	0.078	0.091
$X_{8.6}$	0.218	0.016	0.045	0.181	-0.063	0.023	-0.177	-0.091	-0.028	- 0.053	0.020	0.074
$X_{8.7}$	-0.214	0.842	-0.093	-0.081	-0.058	0.126	0.026	-0.043	-0.101	0.066	0.057	0.194

Source: Own processing.

For all indicators, there is a direct relationship with the business environment of the region. This connection is mostly noticeable in regions with a high level of computer technology use and good road quality ($X_{8.1}$, $X_{8.3}$). The importance of well-organized logistics is shown, in particular, in the relationship between the turnover of automobile transport and the volume of turnover of private enterprises ($r_{y2x8.5}$ = 0,293). However, the state of the environment has a much more significant impact on the financial results of the business - an increase in emissions of harmful substances leads to an increase in the values of the Y_2indicator. This can be explained by the fact that the complex environmental situation is inherent in industrially developed territories, where there are likely to be conditions for business prosperity. Moreover, it is not the number of enterprises (with the performance indicators Y_1 и Y_3 there is an inverse correlation), but their turnover. This block of factors is one of the few where business demography indicators have significant correlations (with the state of fixed assets $X_{8.2}$ and with passenger transport turnover $X_{8.4}$).

4 Results

The next stage of the study is regression analysis. It is intended for an analytical form of expression of the indicators relationship. Preliminary performance indicators are checked for compliance with the normal distribution law. This is important for subsequent analysis and application of certain statistical methods. The instrument is the Kolmogorov-Smirnov criterion, which is justified in a relatively large population. The calculation results are shown in Table 9.

Table 9. Kolmogorov-Smirnov criteria for business environment of the region indicators in the Russian Federation in 2017 (significance level 5 %)

Indicators	The criterion value	P-value	Compliance with the normal distribution law
Y_1	0.994	0.277	Yes
Y_2	1.057	0, 214	Yes
Y_3	2.962	0.000	No

Indicators	The criterion value	P-value	Compliance with the normal distribution law
Y_4	0.784	0.571	Yes
Y_5	1.700	0.006	No
Y_6	1.274	0.078	Yes
Y_7	2.243	0.000	No
Y_8	1.285	0.073	Yes
Y_9	3.005	0.000	No
Y_{10}	1.935	0.001	No
Y_{11}	1.225	0.100	Yes
Y_{12}	1.790	0.003	No

Source: Own processing.

Thus, of the 12 indicators of the Russian business environment of the region, only 6 are in accordance with the normal distribution law (Y_1, Y_2, Y_4, Y_6, Y_8, Y_{11}). Regression models will be built based on them. The regression model for the Y_1 indicator - "The share of private organizations" will have the following form:

$$\hat{Y}_1 = 70{,}107 + 0{,}127x_{2.1} - 1{,}374x_{2.5} + 0{,}024x_{2.7} - 2{,}853x_{3.13} - -0{,}001x_{3.14} - 0{,}004x_{3.16} - 0{,}436x_{5.3} + 0{,}157x_{7.5} + 0{,}260x_{8.1}, \quad (2)$$

where: $X_{2.1}$ – share of the urban population, %; $X_{2.5}$ – infant mortality rate, ‰; $X_{2.7}$ – the ratio of divorces to marriages (there are divorces per 1000 marriages); $X_{3.13}$ – number of gyms, units (per 10,000 inhabitants); $X_{3.14}$ - library fund, copies (per 1000 inhabitants); $X_{3.16}$ – crime rate, units (per 100,000 inhabitants); $X_{5.3}$ – average annual air temperature, degree; $X_{7.5}$ – share of organizations that used special software tools, % of the total number of surveyed organizations; $X_{8.1}$ – number of personal computers, units (per 100 employees).

The multiple correlation coefficient in the model is 0.849. The variation in the share of private organizations in the region by 72.1 % is due to the variation of the factors considered in the model, and by 27.9 % is due to the variation of other factors not included in this model. According to the F-criterion, the model is significant (with a 95 % probability). The above indicates a good quality and adequacy of the built model.

An important element of multiple regression analysis is finding partial coefficients of determination:

$$d_i = \beta_i * r_{yxi}, \quad (3)$$

where: β_i - coefficients of the regression equation on a standardized scale; r_{yxi} - paired correlation coefficients.

The need to calculate partial coefficients of determination is caused by the different scales of factors-arguments. They can be used to estimate the "net" contribution of a factor to the variation of the values of the dependent variable. Summarizing the results of calculations of private coefficients of determination, we can note that the main role in the territorial differences in the share of private organizations is played by social and demographic factors: the region's urbanization, crime rate, development of cultural infrastructure and sports, health status (which is indicated by the infant mortality rate) – in total, they account for 69.3% of the impact on the effective indicator. However, the nature of their influence is different.

Thus, the relative number of private institutions is higher in urban areas with a high level of ICT development, good health, low crime rate, geographically belonging to the North-West, central part of the European part and Western Siberia. In the Russian Federation, the largest share of private organizations is in Saint Petersburg (92.7 % of all organizations), in the Tyumen region (91.5%), and in the Samara region (91.2%).

The regression model for the Y_2 indicator - "Turnover of private organizations" will have the following form:

$$\hat{Y}_2 = 64{,}714 + 0{,}592x_{1.5} + 0{,}008x_{5.1} + 0{,}086x_{6.1} - 0{,}579x_{6.6} - 1{,}113x_{8.1} + +6{,}518x_{8.4} + 6{,}322x_{8.7}, \quad (4)$$

where: $X_{1.5}$ – share of mineral extraction in the gross value added of the region, %; $X_{5.1}$ – population density, people/km^2; $X_{6.1}$ – revenues of the consolidated budgets of the subjects of the Russian Federation per capita, thousand rubles; $X_{6.6}$ – share of unprofitable organizations,%; $X_{8.1}$ – number of personal computers, units (per 100 employees); $X_{8.4}$ – passenger turnover of public buses per capita, thousand pass-km; $X_{8.7}$ – emissions of pollutants into the atmosphere from stationary sources, thousand tons (per 1000 inhabitants).

The multiple correlation coefficient is 0.905. The variation in turnover of private organizations in the region by 82.0 % is due to the influence of factors included in the model. According to this performance indicator, priority belongs to infrastructure factors, which account for 54.6% of the impact. Thus, the turnover per 1 private organization is likely to be higher in a financially secure region with a predominance of extractive industries, a compact settlement system and a complex environmental situation. That is why we see Yamalo-Nenets, Nenets and Khanty-Mansiysk Autonomous okrugs, Kemerovo region, Komi, etc. among the leaders in this Y_2 indicator.

The dependence of small enterprises turnover (Y_4) on various factors is approximated by the multiple regression model:

$$\hat{Y}_4 = 35{,}302 - 0{,}113x_{1.6} - 0{,}014x_{2.7} - 0{,}599x_{4.7} - 0{,}345x_{8.7}, \quad (5)$$

where: $X_{1.6}$ – share of mineral extraction in the gross value added of the region, %; $X_{2.7}$ – the ratio of divorces to marriages (there are divorces per 1000 marriages); $X_{4.7}$ – share of the population with monetary incomes below the subsistence minimum, %; $X_{8.7}$ – emissions of pollutants into the atmosphere from stationary sources, thousand tons (per 1000 inhabitants).

The multiple correlation coefficient is 0.674. All these factors account for 45.5 % of the variation in the birth rate coefficient of organizations by regions. According to the indicator "Turnover of small enterprises", the dominant influence belongs to the factor of living standards – the share of the population with incomes below the subsistence minimum, which causes more than 1/3 (more precisely, 36.5%) of the variation of the effective variable. Other indicators have an important, but significantly less impact. Therefore, we can say that the largest volume of turnover of small businesses is observed in regions that are not industrially developed and, consequently, do not have global environmental problems, and at the same time have a decent standard of living in combination with traditional family lifestyle.

The leaders among the regions in terms of turnover of small enterprises are the Crimea, Astrakhan and Volgograd regions, and Saint Petersburg. In particular, these subjects of the Federation have very low values for emissions of harmful substances (much lower than the arithmetic average).

The dependence of the IP revenue volume (Y_6) on factors is described by the following linear multi-factor model:

$$\hat{Y}_6 = -4090{,}7 + 189{,}1x_{1.4} + 279{,}4x_{2.9} + 27{,}4x_{3.4} - 427{,}2x_{3.15} - -315{,}1x_{4.7} - 329{,}6x_{4.10} + 29{,}7x_{5.2} + 570{,}0x_{6.4}, \quad (6)$$

where: $X_{1.4}$ – share of agriculture, hunting and forestry in the gross value added of the region, %; $X_{2.9}$ – share of the working age population, %; $X_{3.4}$ – number of students enrolled in the programs of mid-level specialists, people (per 10,000 inhabitants); $X_{3.15}$ – number of collective accommodation facilities, units (per 10,000 inhabitants); $X_{4.7}$ – share of the population with monetary incomes below the subsistence minimum, %; $X_{4.10}$ – the total area of residential premises, on average per one inhabitant, m^2/ person; $X_{5.2}$ – woodiness area, %; $X_{6.4}$ – foreign direct investment in fixed assets per capita, dollar; USA.

The factors collected in the models explain more than half (55.3 %) of the regional variation of IP revenue volume. The multiple correlation coefficient is 0.744. Among all the independent variables, the "foreign direct investment in fixed assets per capita" factor clearly stands out, explaining the 19.9 % of Y_6variation. The second most important factor is "the share of the population with monetary incomes below the subsistence minimum" (14.4 %), followed by "the total area of residential premises per inhabitant on average" (7.9%).

The largest revenue of IP is achieved in regions of natural forest area with its agricultural specifics and a good level of training of middle managers that are non-tourist centers while providing its citizens with an acceptable standard of living and provide the economy with investment (including foreign). Private entrepreneurs in Dagestan, the Sakhalin region, the Khabarovsk territory and Buryatia have the largest volume of revenue. Thus, in the Republic of Dagestan, the share of agriculture, hunting and forestry in the region's GVA is 23.5 %, which is more than 2 times higher than the average value in Russia.

We built a multi-factor regression model of the dependence of the organization's mortality rate (Y_8) from several factors. General view of the resulting model:

$$\hat{Y}_8 = -139{,}204 - 3{,}557x_{1.11} + 656{,}389x_{3.5} - 0{,}007x_{3.14} - 0{,}200x_{4.9} - -2{,}173x_{7.2} + 2{,}139x_{7.5} + 2{,}825x_{8.2} + 37{,}423x_{8.4}, \quad (7)$$

where: $X_{1.11}$ – unemployment level, %; $X_{3.5}$ – number of educational institutions of higher education and scientific organizations, units (per 10,000 inhabitants); $X_{3.14}$ – library fund, copies (per 1000 inhabitants); $X_{4.9}$ – availability of cars (number of own cars), units (per 1000 inhabitants); $X_{7.2}$ – share of innovative goods, works, services, % of the total volume of shipped goods, performed works, services; $X_{7.5}$ – share of organizations that used special software tools, % of the total number of surveyed organizations; $X_{8.2}$ – depreciation rate of fixed assets, %; $X_{8.4}$ – passenger turnover of public buses per capita, thousand pass-km.

This model has a good statistical quality, which is expressed by the values of the t-criterion and the F-criterion. Thus, the parameters of the model and the model as a whole can be considered statistically significant with a probability of 95 %. The variation of the factors collected in the model by 61.9 % causes a variation in the mortality rate coefficient of organizations. All other (random) factors account for 38.1 % of the variation. The value of the correlation coefficient is 0.787. In this model, there are 2 factors from the groups "Social factors" and "Innovative factors", but even in total, they caused only 1/5 of the effective feature variation (19.7 %). The other 2 factors, which we previously attributed to the "Infrastructure factors" group (31.8%), had a much greater impact.

Based on the results of the simulation, we can say that a high level of mortality is observed in regions with low unemployment, with a focus on higher education and science, which do not differ in the renewal of fixed assets and a good standard of living. Regions with high business mortality include the Crimea, Astrakhan and Vologda regions, and the Khanty-Mansiysk Autonomous Okrug – Yugra.

The following multiple regression equation was obtained based on the effective indicator Y_{11} – "Number of fast-growing enterprises (estimated growth by the turnover)":

$$\hat{Y}_{11} = -10{,}173 - 1{,}222x_{1.11} + 0{,}521x_{7.5} - 0{,}639x_{7.6} + 8{,}156x_{8.4}, \quad (8)$$

where: $X_{1.11}$ – unemployment level, %; $X_{7.5}$ – share of organizations that used special software tools, % of the total number of surveyed organizations; $X_{7.6}$ – number of active subscribers of fixed and mobile broadband Internet access, units (per 100 inhabitants); $X_{8.4}$ – passenger turnover of public buses per capita, thousand pass-km.

5 Discussion

The quality of this model was not very high – the correlation coefficient was 0.594, and the determination coefficient was 0.352 (factors explain only 35.2 % of the variation in effective breeding). In these conditions, it is difficult to distinguish the dominant factor (a group of factors); we only note that, according to the particular coefficients of determination, the main factor is the "passenger turnover" (14.8% of the influence). In general, the number of fast-growing enterprises (by turnover) increases in regions with high employment, a good level of ICT development and a large volume of passenger traffics. These regions include the Republic of Crimea, Astrakhan region, Krasnoyarsk territory, and Bryansk region. For example, in the Astrakhan region, the passenger turnover of public buses per capita is the largest (4.41 thousand pass-km) in the country. The distribution of 6 performance indicators does not follow the normal law (Y_3, Y_5, Y_7, Y_9, Y_{10}, Y_{12}). Thus, it is incorrect to apply parametric analysis methods to them. We will use nonparametric methods for analyzing relationships – rank correlation methods. The results of the analysis of factor relationships are briefly shown in table 10. We show only the highest values of Spearman's rank correlation coefficient (regardless of the sign in the coefficient).

Table 10. Spearman's rank correlation coefficients (5% level of significance)

Name	Coefficients
Y_3	
$X_{2.1}$ – share of the urban population, %;	0.680
$X_{7.3}$ – internal costs on scientific research and developments per capita, thousand rubles;	0.649
$X_{7.6}$ – number of active subscribers of fixed and mobile broadband Internet access, units (per 100 inhabitants);	0.636
$X_{3.11}$ – number of theater audience, people (per 1000 inhabitants);	0.565
$X_{8.1}$ – number of personal computers, units (per 100 employees).	0.545
Y_5	
$X_{3.15}$ – number of collective accommodation facilities, units (per 10,000 inhabitants);	0.294
$X_{2.9}$ – share of the working age population, %.	-0.236
$X_{3.3}$ – number of students enrolled in training programs for qualified workers, employees (per 10,000 inhabitants);	0.214
The other coefficients are not statistically significant	
Y_7	
$X_{3.11}$ – number of theater audience, people (per 1000 inhabitants);	0.333
$X_{4.10}$ – the total area of residential premises, on average per one inhabitant, m^2/person;	-0.322
$X_{2.3}$ – birth rate, ‰	0.302
$X_{1.4}$ – share of agriculture, hunting and forestry in the gross value added of the region, %;	-0.297
$X_{3.14}$ - library fund, copies (per 1000 inhabitants);	-0.293
Y_9	
$X_{1.6}$ – share of manufacturing in the region's gross value added, %;	0.547

Name	Coefficients
$X_{2.6}$ – migration growth coefficient, ‰	0.490
$X_{3.11}$ – number of theater audience, people (per 1000 inhabitants);	0.469
$X_{7.3}$ – internal costs on scientific research and developments per capita, thousand rubles;	0.460
$X_{2.1}$ – share of the urban population, %;	0.447
Y_{10}	
$X_{1.11}$ – unemployment level, %.	-0.445
$X_{7.5}$ – share of organizations that used special software tools, % of the total number of surveyed organizations;	0.387
$X_{1.7}$ – share of education in the region's gross value added, %;	-0.346
$X_{4.9}$ – availability of cars (number of own cars), units (per 1000 inhabitants);	0.319
$X_{7.4}$ – expenditures on ICT per capita, thousand rubles;	0.307
Y_{12}	
$X_{6.3}$ – fixed capital investment per capita, thousand rubles;	0.648
$X_{4.4}$ – volume of paid services per capita, rubles;	0.629
$X_{4.2}$ – average consumer spendings per capita (per month), rubles.	0.600
$X_{1.2}$ – GRP physical volume index, %;	0.592
$X_{1.1}$ – gross regional product (GRP) per capita, rubles/person;	0.582

Source: Own processing.

According to the obtained results, it is difficult to single out any one enlarged group of factors that significantly affect the above-mentioned performance indicators of the business environment of the region. For example, the number of small businesses and IP (Y_3 и Y_5) is influenced primarily by social and demographic indicators, as well as factors related to the development of IT infrastructure. The birth rate of organizations (Y_7), is also related to the traditional demographic birth rate of the population, and the relationship between them is reversed.

To create and grow a business (Y_7, Y_9 и Y_{10}), the production specialization of the region is important – the birth rate of organizations and their entrepreneurial activity is higher in territories where agriculture and education take a small share in the GRP. Active business growth requires appropriate infrastructure support and investment in research and development.

The final indicator is a multidimensional assessment of the business environment of the region (Y_{12}) that synthesizes the key determinants of business development. Subjects of the Federation with a high level of investment, standard of living and level of economic development are preferred for creating and running a business. Thus, our key hypothesis of the study was generally confirmed.

6 Conclusion

In conclusion, we will bring together the factor relationships of the business environment of the region indicators and identify the most important groups of influence factors (Table 11). The "+" sign indicates the significance of the influence of this factors' group.

Table 11. Significance of the influence of enlarged groups of factors on the business environment of the region

Group of factors	Y_1	Y_2	Y_3	Y_4	Y_5	Y_6	Y_7	Y_8	Y_9	Y_{10}	Y_{11}	Y_{12}
1. General economic							+		+	+		+
2. Demographical	+	+			+		+		+			
3. Social	+	+			+		+	+	+			
4. Factors of living standards				+		+	+			+		+
5. Natural and geographical												
6. Financial						+						+
7. Innovative		+						+	+	+		
8. Infrastructural		+	+					+			+	

Source: Own processing.

The summary data show the key role of socio-demographic, general economic, and innovative infrastructure regions' conditions in the development of the business environment. The financial condition of the regions plays a secondary role, and natural and geographical conditions are almost irrelevant. Thus, if the regional government adds adequate business support programs to the existing starting conditions (including at the municipal level), this will bring the business environment of the region to a qualitatively new level of development.

References

1. A.M. Mikhaylov, G.T. Sabirova, K.V. Shnyakin, Innovation and investment activities in businesses as a determining factor of development of the Russian Federation. In: S. Ashmarina, M. Vochozka (Eds.), Sustainable Growth and Development of Economic Systems (pp. 143-156). *Contributions to Economics*. (Springer, Cham, 2019). DOI: 10.1007/978-3-030-11754-2_11.
2. D. Urbano, D. Audretsch, S. Aparicio, M. Noguera, Does entrepreneurial activity matter for economic growth in developing countries? The role of the institutional environment. *International Entrepreneurship and Management Journal*, **16**(3), 1065-1099. DOI: 10.1007/s11365-019-00621-5. (2020).
3. E.A. Kandrashina, S.I. Ashmarina, D.V. Aleshkova, M.V. Vorotnikova, The level of production localization of automotive industry enterprises in Russia. In J. Horák, J. Vrbka and Z. Rowland (Eds.), Proceedings of Innovative Economic Symposium 2019 – Potential of Eurasian Economic Union (IES2019), *SHS Web of Conferences*, **73**, 01011. (EDP Sciences, Les Ulis, 2020). DOI: 10.1051/shsconf/20207301011.
4. E.V. Rybakova, M.A. Nazarov, Entrepreneurship in digital era: Prospects and features of development. In: S. Ashmarina, V. Mantulenko (Eds.), Current Achievements, Challenges and Digital Chances of Knowledge Based Economy. *Lecture Notes in Networks and Systems*, **133** (pp. 105-112). (Springer, Cham, 2021). DOI: 10.1007/978-3-030-47458-4_13.
5. Federal law "On the development of small and medium-sized enterprises in the Russian Federation" of 24.07.2007 N 209-FZ. (2007).URL: http://www.consultant.ru/document/cons_doc_LAW_52144/. Accessed: 03.08.2020.
6. Forbes, Sly figures: What is hidden by income data in Russia. (2019).URL: https://www.forbes.ru/biznes/372965-lukavye-cifry-chto-skryvayut-dannye-o-dohodah-v-rossii. Accessed: 03.08.2020.
7. G.I. Yakovlev, A.V. Streltsov, E.Y. Nikulina, Reforming issues of entrepreneurship in the digital economy. In: S. Ashmarina, V. Mantulenko (Eds.), Current Achievements, Challenges and Digital Chances of Knowledge Based Economy. *Lecture Notes in Networks and Systems*, **133** (p. 825-834). (Springer, Cham, 2021). DOI: 10.1007/978-3-030-47458-4_95.
8. G.P. Gagarinskaya, A.V. Gagarinskii, E.S. Potokina, T.N. Obuschenko, State support as incentive

for developing enterprises and increasing competitiveness. In: S. Ashmarina, V. Mantulenko, M. Vochozka (Eds.), Engineering Economics: Decisions and Solutions from Eurasian Perspective. Engineering Economics Week 2020. *Lecture Notes in Networks and Systems*, **139** (p. 201-211). (Springer, Cham, 2021). DOI: 10.1007/978-3-030-53277-2_23.
9. J. Váchal, M. Talíř, The development of organizational and management structures in small-scale and mid-scale entrepreneurship in the Czech Republic. In J. Horák, J. Vrbka and Z. Rowland (Eds.), Proceedings of Innovative Economic Symposium 2019 – Potential of Eurasian Economic Union (IES2019), *SHS Web of Conferences*, **73**, 02006. (EDP Sciences, Les Ulis, 2020). DOI: 10.1051/shsconf/20207302006.
10. M. Holienka, Entrepreneurial environment in Slovakia: Multi-perspective comparison with innovation-driven economies. *Procedia Economics and Finance*, **34**, 437-444. DOI: 10.1016/S2212-5671(15)01652-4. (2015).
11. M.R. Efimova, E.A. Dolgikh, T.A. Pershina, L.S. Parshintseva, The methodology for multivariate statistical analysis of innovation potentials: The case of Russia. In: A.V. Bogoviz, A.E. Suglobov, A.N. Maloletko, O.V. Kaurova, S.V. Lobova (Eds.), Frontier Information Technology and Systems Research in Cooperative Economics. *Studies in Systems, Decision and Control*, **316** (p. 239-247). (Springer, Cham, 2021). DOI: 10.1007/978-3-030-57831-2_25.
12. P. Muñoz, W. Naudé, N. Williams, T.Williams, & R. Frías (2020). Reorienting entrepreneurial support infrastructure to tackle a social crisis: A rapid response. *Journal of Business Venturing Insights*, 14. DOI: 10.1016/j.jbvi.2020.e00181.
13. S.I. Ashmarina, G.M. Murzagalina, Entrepreneurial potential of Russian regions: Comparative assessment and development prospects. In: S.I. Ashmarina, J. Horák, J. Vrbka, P. Šuleř (Eds.), Economic Systems in the New Era: Stable Systems in an Unstable World. IES 2020. *Lecture Notes in Networks and Systems*, **160** (pp. 33-39). (Springer, Cham, 2021). DOI: 10.1007/978-3-030-60929-0_5.
14. S.I. Ashmarina, V.V. Mantulenko, E.P. Troshina, Readiness to changes as one entrepreneurial value of the innovation-riented conomy. In: S. Ashmarina, M. Vochozka (Eds.), *Sustainable Growth and Development of Economic Systems. Contributions to Economics* (pp. 157-166). (Springer, Cham, 2019). DOI: 10.1007/978-3-030-11754-2_12.
15. S.I. Ashmarina, V.V. Mantulenko, Regional development of small and medium-sized businesses: Value and behavioral aspects. In *Proceedings of the 13th International Scientific Conference European Forum of Entrepreneurship 2020 "Protectionism – protection or destruction of national economies?"* (pp. 46-52). (NEWTON College, Prague, 2020).
16. V.V. Radaev, The development of small entrepreneurship in Russia. (UNU/WIDER, Helsinki, 2001).
17. V.V. Radaev, The market as an object of sociological investigation. *Sociological Research*, **39**(1), 51-66. (2000).
18. VCIOM, Entrepreneurship in Russia: Trust, barriers and success factors. (2017). URL: https://wciom.ru/index.php?id=236&uid=3605. Accessed: 03.08.2020.

Sustainability of passenger transport as a precondition for fulfilling the 4.0 company concept – analysis of trends in the Czech Republic

Jana Hinke[1] and *Veronika* Komorousová[2]

[1]Faculty of Economics, University of West Bohemia, Univerzitní 22, Plzeň, 306 14, Czech Republic
[2]Faculty of Economics, University of West Bohemia, Univerzitní 22, Plzeň, 306 14, Czech Republic

Abstract. Strengthening public transport can significantly contribute to transport sustainability, in particular to reducing greenhouse gas emissions. This article primarily aims to analyse the long-term trend in the production of carbon dioxide by individual types of transport throughout the Czech Republic and to compare the performance of public passenger transport and passenger car transport. The analysis concludes that the performance of public passenger transport by land in the Czech Republic has not been increasing over a long period; on the contrary, individual transport has been strengthening. There has been an increasing trend in CO_2 emissions from transport for a long period. Thus, the development of subsidies in the public transport clearly shows that to fund public passenger transport is in the public interest.
Key words: Sustainable development, transport, emissions, performance.

1 Introduction

Environmental issues, growing urban population and dwindling crude oil reserves are the main reasons why there is more and more talk about transport sustainability. Gumbo and Moyo [1] warn of this problem especially in developing countries. Lefevre, Briand, Pye, et al. [2] state that "the transport sector represents 25% of global CO_2 emissions" and further point to the fact that "while a range of options exist to decarbonize the passenger transport sector, the detailed sequencing of actions and resulting transformations over time remain largely unexplained in political debates".
The problem of CO_2 emissions is compounded by the decline in organic productive areas, especially in cities and on outskirts. This decline is exacerbated by the lack of capacity to sequester CO_2 on both the microscale and mesoscale, which results in rising temperatures and reduced air quality within urban boundaries [3]. The above-named authors (Lefevre, Briand, Pye, et al.) [2] indicate that the results across countries correspond with the following theses:

- to achieve deep decarbonisation, resolute action for all types of transport is needed, especially for solutions on the demand side;

- deep decarbonisation is compatible with other political priorities, such as meeting the mobility needs at affordable costs; and
- strategies should be adapted to the purpose of mobility, to local contexts and national circumstances, with the framework being adapted to other sectors and further developed in the context of future political processes.

Stefaniec, Hosseini, Xie, et al. [4] propose to introduce a systematic approach to transport evaluation based on the social, economic and environmental dimensions of sustainability in order to support decision makers (in better evaluation of transport sustainability performance). Sultana, Salon and Cuba [5] also admit that there are inherent trade-offs that may hinder win-win-win strategies across this environmental, economic and social domain.
Umar, Ji, Kirikkaleli et al. [6] demonstrate that there is an important cointegration equation between CO_2 emissions, innovations, financial development, transport infrastructure and real GDP, on which individual political decisions should be based - for example, to strengthen the performance of public passenger transport and to encourage restrictions on individual transport or to support pooling of drives - more people travel in fewer vehicles [7].
All over the world governments have introduced sustainability programs and policies that lead to a shift from traditional private car-dependent transport systems to the greater use of public transport, cycling and walking [8] or the use of electric three-wheelers [9]. Many research teams [10, 11] offer tools for decision-making of local governments in order to assess the global and local sustainability of transport systems, mostly based on multi-criteria decision making. The trend of transport inclusion into a sophisticated system of life is growing stronger in all aspects - in the offer of equal access to public services and job opportunities, in environmental protection and sustainable use of natural resources [12]. This article aims to create a time series of data on carbon dioxide production by individual types of transport throughout the Czech Republic, to determine a long-term trend between 2000 and 2018 and to describe the trend. Furthermore, based on the comparison of performance of public passenger transport, a conclusion will be drawn as to whether there are really changes or shifts in favour of public passenger transport and how strong these shifts are. A partial aim is to evaluate the transport output, both by the number of transported persons and by kilometres per person. The last aim is to describe the amount of subsidies provided from the state budget of the Czech Republic, on the basis of which it will be possible to state whether funding and supporting public transport as a whole is in the real interest of the state.
To meet the aforementioned aims, statistical data of the Ministry of Transport of the Czech Republic from individual yearbooks (publicly available) and data of the Ministry of the Environment of the Czech Republic are used. Only the data on the land passenger transport are used from the data set on transported persons. Passenger shipping and air travel are excluded from the survey. This procedure is also applied to the analysis of transport output, which will be supplemented by data on the length of lines and roads. MS Excel is used for data processing.

2 Sustainability in passenger transport – greenhouse gas emissions

European countries have undertaken to meet the commitments of the White Paper, which is the so-called Single European Transport Area Plan, i.e. creating a competitive and resource-efficient transport system. Strengthening public transport can significantly contribute to meeting one of the objectives of the White Paper that reads as follows: "To reduce greenhouse gas emissions by at least 60% by 2050 compared to 1990" [13, p.3].
In order to determine the trend in the production of greenhouse gases in the Czech Republic, it is necessary to create a time series of carbon dioxide emissions for individual types of

transport. Data from the Ministry of the Environment are used for a longer time series (Table 1). It is not possible to compile a sufficiently long time series from the data of the Ministry of Transport, where the data for 2019 are also given. This is due to the fact that in 2019 there was a change in the methodology and only the data for the last five-year period were revised retrospectively. If we compare the data from both sources, the total emissions from transport are the same, but there are noticeable deviations between individual categories.

Table 1. Carbon dioxide (CO_2) emissions between 2000 and 2018 (in thousands of tonnes) in the Czech Republic

	2000	**2001**	**2002**	**2003**	**2004**	**2005**	**2006**	**2007**	**2008**	**2009**
Passenger car transport	7,391	7,807	8,295	9,596	9,997	10,579	10,822	11,545	11,365	11,295
Public passenger transport by land	1,143	1,152	1,120	1,128	1,105	1,162	1,169	1,186	1,275	1,271
Other transport	4,031	4,433	4,526	5,176	5,677	6,303	6,475	6,703	6,909	6,543
Transport in total	12,565	13,392	13,940	15,900	16,779	18,044	18,467	19,434	19,550	19,109

	2010	**2011**	**2012**	**2013**	**2014**	**2015**	**2016**	**2017**	**2018**
Passenger car transport	10,761	10,887	10,622	10,690	10,978	11,461	12,068	12,420	12,751
Public passenger transport by land	1,277	1,287	1,292	1,219	1,355	1,304	1,293	1,268	1,189
Other transport	6,104	6,004	5,912	5,762	5,997	6,291	6,605	6,811	6,898
Transport in total	18,143	18,178	17,825	17,671	18,330	19,056	19,966	20,499	20,838

Source: Own processing according to the Ministry of the Environment of the CR, 2020.

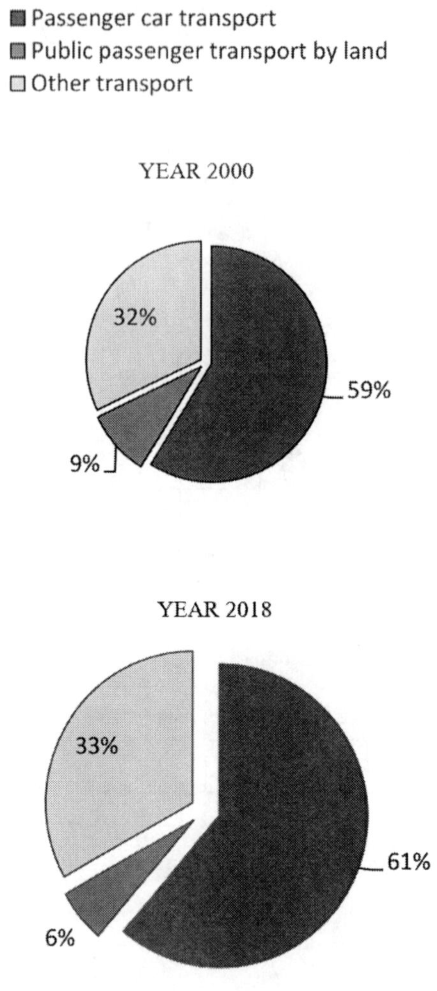

Fig. 1. Proportion of carbon dioxide (CO_2) emissions in the total emissions from transport in % - year 2000
Source: own processing according to the Ministry of the Environment of the CR, 2020
Fig. 2. Proportion of carbon dioxide (CO_2) emissions in the total emissions from transport in %. – year 2018
Source: Own processing according to the Ministry of the Environment of the CR, 2020.

Figure 1 and figure 2 clearly show that that the individual car transport accounted for 59% of total CO_2 emissions in 2000 and by 2018 this proportion rose to 61%. It is necessary to realize that the minority proportion of emissions includes not only public passenger transport, but also freight transport and air and water transport.
We can derive the trend of emissions in the time series:

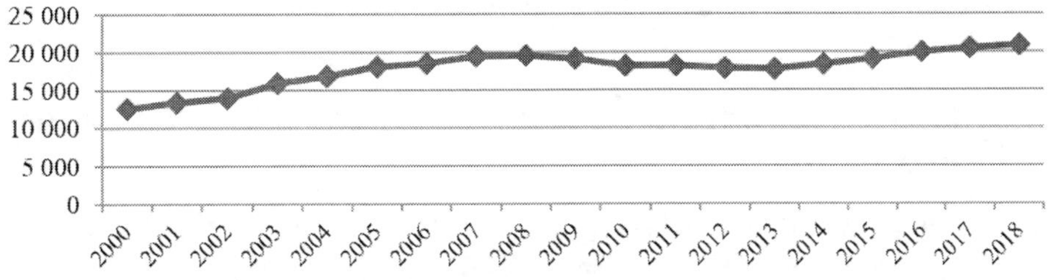

Fig. 3. Trend in the development of the total carbon dioxide emissions (CO_2) from transport between 2000 and 2018 (in thousands of tonnes)
Source: Own processing according to the Ministry of the Environment of the CR, 2020.

It can be observed from the aforementioned that the total CO_2 emissions have not been reduced yet. The growing trend is confirmed by the fact that between 2000 and 2018 CO_2 emissions from transport increased by 66%. The trend in the development of total emissions from transport (Figure 3) clearly shows that there was a linear growth between 2000 and 2008. Between 2008 and 2013, there was a slight decrease and stagnation in total CO_2 emissions from transport and since 2014 emissions were growing linearly again.

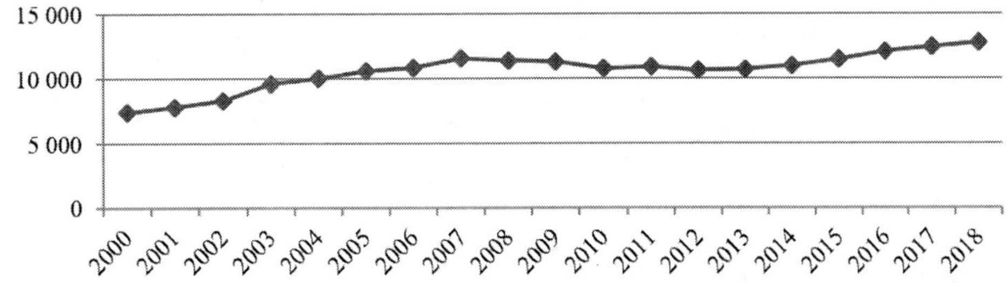

Fig. 4. Trend in the development of carbon dioxide emissions (CO_2) from the passenger car transport between 2000 and 2018 (in thousands of tonnes).
Source: Own processing according to the Ministry of the Environment of the CR, 2020.

CO_2 emissions from the passenger car transport (Figure 4) more or less copy the curve of the total emissions from transport.

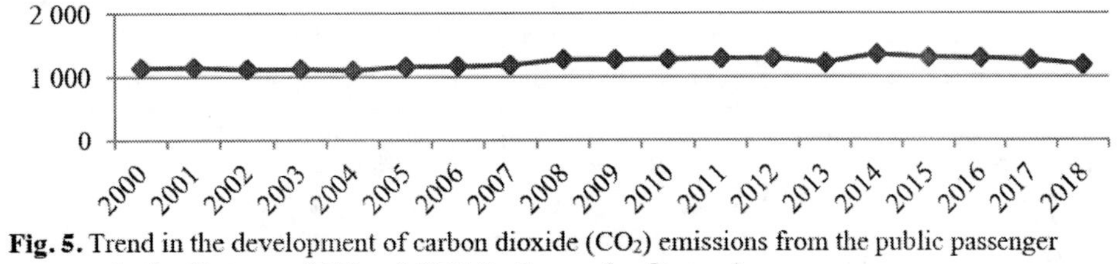

Fig. 5. Trend in the development of carbon dioxide (CO_2) emissions from the public passenger transport by land between 2000 and 2018 (in thousands of tonnes)
Source: Own processing according to the Ministry of the Environment of the CR, 2020.

The curve of emissions from the public passenger transport by land (Figure 5) is rather different. There was an evident stagnation throughout the monitoring period. Between 2007 and 2008 there was a certain growth. On the contrary, in 2013 there was a more significant decline, which was followed by a steeper growth, and then the curve had a declining trend until 2018. It can be stated from the above that it is really necessary to start motivating the inhabitants to reduce the individual car transport and to strengthen the public transport in order to reduce greenhouse gas emissions.

3 Transport performance in passenger transport, passengers transported

Passenger transport performance is usually expressed in two ways - on the one hand in the number of transported passengers and on the other hand in kilometres per person. The kilometre per person represents the transport of one person over a distance of one kilometre. It is calculated as the product of the transport performance (i.e. the distance a passenger travels on average) and the number of transported persons. The Ministry of Transport of the Czech Republic regularly collects data from transport companies and summarises them in yearbooks. One set of the data is the transported persons and kilometres per person (according to the data of the Ministry of Transport of the Czech Republic, 2020).

3.1 Evaluation of transport performance by the number of persons transported

First, the development trend in the number of transported persons is described on the basis of comparing the data from the yearbooks of the Ministry of Transport of the Czech Republic:

Table 2. Passenger transport (in millions) between 2005 and 2019 in the Czech Republic

	2005	2006	2007	2008	2009	2010	2011	2012
Roads in total	4,967.42	4,968.74	5,037.65	5,124.62	5,034.57	4,682.10	4,701.00	4,732.02
Railway transport	180.27	183.03	184.23	177.42	164.96	164.80	167.93	172.80
Bus transport	388.26	387.71	375.02	373.40	367.65	372.60	364.62	344.99
Municipal public transport	2,268.89	2,238.01	2,258.39	2,323.80	2,261.96	2,174.70	2,138.46	2,224.23
Public transport in total	2,837.42	2,808.75	2,817.65	2,874.62	2,794.57	2,712.10	2,671.00	2,742.02
Individual car transport	2,130.00	2,160.00	2,220.00	2,250.00	2,240.00	1,970.00	2,030.00	1,990.00

	2013	2014	2015	2016	2017	2018	2019
Roads in total	4,695.72	4,728.50	4,849.20	4,954.10	5,016.70	5,203.40	5,396.30
Railway transport	174.49	176.05	176.60	179.20	183.00	189.50	193.80
Bus transport	337.98	349.52	350.90	332.80	329.70	340.20	354.70
Municipal public transport	2,173.25	2,142.93	2,146.30	2,168.80	2,135.20	2,184.10	2,231.20
Public transport in total	2,685.72	2,668.50	2,673.80	2,680.80	2,647.90	2,713.80	2,779.70
Individual car transport	2,010.00	2,060.00	2,175.40	2,273.30	2,368.80	2,489.60	2,616.60

Source: Own processing according to the Ministry of Transport of the CR, 2020.

There was a decline in the proportion of passengers transported by the land public transport in the total transport (Table 2). In 2005, this percentage was 57% and by 2019 it dropped to 52%. This decrease was caused by a faster growth in the number of persons transported by the individual car transport.
The municipal public transport accounted for the highest percentage in the land public transport, with its stable proportion in public transport remaining at 80%.

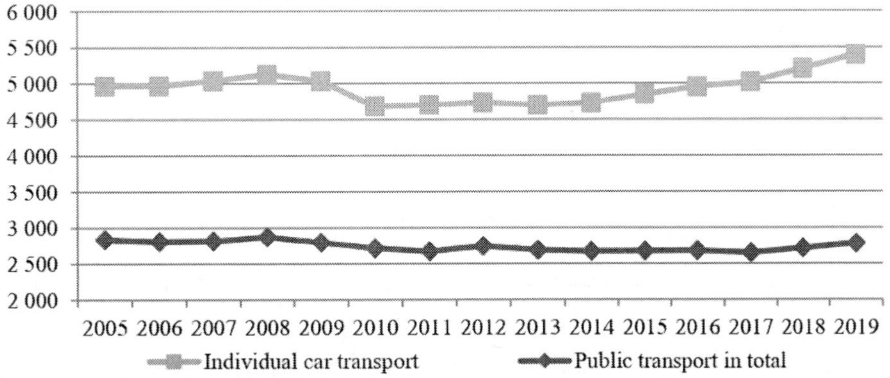

Fig. 6. The development trend in transported persons (in millions) in the individual car transport and the land public transport between 2005 and 2019
Source: Own processing according to the Ministry of Transport of the CR, 2020.

It is evident from Figure 6 that the development curve of the number of transported persons in the land public transport had a rather stagnant character between 2005 and 2019. There were no significant changes in the development curve, since 2005 there was a slight decrease and only in the last two years a slight increase was recorded. On the contrary, the development curve of transported persons in the individual car transport was significantly growing since 2013. The growth trend of transported persons in the individual car transport had been preceded by a stagnation and significant decline in 2010.

3.2 Evaluation of transport performance by kilometres per person

The second way to evaluate the transport performance is the method of describing kilometres per person. Based on the statistical data of the Ministry of Transport of the Czech Republic, the data of registered kilometres per person in the individual types of transport were determined and a time series was created from them:

Table 3. Transport performance (in millions of kilometres per person) between 2005 and 2019 in the Czech Republic

	2005	2006	2007	2008	2009	2010	2011	2012
Roads in total	98,849.00	100,366.00	102,310.00	104,279.00	103,842.00	96,113.80	96,753.00	97,164.86
Railway transport	6,667.00	6,922.00	6,898.00	6,803.00	6,503.00	6,590.70	6,714.00	7,264.70
Bus transport	8,607.00	9,501.00	9,519.00	9,215.00	9,494.00	10,335.70	9,267.00	9,015.41
Municipal public transport	14,935.00	14,313.00	14,353.00	15,881.00	15,555.00	15,617.40	15,282.00	16,624.75
Public transport in total	30,209.00	30,736.00	30,770.00	31,899.00	31,552.00	32,543.80	31,263.00	32,904.86
Individual car transport	68,640.00	69,630.00	71,540.00	72,380.00	72,290.00	63,570.00	65,490.00	64,260.00

	2013	2014	2015	2016	2017	2018	2019
Roads in total	97,552.36	100,336.89	104,099.00	108,742.60	112,826.60	117,113.50	121,176.80
Railway transport	7,600.60	7,796.50	8,298.10	8,843.40	9,497.60	10,286.00	10,930.60
Bus transport	9,025.57	10,010.20	9,995.90	10,257.10	11,177.80	10,950.40	10,547.00
Municipal public transport	16,276.19	16,270.19	16,100.00	17,387.10	17,824.20	17,906.10	18,520.20
Public transport in total	32,902.36	34,076.89	34,394.00	36,487.60	38,499.60	39,142.50	39,997.80
Individual car transport	64,650.00	66,260.00	69,705.00	72,255.00	74,327.00	77,971.00	81,179.00

Source: Own processing according to the Ministry of Transport of the CR, 2020.

Table 3 gives an overview of the transport performance in kilometres per person. With this evaluation method of the transport performance, there was a slight increase in the proportion of the land public transport, when in 2005 it accounted for 31% which increased to 33% within a year. This was due to the fact that the average transport distance in the public transport increased from 10.6 km in 2005 to 14.4 km in 2019. This phenomenon is positive, it informs us that people started using the public transport more even for longer journeys. This increase in the average transport distance is evident especially in the railway transport, where the average transport distance increased from 37 km to 56 km. On the contrary, there was a slight decrease in the average transport distance in the individual car transport from 32.2 km to 31 km.

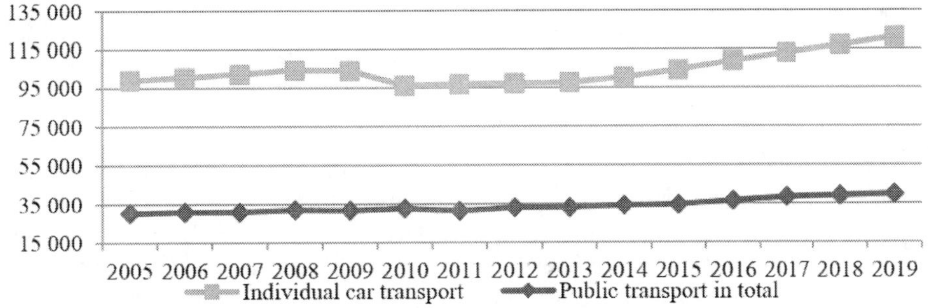

Fig. 7. The development trend in the transport performance (in millions of kilometres per person) of the individual car transport and the passenger public transport by land between 2005 and 2019
Source: Own processing according to the Ministry of Transport of the CR, 2020.

The development trend in the transport performance (Figure 7) clearly shows that there was a slight increase in the public passenger transport throughout the monitoring period. However, there was a relatively steep growth in the performance of passenger car transport since 2013 after a phase of stagnation and decline between 2005 and 2012.

3.3 Subsidizing the land public transport from the state budget

The analysis of the public transport use can be supplemented with a time series on the development of subsidies as a state aid. This description provides information on whether it is in the real interest of the state to support public transport as a whole.

Table 4: The amount of subsidies in the scheduled public passenger transport (in millions of CZK)

	2010	2011	2012	2013	2014	2015	2016	2017	2018
Line bus transport (without municipal public transport)									
Budgets of regional (district) authorities	4,823	4,712	5,054	5,184	5,275	5,251	5,422	6,180	6,642
Municipality budgets	370	368	462	470	550	470	658	632	376
Total	5,193	5,080	5,516	5,654	5,825	5,721	6,080	6,812	7,018
Railway passenger transport									
State budget	4,070	4,074	4,009	4,043	4,233	4,414	4,497	4,645	4,802
Subsidies from regional (district) authorities	8,458	8,672	8,864	9,245	9,290	9,413	9,584	9,794	10,180
Total	12,528	12,746	12,873	13,288	13,523	13,827	14,081	14,439	14,982
Subsidies in total	17,721	17,826	18,389	18,942	19,348	19,548	20,162	21,251	22,000

Source: Own processing according to the Ministry of Transport of the CR, 2020.

Subsidies (Table 4) are listed only for the period up to 2018 (2019 is not available yet) and only for the line bus and railway transport. The table clearly shows that transport subsidies were growing over time. Between 2010 and 2018 there was an increase in subsidies by 24%. The ratio of railway transport subsidies to total subsidies ranges from 68% to 70%.

4 Conclusion

At first glance, the analysis clearly shows that the overall performance of the land public transport has not significantly improved since 2005; on the contrary, the use of the individual car transport seems to be strengthening and intensifying. The development trend in the transport performance clearly shows that there was a slight increase in the public passenger transport throughout the monitoring period. However, there was a relatively steep growth in the performance of passenger car transport since 2013 after a phase of stagnation and decline between 2005 and 2012.
The total CO_2 emissions have not been reduced yet. The growing trend is confirmed by the fact that between 2000 and 2018 CO_2 emissions from transport increased by 66%. Given the longer-term efforts and goals to substantially reduce CO_2 emissions (which was illustrated with the increasing trend in provided subsidies), this signal is alarming and it would certainly be appropriate to address this topic in more depth – to conduct more detailed research both at the level of individual regions or at the level of individual transport companies and to look for problematic aspects that prevent the land public transport being stronger pushed through at the expense of the individual road transport. Unfortunately, the current situation in connection with the coronavirus pandemic may, on the contrary, exacerbate the transport crisis and further reduce the use of public transport.

Acknowledgement

This article was elaborated within the project called "TL02000136 - Adaptation of the knowledge-intensive services sector to the conditions of Society 4.0" addressed by the University of West Bohemia in Pilsen, Faculty of Economics, with the state support of the Technology Agency of the Czech Republic.

References

1. T. Gumbo, T. Moyo, T, Exploring the Interoperability of Public Transport Systems for Sustainable Mobility in Developing Cities: Lessons from Johannesburg Metropolitan City, South Africa, Sustainability, 12(5). DOI: 10.3390/su12155875, (2020)

2. J. Lefevre, Y. Briand, S. Pye, et al., A pathway design framework for sectoral deep decarbonization: the case of passenger transportation, Climate Policy, DOI: 10.1080/14693062.2020.1804817, (2020)

3. S. M. Labib, M. N. Neema, Z. Rahaman, et al., Carbon dioxide emission and bio-capacity indexing for transportation activities: A methodological development in determining the sustainability of vehicular transportation systems. Journal of Environmental Management. 223. 57-73. DOI: 10.1016/j.jenvman.2018.06.010, (2018)

4. A. Stefaniec, K. Hosseini, J. Xie, et al., Sustainability assessment of inland transportation in China: A triple bottom line-based network DEA approach. Transportation Research

Part D-Transport and Environment. 80. DOI: 10.1016/j.trd.2020.102258, (2020)

5. S. Sultana, S. Salon, M. Kuby, Transportation sustainability in the urban context: a comprehensive review. Urban Geography. 40(3). 279-308. DOI: 10.1080/02723638.2017.1395635, (2019)
6. M. Umar, X. Ji, D. Kirikkaleli, et al., COP21 Roadmap: Do innovation, financial development, and transportation infrastructure matter for environmental sustainability in China? Journal of Environmental Management. DOI: 271. 10.1016/j.jenvman.2020.111026 (2020)
7. L.A. Merlin, L.A., Transportation Sustainability Follows from More People in Fewer Vehicles, Not Necessarily Automation, Journal of the American Planning Association, 85(4). 501-510. DOI: 10.1080/01944363.2019.1637770, (2019)
8. L.E. Karjalainen, S. Juhola, Framework for Assessing Public Transportation Sustainability in Planning and Policy-Making. Sustainability, 11(4). DOI: 10.3390/su11041028, (2019)
9. R.A. de Mello Bandeira, G.V. Goes, D. N. Schmitz Goncalves, et al, Electric vehicles in the last mile of urban freight transportation: A sustainability assessment of postal deliveries in Rio de Janeiro-Brazil, Transportation Research Part D-Transport and Environment, 67. 491-502. DOI: 10.1016/j.trd.2018.12.017 (2019).
10. U. Oses, E. Roji, J. Cuadrado, et al., Multiple-Criteria Decision-Making Tool for Local Governments to Evaluate the Global and Local Sustainability of Transportation Systems in Urban Areas: Case Study. Journal of Urban Planning and Development. 144(1). DOI: 10.1061/(ASCE)UP.1943-5444.0000406, (2018)
11. P. Zhao, Z. Yu, Investigating mobility in rural areas of China: Features, equity, and factors. Transport Policy. 94. 66-77. DOI: 10.1016/j.tranpol.2020.05.008, (2020)
12. Z. Liu, M. de Jong, F. Li, et al., Towards Developing a New Model for Inclusive Cities in China-The Case of Xiong'an New Area, Sustainability, 12(5), DOI: 10.3390/su12156195, (2020)
13. European Commission, White Paper - Roadmap to a Single European Transport Area - Towards a competitive and resource efficient transport system. Retrieved August 17, 2020 from https://www.mdcr.cz/getattachment/Dokumenty/Evropska-unie/Zakladni-dokumenty/Bila-kniha-Plan-jednotneho-evropskeho-dopravniho/Bila-kniha-Plan-jednotneho-evropskeho-dopravniho-prostoru-%E2%80%93-vytvoreni-konkurenceschopneho-dopravniho-systemu-ucinne-vyuzivajiciho-zdroje.pdf.aspx, (2011)
14. Ministry of Transport, Transport Yearbooks of the Czech Republic, Retrieved July 15, 2020, from https://www.sydos.cz/cs/rocenky.htm, (2020)
15. Ministry of the Environment, Environmental Yearbook of the Czech Republic. Retrieved September 10, 2020, from https://www.mzp.cz/cz/statisticka_rocenka_zivotniho_prostredi_publikace, (2020)

IoT - Tool of Society 4.0 - Literary Review

Marie Černá[1,*] and *Lucie* Vallišová[2]

[1]University of West Bohemia, Faculty of Economics, Univerzitní 8, 301 00 Plzeň
[2]University of West Bohemia, Faculty of Economics, Univerzitní 8, 301 00 Plzeň

Abstract. Society 4.0 and its tools influence significantly current human activities. Approach of people to use of information have changed, information technologies became part of our daily life. Data are shared and processed using different processing procedures than were used 20 years ago. It has many impacts on lives of individuals and interpersonal relations. This contribution deals with the issue of Internet of Things (IoT) as the tool enabling people more efficient information management. The objective of this contribution is to outline the significance of IoT and provide the overview of this technology. The contribution is based on the literature research and contains bibliometric analysis confirming the IoT significance for current and future human activities.
Keywords: automation, data sharing, digitization, Industry 4.0, IoT, Society 4.0

1 Introduction

Current era is typical by constant changes of practices established in all areas of human activity. It is closely associated with rapid development of many scientific areas and tools used by them. The most visible changes caused by such development has occurred in the field of information technologies. Information technologies represent one of the basic preconditions for the human society development. This can be documented by the following table showing the stages of society development defined by Roser [1], Veza, Mladineo, Peko [2] and Schwab [3], Table 1.

* Corresponding author: macerna@kfu.zcu.cz

Table 1. Four stages of the industrial revolutions

No.	Period	Classification	Brief description
1.	18th century	The first industrial revolution	Mechanization, water (use of water power), steam.
2.	19th century	The second industrial revolution	Mass production, assembly lines, use of electricity.
3.	2nd half of the 20th century	The third industrial revolution	Use of computers and automation.
4.	2011 (resp. 2013)	The fourth industrial revolution	Information systems (digitization, advanced automation, cognitive technologies, high-capacity networks).

Source: Own processing based on [1], 2020.

Currently used concept of bringing new technology, influencing work and way of human behaviour itself, is called Society 4.0 - Industry 4.0.

The fourth industrial revolution is currently accepted as the concept of bringing technology (internet of Things - IoT, robotics, virtual reality, cyber-physical systems, artificial intelligence, big data) influencing significantly work and life generally. Original concept of Industry 4.0 was related to the production area. What comes out, was the fact that this concept will influence the overall design of today´s society, therefore the designation of this era was widened. Currently are used terms Industry 4.0 and Society 4.0.

Individuals start to come to terms with the need to use new or improved tools to ensure smooth running of individual work activities. New conditions represent for people challenge and shift to greater efficiency in its actions (time optimization of performed activities). On the other hand, it is simultaneously seen as limitation [4] and people are dealing with the question: 'What is really important in real life?'

Those important things should be ensured by the use of tools provided by Society 4.0. Tools of Society 4.0 that can be used during optimization of business activities (activities ensured by the corporation) are summarized in the following tables, Table 2a and Table 2b.

Table 2a. Tools of Society 4.0 that can be used to optimize the business activities (A-P)

Tool	Abbreviation	Use of tool (selected options)	Positives	Negatives
Artificial Intelligence	AI	✓ Predictive management of inventories (ERP systems). ✓ Predictive maintenance. ✓ New business models development.	✓ Automation. ✓ Focus on higher value added cases solving. ✓ Lack of workforce solving. ✓ Acceleration of activities.	✓ Financial demands of research. ✓ Misuse of research results. ✓ The need for reorganization of business activities.
Blockchain		✓ Reliable and anonymized transaction recording and verification.	✓ Use in logistic chains. ✓ Protected communica-tion.	✓ Transmission capacity and storage media requirements. ✓ Lack of competencies.
Drones		✓ Consignment deliveries. ✓ Insurance claims solving. ✓ Real estate activities. ✓ Construction inspections. ✓ Mapping.	✓ Inspection and mapping of various activities.	✓ Ensuring safety. ✓ Regulation of users.
Internet of Things	IoT	✓ Processing and analysing data gathered by sensors.	✓ Process optimization. ✓ Preventive maintenance. ✓ Remote diagnostics	✓ Building of sufficient infrastructure.
3-D printing Additive production		✓ 3D objects. ✓ Nano-technology.	✓ Production anywhere. ✓ Data transfer. ✓ Fast prototyping. ✓ Waste-free technologies.	✓ Necessary transformation of development and production processes. ✓ Lack of specialists.

Source: Own processing based on [5], 2020.

Table 2b. Tools of Society 4.0 that can be used to optimize the business activities (Q-Z)

Tool	Abbreviation	Use of tool (selected options)	Positives	Negatives
Robotics		✓ Automation of activities. ✓ More possibilities.	✓ Use in health care, social services. ✓ Possibility to work 24 hours/week, no illnesses, conflicts, etc.	✓ Initial investment. ✓ Staff training. ✓ Socio-economic consequences.
Virtual and Augmented Reality	VR, AR	✓ Visualization of industrial equipment. ✓ Better diagnostics. ✓ Navigation, easier orientation.	✓ Real time operations. ✓ New ways of carrying out "experienced" activities. ✓ Unlimited learning opportunities. ✓ Greater emotional experience for consumer.	✓ Market penetration barriers. ✓ High price.

Source: Own processing based on [5], 2020.

1.1 Internet of things (IoT)

IoT (Internet of Things) is often presented as the new paradigm becoming very strong in the field of telecommunication [6], [7]. Basic thought of this concept is pervasive identification of various things or objects using RFID chips, sensors, drivers, mobile phones, etc. [8]. These can communicate and cooperate one with each other, because of the use of unique addressing scheme [9].

Pervasive systems are developed taking into account the necessity to offer open, extensible, evolving portfolio of services integrating sensor data from various sources. It is necessary to ensure suitable and consistent adaptive behaviours during provision of such services even if the huge amount of data gathered by the sensors report different degree of precision, accuracy and dynamism [10].

Internet of things brings together two terms, internet and things. This tool may be accepted as the global network enabling communication between any objects (physical and virtual), between people (human-human), between physical person and thing (human-thing) or between things (thing-thing) [11]. Each object has to be clearly identifiable [12].

IoT does not mean only 'connection', in the sense of connecting electronic facilities as servers, PCs, tablets and smartphones. IoT also means sensors, codes, or embedded systems using the same IP address connecting internet. These networks produce huge amounts of data that have to be further analysed by PCs [13]. If these objects are able to perceive their environment because of their intercommunication abilities, they become more or less autonomous tools that can operate largely without human intervention (act without direct supervision of a human) [14].

Internet of Things is taken as the coding and networking of daily used objects to let them become individually machine-readable and traceable using the internet. Future tendencies are to ensure the ability to identify, automate, monitor and control each object [14], [15].

IoT is listed by the US National Intelligence Council as one of the six most important 'disruptive' civil technologies for the potential strengthening of US power [16]. In the following parts of this contribution will be analysed the current level of IoT significance using the bibliometric analysis of this term.

2 Methodology

With respect to the set objective of this article '... to outline the significance of IoT and provide the overview of this technology', was conducted bibliometric analysis of the term IoT. The plan of systematic literature review consists of basic steps that the authors followed in preparing this paper as visible in the next table, Table 3.

Table 3. Plan of systematic literary review

Stage	Description
1	An overview of publishing activities in a given area by searching for specified phrases.
2	Application of selection criteria as well as exclusion criteria.
3	Identification of key terms, authors and journals publishing on this topic.
4	Conclusions, evaluation.

Source: Own processing, 2020.

Authors of this contribution used as the tool quantitative analysis, systematic literary review from the general point of view, focused on selecting basic areas associated with the term IoT and IIoT, as a specific area of IoT. As the database used for the records selection was taken the Scopus database. Within the bibliometric analysis were the terms 'Internet of Things' and 'Industrial Internet of Things' and their abbreviations, 'IoT' and 'IIoT', searched simultaneously using the logical operator AND.

Research took into account headings, key words and also abstracts of the sources. Other precondition set before the research was identification of selected sectors of scientific interest:

- Business, Management and Accounting,
- Social Sciences,
- Engineering,
- Computer Science,
- Economics,
- Econometrics and Finance,
- Decision Sciences.

As the potential sources were used articles published in reviewed professional journals provided in English, without taking into account the place of publication. Next set limitation was the time horizon. Analysed were only publications published since 2000 till 2020. All set criteria are summarized by the following tables, Table 4 and Table 5.

Table 4. Selection criteria

Criteria	Explanation
Source	Database Scopus.
Sectors covered	Scopus: Business, Management and Accounting; Social Sciences; Engineering; Computer Science; Economics, Econometrics and Finance; Decision Sciences.
Type of studies	Theoretical and empirical character.
Language	English.
Time period	2000 - 2020.
Type of publication	Magazine articles.
Geographical coverage	Worldwide.

Source: Own processing, 2020.

Table 5. Exclusion criteria

Criteria	Explanation
Time period before 2000	Only a horizon since 2000 has been included. This is a period of increased interest in this issue.
Sectors	Excluded are all other sectors that did not correspond to the monitored areas. In these sectors, the occurrence of articles was minimal.
Language	Much of this issue is discussed in English.
Type of publication	Focused on high quality magazine articles.

Source: Own processing, 2020.

3 Results and discussion

As the limits of the research can be seen focusing on strictly defined areas of scientific interest. Authors tried to avoid using as the sources those areas of scientific interest where occurrence of the articles dealing with IoT and IIoT is minimal, because of the objective reasons. Therefore, it was decided to take into account the association of IoT with the area dealing with the issue of Industry 4.0. As written above, the twenty-year time horizon was selected. The reason for this represents the fact that the amount of articles dealing with the analysed issue before the year 2000 was very low.

After taking into account all defined prerequisities were, by the Scopus database, searched 4 297 records.

If analysing the individual records in time, it is possible to identify significant increase of publications dealing with this issue, since 2015 can be this increase defined as exponential. In 2020 was the amount of selected articles dealing with the analysed issue 1 667. The timeline of the records found in each year can be seen in the following figure, Fig. 1.

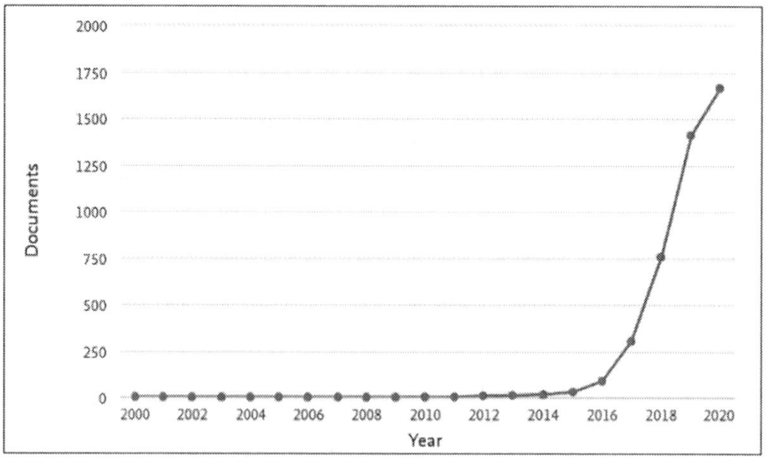

Fig. 1. Number of records found in individual years

Source: Own processing based on [17], 2020.

Increased interest in this issue, visible in Fig. 1, is very likely caused by focusing on simplification of human activities realized by the use of new technologies. Development of these technologies arose as the reaction to turbulent environment and constant changes in current human activities. Technologies belong among the areas of interest of concept Industry 4.0 that is still discussed, even if scientists currently work also with new concepts like Society 5.0 [18] bringing to the previous concept added value in the form of focusing back on people and their needs (even immaterial). This concept takes new technologies more as the tool to reach individual people goals, not as the objective itself. IoT represents one of those upcoming technologies [14].

Among the most frequently mentioned keywords in the found articles were the phrases:

- internet of things,
- network security,
- industry 4.0,
- blockchain,
- wireless sensor networks,
- industrial internet of things,
- or cloud computing.

Similarly, the distribution of records by country of origin is not surprising. This topic is most often published in China, followed by the United States and India. Such result fully corresponds to the number of 'Internet users' in each country. The number of found records according to individual territories can be seen in the following figure, Fig. 2.

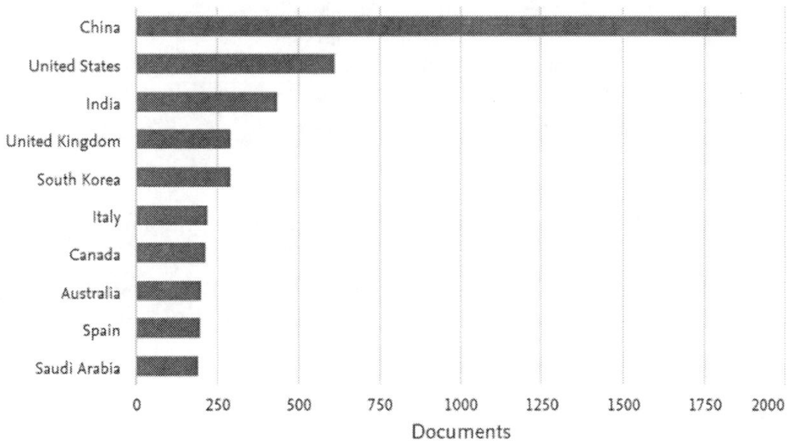

Fig. 2. Number of records found by territory

Source: Own processing based on [17], 2020.

The given phrases appear most often under the heading of the scientific field of computer science (3 095 records) and engineering (2 822 records). In other areas of scientific interest, the number of records is, compared with the two mentioned fields, much lower.

Set prerequisities can be further discussed, but in this case (compilation of this contribution), represent thoughtful limitations arising from the research objective.

4 Conclusion

In conclusion, it can be stated that the bibliometric analysis confirmed the growing interest in the analysed issue. In the period under review, between 2015 and 2020, can be talked about an exponential increase in searching for information dealing with this issue on the Internet.

IoT (or IIoT) belongs to the tools that organizations (and individuals) use to achieve the optimization of business activities. This is one of the most frequently mentioned tools of the Society 4.0. In addition to IoT, such tools most often include artificial intelligence, augmented reality, blockchain, drones, robotics, virtual reality, 3D printing and additive manufacturing as mentioned in the first part of this contribution.

All above mentioned tools of Society 4.0 will (already have) have an impact on the changes in standard human activities (labour market, provision of services, care for the elderly people, etc.) and interpersonal relations. It starts to be more visible also thanks to the current global pandemic situation.

'What is really important in real life?' Question asked in the introduction can't be answered easily. It depends on many external factors (political situation, environmental changes, etc.). It is also associated with subjective feeling of each person and perception of values (morality, ethics). Future social and technological changes that will be brought by the current pandemic situation probably more earlier are inevitable. We can just estimate whether or not mankind will use this negative experience (may be opportunity from opposite point of view) to its own welfare.

This contribution was created within the project 'TL02000136 - Knowledge-intensive services sector adaptation to the conditions of Society 4.0', solved by the University of West Bohemia in Pilsen and the Institute of Technology and Business in České Budějovice, with financial support of the Technology Agency of the Czech Republic.

References

1. Ch. Roser. *AllAboutLean. Organize Your Industry* [online], Available at: https://www.allaboutlean.com/christoph-roser/ (2020)
2. I. Veza, M. Mladineo, I. Peko, Analysis of the Current state of Croatian Manufacturing Industry with Regard to Industry 4.0. *15th International Scientific Conference on Production Engineering (CIM 2015)*, **15** (2015)
3. K. Schwab. *The Fourth Industrial Revolution: what it means, how to respond* [online], Available at: https://professionallearning.education.gov.scot/media/1352/the-fourth-industrial-revolution-what-it-means-and-how-to-respond-world-economic-forum.pdf (2016)
4. R.H. Weber, Internet of Things - New security and privacy challenges. *Computer Law & Security Review*, **26**(1), 23-30 (2010)
5. L. Dvořáková et al., *Methodics of SMEs enterprises in tertiary sector adaptation on implementation of principles, procedures, methods and tools of Society 4.0*. Pilsen: UWB and ITB (2020)
6. A. Saint, Where next for the internet of things? *Engineering & Technology*, **10**(1), 72-75 (2015)
7. H. Tran-Dang, N. Krommenacker, P. Charpentier, D.S. Kim, Toward the Internet of Things for Physical Internet: Perspectives and Challenges. *IEEE Internet of Things Journal*, **7**(6), 1 (2020)
8. D. Miorandi, S. Sicari, F. De Pellegrini, I. Chlamtac, Internet of things: Vision, applications and research challenges. *Ad Hoc Networks*, **10**(7), 1497-1516 (2012)
9. A. Jules, RFID security and privacy: a research survey. *IEEE Journal on Selected Areas in Communication*, **24**(2), 381-394 (2006)
10. J. Ye, S. Dobson, S. McKeever, Situation identification techniques in pervasive computing: A review. *Pervasive and Mobile Computing*, **8**(1), 36-66 (2012)

11. L. Tan, N. Wang, Future internet: The Internet of Things. *ICACTE (3rd International Conference on Advanced Computer Theory and Engineering*, **3** (2010)
12. R. Aggarwal, M. Las Das, *RFID Security in the Context of 'Internet of Things'. First International Conference on security of Internet of Things*, **1** (2012)
13. R. Yugha, S. Chitra, A survey on technologies and security protocols: Reference for future generation IoT. *Journal of Network and Computer Applications*, **169**, 1084-8045 (2020)
14. M. Mahbub, Progressive researches on IoT security: An exhaustive analysis from the perspective of protocols, vulnerabilities, and preemptive architectonics. *Journal of Network and Computer Applications*, 168, 1084-8045 (2020)
15. A. Luigi, I. Antonio, M. Giacomo, The Internet of Things: Survey. *Computer Networks*, **54**(15), 2787-2805 (2010)
16. L. Huichen, N.W. Bergmann, IoT Privacy and Security Challenges for Smart Home Environments. *Information*, **7**(3), 44 (2016)
17. Elsevier B.V. *Scopus database* [online], Available at: https://www.scopus.com/ (2020)
18. S. Serpa, C. M. Ferreira, Society 5.0 and Sustainability Digital Innovations: A Social Process. *Journal of Organizational Culture, Communications and Conflicts*, **23**(2), 1-14 (2019)

Methodology for Calculating the Reproduction Value of Buildings

Eduard Hromada[1], *Renáta* Schneiderová Heralová[1], *Petr* Dlask[1], and *Tomáš* Krulický[2]

[1]Czech Technical University in Prague, Faculty of Civil Engineering, Thákurova 7, Prague 6, Czech Republic
[2]VŠTE, Okružní 517/10, České Budějovice, Czech Republic

Abstract. The article deals with the description of the methodology for calculating the reproduction value of buildings. It is a mathematical model that uses data obtained from a statistical survey and calculates the total reproduction value of buildings from a large data set. The model works with many variable inputs that describe the valued building. These include, for example, the use of the building, the enclosed space of the building, the floor area of the building, the building material of the supporting structures of the building, the age of the building, the technical condition of the building, the year the building was acquired and others. Based on the results, the model was calibrated and its parameters optimized to achieve the most accurate values.
Keywords: real estate market, real estate valuation, building wear, statistics

1 Introduction

Real estate valuation is one of the basic building blocks of the valuation issue which by its nature lies on the border between economics and other technical fields (construction, engineering, ...). Many authors focus on real estate valuation individually and deal with methods and approaches of valuing just one specific real estate. However, these methods fail in the valuation tasks of a statistically significant and internally inhomogeneous set of real estate, numbering several tens of thousands of units. When valuing such large sets of real estate, it is necessary to slightly modify the methods focused on the valuation of individual real estates and to adapt them to the possibilities of the appraiser. In the case of a valued group of several tens or hundreds of thousands of properties, it is basically unthinkable to perform selected valuation acts, such as local surveys etc. The presented article withdraws from this established practice and aims to determine the reproduction value of a large internally inhomogeneous set of real estate.

For this purpose, a set of a total of 50,000 properties will be examined, containing only buildings (excluding land) in the Czech Republic. The actual calculation of the reproduction value will be performed in two variants, both considering and not considering the ancillary budget costs, i.e. the costs expected to be invested for the acquisition of the construction.

2 Literary Research

Real estate valuation is still a very topical issue to be solved in the field of expertise and valuation. This is an area where innovation is constantly taking place, from every conceivable point of view. This manifests itself not only in constant shifts in trends in construction, which are made possible by the development of new construction technologies, but also in the constant development of real estate valuation procedures. The ever-increasing number of expert opinions on real estate valuation issues can also serve as a proof of the growth of the given sector. In practice, it is also possible to encounter a situation in which the contracting authority places requirements on the assessor regarding the method, methodology or even the result of the valuation. This may lead to lower objectivity and a reflection of the assessor's disproportionately strong subjective feeling in the valuation itself. One of the tools to eliminate these adverse effects is the development of uniform methodologies and procedures for valuation, which on the one hand sufficiently balances the specificity of the property under consideration and on the other hand ensures a maximum objectivity of the result. This can be achieved by leveling the real estate market environment, suppressing subjective influences or eliminating errors that may have been unintentionally caused in preceding expert opinions due to insufficient research in the entire field. At this moment, the effect of the scientific basis behind the whole issue of real estate valuation is evident, which, in this respect, is a deterministic parameter of the final value of the valued real estate [1].

Usually, units of measure are used to determine the value of specific buildings putting their size in concrete terms. In some cases, they also supply a determinant of specificity to the buildings being valued. This is mainly the size of the built-up floor area in m^2 (CZK / m^2) or the number of rooms (CZK / pc). However, today's technologies allow the use of other parameters of buildings as units of measure which distinguish individual buildings very precisely from other buildings [2]. The development of new technologies also allows, provided the cooperation of all parts of the real estate industry (construction, marketing, work of real estate agencies, etc.) to significantly accelerate all transactions related to the sale and rental of real estate, their valuation, etc. [3]. The value of buildings is also affected by their location in the landscape as such. If the valued building is located, for example, in an ecologically rich area, a higher value of buildings and adjacent land can be assumed [4]. To provide a value estimate of a building, it is also possible to use the hedonic regression model, which has already been examined many times, or also for specific buildings, data on their repeated sales [5]. The random forest method has proven to be a useful complementary tool to capture the complexity and nonlinearity of the entire set of buildings that is the subject of the valuation [6].

Using an interoperable data model for the valuation of land intended for building, national geographic data infrastructures can be extended, which could lead to mass valuation when using machine learning as a numerical processor. The prediction robustness of the values of this land increases with the quality and reliability of the input data [7].

Most often, the real estate valuation expert community uses traditional valuation methods, such as comparative, reproduction and yield methods. Modern valuation methods are not currently used to such an extent [8]. When comparing econometric and non-econometric valuation methods of buildings, the market value of the building used in determining the value of another building in a comparative econometric way was discovered to play an important role. However, when using a comparative econometric valuation method, there is a high demand for the amount of data that can be used for comparison. In practice, this can only be achieved with a sufficiently large database of data which is freely available, for example on the Internet [9].

Competition has a negative effect on the price range [10]. In such cases, the market price of the buildings was to be regulated by the state through its intervention [11]. With the help of statistical tests of data on the development of building prices, the effectiveness of individual buildings was examined according to the way they are used on the real estate market. There is a different level of risk of return on investment from individual buildings according to the method of their use. This is one of the consequences of the inhomogeneity of the real estate market [12]. Another determinant of the value of buildings is also the length of their service life which is already taken into account during their construction. This parameter is considered in connection with the intended use of the building, and, based on this deliberation, such construction materials are chosen that can withstand years of intensive use (e.g. floors of assembly halls, which are driven on by heavy equipment). Usually, in connection with this viewpoint, individual intervals of necessary reconstruction interventions to prolong the service life of the buildings are also planned. This also determines the life cycle costs of buildings (Building Information Modeling - BIM) (In the case of historic buildings, the relevant model is Historic Building Information Modeling - HBIM [13]). Nowadays, digital technologies can also be used for the evaluation of BIM itself, which can be used to scan a building from the interior and the exterior, and to determine its technical condition on the basis of the evaluated parameters [14]. BIM can also be used to evaluate the thermal properties of buildings. A three-dimensional model of buildings with specific heat leakage locations can be created for this purpose [15]. Therefore, it is also necessary to take into account the service life of the individual building materials used. As a consequence, it is advantageous to focus on the creation of a model, which could be used to determine the service life of individual structural elements of buildings [16] or to focus directly on the evaluation of the life of buildings as a whole on the basis of a visual assessment of the current state of a particular building which also takes into account the mechanisms of degradation of individual materials [17]. From an economic point of view, the best solution is to always determine the most optimal moment to carry out the reconstruction and extend the life of the buildings. This moment should respect the economic aspect so that the required investment is made with the greatest possible efficiency in terms of its necessary financial amount and also the greatest efficiency in terms of technical reconstruction of the building [18]. An appropriate prediction of construction life cycle costs can also reduce future economic costs of its operation [19]. The methodology of determining BIM has already been discussed by many authors and some have contributed to its modernization in the form of individual modifications etc. [20].

3 Methods and data

The examined set contains data on 50,000 constructions realized and executed on the territory of the Czech Republic beginning from long-time history to the present day, in a diverse construction-technical condition. The set thus contains new constructions, carried out after 2010 in excellent condition, as well as constructions carried out decades or hundreds of years ago in good maintained, but also in desolate condition intended for demolition or general reconstruction. From the point of view of the function of the buildings, the set is based on buildings with a predominant use other than residential (non-residential premises). The file does not include information on ground line constructions, but there are constructions established, for example, for educational and cultural purposes, historical buildings and monuments, church buildings, hospitals, administrative buildings, halls, chimneys, buildings that are part of higher technological units (wastewater treatment plant, etc.).

The character of the building will be determined on the basis of the classification of the Uniform Classification of Buildings. The basic element of the valuation is the unit price of Technical and Economic Indicators (THU) which are determined on the basis of long-term statistics and price developments in construction, and are regularly published by specialized companies engaged in processing price systems in the field of construction. THU indicators are related to the unit price, i.e. CZK / unit where a unit can be represented according to its character by a cubic meter of the basic enclosed space of the building, a square meter for civil engineering or a length meter for linear structures. Unit prices of THU indicators are regularly published on the website www.stavebnistandardy.cz.

The initial data obtained this way will be further adjusted by the coefficients of the respective tenders. This coefficient represents another degree of objectification of the entire calculation, and seeks to bring unit prices as close as possible to the market reality in the Czech Republic. The coefficient is based on the survey of previous tenders for contractors / suppliers of the construction. It included all tenders with a construction theme, published in the Czech Republic in the period 2016-2018 with an expected scope of > CZK 1 million for supplies and services and > 3 CZK million for construction work; a total of 1,161 tenders were included in the survey. The coefficient was further divided into sub-coefficients according to the nature and essence of the solved contract in relation to the definition of JKSO (Uniform classification of construction objects). The coefficient thus expresses the ratio between the originally planned (budget) price of the contract and the actual winning price in the given procedure.

3.1 Calculation variants of the reproduction value

The calculation of the reproduction value of all buildings can be calculated in two variants:

- Reproduction and time value of the building.
- Reproduction and time value of the building including ancillary budget costs.

The first variant of the calculation takes into account the reproduction purchase price of the building and the wear and tear of the building. In addition, the second variant of the calculation includes the relevant ancillary budget costs (costs of design work and engineering activities, costs of site location, costs of technical infrastructure and risk reserve).

Table 1. Structure for calculating the reproduction value of all buildings

Calculation variant 1	Calculation variant 2
Reproduction purchase price of the building	Reproduction purchase price of the building
→ taking into account the wear of the building	→ taking into account the wear of the building
	+ costs of design work and engineering activities
	+ costs of site location
	+ costs of technical infrastructure
	+ risk reserve
= **Reproduction and time value of the building**	= **Reproduction and time value of the building including ancillary budget costs**

Source: Own processing.

3.2 Technical and economic indicators

The starting point for calculating the reproduction value of buildings is based on the technical and economic indicators of RTS, a.s., which are publicly available free of charge on the website: http://www.stavebnistandardy.cz/default.asp?Bid=6&ID=6.

Technical and economic indicators are commonly used for valuing construction plans at the planning stage and calculating construction costs. The values of indicators were compiled on the basis of long-term statistics on prices of buildings and constructions from representative item budgets and indexation. Each price indicator has an assigned unit of measure, i.e. a unit of measure to which construction costs are related (it's just about only basic budget costs - they do not include ancillary and other costs). For buildings and halls, these are usually cubic meters of enclosed space.

This default data of RTS, a.s. (technical and economic indicators) was further adjusted by conversion coefficients of tenders based on the research team's own survey. The coefficient of tenders expresses the ratio between the expected (budget) price of the contract and the final winning (lowest bid) price of the selected tenderer. It is implicitly assumed that the tenderer submitted the tender to the tender freely, voluntarily, and is willing to implement the subject of performance for the tender price, while the tenderer behaves with due diligence and the offered price brings sufficient value from the work (covers all anticipated costs and reasonable profit). The calculation does not take into account the final price for the implementation of the contract, which may be affected by other costs during construction (more / less work), as it can be assumed that these costs were not expected and foreseeable at the time of awarding the contract and would occur another supplier.

Information on published tenders was drawn from a database managed by the Ministry of Regional Development of the Czech Republic, available online: https://www.vestnikverejnychzakazek.cz/. The main advantages of this database include a large number of orders with all files and attachments owned by the home domain and no redirection to other websites during the search or viewing of documents, as is the case with some competing databases. The Public Procurement Bulletin records public contracts with an expected scope of more than CZK 1 million for supplies and services and more than CZK 3 million for construction work.

An extensive table was created for the purpose of recording values and information from suitable public contracts. Its content consists of a horizontal axis consisting of columns: order number, region, city, contracting authority, building code, type of building, project name, operation, project description, closing date, contract closing date, estimated price (excluding VAT), final price, ratio of final and estimated price, supplier, registered office of the supplier, duration in months and source. The vertical axis then consists of individual public contracts.

Table 2. Calculated coefficients of tenders according to the use of the building

Use of the building	The number of records	Coefficients of tenders
Administrative buildings	70	0,9279
Other residential buildings	122	0,8949
Buildings for religious activities	41	0,9451
Buildings for industry	43	0,8743
Buildings for social purposes	66	0,9014
Sports buildings	109	0,9504

Healthcare buildings	175	0,9432
Garages	41	0,9586
Historical or cultural monuments	47	0,9288
Hotels	36	0,8802
Museums and libraries	73	0,9284
Schools, universities and research buildings	338	0,9116
Total / weighted average	**1161**	**0,9209**

Source: Own processing.

3.3 Determination of building wear

The linear wear method was used to determine the wear of the buildings. For the linear method, wear increases in direct proportion to time. It is assumed that the building as a whole degrades over time. We can express this dependence by a formula:

$$W = 100 * \frac{A}{L} \quad (1)$$

where
W ... wear of the building calculated by the linear method [%],
A ... age of construction [years],
L ... lifetime of the building [years].

A service life of 200 years is considered as the limit value. The reason is the maximum permissible service life of foundations, vertical structures, ceilings and stairs in relation to Table No. 7, Annex No. 15, Decree No. 3/2008 Coll. (so-called valuation decrees) on the implementation of certain provisions of Act No. 151/1997 Coll. on property valuation, as amended. This period is just 200 years.

In determining the service life of buildings, it was based on Annex No. 15 to Decree No. 3/2008 Coll. (so-called valuation decrees), other information sources and the experience of the research team. The table of building lifetimes distinguishes JKSO (uniform classification of construction objects, Czech national regulation) in the level - the field of the building object (1st to 3rd place of the numerical code) and at the same time the building material of the load-bearing structures.

Table 3. Lifespan of buildings in years

JKSO (uniform classification of construction objects)		Structural and material characteristics of load-bearing structures								
		1	**2**	**3**	**4**	**5**	**6**	**7**	**8**	**9**
801	Civil engineering buildings	200	150	200	100	100	75	70	80	80
802	Civil engineering halls	150	150	150	100	100	75	70	80	80
803	Residential buildings	200	150	200	100	100	75	70	80	80
812	Buildings for production and services	150	150	150	100	100	75	70	80	80

Source: Own processing, related legal and technical regulations.

For a complete understanding of the above table, it is further stated that the JKSO classification distinguishes a total of 9 types of building materials of vertical load-bearing structures of buildings:

- 1 - vertical load-bearing structure made of bricks and blocks,
- 2 - vertical load-bearing structure monolithic concrete bar,
- 3 - vertical load-bearing structure monolithic concrete surface,
- 4 - vertical load-bearing structure assembled from concrete bar parts,
- 5 - vertical load-bearing structure assembled from concrete surface parts,
- 6 - vertical load-bearing structure mounted from spatial cells,
- 7 - vertical metal supporting structure,
- 8 - vertical wooden and wood-based load-bearing structure,
- 9 - vertical load-bearing structure made of other materials.

3.4 Overall assessment of the condition of the building

A parameter that evaluates the technical condition of the building is also included in the calculation. A total of 5 categories of the overall assessment of the condition of the building are distinguished, from the new building to the state of emergency. For each category, a building condition coefficient is assigned, which multiplies the basic technical and economic indicator.

Table 4. Overall assessment of the condition of the building and risk reserve

Overall assessment of the condition of the building	Coefficient of building condition	Risk percentage [%]
Brand new building, max. age 5 years	1,00	10,00
Excellent	0,95	11,25
Quite satisfactory	0,90	12,50
Requires renovation	0,80	13,75
Emergency (ruin)	0,70	15,00

Source: Own processing.

3.5 Risk reserve

The risk reserve is used to cover unexpected expenses during construction. It is considered in the range of 7 - 10% for new buildings, 10 - 15% for reconstructions. The basis for the calculation is the reproduction purchase price of the building. The previous table is used for the calculation, which determines the risk percentage depending on the overall assessment of the condition of the building.

3.6 Costs of design work and engineering activities

To calculate the costs of design work and engineering activities, it is necessary to include the following three performance phases:

- construction management phase (planning),
- construction phase,
- phase after construction completion.

According to the website http://cenyzaprojekty.cz/sazebnik.html, these three phases represent 85% of the total costs of project work and engineering activities. This percentage is the same for all construction categories and all performance zones.

Costs of design work and engineering activities are calculated using the appropriate performance zone for each building. From the following table, the appropriate percentage will be used to calculate the fee.

Table 5. Cut-out from the percent table for the calculation of the costs of design work and engineering activities

Eligible costs	**Performance zone by category of building**				
CZK	1	2	3	4	5
	Percentage for calculating the fee				
0	8,87	10,71	13,25	15,46	17,14
1 000 000	8,24	9,91	11,99	14,03	15,75
2 000 000	7,88	9,51	11,49	13,38	15,00
3 000 000	7,64	9,15	11,05	12,93	14,47
4 000 000	7,46	8,92	10,77	12,59	14,08
5 000 000	7,31	8,74	10,54	12,33	13,77
...	...	...	...	...	...

Source: Performance and fees regulations (ČKA, ČKAIT), own processing.

3.7 Costs of site location

The costs associated with the location of the building include the costs of operational and social equipment of the construction site, territorial influences, extremely difficult working environment. This item includes costs that are not valued within the costs of construction objects. The costs associated with the location of the building can be, for example:

- site equipment costs,
- costs caused by operational influences (for example, the operation of the investor, third parties, road, rail, urban traffic near the construction site, harmful environment),
- costs caused by areas with difficult production conditions,
- costs related to the impact of extreme climatic conditions,
- costs caused by extremely difficult traffic conditions,
- costs of necessary transport of the supplier's employees to the workplace and back,
- off-site transport costs (above the average costs included in the cost of construction sites or for selected materials),
- costs arising from the title of work on protected monuments (reduction of staff performance due to slower and more careful work).

The costs of site location is determined by a percentage rate that ranges from 2 to 5% of the value of the basic technical and economic indicator and depends on the size of the building. The larger the building, the smaller the percentage.

The function is used for the calculation, which determines the percentage of costs of site location depending on the size of the enclosed space of the building. For individual categories of the enclosed space of the building in the interval <0 m^3; 100,000 m^3>, the percentage rate of costs of site location is determined using the linear function:

$$y = -ax + 5 = \frac{3}{100\,000} + 5 \qquad (2)$$

where
100,000… is the upper limit of the built-up area of the building.

In the case of a built-up area of a building larger than 100,000 m^3, a constant percentage rate of 2% is considered.

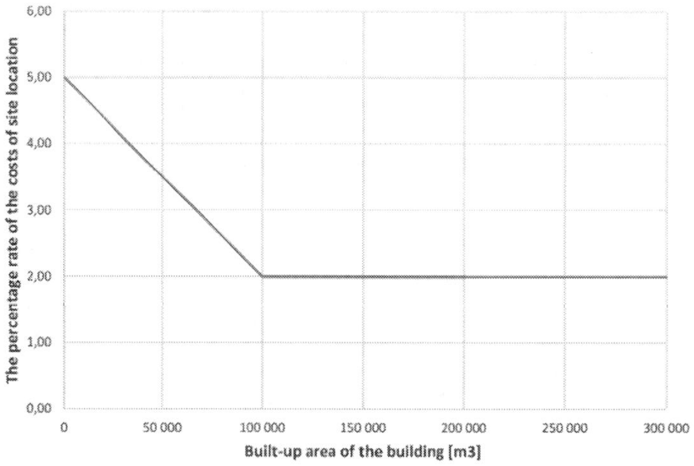

Fig. 1. Dependence of the percentage rate of the costs of site location on the size of the enclosed space of the building

3.8 Costs of technical infrastructure

The costs of the technical infrastructure of the building include the costs of creating the electricity connection, water supply, sewerage and gas, communications, fencing and landscaping. The value in the range of 2% to 3.5% of the reproduction value of the building is considered (even the connections and other technical infrastructure wear out over time) and depends on the size of the building. The larger the building, the smaller the percentage.

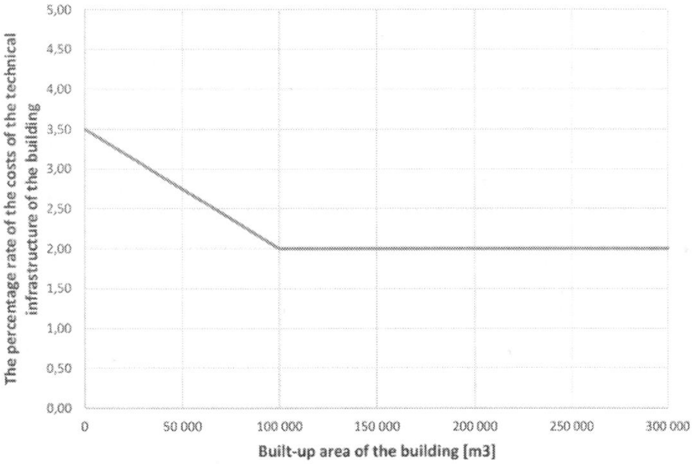

Fig. 2. Dependence of the percentage rate of the costs of the technical infrastructure of the building on the size of the enclosed space of the building

4 Results

By performing the procedure described above, the results shown in the table below were obtained:

Table 6. Determination of the reproduction price of buildings for individual groups

Group of construction objects	Amount of constructions in the group	Average wear (%)	Determined reproduction price excluding ancillary budgetary costs	Determined reproduction price including ancillary budgetary costs
Civil engineering buildings	9015	56,3	19 819 942 918 CZK	33 390 530 687 CZK
Residential buildings	11967	49,5	80 631 263 420 CZK	122 543 340 070 CZK
Healthcare buildings	642	35,1	17 335 769 709 CZK	25 054 084 116 CZK
Educational buildings	5463	37,2	116 915 843 702 CZK	168 757 599 414 CZK
Science, culture and research buildings	5373	46,6	37 093 549 320 CZK	57 493 214 049 CZK
Sports buildings	2031	26,0	31 269 541 757 CZK	43 443 075 511 CZK
Administrative buildings	3234	43,7	31 627 415 090 CZK	48 677 791 405 CZK
Buildings for accommodation and recreation	1334	41,6	4 098 573 291 CZK	6 164 947 634 CZK
Buildings for trade and catering	1334	34,8	6 383 892 756 CZK	9 623 503 822 CZK
Industrial buildings	2393	30,8	8 711 032 908 CZK	12 840 100 740 CZK
Buildings for agricultural production and animal husbandry	478	49,4	1 107 273 661 CZK	1 988 613 494 CZK

Buildings for transport and communication	1141	34,2	1 280 144 515 CZK	1 987 182 872 CZK
Garages, buildings for repair and maintenance of vehicles, machinery and equipment	2704	35,1	4 832 890 636 CZK	6 998 214 028 CZK
Storage buildings	2 891	43,6	7 639 371 538 CZK	12 328 699 922 CZK
Total	**50 000**	--	**368 746 505 228 CZK**	**551 290 897 771 CZK**

Source: Own processing.

From the table above, it is obvious that the total reproduction value of the examined real estate sample excluding ancillary budgetary costs is 368 746 505 228 CZK and including the ancillary budgetary costs then 551 290 897 771 CZK.

5 Conclusion

The presented paper introduced its own comprehensive methodology for determining the reproductive value of a significant internally inhomogeneous set of buildings. The methodology introduced a method of approximating budget costs and list prices to market reality, while this coefficient may be applicable for other contracting parties as well. Furthermore, the authors introduced a unique way of determining the ancillary costs of construction, as a necessary cost element in construction activities. This method can also be further reproduced and thus contribute to the improvement of the initial estimates of construction costs. The application of this element does not depend on local conditions, and thus the given method can be applied across markets and depends on the actual mass of the building (enclosed space). The proposed methodology was subsequently applied to a set of 50,000 randomly selected buildings in the Czech Republic. The goal of the paper was fulfilled. The authors see a further research possibility in the focus of the scientific community's interest on the pricing factor of location, resp. locality, i.e. a stronger emphasis on regional influence in the set reproduction price. The outputs of the paper can also be used by the public sector in defining government housing policy.

References

1. U. Ozdilek, Scientific basis of value and valuation. *Journal of revenue and pricing management*, **18**(3), 266-277 (2019)
2. J. Bin, B. Gardiner, E. Li, Z. Liu, Multi-source urban data fusion for property value assessment: A case study in Philadelphia. *Neurocomputing*, **404**, 70-83 (2020)
3. A. Saull, A. Baum, F. Braesemann, Can digital technologies speed up real estate transactions? *Journal of property investment & finance*, **38**(4), 349-361 (2020)
4. A. Binter, K. Krol, M. Frosik, M. Furczon, Ecological Considerations in Real Estate Valuation. *Journal of ecological engineering*, **21**(5), 47-55 (2020)
5. A. Oust, S. N. Hansen, T. R Pettrem, Combining Property Price Predictions from Repeat Sales and Spatially Enhanced Hedonic Regressions. *Journal of real estate finance and economics*, **61**(2), 183-207 (2020)
6. J. Hong, H. Choi, W.S. Kim, A house price valuation based on the random forest approach: the mass appraisal of residential property in South Korea. *International Journal of Strategic Property Management*, **24**(3), 140-152 (2020)

7. A. C. Aydinoglu, R. Bovkir, I. Colkesen, Implementing a mass valuation application on interoperable land valuation data model designed as an extension of the national GDI. *Survey review*, [*Early Access*] (2020)
8. R. B. Abidoye, M. Junge, T. Y. M. Terence, T. B. Oyedokun, M. L. Tipping, Property valuation methods in practice: evidence from Australia. *Property management*, **37**(5), 701-718 (2019)
9. J. Zyga, Object and objective of property appraisal and their effects on valuation methods and databases. *Geodetski vestnik*, **63**(1), 92-103 (2019)
10. K. S. Gerardi, A. H. Shapiro, Does Competition Reduce Price Dispersion? New Evidence from the Airline Industry. *Journal of Political Economy*, **117**(1), 1-37 (2009)
11. J. B. H. Yap, X. H. Ng, Housing affordability in Malaysia: perception, price range, influencing factors and policies. *International Journal of Housing Markets and Analysis*, **11**(3), 476-497 (2018)
12. F. Almudhaf, A. J. Hansz, Random Walks and market efficiency: Evidence prom real estate investment trusts (REIT) subsectors. *International Journal of Strategic Property Management* **22**(2), 81-92 (2018)
13. A. Mol, M. Cabaleiro, H. S. Sousa, J. M. Branco, HBIM for storing life-cycle data regarding decay and damage in existing timber structures. *Automation in Construction*, **117**, Art. No. 103262 (2020)
14. A. G. Mohamed, M. R. Abdallah, M. Maryouk, BIM and semantic web-based maintenance information for existing buildings. *Automation in Construction*, **116**, Art. No. 103209 (2020)
15. D. Utkucu, H. Sozer, Interoperability and data exchange within BIM platform to evaluate building energy performance and indoor comfort. *Automation in Construction*, **116**, Art. No. 103225 (2020)
16. C. Chai, J. De Brito, P. L. Gaspar, A. Silva, Statistical modelling of the service life prediction of painted surfaces. *International Journal of Strategic Property Management*, **19**(2), 173-185 (2015)
17. C. Marques, J. de Brito, A. Silva, Application of the factor method to the service life prediction of etics. *International Journal of Strategic Property Management*, **22**(3), 204-222 (2018)
18. A. Farahani, H. Wallbaum, J. O. Dalenback, Optimized maintenance and renovation scheduling in multifamily buildings - a systematic approach based on condition state and life cycle cost of building components. *Construction Management and Economics*, **37**(3), 139-155 (2019)
19. J. Lee, H. Yang, J. Lim, T. Hong, J. Kim, K. Jeong, BIM-based preliminary estimation method considering the life cycle cost for decision-making in the early design phase. *Journal of Asian Architecture and Building Engineering*, **19**(4), 384-399 (2020)
20. D. Wieczorek, E. Plebankiewicz, K. Zima, Model estimation of the whole life cost of a building with respect to risk factors. *Technological and Economic Development of Economy*, **22**(1), 20-38 (2019)
21. D. Macek, M. Kosina, Specifications of M&E requirements in the BIM model in the context of use for facility management. *3rd International Conference on Engineering Sciences and Technologies, ESaT 2018*, 425-430 (2018)
22. T. McGough, J. Berry, Pricing risk and its use in modelling real estate market yields. *Journal of Property Investment and Finance*, **38**(5), 419-433 (2020)
23. W. Kuang, C. Chen, Q. Wang, Home purchase restriction, real estate investment, and corporate innovation. *Frontiers of Business Research in China*, **14**(1) (2020)

Stable development of travel business in the Odessa region: Estimate using multivariate statistical analysis

Valerii Matskul[1,*], *Anatoliy* Kovalyov[1], *Valentyn* Chernyshev[1], and *Diana* Okara[2]

[1]Odessa National Economic University, 65082 Odessa, Ukraine
[2]Odessa State Academy of civil engineering and architecture, 65029 Odessa, Ukraine

Abstract. The article examines the level of sustainable development of the travel business in the Odessa region of Ukraine. This level is evaluated as a latent indicator, which is characterized by a certain set of signs-symptoms. The methodology used for modeling is Partial Least Squares-Path Modelling (or PLS-PM) and the method Time-Wise Multi-Way Principal Component Analysis. This approach allowed us to identify and evaluate latent and explicit factors affecting the sustainable development of travel business in the Odessa region.
Keywords: travel business, sustainable development, Time-Wise Multi-Way Principal Component Analysis, Partial Least Squares-Path Modelling

1 Introduction

In recent years, there has been a reorientation of the main goals of the business. Previously, the main task of business was to obtain the greatest possible profit, now stable development is taking priority. A lot of research has been devoted to the issues of sustainable development of the travel business, among which we mention works [1-8]. The problem of quantitative assessment of the level of stable development of the tourism business is especially urgent. To solve it, it is necessary to apply the methods of mathematical modeling. A number of works are devoted to these questions [9-17]. Many economic indicators (such as development stability, competitiveness, investment attractiveness, efficiency, etc.) are latent, clearly unmeasured. Therefore, it becomes necessary to use special methods for their evaluation. This work is devoted to assessing the level of sustainable development of the tourism business in the Odessa region using special methods of multivariate statistical analysis. It should be noted that the Odessa region has strong potential for sustainable development. Among the many factors influencing this, we note the following: the presence of an extended line of the Black Sea coast, the presence of a developed infrastructure (sea and river ports, an international airport), sanatoriums, rest houses, cultural attractions, etc. Unfortunately, in 2020, the tourism business in the Odessa region (due to COVID-19) practically ceased to function normally (like all over the world).

[*] Corresponding author: valerii.matskul@gmail.com

2 Materials and methods

Data for analysis is available on the website of the State Statistics Service of Ukraine (regional statistics) and Main Department of Statistics in Odessa region [18, 19]. The time period for which the data will be available starting from 2010 and ending in 2019. Preliminary data processing was carried out in MS Excel spreadsheets. When modelling and computing was used the following tools: a package for component analysis (PCA-PM) of Statistica12 [8], and program XLSTAT (a package XLSTAT-PLSPM for analysis using the least partial square method PLS-PM: Partial Least Squares Path Modelling or Projection on the Latent Structures Path Modelling). PLS-PM methods started to be widely used in a variety of applied researches of the 70-s years of the twentieth century because of the work of Herman Wold and his son Sven Wold, in this works the basic principles of the technique of modelling PLS–PM were established. PLS-PM is a tool for modelling of interconnections between latent variables. The method PLS-PM is used for analyses of high dimensional data in the conditions of ill-structured environment. It is widely applied, in particular, in the economy to evaluate such latent indexes as a utility, level of economic development, etc. Following, the task of modelling by the method of PLS-PM can be described as follows: let X is a $n \times p$ data matrix:

$$X = \begin{pmatrix} x_{11} & x_{12} & \cdots & x_{1p} \\ x_{21} & x_{22} & \cdots & x_{2p} \\ \vdots & \vdots & \cdots & \vdots \\ x_{n1} & x_{n2} & \cdots & x_{np} \end{pmatrix}$$, where n is the number of observations (objects), and p is the number of variables (characteristics). The matrix X can be divided into l blocks $X_j, j=\overline{1,l}$, each of which X_j is associated with a latent variable LV_j, the estimate of which is denoted $Y_j = LV_j$. All interconnections between variables can be divided into two types: interconnections between latent variables LV_j and blocks X_j (these interconnections form the outer model) and interconnections between the latent variables with each other (these interconnections form the internal model). Internal model can be written in the form of a system of linear equations:

$$LV_j = \beta_0 + \sum_{i \to j} \beta_{ji} LV_i + \varepsilon_j^{in}; \tag{1}$$

where LV_i is all latent variables that affect the latent variable LV_j; β_{ji} are structural coefficients that characterize the force and the direction of connection between latent variables LV_i and LV_j; β_0 - free member; ε_j^{in} - a random deviation of the internal model. The internal model imposed by the following requirements:

1. The system of latent equations should be recursive so it has to correspond to loop-free graph.
2. Internal model is recursive, so the following requirement is true:

$$E(LV_j \setminus LV_i) = \beta_{0i} + \sum_{i \to j} \beta_{ji} LV_i \,.$$

3. Random deviation are independent of the explanatory variables, that is, the following requirement is satisfied: $\mathrm{cov}(LV_j, \varepsilon_j) = 0$.

Thus, at the simulation by using the method of projection on the latent structure the requirements for the statistical distributions of variables and random deviations are absent. The external model describes two type of interconnection between latent and explicit variables: reflective and formative. Reflective type is the most widespread type of external models in which the latent variable is the cause of implicit variables, that is, explicit variables reflect latent. The external model can be written in the form of the system of linear equations:

$$x_{jk} = \lambda_{0jk} + \lambda_{jk} LV_j + \varepsilon_{jk}^{out}\,; \tag{2}$$

where λ_{jk} are load coefficients, λ_{0jk} are free members, ε_{jk}^{out} is random deviation of the external model. Formative type is a kind of external model when explicit variables are "cause" of latent variables, i.e. they «form» latent variables. External formative model can be represented in the form of a system of linear equations:

$$LV_j = \lambda_{0j} + \lambda_{jk} x_{jk} + \varepsilon_j^{out}\,; \tag{3}$$

where λ_{jk} are load coefficients, λ_{0j} are free members, ε_j^{out} is random deviation of the external model. Latent variables can not be measured directly. Therefore we enter the notion of evaluation of latent variable $Y_j = LV_j$, which is a linear combination of the corresponding implicit variables: $Y_j = \sum_k w_{jk} x_{jk}$, where w_{jk} - external weight (weight multipliers) of the model. Thus, variables LV_j and $Y_j = LV_j$ characterize one and the same entity, but are used in different purposes: the first is for theoretical, the other is for practical.

PLS-PM modelling can be represented as a three-steps algorithm:

a) calculating of external weights w_{jk} to obtain estimates of the latent variables;

b) computation of structural coefficients β_{jk} of internal model;

c) calculation of loads λ_{jk} of external model.

The first step is the main and the most complicate step of PLS-PM methodology, that is iteration, its purpose is to calculate the values of external measures and, as a result, ratings Y_j. The result of this step is that abstract latent variables materialize. The second step is solution of a system of linear equations by the method of least squares, the third is calculation of correlations $\lambda_{jk} = cor(\mathrm{x}_{jk}, Y_j)$.

3 Results and Discussion

Building PLS-PM for the model of sustainable development of travel business for Odessa region. The following table presents indicators of sustainable travel business development:

Table 1. Indicators of sustainable travel business development

Indicator	Description of the indicator	Notation
Number of tourists	Total number of tourists per calendar year	x_{11}
Wage of employees in the service sector	Total wage of employees in the service sector per calendar year	x_{12}
Number of travel company	Number of travel company in region	x_{13}
Tax revenues from the travel industry	Total tax revenues from the travel industry for calendar year	x_{14}
Subsidies for the population	Total subsidies for the population per calendar year	x_{21}
State cost on social services for the population	The sum of state cost on the protection of public safety, health and other social services for a year	x_{22}
The unemployment rate	The proportion of the total number of unemployed persons to the economically active population	x_{23}
Emissions of pollutants into atmospheric air	Quantity of the emissions of pollutants into atmospheric air in region	x_{31}
Wastes generation	Quantity of the wastes generation in region	x_{32}
State cost on the protection of the environment	The sum of state cost for the protection of the environment per calendar year	x_{33}
Current expenditures on environmental	Costs on environmental per calendar year	x_{34}
Gross Regional Product	Value of the GRP per calendar year	x_{41}

State spending on economic development	State spending on economic development per calendar year	x_{42}
Utility Costs	The amount of the state cost for utility bills populated	x_{43}
Total income	Total population income for a calendar year	x_{44}

Source: Own processing.

Note that these indicators of sustainable development of the tourism business were selected from a large number of factors affecting the stable development of the region.

Following the works [20-22], we build a hierarchical structural model. The original model is represented in the following Figure 1. Latent variables: LV_1 - level of sustainable development of the tourism business, LV_2 – level of social condition, LV_3 - level of the environmental protection, LV_4 – condition of economics are denoted by ovals, and explicit variables (indicators from a table) by rectangles.

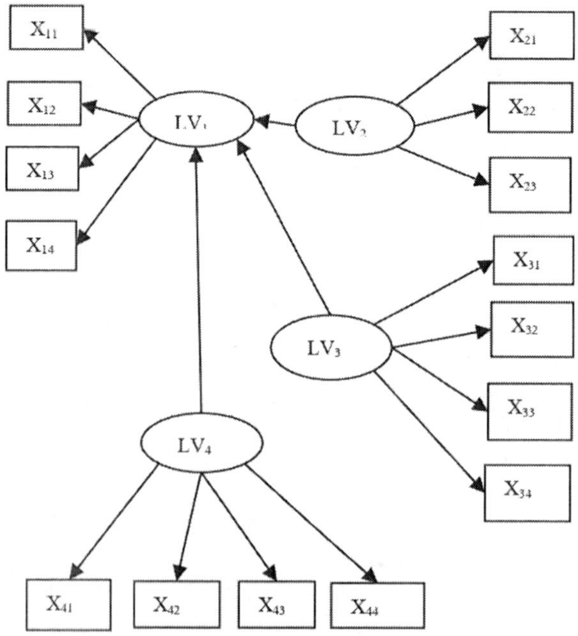

Fig. 1. Schematic representation of the basic model.
Source: Own processing with XLSTAT-PM.

The relationship of the latent variables is the internal model, and the relationships between latent and explicit variables represent an external model. Let's write these models in analytical form. The internal model: $LV_1 = \beta_0 + \beta_2 LV_2 + \beta_3 LV_3 + \beta_4 LV_4 + \varepsilon_1$, where

LV_j - latent variables; $\beta_2, \beta_3, \beta_4$ - structural coefficients, which characterize the strength and direction of the interconnection between latent variables, β_0 - free member and ε_1 is residual. The external model looks like the following system:

$$\begin{cases} x_{11} = \lambda_{01} + \lambda_{11} LV_1 + \varepsilon_{11} \\ \qquad \dots \\ x_{44} = \lambda_{04} + \lambda_{44} LV_4 + \varepsilon_{44} \end{cases}; \quad (4)$$

where $x_{11},...,x_{44}$ are explicit variables, $\lambda_{11},...,\lambda_{44}$ are coefficients of loads, $\lambda_{01},...,\lambda_{04}$ - free members, $\varepsilon_{11},...,\varepsilon_{44}$ - residuals. Note that at structural coefficients simulating by means of the partial least squares (PLS) we use the estimations of latent variables: $Y_j = LV_j = \sum_k w_{jk} x_{jk}$. For our model, these estimations can be presented in the form of the following system of equations:

$$\begin{cases} Y_1 = w_{11}x_{11} + w_{12}x_{12} + w_{13}x_{13} + w_{14}x_{14} \\ Y_2 = w_{21}x_{21} + w_{22}x_{22} + w_{23}x_{23} \\ Y_3 = w_{31}x_{31} + w_{32}x_{32} + w_{33}x_{33} + w_{34}x_{34} \\ Y_4 = w_{41}x_{41} + w_{42}x_{42} + w_{43}x_{43} + w_{44}x_{44} \end{cases}; \quad (5)$$

where $w_{11},...,w_{44}$ are external weights of the model.

Let's check correctness of breaking the X matrix of raw data into 4 blocks X_1, X_2, X_3, X_4 (which corresponds to latent components LV_1, LV_2, LV_3, LV_4). The method Time-Wise Multi-Way Principal Component Analysis (TMPCA) is applicable. We calculate using the appropriate module of the Statistica. Note that variables should be standardized before analysis ($z_{ij} = \frac{x_{ij} - \overline{x_{ij}}}{\sigma_j}$, $i = \overline{1,10}; \ j = \overline{1,15}$). This brings all variables to dimensionless values with mean values of 0 and standard deviations of 1. For the analysis, we use the following results of the module TMPCA:

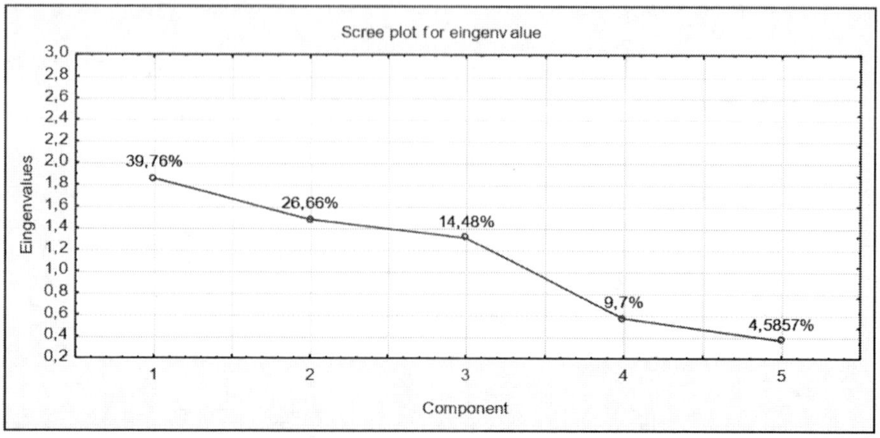

Fig. 2. Screen plot for eigenvalues.
Source: Own processing with Statistica.

Table 2. Eigenvalues and percentage contribution of PCA at the total variation of components LV_1, LV_2, LV_3, LV_4.

Components	eigenvalue	% total variance	accumulated %
LV_1	3,67872	39,76	39,76
LV_2	2,499577	26,66	66,42
LV_3	2,3032	14,48	80,9
LV_4	1,8732	9,7	90,6

Source: Own processing with Statistica.

Figure 2 and Table 2 show that the components-factors LV_1, LV_2, LV_3, LV_4 selected for modeling provide more than 90% of the total variation.

Analyses of the model by PLS-PM includes the following items: a) verification of the internal consistency of the blocks; b) verification of significance of variables in the external model; c) verification of cross correlations of variables in block with latent variables of other blocks; d) verification of the internal model; e) verification of the model qualification by index of conformity of model to data.

Step a). The following criteria are used to verify internal consistency in blocks: Cronbach-alpha (α) and Dillon-Goldstein (ρ) coefficients (see, for example, [20-22]), as well as eigenvalues λ_i of the correlation matrix of explicit variables (indicators).

Table 3. Checking internal consistency in blocks

Block	α	ρ	λ_1	λ_2
LV_4	0.8843	0.9123	1.87	0.72
LV_3	0.1009	0.0912	2.3	0.73
LV_2	0.2012	0.5809	2.50	0.38
LV_1	0.8547	0.9035	3.68	0.70

Source: Own processing with XLSTAT-PM.

The Table 3 shows that the two blocks (LV_4 and LV_1) have good $\alpha > 0,7$ and $\rho > 0,7$ coefficients. But the LV_3 and LV_2 blocks have poor internal consistency ($\alpha, \rho < 0,7$). The following Table 4 shows the correlation coefficients of explicit and latent variables for each of the blocks.

Table 4. The correlation coefficients of explicit and latent variables for each of the blocks

Block	Variables			
	x_{31}	x_{32}	x_{33}	x_{34}
LV_3	-0,8234	-0,9123	0,9546	0,8756
	x_{21}	x_{22}	x_{23}	
LV_2	0.8947	0.9035	-0,8759	

Source: Own processing with XLSTAT-PM.

In the block LV_3 two variables (x_{31} and x_{32}) have a negative correlation with latent variable. In the LV_2 also there is a variable with negative correlation (x_{23}). This leads to internal inconsistencies in blocks, since these indicators are so-called de-stimulators (the higher their values, the worse). For the correct application of multivariate statistical analysis methods, it is recommended to make all indicators unidirectional (as a rule, stimulants). Therefore, we will make a replacement: $x_{31}'=1- x_{31}$, $x_{32}'=1- x_{32}$, $x_{23}'=1- x_{23}$. After modification of variables we obtain the index of cooperation of external model, are presented in the following Table 5:

Table 5. Cooperation indexes of the external model

Block	α	ρ	λ_1	λ_2
LV_4	0.8802	0.9113	1.87	0.72
LV_3	0.8854	0.9463	2.3	0.73
LV_2	0.8567	0.9347	2.50	0.38
LV_1	0.8573	0.9086	3.68	0.70

Source: Own processing with XLSTAT-PM.

All three of the cooperation criterions satisfy the necessary ($\alpha, \rho > 0,7 ; \lambda_1 > 1, \lambda_2 < 1$).

Step b). The second phase is verification of the external model:

Table 6. The value of the coefficients of the external model

Block	Variable	External weight w_j	Load λ_{1j}
LV_4	x_{41}	0.5347	0.2678
	x_{42}	0.4342	0.8473
	x_{43}	0.3245	0.4565
	x_{44}	0.3641	0.3172
LV_3	$1-x_{31}$	0.3274	0.9163
	$1-x_{32}$	0.3047	0.8132
	x_{33}	0.2679	0.9024
	x_{34}	0.2473	0.8271
LV_2	x_{21}	0.3274	0.9045
	x_{22}	0.4382	0.9134
	$1-x_{23}$	0.3645	0.8356
LV_1	x_{11}	0.3468	0.9768
	x_{12}	0.2173	0.7688
	x_{13}	0.1012	0.5467
	x_{14}	0.4037	0.9736

Source: Own processing with XLSTAT-PM.

Variables are considered significant if the load factor $\lambda_{1j} > 0,7$. By the results of the model validation it is necessary to eliminate variables x_{13}, x_{41}, x_{43}, x_{44}. After exclusion of nonsignificant variables we get the following test results of external model:

Table 7. The value of the coefficients of the external model after modifications

Block	Variable	External weight w_j	Load λ_{1j}
LV_4	x_{42}	0.3769	0.9151
LV_3	$1-x_{31}$	0.3453	0.9273
	$1-x_{32}$	0.3048	0.8601
	x_{33}	0.2869	0.9041
	x_{34}	0.2283	0.8171
LV_2	x_{21}	0.3394	0.9032
	x_{22}	0.4192	0.9081
	$1-x_{23}$	0.3586	0.8576
LV_1	x_{11}	0.4071	0.9879
	x_{12}	0.2412	0.7697
	x_{14}	0.4394	0.9677

Source: Own processing with XLSTAT-PM.

After checking the load factors λ_{1j}, that is, after checking the interconnection of explicit variables with latent variable of the corresponding block, you need to check the cross-loading, i.e. to determine the strength of the connection between explicit variables and latent variables of other blocks. This will exclude such indicators whose connection with latent variable of another block is greater than with latent variable of the corresponding block. The Table 8 shows that the strength of the connection of all explicit variables with latent variables of corresponding blocks is greater than with latent variables of other blocks, that is, all variables are "loyal" to their blocks.

Table 8. The value of the coefficients of cross load

Block	Variable	Block			
		LV_4	LV_3	LV_2	LV_1
LV_4	x_{42}	0.9071	-0.0750	0.0934	0.4374
LV_3	$1-x_{31}$	0,3149	0.9273	0.5768	0.8091
	$1-x_{32}$	0.2133	0.8081	0.5638	0.7283
	x_{33}	-0.5463	0.9023	0.6475	0.7205
	x_{34}	-0.3148	0.8087	0.6405	0.5586
LV_2	x_{21}	-0.0384	0.5768	0.9120	0.6879
	x_{22}	-0.1061	0.7495	0.9071	0.8412
	$1-x_{23}$	-0.0834	0.5205	0.8586	0.7225
LV_1	x_{11}	-0.2412	0.8364	0.8031	0.9879
	x_{12}	0.0945	0.4273	0.6814	0.7907
	x_{14}	-0.2927	0.2489	0.8506	0.9687

Source: Own processing with XLSTAT-PM.

The next step is to check the quality of the internal model. In the Table 9 are estimates β_i of the equations of structural model, as well as the results of the criterion-statistics.

Table 9. Estimation of the parameters β_i of a structural model

	Estimation	Standard error	t-statistics	Pr(>|t|)
Free member	2,1628E-17	0.0354054	2,53E-16	1
LV_4	0.29028279	0.1157102	1.598248	0.12
LV_3	0.67353833	0.1663135	3.849672	0.009
LV_2	0.51436382	0.1306488	2.968084	0.003

Source: Own processing with XLSTAT-PM.

Criterion of t-statistics is held for blocks LV_3 and LV_2 ($Pr(>|t|)<0.05$), but not to block LV_4 ($Pr(>|t|)>0{,}05$). Following the criterion of t-statistics, the latent factor LV_4 should be excluded from consideration. But we decided to leave it in the model. After all, the impact of the level of economic development of the region on the environment and the level of social development is obvious. Our hierarchical model did not take into account the relationship between latent factors LV_4, LV_3 and LV_2. The following Table 10 presents summary statistics for the structural model.

Table 10. Total statistic of internal model

Block	R^2	Ratio explanation variation	Ratio rejecting variation AVE
LV_4	0.00	0.79	0.89
LV_3	0.00	0.85	0.75
LV_2	0.00	0.89	0.79
LV_1	0.87	0.85	0.67

Source: Own processing with XLSTAT-PM.

Showing R^2 for target block LV_1 more than 87%. The proportion of variation characterizes share variations of a block, which reproduced latent variable this block. This characteristic for the all blocks far exceeds 50%, that positively characterizes the model. In the last column of a given indicator, which characterizes the dispersion excluded share (average particle dispersion indicators unit that explains the latent variable dispersion in the block that contains the error of measurement). Fashion AVE for the all blocks exceeds 50%, therefore, this criterion of the internal model is also considered satisfactory. The fifth stage is the calculation of the single factor of quality conformity model data - GoF (Goodness-of-Fit [21]). Coefficient characterizes the quality as an internal model of the system, and external, and serves as an indicator of the reliability of the forecast model (predictive reliability model is considered to be high, if the coefficient of GoF > 70%). Our model coefficient GoF=82%. Therefore, the constructed model can be used for forecasting. Finally, let's analyze the final version of our model. The internal model can be written as the following equation:

$$LV_1 = 0,5143LV_2 + 0,6735LV_3 + 0,2903LV_4 + \varepsilon_1^{in} . \quad (6)$$

Estimates of latent variables are a system of equations:

$$\begin{cases} LV_1 = 0,41x_{11} + 0,24x_{12} + 0,44x_{14} \\ LV_2 = 0,34x_{21} + 0,42x_{22} + 0,36(1 - x_{23}) \\ LV_3 = 0,35(1 - x_{31}) + 0,30(1 - x_{32}) + 0,29x_{33} + 0,23x_{34} \\ LV_4 = 0,38x_{42} \end{cases} . \quad (7)$$

4 Conclusion

Two factors have the most significant direct impact on the level of stable development of the tourism business in the Odessa region (LV_1): LV_2 – level of social condition (comfort) and LV_3 - level of the environmental protection (ecological state of the region). The strength of their influence is 0,5143 and 0,6735 respectively. A weaker direct effect (with a strength of 0,2903) on the endogenous factor LV_1 is exerted by the exogenous factor LV_4 - condition of economics. Obviously, its influence is direct (through factors LV_2 and LV_3). According to the results obtained, the level of social comfort (LV_2) can be assessed using three factors: x_{21} - total subsidies for the population; x_{22} - the sum of state cost on the protection of public safety, health and other social services; x_{23} - the proportion of the total number of unemployed persons to the economically active population (de-stimulating effect). The level of the environmental protection (LV_3) can be assessed using 4 factors: x_{31} - emissions of pollutants into atmospheric air and x_{32} - wastes generation (de-stimulating effect); x_{33} - state cost on the protection of the environment and x_{34} - current expenditures on environmental. On improving the state of the economy (factor LV_4) is significantly influenced by x_{42} - state spending on economic development. The level of stable development of tourism business (LV_1) determined by the direct impact of three factors: x_{11} - the number of tourists; x_{12} - wage of employees in the service sector; x_{14} - tax revenues from the travel industry. On the model you can count latent indexes for each year and create a stimulant to predict future values. Sustainability of development can be assessed by comparing the values of estimates for different periods. For example, if the estimation of the latent variables (LV_2, LV_3, LV_4) current period previous period ratings are smaller, then the corresponding block is considered to be sustainable. Or for every variables you can define a baseline with which to conduct the comparison.

References

1. L. Butowski, *Sustainable Tourism – A Model Approach, Visions for Global Tourism Industry – Creating and Sustaining Competitive Strategies* (2012) [online], Available at: http://www.intechopen.com/books/visions-for-global-tourism-industry-creating-and-sustaining-competitive-strategies/sustainable-tourism-a-modeling-approach (2019)
2. R. Casagrandi, S. Rinaldi, A theoretical approach to tourism sustainability, *Conservation Ecology*, **1**, 6, (2002)
3. J. B. Greenwood, *Sustainable development in a tourism destination context: a Plimsoll model of sustainability in Tyrell Country, North Carolina*, Raleigh, NC. (2006)
4. J. Steele, D. Dredge, P. Scherrer, Monitoring and evaluation practices of volunteer tourism organizations, *Journal of Sustainable Tourism*, **25** (11), 1674-1690 (2017)
5. E. Wilkins, S. de Urioste-Stone, A. Weiskittel, T. Gabe, Effects of Weather Conditions on Tourism Spending: Implications for Future Trends under Climate Change, *Journal of Travel Research*, **57** (8), 1042-1053 (2018)
6. F. Higgins-Desbiolles, S. Carnicelli, C. Krolikowski, G. Wijesinghe, K. Boluk, Degrowing tourism: rethinking tourism, *Journal of Sustainable Tourism*, **27** (12), 1926-1944 (2019)
7. D. Zeng, Z. Liang, B.W. Ritchie, Residents' social dilemma in sustainable heritage tourism: the role of social emotion, efficacy beliefs and temporal concerns, *Journal of Sustainable Tourism*, **28** (11), 1782-1804 (2020)
8. R. Sharpley, Tourism, sustainable development and the theoretical divide: 20 years on, *Journal of Sustainable Tourism*, **28** (11), 1932-1946 (2020)
9. H. Liu, H. Song, New Evidence of Dynamic Links between Tourism and Economic Growth Based on Mixed-Frequency Granger Causality Tests, *Journal of Travel Research*, **57** (7), 899-907 (2018)
10. V.S. Lin, Y. Yang, G. Li, Where Can Tourism-Led Growth and Economy-Driven Tourism Growth Occur?, *Journal of Travel Research*, **58** (5), 760-773 (2019)

11. Z. Li, H. Shu, T. Tan, S. Huang, J. Zha, Does the Demographic Structure Affect Outbound Tourism Demand? A Panel Smooth Transition Regression Approach, *Journal of Travel Research*, **59** (5), 893-908 (2020)
12. C. Wang, G. Li, H. Xu, Impact of Lifestyle-Oriented Motivation on Small Tourism Enterprises' Social Responsibility and Performance, *Journal of Travel Research*, **58** (7), 1146-1160 (2019)
13. S. Petit, N. Seetaram, Measuring the Effect of Revealed Cultural Preferences on Tourism Exports, *Journal of Travel Research*, **58** (8), 1262-1273 (2019)
14. A. Fernandez-Morales, J.D. Cisneros-Martinez, Seasonal Concentration of Cruise Tourism Demand in Southern Europe, *Journal of Travel Research*, **58** (8), 1389-1407 (2019)
15. B. Peng, H. Song, G.I. Grouch, S.F. Witt, A Meta-Analysis of International Tourism Demand Elasticities, *Journal of Travel Research*, **54** (5), 611-633 (2015)
16. G.E.D.O. Santos, V. Ramos, J. Rey-Maquieira, Length of Stay at Multiple Destinations of Tourism Trips in Brazil, *Journal of Travel Research*, 54 (6), 788-800 (2015)
17. O. Ghorbanzadeh, S. Pourmoradian, T. Blaschke, B. Feizizadeh, Mapping potential nature-based tourism areas by applying GIS-decision making systems in East Azerbaijan Province, Iran, *Journal of Ecotourism*, 18 (3), 261-283 (2019)
18. State Statistics Service of Ukraine [online], Available at: http://www.ukrstat.gov.ua (2020)
19. Main Department of Statistics in Odessa region [online], Available at: http://www.od.ukrstat.gov.ua (2020)
20. M. Tenenhaus, V.E. Vinzi, Y.M. Chatelin and C. Lauro. PLS Path Modeling. *Computational Statistics & Data Analysis*, **48**, (2005)
21. M. Tenenhaus, S. Amato and V.E. Vinzi A Global Goodness-of-Fit Index for PLS Structural Equation Modelling, *Proceeding of the XLII SIS Scientific Meeting*, **1** (2004)
22. M. Wtzels, G. Odekerken-Schroder and C. Van Oppen. Using PLS Path Modeling for Assessing Hierarchical Construct Models: Guidelines and Empirical Illustration. *MIS Quarterly*, **31** (2009)

Big Five in University Students in Relation to Psychological and Entrepreneurial Competences

Iveta Kmecová[1] and *Alena* Kajanová[2]

[1]Department of Human Resource Management, Institute of Technology and Business in Ceske Budejovice, Czech Republic
[2]Institute of Technology and Business in Ceske Budejovice, Czech Republic

Abstract. The objective of the research was to find and analyse the dependence factors of the individual personality. For the purpose of the analysis, students of two higher education institutions in the South Bohemian Region, the Czech Republic, were chosen. The students were from different fields of study, men and women, of different age. The research further dealt with the dependency of the Big Five factors on the positive and negative stress management strategies. The research results can be used for further studies that would be a contribution to theory as well as for the individualization of the approach to concrete groups or individual students.
Keywords: NEO-FFI, entrepreneurial competences, big five theory, stress management strategies

1 Introduction

The aim of our paper is to analyze the relationship a) between basic personality characteristics according to Big Five measured using the NEO-FFI questionnaire and selected sociodemographic variables and then b) between Big five personality characteristics and coping strategies according to SVF-78 and all students of the University of Technology and economic in Č. Budějovice (Czech Republic). We proceeded from the following hypotheses: H1: There is a relationship between the NEO-FFI dimensions and the SVF-78 coping strategies. H2: There is a difference in the dimensions of NEO-FFI according to the field studied. H3: There is a difference in the dimensions of NEO-FFI according to age.

2 Theoretical Background

NEO Five Factor Inventory (NEO-FFI) is a convenient tool for many industrial and organizational applications, including candidate screening and placement, performance prediction – as well as wide applications in research. NEO-FFI is a quick and efficient tool that has found successful applications in a number of applied and research field. The study defines term NEO Five-Factor Inventory (NEO-FFI) as a shortened version of a solid empirically-validated NEO Personality Inventory-Revised, an implementation of the five-factor model (FFM) of personality. Originally intended to provide a quick measure for screening and introductory purposes, NEO-FFI gained a position in a number of applied and research fields [1]. The Topic NEO Big Five has long been a concern for many researchers; NEO Five-factor inventory has been published in a number of books and journals, for example in [1] and in others. NEO Five-Factor Inventory (NEO-FFI) test is one of the latest developments in the field of personality measurement, intended to assess emotional, interpersonal, experiential, attitudinal, and motivational styles of individuals. These styles represent the empirically discovered five-factor model (FFM) of the human personality. NEO-FFI being a much shorter version of a mainstream NEO Personality Inventory-Revised (NEO PI-R), has its specific advantages and disadvantages that are discussed in this paper [1]. A meta-analysis reports a number of studies where NEO-FFI was used to predict academic performance (GPA, course grade, and average exam grade) with Conscientiousness scale most consistently linked to post-secondary academic success. NEO-FFI is a shortened version of the NEO Personality Inventory-Revised, an implementation of a successful, and extensively empirically-validated five-factor model of human personality. In accordance to the two methods of personality assessment outlined below—personality questionnaire and peer rating—NEO-FFI offers two observer-rating forms (for rating men and women) along with the standard self-report form, as well as a college-age self-report form. All 60 items of the NEO-FFI test require a response rated on the five-point Likert scale. The published expects most people to complete NEO-FFI within 15 minutes with scoring taking only 1 or 2 minutes [2]. The study holds the idea that the NEO-FFI-3 can help understand respondent's basic emotional, interpersonal, experiential, attitudinal, and motivational styles. It can also help quickly develop rapport with respondent, provide meaningful feedback and insight that will help our respondent/client develop greater self-understanding. The NEO Five-Factor Inventory-3 (NEO-FFI-3), is the updated version of the NEO-FFI — a 60item version of the NEO-PI-3. It provides a quick, reliable, and accurate measure of the five domains of personality and is particularly useful when time is limited and when global information on personality is needed [3]. Another result outlines in their other studies: In vocational counselling, the NEO PI-R can supplement measures of vocational interests and abilities, especially by calling attention to the client's strengths and weaknesses in adjustment and motivation. Use of the NEO Job Profiler a tool designed to help identify the personality requirements of different occupations is illustrated in the police selection sample. Together, the NEO Job profiler and NEO PI-R can help determine the optimal match between person and occupation. The NEO-FFI test is published in 9 foreign languages (including British English) and has validated versions available in other 25 languages [3]. Author describes the use of NEO-FFI test as a tool for quick individual personality testing in counselling and clinical setting but there is already a large body of research reporting use of this test in educational, organizational domains. [3, 4] Another study informs about results of research on a sample of 335 adult respondents. This study deals with the analyse of NEO Five-Factor Inventory too [5, 6]. Next study reported the significant correlations between NEO-FFI Extraversion and four different measures of creativity. Almost 50% of the variance in divergent thinking could be accounted by the NEO-FFI factors, while intelligence measured by Wonderlic Personnel Test provided no incremental variance in creativity scores [6]. Interesting is the studies [7, 8] is that the personality traits of treatment-seeking problem gamblers have been compared to healthy control groups in several studies and, although there is consistent evidence for high Neuroticism and low Conscientiousness in problem gamblers, past results may have been affected by selection bias. This study replicated these findings in a correlational design. The participants were nontreatment-undergraduate students who were screened for excessive and potentially addictive self-defeating behaviours. The Shorter Promis Questionnaire and the NEO PI-R were completed by undergraduate students at two Canadian universities (N = 369). Scores on the gambling subscale showed modest but statistically significant correlations with high Neuroticism, low Agreeableness, and low Conscientiousness [9]. Another study evaluated the latent structure of the NEO Five-Factor Inventory (NEO FFI) and relations between the five-factor model (FFM) of personality and dimensions of DSM-IV anxiety and depressive disorders (panic disorder, generalized anxiety disorder [GAD], obsessive—compulsive disorder, social phobia [SOC], major depressive disorder [MDD]) in a large sample of outpatients (N = 1,980) [10]. Other study reports large positive correlation between NEO-FFI Extraversion and emotional intelligence, an array of social and emotional skills that help people understand emotions of other people, as well as monitor their own feelings [11]. Goal of the next study was investigated with the change and stability of the Big-Five personality factors measured on a short scale. Fifteen similar studies using longer instruments were reviewed. In this study 7554 participants aged between 16 and 92 years completed a short 15 item FFM inventory twice six years apart in a large British sample. Participants were divided into six age groups and the results were broadly similar. Correlational analysis showed all five personality factors were considerably stable over six years after controlling for gender and age ($r = 0.47$ to $r = 0.60$, $p <.001$). Implications and limitations of using short scales particularly problems of measurement invariance, are acknowledged [12]. The next paper deals with the popular NEO Personality Inventory-Revised (NEO-PI-R) has a short form—the NEO Five-Factor Inventory (NEO-FFI) — that taps the five broad factors with fidelity and reliability. In this study, 13 item clusters were found to replicate across halves of a sample of self-descriptions by adults (N = 732). Thirteen factor-analytically derived scales were developed for the item clusters. The scales demonstrated reliability and factor structure comparable to that of the 30 facet scales of the NEO-PI-R. Correlation and multiple regression analyses showed that the content coverage of the 13 scales has high overlap with that of the NEO-PI-R facet scales [13]. Analyses of variants of the FFM revealed that only the unrestricted exploratory factor model showed acceptable fit as well as replicability across gender and educational level the goal of the paper was to identify changes in self and peer evaluations of personality among team members using the Five Factor Model. Multidisciplinary teams of five students in an undergraduate research design project-based course were used to evaluate their own and their peer's personalities over the course of one semester. Results showed that team members' evaluations of their own personalities did not change significantly through four iterations. Team member's evaluations of their peers did change for Neuroticism, Agreeableness, and Extraversion [14]. The study reports result of a research of coping strategies identified in students in two different study modes: full-time students and part-time students. The SVF-78 stress coping inventory was administered. The relationship of SVF-78 subscales to the variable study mode (full-time, part-time) was investigated. As for the statistical procedure, Levene's Test for Equality of Variances and two independent sample t tests were used. The most important finding shown by results of this study is a higher usage of the tree optimal coping strategies - situation control, reaction control and positive self-instructions - by part-time students. Part-time students also use the minimization strategy more frequently and the resignation strategy less frequently than full-time students. The study has some limitations, because the sample is not representative - subjects were students of the Faculty of Economics and Management [15]. Authors of the next paper are concerned with the efficiency of the lessons with the special focus on an effect of personality variables, namely extraversion and neuroticism, on preference of stress coping strategies. The issue of relationship between extraversion resp. neuroticism and use of stress coping strategies by undergraduate students at Czech University of Life Sciences (CULS) (n=109) was investigated. The data were collected with the use of standardized questionnaires SVF 78 and EOD. The mutual relationship of the variables was tested by Spearman's correlation coefficient. The results indicate association of negative coping strategies in general with neuroticism and positive strategies in general with extraversion. Practical implications of the findings are discussed and interventions for mental health reinforcement are recommended. The questionnaire - Stress Management Strategy SVF 78 and its detailed results will be the subject of the forthcoming contribution [16].

3 Methodology

The research tool was the NEO Five-factor personality inventory questionnaire [17]. NEO-FFI test represents one of the latest trends in the field of measuring personality traits, aiming at assessing emotional, interpersonal, experience, attitude and motivational styles of individuals [1]. The Five Factor Model (FFM) of personality has become a popular benchmark in terms of personality metrics [3]. This five-factor model measures a personality using five traits: openness to experience, conscientiousness, extraversion, agreeableness, and neuroticism.

Next, we examined the correlation relationship (the statistical dependence of two quantitative variables), specifically the relationship of two main domains of the SVF-78 questionnaire – positive and negative coping strategies and dimensions of the NEO-FFI. The SVF-78 questionnaire (Stress coping style) was developed by Wilhelm Janke and Gisela Erdmann (2002) It is a multi-dimensional self-observation inventory capturing the individual tendencies to use various methods of coping with stress in stressful situations. Together with the NEO-FFI questionnaire, the SVF-78 was a part of a research project implemented within an internal grant competition. The interpretation of the detailed results obtained from the analysis of SVF 78 (Stress coping style) questionnaire will be a subject of another study [18].

The sample consisted of 277 students of the Institute of Technology and Business (ITB) in Ceske Budejovice (South Bohemian Region) in both full-time and part-time forms of study (at the age ranging between 19 and 54), with m = 27 years, studying various fields of study: Mechanical Engineering (24.6 % respondents), Logistics (10.9 %), Economics (59.4 %) and Civil Engineering (5.1 %). The questionnaire was anonymous and it was completed in accordance with ethical principles and its completing was voluntary. The questionnaire was supplemented with several more questions focusing on socio-demographic indicators – sex, age, and field of study of the students. The data was then statistically processed by the SPSS programme at 95% significance level. After performing the normality test and ensuring the normality of the data, the following statistical parametric tests were used: t-test for two independent groups, ANOVA and correlation.

4 Results

In the sample mentioned above, the highest mean values were obtained in the case of the "conscientiousness" dimension (32.509). On the other hand, the lowest value was for the dimension "openness to experience" (22.054). The highest standard deviation was detected in the case of the dimension "extraversion", which indicates a large diversity of the sample in terms of this dimension (Table 1).

Table 1. NEO-FFI dimensions, basic descriptive statistics

	Minimum	Maximum	Mean	STD
Neuroticism	.00	46.00	22.054	8.122
Extraversion	.00	48.00	31.271	8.195
Openness to	.00	44.00	25.433	6.646

experience				
Agreeableness	.00	42.00	29.404	6.701
Conscientiousness	.00	48.00	32.509	7.680

Source: Own processing.

Table 2 shows testing the NEO-FFI dimensions in relation with the sex of the respondents. For testing, an independent sample t-test was used, which showed that there is a significant difference between the sexes in the case of "extraversion" (p=0.004), "agreeableness" (p <0.000) and "conscientiousness" (p <0,000). In the case of "extraversion", women achieved a higher mean score than men (32.631 for women and 29.797 for men). Also, in the case of "agreeableness", women´s score was higher than men´s score (31.319 to 27.331), as well as in the case of "conscientiousness" (34.722 for women and 30.113 for men).

Table 2. Relationship between NEO-FFI dimensions and sex

Dimension	Sex	Mean	Std. Deviation	t	Sig. (2-tailed)
Neuroticism	Male	21.278	8.384	-1.532	.127
	Female	22.771	7.834	-1.528	.128
Extraversion	Male	29.797	8.489	-2.915	**.004**
	Female	32.632	7.695	-2.904	**.004**
Openness to experience	Male	24.902	7.025	-1.279	.202
	Female	25.923	6.261	-1.273	.204
Agreeableness	Male	27.331	7.036	-5.175	**.000**
	Female	31.319	5.769	-5.135	**.000**
Conscientiousness	Male	30.113	8.240	-5.222	**.000**
	Female	34.722	6.395	-5.171	**.000**

Source: Own processing.

The relationship between the age of the respondents and the NEO-FFI dimensions was tested using the Analysis of Variance (ANOVA) test. In this case, no statistically significant dependence has shown (see Table 3). It follows from the results that according to the NEO-FFI, personal traits do not depend on age.

Table 3. Relation between NEO-FFI dimensions and age

Dimension	F	Sig.
Neuroticism	1.524	.043
Extraversion	1.275	.159
Openness to experience	.913	.603
Agreeableness	1.087	.352
Conscientiousness	1.748	**.011**

Source: Own processing.

In terms of identifying the relation between the dimensions and the studied field (tested using ANOVA), the only significant relationship was detected in the case of the "conscientiousness" dimension (p=0.005; F=4.387). As Table 4 shows, the mean value for mechanical engineering was 31.779, while it was 30.3 for logistics, 29.433 for economics, and 30.429 for civil engineering. The highest score was thus achieved in the case of mechanical engineering, while students of economics appear to be least conscientious.

Table 4. Relation between NEO-FFI dimensions and studied field

Dimension	F	Sig.
Neuroticism	1.250	.292
Extraversion	3.725	.012
Openness to experience	.707	.549
Agreeableness	3.832	.010
Conscientiousness	4.387	**.005**

Source: Own processing.

Table 5. Correlation of two main domains of SVF-78 questionnaire and NEO_FFI dimensions

	SVF POZ		SVF NEG	
NEO-FFI dimension	P	r	p	r
Neuroticism	**0.000**	-0.323	**0.000**	0.703
Extraversion	**0.000**	0.408	**0.000**	-0.363
Openness to experience	**0.000**	0.254	0.156	-0.086
Agreeableness	0.028	0.132	0.562	0.562
Conscientiousness	**0.000**	0.287	**0.000**	-0.308

Source: Own processing.

Table 5 shows the correlation of the two main SVF-78 questionnaire domains results – positive and negative coping strategies and NEO-FFI dimensions. Significant correlations have been detected in the case of positive strategies and all NEO-FFI dimensions. In terms of negative strategies, only neuroticism, extraversion, and conscientiousness correlated. In all cases, these were rather weak correlations.

5 Discussion

The results show certain trends. University students achieved a high number of points in the questionnaires especially for conscientiousness, while the lowest scores were achieved in openness to experience. In Czech context, these results can be interpreted among the university students by the fact that university studies are a certain traditional basis for entry into life, to which the students are motivated both by their parents and people around them. Thanks to free studies, higher education is perceived by society as a standard way for the students of secondary schools. Lower scores for openness to experience can thus be caused by the above-mentioned "standard way" [19]. However, for successful study, the ability of students to conscientiously fulfil the requirements is necessary. Therefore, high scores for conscientiousness are expected. It is also necessary to take into account time, financial and psychic (stress) burden placed on university students, but the students did not show higher scores or neuroticism associated with stress. Although the research does not deal with the personality traits, the results of the questionnaire may affect these negative factors [20].

The dependence of the individual personality traits on the sex was confirmed for conscientiousness, agreeableness, and extraversion. In this case, women an achieved higher score but also a lower standard deviation [21]. These results partially correspond to where the female students achieved higher scores in conscientiousness and agreeableness compared to male students. However, correlation with neuroticism was not detected. In the case of extraversion, the female students showed a lower score that the male students. This could be due to different types of universities, as included also medical and social-type of schools and pedagogical schools, which are popular rather with female students in the Czech Republic [19, 21]. The results of the individual studies show different outcomes (Rubinstein, 2005), which can point to the differences between the male and female students of the individual fields of study in higher education institutions [19].

In terms of the age of the students, no significant dependence was confirmed, which corresponds with the conclusion that the age factor does not influence the individual personality traits of students. However, it is questioned whether a higher score in conscientiousness and extraversion would be achieved if a higher number of older students were included [22].

When investigating the dependence of personality traits on a field of study, dependence was confirmed only for conscientiousness. In this case, the highest score was achieved by mechanical engineering students. Civil engineering and logistics students achieved a lower score, while the lowest number of points was achieved by economics students. This indicates that students of economics are less conscientious than students in technical fields. However, it is necessary to take into account that the representation of men and women in individual fields of study is not even. In the Czech Republic, students of technical fields are mostly men [23]. Given that women showed higher score in conscientiousness than men but men showed higher standard deviation, it can be concluded that male students studying economic fields achieved lower score than male students in technical fields, which can be influenced by the expected roles of these student's bone.

For verification of this conclusion, further investigation is necessary. This research could help create learning processes that would make use of the strengths of students in individual fields of study [24].

Finally, when comparing the results of Big Five personality traits with the stress management strategy (SVF-78) questionnaires, there was a correlation between positive strategies with all kinds of personality traits, which does not have to be considered decisive, as these results can be distorted by individual students' self-assessment that does not necessarily correspond with the reality. In the case of negative stress management strategies, there was a correlation with neuroticism, extraversion, and conscientiousness. This partly corresponds with the specific personality traits. Neuroticism can be seen as a negative trait; its correlation with negative stress management strategies was thus expected. Extraversion can be seen as neutral, but pointing out to a greater need for being with people and thus exposure to greater stress [20]. Correlation of conscientiousness with negative stress management strategies might be caused by a higher degree of stress acting on a student with higher demands on his or her success [25].

6 Conclusion

This Big Five research realized among university students seeks correlations between individual personality traits and age, sex, field of study, and stress management strategies (SVF-78). Big Five research shows correlations that differ from each other within specific groups of people. The aim is to analyse such correlations and propose further research among university students.

The research has shown that students show a higher score in conscientiousness and the lowest score in openness to experience. Women achieved a higher number of points in conscientiousness, agreeableness, and extraversion. Men showed greater differences in all traits (greater standard deviation) than women. No dependence on age has been detected. Students of technical fields showed higher score in conscientiousness compared to economics students. Positive stress management strategy showed a correlation with all five personality traits, while negative stress management strategies showed correlation with neuroticism, extraversion, and conscientiousness.

Acknowledgements

The subject-matter article was published as a part of IGS Project– Internal grant competition); Project Number: 8110-009, Order Number: 05IRP014; Project Name: Psychological-Entrepreneurial competences of university students at IT.

References

1. Chadyuk, O. NEO Five-Factor Inventory and its Applications in Industrial /Organizational Psychology [online], Available from: https://www.researchgate.net/publication/278026898_NEO_Five-Factor_Inventory_and_its_Applications_in_Industrial_Organizational_Psychology (2009)
2. Costa, P. T. & McCrae R. R. NEO Five-Factor Inventory [online], Available from: https://www.researchgate.net/publication/278026898_NEO_FiveFactor_Inventory_and_its_Applications_in_Industrial_Organizational_Psychology (2005).
3. Costa P. T. & McCrae R. R. NEO-FFI-3. NEO Five-Factor Inventory-3, [online], Available from: https://www.sigmaassessmentsystems.com/assessments/neo-five-factor- inventory-3/ (2010).
4. C. Alonso, E. Romero, Study of the Domains and Facets of the Five-Factor Model of Personality in Problematic Internet Use in Adolescents. International Journal of
5. Mental Health and Addiction. 18, 293–304 (2020).
6. Murray, G., Rawlings, D., Allen, N. & Trinder, J. NEO Five-Factor Inventory Scores: Psychometric Properties in a Community Sample. Journal Measurement

and Evaluation in Counseling and Development. 36 (3), 140-149 (2003).

7. J. W. Lace, L. N. Evans, Z. C. Merz, P. J. Handal, Five-Factor Model Personality Traits and Self-Classified Religiousness and Spirituality. Journal of Religion and Health. 59, 1344–1369 (2020).

8. Furnham, A. & Bachtiar, V. Personality and intelligence as predictors of creativity. Personality and Individual Differences 45, 613–617 (2008).

9. K. A. Lyon, G. Juhasz, L. J. E. Brown, R. Elliott, Big Five personality facets explaining variance in anxiety and depressive symptoms in a community sample. Journal of Affective Disorders. 274, 515–521 (2020).

10. MacLaren, V., Best, L., Dixon, M. & Harrigan, K. Problem gambling and the five-factor model in university students. Personality and Individual Differences, 50 (3), 335-338 (2011).

11. Rosellini, A. J. & Brown, T. A. The NEO Five-Factor Inventory: Latest Structure and Relationships Wth Dimensions of Anxiety and Depressive Disorders in a Large Clinical Sample. SAGE yournal, 18 (1), 27-38 (2010).

12. Craig, A. et al. Psychological and neural correlates of emotional intelligence in a large sample of adult males and females. Personality and Individual Differences 46, 111–115 (2009).

13. Furnham, A. & Cheng, H. The change and stability of NEO scores over six-years: A British study and a short review. Personality and Individual Differences 144, 105–110 (2019).

14. Saucier, G. Replicable Item-Cluster Subcomponents in the NEO Five-Factor Inventory. Journal of Personality Assessment, 70(2), 263-276 (2010).

15. Stidham, H., Summers, J. & Shuffler, M. Using the Five Factor Model to Study Personality Convergence on Student Engineering Design Teams. Proceedings of the DESIGN 2018 15 th International Design Conference (2018).

16. Natovova, L., Chylova, H., Michalek, P. & Natov, P. Which coping strategies are preferred by students in different modes of study. Efficiency and responsibility in education 2012, 416-424 (2012).

17. Chylova, H. & Natovova, L. Personality and stress coping at Czech University of Life sciences students. Proceedings of the 15th international conference efficiency adn responsibility in education (ERIE), 125-131 (2018).

18. Hrebickova, M. & Urbanek, T. NEO Five-factor personality inventory (according to NEO Five-factor Inventory P.T. Costa and R.R. McCraee. Prague: Testcentrum, s.r.o. (2001).

19. Janke, W. & Erdmannova, G. Stress management strategies – SVF 78. Praha: Testcentrum, s.r.o. (2003).

20. Prudky, L., Pabia,n P. & Sima, K. Czech higher education: on the way from elite to universal education 1989-2009. Grada Publishing as (2010).

21. Bunevicius, A., Katkute, A. & Bunevicius, R. Symptoms of anxiety and depression in medical students and in humanities students: Relationship with big-five personality dimensions and vulnerability to stress. International Journal of Social Psychiatry 54, 494–501 (2008).

22. Mlcak, Z. Five-factor model of personality, prosocial behavior and empathy in high school and university students. Social processes and personality (2010).

23. Noftle, E. E. & Fleeson, W. Age differences in big five behavior averages and variabilities across the adult life span: Moving beyond retrospective, global summary accounts of personality. Psychology and Aging, 25(1), 95 (2010).

24. Jarkovska, L. & Liskova, K. Gender aspects of Czech education. Sociological journal 44, 683–701 (2008).

25. Komarraju, M., Karau, S. J. & Schmeck, R. R. Role of the Big Five personality traits in predicting college students' academic motivation and achievement. Learning and individual differences, 19(1), 47-52 (2009).

26. Trapmann, S., Hell, B., Hirn, J. O. W. & Schuler, H. Meta-analysis of the relationship between the Big Five and academic success at university. Zeitschrift für Psychologie/Journal of Psychology, 215(2), 132-151 (2007).

The state's responsibility towards entrepreneurs in the Czech Republic for the measures adopted in relation to the epidemic of coronavirus

Stanislav Bílek[1,*], *Zdeněk* Caha[1], and *Vendula* Velková[1]

[1] Institute of Technology and Business in České Budějovice, Faculty of Corporate Strategy, Okružní 517/10, 370 01 České Budějovice, Czech Republic

Abstract. This paper briefly analyses the legislation of the state's liability for the measures adopted in relation to the epidemic of coronavirus in the period of the state of emergency, it especially deals with the state's liability for damage accrued upon entrepreneurs in a casual relation to the restriction of the right to engage in enterprise. The aim of the paper is to carry out a legal analysis of liability for damage according to the Crisis Act on one hand, and of liability for damage caused by the terminated measures of the state according to the Act referring to the protection of public health, on the other hand. The paper provides a practical guideline for claiming damages and draws attention to the limits of such claims.
Keywords: state of emergency, business, measures of general nature, liability for damage

1 Introduction

Shortly after the pandemic outbreak numerous studies that reflect the impact of the corona virus crisis from the medical, economic, social, environmental as well as legal points of view started to appear. Scientists started to quantify the impact of the ordered restrictions just after the pandemic outbreak. Firano and Fatine [1] model the economic consequences of the epidemic in Morocco in their research, and they emphasize that the choice of the restrictive policy has enabled reduction of the negative medical impacts of the pandemic. A study by Donthu, Gustafsson [2] points out serious economic consequences in the global scale, which lead to dramatic changes in the behaviour of businesses as well as consumers. Farias and de Araujo [3] point out the influence of the pandemic on the global food supplies and consequent grow of consumer prices. A study by Malinovsky et al. [4] observes that Russian economy has been strongly negatively affected by the restrictions and worsened macroeconomic indicators are also expected in the future. In relation to individual industries Varelas and Apostolopoulos [5] for example analysed the strategic approaches to management among entities involved in tourism, which were focused on the mitigation of the economic impacts. Restrictions introduced to prevent the virus from spreading have changed the life style from the roots, which has also accented social inequality. Countries as well as individual regions

* Corresponding author: stanislav.bilek@mail.vstecb.cz

and municipalities had to seek quick solutions how to reduce the social-economic impacts and had to rely on their own abilities to respond flexibly and adapt themselves to the situation that has arisen [7]. Another study [8] emphasizes that the COVID-19 pandemic has forced the society worldwide to resort to social distancing to slow down the spreading of the SARS-CoV-2 virus. Aldaco et al. [8] deals with the environmental impacts shortly after the pandemic outbreak. Studies also exist that point out the fact that the negative economic impact of the Covid 19 virus spreading might have political consequences and give rise to the strengthening of extremist groups and opposition movements and escalate into armed conflicts [9]. A study by Neumeyer et al. [10] on the other hand points out that Covid 19 also represents a great challenge to entrepreneurs and also an opportunity towards a permanently sustainable and circular economy.

It is obvious that the situation that has arisen is completely unique, where unexpected economic and social impacts are among others influenced by targeted legislative interventions of the state in search for the prevention of the system overloading.

2 Methods and data

The author uses the method of data collection based on analysing the valid legislation, above all, the constitutional legislation, the resolutions of governments, measures of general nature, and analysing judgements, academic papers. In terms of proposed measures, they are based on author's private practice.

The government of the Czech Republic adopted a resolution to declare an state of emergency for the territory of the Czech Republic for the reason of endangering health in relation to proving the incidence of coronavirus /designated as SARS CoV-2/ on the territory of the Czech Republic for the period of 30 days on 12th March 2020. The resolution was published in the Collection of Laws No. 69/2020. The following, inter alia, results from the text: 'the government order pursuant to sec. 5 paragraph a) to e) and sec. 6 Act No. 240/2000 sb. [11], Crisis Management and the Amendments of Certain Statues (the Crisis Act), as amended, crisis measure for dealing with the developed crisis situation whose specific implementation will be determined by the separate resolution of the government.'

On 09.04.2020 the government, in resolution published in the Collection of Laws No. 156/2020, extended the state of emergency till 30.04.2020, with the prior consent of the Chamber of Deputies. Further, the Chamber of Deputies extended the state of emergency till 17.05.2020.

In relation to the declaration of state of emergency, the government, individual ministries and regional bodies have adopted dozens of measures. The adopted measures resulted in a significant limitation of fundamental human rights formulated in the Charter of Fundamental Rights and Freedoms, inter alia, the significant limitation of the right to do business and economic activities, above all, in the connection to the limitation or ban on the retail sale of goods and services.

The purpose of the following legal interpretation is to focus on the state's liability for damage accrued upon an entrepreneur in casual relation to the measures adopted by the bodies of public authority in the period of state of emergency, and the reviewability of these measures by courts.

Two lines of legal defence are considered; first, the state's liability for damage amended directly in Act No. 240/2000 Sb. [11], Crisis Management and the Amendment of Certain Laws (the Crisis Act), second, the states liability for damage caused by unlawful decisions according to Act No. 82/1998 Sb. [12], Liability for damage caused in the exercise of official authority by a decision or maladministration. The chosen way of defence is based on the legal form of adopted measures and on the body of the state having adopted the measure. The author focuses on the adopted resolutions of the government and on the measures of the Ministry of Health, as they encroached upon the rights of entrepreneurs the most significantly.

The regime of both acts is different, as damage, according to the Crisis Act, can be compensated without the state's divergence from prescribed rules or standards (the prerequisite of liability for damage will be the mere existence of an adopted measure), on the contrary, when the state is liable to unlawful decisions, the state must revoke the measure on the grounds of its unlawfulness. During legal analysis, the general amendment of liability for harm, according to the Civil Code, cannot be omitted. It cannot be excluded that the injured parties will assert their claims according to both legislations concurrently.

In relation to this issue, it is necessary to mention, whether and to what extent the adopted measures are the subject of the control of judicial power, either within general administrative judiciary or constitutional judiciary.

3 Results

3.1 The analysis of legislation

3.1.1 Liability for damage according to the Crisis Act

Article 26 of the Charta of Fundamental Rights and Liberties, which forms the constitutional order of the Czech Republic, states that 'Everybody has the right to the free choice of her profession and the training for that profession, as well as the right to engage in enterprise and pursue other economic activity'.

The piece of constitutional legislation No. 110/1998 Sb. [13], Regarding the security of the Czech Republic, modifies the rules for the limitation of fundamental human rights. Regarding the institute of state of emergency, it provided in Article 5 as follows: '*The government may declare a state of emergency in the event of natural disasters, environmental or industrial accidents, incidents or other dangers which pose a significant threat to lives, health or property, or internal order and safety. The Government will immediately inform the Chamber of Deputies of the declaration of an state of emergency, who may cancel the declaration.'* Article 6 of the same piece of legislation provides further rules for the declaration of state of emergency: *State of emergency can only be declared upon reasons given for a specific time period and specific area. Concurrently with the declaration of state of emergency the government must specify which rights, established in in a separate piece of legislation, and to what extent, are restricted in compliance with the Charter of Fundamental Rights and Liberties, and which liabilities, and to what extent, are imposed. Details are determined by law. State of emergency can be declared for the period not exceeding 30 days. The given period can only be extended after an earlier consent of the Chamber of Deputies. The state of emergency ends by the expiration of the period that it has been declared for unless the government or the Chamber of Deputies decides to terminate it before the expiration of this period.'*

The assumption of accuracy to declare state of emergency is, inter alia, the determination of the grounds for a declaration, temporality and the determination of territorial extent and validity. It is especially a constitutional piece of legislation that imposes on the government of the Czech Republic which specific rights determined in the Charter of Fundamental

Human Rights and Liberties are restricted by such measure. It results from the text of the resolution to declare an state of emergency by the government that the list of the restricted rights is absent, although the constitutional piece of legislation requires to determine it 'concurrently' with the declaration of state of emergency.

The details of the state of emergency are determined by Act No. 240/2000 Sb. [11], Crisis Management and the Amendment of Certain Laws (the Crisis Act). Our interest is primarily concerned with the issue of damages caused by the casual relation to the adopted crisis measures. This law determines the applicability and powers of statutory bodies and territorial self-governing units, the rights and liabilities of artificial legal persons and natural persons during preparing for crisis situations which are not related to establishing the defence of the Czech Republic against external attack, the dealing with them and the protection of critical infrastructure and liability to the violation of these obligations. Before analysing the issue of the state's liability for damage, we must define the concepts of crisis management and crisis measures. Crisis management is defined in section 2 of the law as: *'the sum of operating activities of the bodies of crisis management concentrated on the analysis and assessment of security risks and the planning, organising implementing and controlling of activities conducted in relation to 1st preparation for crisis situations and the solutions therein or 2nd protection of critical infrastructure. Crisis measure is an organisational measure or technological measure for the solution of crisis situation and removing its consequences, including measures infringing the rights and liabilities of persons'.* For the purpose of this paper, it is important to conclude that crisis measures are a sum of all measures adopted in relation to the state of emergency by various competent bodies of public authority. The form of crisis measures can be a resolution of the government, a piece of legislation, regulation, measure of general nature, administrative decisions, factual acts enforced by the bodies of public authority.

State compensations are determined in the Crisis Act in the following provisions, sections 35 to 37. The author focuses on damage accrued upon entrepreneurs in relation to the restriction of free movement of persons and the restriction or complete ban on retail sale of goods and services, therefore, our interest is in the provision of section 36, called the compensation of damage. According to the provisions of sections 5 and 6 of the Crisis Act it is possible: *'for the shortest time possible and to the minimum necessary extent, that natural persons and artificial legal persons have the right of ownership and the right to use and enjoy restricted, inter alia, in terms of assets, free movement and residence in the determined territory endangered or inflicted by crisis situation, and the restriction also applies to the right to do business which would threaten the implemented crisis measures or infringe or make impossible their implementation.'*

Liability for damage resulting from the adopted crisis measures is determined by the law as follows: *'The state is liable to compensate damage accrued upon artificial legal persons and natural persons in a casual relation to crisis measures and exercises (section 39, subsection 4) conducted according to this law. The state can be relieved from liability only where it proves that the injured party caused damage by himself or herself. Monetary compensation is provided by the body of crisis management that ordered the crisis measure or exercise, during which or as a result of which the damage or harm was caused. The claim for damages with provided reasons is put forward by an artificial legal person or natural person in writing to the competent body of crisis management within 6 months from the moment it became known, not more than 5 years from the moment of the occurrence of damage, otherwise the right extinguishes. The body of crisis management is entitled, in the cases deserving special attention, to adjudicate a claim even after the expiration of term for application or without application, however, not longer than 5 years from the moment of the occurrence of damage. The compensation of damage is not provided for artificial legal*

 https://doi.org/10.1051/shsconf/20219101037

persons and natural persons responsible for the origin of harmful event. The body of crisis management is entitled to require to cover the costs that were incurred as the compensation for damage from the originator of accident or other event resulting in a crisis situation that required the imposition of crisis measures.'

Apart from that, the law determines reimbursements for the restriction of ownership right, providing material means and duty to work and work assistance.

Especially these practical conclusions for an entrepreneur incurred with crisis measures can be deduced from the above quoted provisions:

The state's liability for damage incurred upon by crisis measures is objective, therefore, the claim arises regardless fault

The prerequisite of liability is not the unlawfulness or erroneousness of adopted crisis measure, but its correctness. Therefore, it is not appropriate to require the measure be revoked

Actual damage and a loss of profit are reimbursed (as for a loss of profit note the judicial decision of the Supreme Court (NS) sp.zn. 25 Cdo 3798/2007, sp.zn. 25 Cdo 1649/2007 and sp.zn. 25 Cdo 1422/2016)

Damage may not occur directly in a casual relation to the declaration of the state of emergency, but 'in relation to' the subsequent crisis measure adopted according to the law or in the summary

The subject-matter of reimbursement is both damage to property and other than proprietary loss

Damage must occur in a casual relation to a crisis measure, while the fulfilment of this condition is proved by the injured party. It may be assumed that this aspect will be the most problematic in court disputes. The subject-matter of reimbursement will only be so called objectively anticipated damage. In relation to this, it is not possible to make a simple direct proportion between an adopted measure to restrict retail sale and a loss of profit. The adopted measure is not necessarily the only cause of damage. It concerns the issue of pandemic itself, what its course was like, what other measures it required and whether their actual or possible consequence was a fall in demand for goods or services of an entrepreneur. Therefore, the calculated loss of profit can relate to the fall in demand from different causes, such as the fear of customers, ordered quarantine and other measures concerning health of customers. It concerns the possible situation, what the demand for goods would have been if the crisis measure had not been adopted, or had been adopted to a different extent, and the rise of infected individuals had been exponential. It is certainly relevant to suggest to courts considering casual relations so that they would assess the effects of the measures of neighbouring countries and their bodies, especially the closures of state boundaries, imposition of curfews, etc. The decrease of foreigner customers has certainly an impact on the loss of profit of many entrepreneurs. Czech state, however, is not liable to the measures adopted by other states. Many measures were adopted by Czech and foreigner companies from their own initiative without being compelled by the crisis measures of the state to close a business or restrict production or sale. It is also necessary to note the obligation of prevention imposed by the Civil Code, i.e. to conduct so as to avoid damage to health and property of third parties. Therefore, in the course of doing business entrepreneurs must adopt measures that may have an impact on the scope of offered goods or services, in fact, the prevention necessarily results in the rising costs of hygiene matters. There is a fact that needs to be taken into account, whether an entrepreneur made use of the possibility to sell goods or services in a way other than direct sale in venues, as the ban did not apply to a shop window sale in certain period and a form of e-shop was a reasonable alternative in case of many entrepreneurs.

As for the burden of proof, the injured is in the position to prove the actual damage or loss of profit. Courts will probably resort to using such procedural institutes that will result in reducing the degree of evidence against practical certainty. According to the Civil Procedure Code: *'where it is not possible to exactly determine the amount of the compensation of damage, a court will determine it by considering the individual circumstances of the case fairly. Where it is possible to find out the amount only with disproportional difficulties or it is not possible to ascertain it at all, the court will determine it upon its consideration'*

It may not be possible to make claims for the wages paid to employees as they are so called wasteful expenditures, which are not reimbursed

If the injured person inflicted the damage on himself or herself, the claim for compensation of damages is not established. Further, if an entrepreneur neglected to adopt such measures that would result in mitigating the effects of crisis measure, moreover, if he or she failed to apply for compensations paid by the state to mitigate the effects of crisis measures, or contractually excluded himself or herself from protection provided by the Civil Code within contractual law. However, it will not be possible to account for public benefit from the adopted measures. It would exclude the state's liability as such

A claim for compensation will not be established in the cases when the adopted measures aim at something different than purely reacting to an occurred crisis situation. If, for instance, the emerged situation results in creating stricter legislation for the protection of health with consequences after the end of epidemic, there is not a casual relation of such measure and occurred damage.

A claim must be asserted in subjective, in fact objective time-limit, otherwise, it becomes extinguished, with regards to the body of authority that has issued the measure. The petition would be against the state or region

3.1.2 Liability for damage according to Act referring to the state's liability for damage incurred upon during the enforcement of public authority by decision or maladministration

The claim for compensation of damage is formulated in this legislation (section 7 and 8) as follows: 'the right to compensation of damage incurred by unlawful decision belong to the parties to procedure in which there was issued a decision incurring damage upon them. The claim for compensation incurred by unlawful decision can, unless provided otherwise, be asserted only then where the final and conclusive decision was reversed for unlawfulness or modified by a competent body. The decision of this body is binding for a court involved in deciding on compensation of damage. Where the extraordinary cases are not concerned, the claim for damage incurred by unlawful decision can only be asserted, when the injured person has used all procedural measures within time-limits determined by the state, provided by the state for the injured person to protect his or her rights; such a measure is a regular remedial measure, extraordinary relief, except a proposition to a new trial, and other procedural measure to protect law with the enforcement of which it is connected a commencement of civil proceedings, administrative proceedings or legal proceedings, or a motion to discontinue an enforcement of the judgement of the court.'

From our perspective of an entrepreneur affected by the ban on or restriction of sale we move our focus of attention to a measure of general nature issued by the Ministry of Health according to Act 258/2000 Sb. [14], Regarding the protection of public health (section 69, paragraph i). However, it is necessary to acknowledge that even the measures adopted by the Ministry of Health are crisis measures in the sense of the Crisis Act, although they are not issued with a reference to this law.

The measures of general nature are defined by the Rules of Administrative Procedure (Act No. 500/2004 Sb. [15], as amended) in section 171. It is an administrative act issued according to special legislation that is not a legal regulation or an administrative decision (it aims at a larger range of persons concerned, but it is not generally binding for everyone). The remedial measure for such a provision is an 'administrative action' according to section 101a and the following Act No. 150/2002 Sb. [16], the Rules of Administrative Procedure, as amended. The motion to terminate the measure is put forward to a regional court. The person who claims that his or her rights have been violated by the measures of general nature issued by administrative body is entitled to put forward a motion to terminate the measures of general nature or its parts. If he or she is entitled in compliance with the law in a case, in which the measure of general nature was enforced, bring an administrative action or other motion within administrative judiciary, he or she can suggest terminating the measure of general nature only concurrently with such a motion. The motion can be put forward within one year period starting on the day when the measure challenged by the motion came into effect.

Therefore, it will be essential, who can assert a claim for the compensation of damage according to Act No. 82/1998 Sb. [12], regarding the measures of general nature, and whether such a measure can be a ground for asserting a claim for the compensation of damage, when the text of the law uses the term 'decision', but a measure of general nature is not a decision.

The decision of Municipal Court in Prague, which decided on the motion to terminate the measure of the Ministry of Health banning or restricting retail sale, can be an appropriate initial guideline. The motion was satisfied on the merits by a judgement, which is not final and conclusive, on 23.04.2020 c.j. 14 A 41/2020. Is it possible to conclude that the terminated measures of the Ministry of Health give rise to a claim for the compensation of damage according to the above mentioned law?

The municipal Court in Prague allowed the unlawfulness of these measures. Let us quote the most relevant reasons in short:

Crisis Act is a special regulation in relation to Act Regarding the protection of public health, with regard to the state of emergency, only the government, not the Ministry of Health, was authorised to restrict the fundamental rights and freedoms to such a massive extent.

Failure to adopt the measure by the government according to the Crisis Act, but adopting it by the Ministry of Health according to the Act Regarding public health is considered by the court as a violation of constitutional guarantees of the division of power. By adopting the measures by the Ministry of Health according to the Act Regarding the protection of public health, the control by the Chamber of Deputies was eliminated. The Ministry violated its subject-matter competence.

Considering the possibility to assert a claim for the compensation of damage caused by the terminated measure of general nature, I tend to accept the conclusion that legislation allows such interpretation [17]. Also note the reasoning of the governmental bill No. 160/2006 Sb. [18], or the judgement of the Supreme Court (NS) on 08.08.2017, sp. Zn. 30 Cdo 3292/2015, NS judgement on 27.10.2015, sp. zn. 25 Cdo 3444/2013, NS judgement on 08.08.2017, sp. zn. 30 Cdo 3292/2015; NS judgement on 11.09. 2018, sp. zn. 30 Cdo 3079/2016; NS judgement on 11.09.2019, sp. zn. 30 Cdo 1642/2018 [19-22].

Whereas the measures in question are not issued on the basis of a previous motion, then it is not conditional that the person concerned makes objections according to the Rules of Administrative Procedure. The ground giving rise to the claim is proposing a successful motion to terminate the measure of general nature of the Ministry of Health.

As for the right to sue (locus standi) and justiciability, it is conclusively presumed that 'in this situation, when it is not possible to ground the definition of the injured person by an unlawful decision on the position of the party to the proceedings, every person who suffered damage caused by this decision is entitled to sue' [23]. Such a person will be the one whose motion caused the termination of the measure, and probably all the others suffering from this terminated measure (the entrepreneurs concerned) with regard to specific causes, as these persons cannot make motions to terminate a measure that has already been terminated.

Let us summarise the prerequisites giving rise to a claim according to Act No. 82/1998 Sb. [12]: - it must be a decision (a measure of general nature), - it must be terminated for unlawfulness, - it must be final and conclusive or enforceable before it is declared final and conclusive, - an infliction of harm, - a casual interrelation between making unlawful decision and inflicting harm, - exhausting administrative remedies, - right to sue (locus standi) and justiciability. If the government substitute the terminated measure with a resolution identical in content issued according to the Crisis Act, then we may conclude that the claim for the compensation of damage is established only for the period, when the ban was based on the measure of the Ministry.

Considering the fact that the measures of the Ministry are based on the existence of the state of emergency, we suppose that the claims may be based on the Crisis Act as well. Then, the conditions according to the Act No. 82/1998 Sb. [12] would not be met.

There are relevant objections to an automatic conclusion that the termination of the measure of general nature gives grounds for a claim for the compensation of damage according to Act No. 82/1998 Sb. [12]. It is argued that there needs to be a decision (individual act) issued on the grounds of the tackled extraordinary measures of the Ministry of Health, only that can subsequently be terminated on the grounds of unlawfulness (e.g. decision on administrative delict). According to the judgement of the Supreme Court on 26.08.2014, sp. Zn. 30 Cdo 2767/2013 [24] 'the fact that a person believes to have suffered damage or harm as a consequence of violating the rights of another on the grounds of a previous unlawful decision, does not constitute the person to be naturally a party to the proceedings and the person does not obtain a right to sue (locus standi) according to Act No.82/1998 sb.)' [12].

4 Discussion and conclusion

Within discussion, the authors highlight two, in their opinion, key issues.

First, what legal regime will be imposed upon the cases, when a person suffered harm by a legal act of the state, however, it is not possible to apply Act No. 82/1998 Sb. [12], or any other particular piece of legislation. In general, the legal regulation of the compensation of harm according to the Civil Code, unfortunately, there is lack of sufficient specifying judicial decisions. The Supreme Court allowed in its judgement on 31.07.2019, c.j. 30 Cdo 2301/2017-144 [25], in compliance with the Charter of Fundamental Rights and Freedoms, that: *'if Act No. 82/1998 Sb. does not establish liability to damage during the performance of public authority (and its reimbursement), as it was in the reviewed case, it does not mean that the liable entities are without liability, but 'only' that their liability is not governed by*

Act No. 82/ 1998 Sb. [12], but it is governed by general provisions for 'the compensation of harm'.

The second issue relates to the problem indicated in the introduction, which bodies of public authority decide on constitutionality and legality of crisis measures. The way of administrative judiciary has already been indicated. In the case of exhausting of all remedial measures including an appeal before the Court of Appeal, the Constitutional Court may decide the case.

As far as the Constitutional Court is concerned (resolution on 22.04.2020, sp. zn. Pl. US 8/20 [26]), it clearly referred to the resolutions of the government, that : *'the declaration of the state of emergency is primarily an act of the application of constitutional law; it represents an 'act of governing' that has a normative effect, it is not principally a subject of the control of the Constitutional Court and it is 'reviewable' by primarily democratically elected political (non-judiciary) body that is the Chamber of Deputies. The act of declaring the state of emergency could be terminated by the Constitutional Court, if it were in conflict with the fundamental principles of a democratic rule of law. The government determined, in the resolution on the declaration of the state of emergency, the restricted rights by referring to section 5 paragraph a) to e) and section 6, the Crisis Act and to a further specifying resolution of the government. Although, such implicit determination is general and vague, it cannot be regarded as an excessive violation or misuse of constitutional law. It cannot be concluded from the given defect of form that the decision to declare the state of emergency is invalid, as excessively strict assessment of formal elements would completely omit the seriousness and extraordinary quality of the situation leading to the declaration of the state of emergency'.*

Upon analysing the above given documents, we may conclude that the entrepreneurs concerned need to assemble all the relevant sources for asserting possible claims for the compensation of damage from the state. In the first phase it should be considered to what extent a particular entrepreneur prevented the infringement of harm, whether all appropriate measures were adopted to prevent of minimalize it.

With respect to the missing judicial decisions on the specifics of measures in the period of this state of emergency, every entrepreneur must consider a regime for asserting claims, whether by the way of the Crisis Act or Act No. 82/1998 Sb. [12]. Cautiously, both ways are recommended concurrently.

Asserting a claim must be duly and timely to the statutory body specified in legislation. The claim must be determined as accurately as possible and it must be attached with evidence in writing or other form. In this sense, it is necessary to have a sufficient record of every internal decision and act of the entrepreneur (instructions to employees, closure of business, etc.), register all communication with business partners and customers, contracts and amendments of contracts reacting to the crisis situation. Even photography or audio recordings proving the conduct of the entrepreneur are recommended. It is always necessary to distinguish which procedures the entrepreneur adopted for prevention and which, on the contrary, were adopted for the purpose of mitigating the effects of state crisis measures.

As for the calculation of real damage and loss of profit, the claims must be certified with particular documentation – contracts, the amendments of contracts, bank statements, the records of meetings, withdrawals from contracts, annual balance-sheet reports and further accounting documents, ideally an expert's report to prove a loss of profit.

A particular application (notice, assertion) must be specific and comprehensible.

Acknowledgment

The paper was created within the project of the Internal Grant Competition 2020 of the Institute of Technology and Business in České Budějovice Nr. 8110-003.

References

1. Z. Firano, F. A. Fatine, The COVID-19: macroeconomics scenarii and role of containment in Marocco. *One Health.* **10** (2020)
2. N. Donthu, A. Gustafsson, Effect of COVID-19 on business and research. *Journal of Business Research.* **117**, pp. 284-289 (2020)
3. D. D. Farias, F. F. de Araujo, Will COVID-19 affect food supply in distribution centers of Brazilian regions affected by the pandemic? *Trends in Food Science & Technology.* **103**, pp. 361-366 (2020)
4. A. A. Malinovsky, D. M. Osina, E. N. Trikoz, State support for the Russian economy affected by covid-19 consequences. *Lecture Notes in Network and Systems.* **139**, pp. 283-287 (2020)
5. S. Varelas, N. Apostolopoulos, The Implementation of Strategic Management in Greek Hospitality Businesses in Times of Crisis. *Sustainability.* **12**(17) (2020)
6. N. Caruso, S. Mela, E. Pede, A resilient response to the social-economic implications of coronavirus. The case of Snodi Solidali in Turin. *Urban Research & Practice.* (2020)
7. T. A. Perkins, G. Espana, Optimal Control of the COVID-19 Pandemic with Non-pharmaceutical Interventions. *Bulletin of Mathematical Biology.* **82**(9) (2020)
8. R. Aldaco, D. Hoehn, J. Laso, M. Margallo, J. Ruiz-Salmón, J. Cristobal, R. Kahhat, P. Villanueva-Rey, A. Bala, L. Batlle-Bayer, P. Fullana-i-Palmer, A. Irabien, I. Vasquez-Rowe, Food waste management during the COVID-19 outbreak: a holistic climate, economic and nutritional approach. *Science of the Total Environment.* **742**(10) (2020)
9. M. Mehrl, P. W. Thurner, The Effect of the Covid-19 Pandemic on Global Armed Conflict: Early Evidence. *Political Studies Review.* (2020)
10. X. Neumeyer, W. S. Ashton, N. Dentchev, Addressing resource and waste management challenges imposed by COVID-19: An entrepreneurship perspective. *Resources, Conservation and Recycling.* **162** (2020)
11. ČESKO, 2000. Zákon č. 240 ze dne 9. srpna 2000 zákon o krizovém řízení a o změně některých zákonů (krizový zákon). In: Sbírka zákonů České republiky. Částka 73, s. 3475-3487. ISSN 1211-1244.
12. ČESKO, 1998. Zákon č. 82 ze dne 15. dubna 1998 zákon o odpovědnosti za škodu způsobenou při výkonu veřejné moci rozhodnutím nebo nesprávným úředním postupem a o změně zákona České národní rady č. 358/1992 Sb., o notářích a jejich činnosti (notářský řád). In: Sbírka zákonů České republiky. Částka 31, s. 5058-5063. ISSN 1211-1244.

13. ČESKO, 1998. Ústavní zákon č. 110 ze dne 29. května 1998 Ústavní zákon o bezpečnosti České republiky. In: Sbírka zákonů České republiky. Částka 39, s. 5386-5387. ISSN 1211-1244.

14. ČESKO, 2000. Zákon č. 258 ze dne 11. srpna 2000 zákon o ochraně veřejného zdraví a o změně některých souvisejících zákonů. In: Sbírka zákonů České republiky. Částka 74, s. 3622-3662. ISSN 1211-1244.

15. ČESKO, 2004. Zákon č. 500 ze dne 24. června 2004 zákon správní řád. In: Sbírka zákonů České republiky. Částka 174, s. 9782-9844. ISSN 1211-1244.

16. ČESKO, 2002. Zákon č. 150 ze dne 21. března 2002 zákon soudní řád správní. In: Sbírka zákonů České republiky. Částka 61, s. 3306-3330. ISSN 1211-1244.

17. F. Korbel, In: F. Ištvánek, P. Simon, F. Korbel, *Zákon o odpovědnosti za škodu způsobenou při výkonu veřejné moci rozhodnutím nebo nesprávným úředním postupem: Komentář*. Praha: Wolters Kluwer. Komentáře (Wolters Kluwer ČR) (2017)

18. ČESKO, 2006. Zákon č. 160 ze dne 27. dubna 2006 zákon, kterým se mění zákon č. 82/1998 Sb., o odpovědnosti za škodu způsobenou při výkonu veřejné moci rozhodnutím nebo nesprávným úředním postupem a o změně zákona České národní rady č. 358/1992 Sb., o notářích a jejich činnosti (notářský řád), ve znění pozdějších předpisů, zákon č. 201/2002 Sb., o Úřadu pro zastupování státu ve věcech majetkových, ve znění pozdějších předpisů, a zákon č. 40/1964 Sb., občanský zákoník, ve znění pozdějších předpisů. In: Sbírka zákonů České republiky. Částka 55, s. 1950- 1952. ISSN 1211-1244.

19. Rozsudek NS ze dne 8. 8. 2017, sp. zn. 30 Cdo 3292/2015.

20. Rozsudek NS ze dne 27. 10. 2015, sp. zn. 25 Cdo 3444/2013.

21. Rozsudek NS ze dne 11. 9. 2018, sp. zn. 30 Cdo 3079/2016.

22. Rozsudek NS ze dne 11. 9. 2019, sp. zn. 30 Cdo 1642/2018.

23. P. Vojtek, *Odpovědnost za škodu při výkonu veřejné moci*: *Komentář*, 3. vydání, C. H. Beck, Praha (2012).

24. Rozhodnutí NS ze dne 26. 8. 2014, sp. zn. 30 Cdo 2767/2013.

25. Rozsudek NS ze dne 31. 7. 2019, č. j. 30 Cdo 2301/2017.

26. Usnesení Ústavního soudu sp. zn. Pl. ÚS 8/20 ze dne 22. 4. 2020.

Subjective evaluation of socio-economic situation of residents of the South Bohemian Region and factors impacting it

Alena Kajanova[1,*] and *David* Urban[2]

[1]Institute of Technology and Business, Faculty of Corporate Strategy, Okruzni 517/10, 37001 Ceske Budejovice, Czech Republic
[2]Department of Social Work, College of Polytechnics Jihlava, Tolsteho 16, Jihlava, Czech Republic

Abstract. In our entry we focus on subjective evaluation of the socio-economic situation of residents of the South Bohemian Region. We explore and describe factors as well as contexts contributing to such subjective evaluations. The research was conducted via a survey, which made use of a self-designed questionnaire consisting of primarily closed scale questions focusing on areas including evaluating the socio-economic situation, the ability to manage own income. The sample consisted of 700 questionnaires. The results of the research show that 43.4% of respondents have average income management skills and that the ability to manage income is statistically independent from age, education level, or sex of the respondent. Employment also does not statistically influence the ability. As part of subjective evaluation of poverty/wealth, the respondents feel to be middle class (29.1%), the extremes were selected seldomly. Only one percent evaluated their situation as poverty; 0.1% saw perceived themselves as wealthy.

Keywords: socio-economic situation, poverty, social status, economic dimension

1 Introduction

In our entry, we focus on subjective evaluations of socio-economic situations. Such subjective evaluations are impacted by several factors (education level, employment, etc.) [1;2; 3]. On the other hand, it is obvious that the socio-economic situation of an individual reflects into other areas of life – health, free time activities, etc. Wilkinson, Marmot [4] state that e.g. poor social and economic circumstances have a life-long impact on health. Persons who are lower on the social ladder, are usually twice more at risk of a severe illness and premature death than persons at the top of the ladder [4]. Results from other published works also point out that population groups with lower socio-economic status are proven to be in worse health, which was not only subjectively stated but also objectively research – their risk of morbidity and mortality is higher. In deprived areas, health of the population is worse but

* Corresponding author: ali.kajanova@email.cz

the use of heath care is higher despite the quality of the health care services in the area being lower than is areas with high socio-economic stats [5]. Considering these information, we must focus on the question of socio-economic status of individuals and reflect on it.

9.6% of the inhabitants of the Czech Republic were at risk of income poverty in 2018 [6]; in 2019, this rate was at 10.1% and therefore impacts one in ten Czech residents [7]. Average annual net income per person in a household was 182,000 CZK in 2017, which was 10,000 CZK more than in 2016. The majority of households can manage their income relatively easily or only with small complications [6]. The income situation of households has improved over the past few years – evidenced by the positive value of the average annual growth rates of their net monetary income per person. This trend is also supported by older data on household income from 2015 that clearly show that the average annual growth coefficient for the period 2011-2015 was 102.8%. On average, household incomes nominally increased each year by 2.8%. After adjusting for average inflation, the income increased by 1.5% [8].

In October 2019, 52% Czech residents evaluated their current economic situation as good, 12% as bad, and 35% as neutral. The data was published by the Public Opinion Research Centre of the Czech Academy of Sciences. The optimism in evaluating current economic situation increases according to researchers alongside increasing satisfaction with own life, political situation, and level of highest achieved education. Highly qualified experts or managers, residents of Prague, residents of the South Moravian Region, and residents of cities with population higher than 80,000 evaluate their economic situation more positively [9].

On the other hand, we must address risks and fears – 17.6% of households have large issues with managing their income. Single mothers and women older than 65 who live alone struggle with their income [6].

The highest percentage of households that struggled with managing their income were unemployed persons (49.5%); however, this percentage also decreased by up to 6.3%. Households of self-employed persons struggled with their income only in 8.5% of cases in 2019, which is a 3.3% decrease compared to 2018 [4]. Fear of the global economic crisis is perceived as a real threat for the Czech Republic by 39% of Czech residents [10].

When it comes to subjective perception of socio-economic situation, 32% of Czech residents perceived their family as poor in 2016 according to a STEM study, 7% were entirely sure that their family was poor. 49% of persons with primary education and 56% of unemployed persons perceived their family to be "poor", as well as 38% of persons with vocational education and 35% of senior citizens. 49% of divorced respondents and 44% of one-person households more often also see themselves as poor. The feeling of poverty depends on the household's assets. Regardless, it is interesting that more than a third of persons with the overall value of assets under 300,000 CZK does not regard themselves as poor. On the other hand, almost a fifth of persons with assets worth more than two million CZK considers themselves poor. These facts prove the subjectivity of poverty [11].

When reflecting on the perception of the socio-economic situation, we must also consider the rate of debt of Czech households. According to the Czech National Bank (information from the system of ARAD timelines [12]), the rate of debt of Czech households in three researched indicators (consumption, living, other) keeps growing.

Whereas the household indebtedness compared to household income was only 55% in 2009 in the Czech Republic in comparison to 96% in the EU-27, the debt of Czech households when compared to the volume of their net financial actives was higher that the European average (56% in comparison to 53% for EU-27). The indebtedness can also be compared with the overall wealth of Czech households defined not only by the value of their financial and tangible fixed assets in the form of apartments and houses (mostly acquired via a loan). In comparison with such wealth, the rate of indebtedness of Czech households was approximately 17 percent in 2010. The comparison with European average is not possible in this case since the state of tangible fixed assets is calculated for only five more European countries [13].

The objective of the entry is to describe the subjective evaluation of the socio-economic situation of residents of the South Bohemian Region and the factors impacting it.

2 Methodology

The research was conducted via a survey, which made use of a self-designed questionnaire consisting of primarily closed scale questions focusing on areas including evaluating the socio-economic situation and the ability to manage income.

The representative research sample consists of quota sampling [14]. The quotas were determined based on sociodemographic indicators, i.e. gender and age, by random selection from among approached residents of the city České Budějovice and neighboring municipalities who were older than 15. The amount of the respondents was N=700. In České Budějovice, there were 77,000 inhabitants in the age group in question at the time of the study. Thus, the research sample comprised of 1% of the population. The data matrix was statistically processed in the SPSS program. Social support was processed via the Chi-squared test, the p-value was 95%.

3 Results

One of the methods of determining the subjective evaluation of socio-economic situation is the question of managing income. Respondents evaluated their ability to do so as average (43.4%); 22.1% of respondents managed their income well; and 14.7% struggled to manage their income. Only 12% of respondents stated that they can save some of their income (Table 1). The income (Table 2) does not correlate with the ability to manage income (p=0.263; r=0.422).

Table 1. How do you manage your income?

	Frequency	%
Not at all	7	1.0
Poorly	103	14.7
Average	304	43.4
Well	155	22.1
I can save up	84	12.0
I do not know	6	0.9
No answer	33	4.7
In total	692	98.9

Source: Own processing.

Table 2. Net monthly income of the household of the respondent

	Valid	Frequency	Percent
	did not answer, does not know	69	9,9
	under 850 €	77	11,0
	851 – 1249 €	47	6,7
	1250– 1649 €	115	16,4
	1650 € and more	81	11,6
Total		389	55,6
Missing System		311	44,4
Total		700	100,0

Source: Own processing.

How respondents managed their income statistically does not correlate with age (p=0.097) or education level (p=0.451). It does also not correlate with sex (p= 0.747) or even with the un/employment of the respondent (p=0.430).

Another way to determine the subjective socio-economic status is a self-evaluation scale on which respondents can choose values between 1 to 10 – 1 symbolizing "very poor" and 10 symbolizing "very wealthy". The most common answer was the average value 5 (29.1%), as well as the values 6 (17.7%) and 4 (17.6%). The average vale was 5.49; the standard deviation was 21.8. Extreme values were seldomly selected – extreme poverty in 1% of cases, and extreme wealth in 0.1% of cases. Such evaluations statistically do not correlate with age (p=0.454) or education level (p=0.236), or (as in the case of previously tested variable) with sex (p=0.788), and un/employment (p=0.688).

Such a self-evaluation scale statistically correlates with how respondents manage their income. It is connected to their subjective evaluation of their asset situation – the respondent could choose on a scale from 1 to 10. 1 symbolized "very poor" and 10 symbolized "very wealthy". The better the self-evaluation of the status of the respondents, the better they manage their income (p=0.000; r=0.229). Also, the better the management of income, the higher the satisfaction with their financial situation (p=0.000; r=0.681). It is interesting that the variable does not correlate with the un/employment situation (p=0.636), see Table 3, since both – the financial situation and the employment situation – correlate (p=0.000; r=0.211).

Table 3. Correlation matrix – relation between subjective status evaluations and selected variables

		How do you get along with from that income?	which describes to your asset situation	Employment situation	Financial situation
How do you get along with from that income?	Pearson Correlation	1	.229**	.018	.232**
	Sig. (2-tailed)		**.000**	**.636**	**.000**
	N	692	675	676	681
which describes to your material situation	Pearson Correlation	.229**	1	.078*	.423**
	Sig. (2-tailed)	.000		.044	.000
	N	675	679	669	674
Employment situation	Pearson Correlation	.018	.078*	1	.211**
	Sig. (2-tailed)	.636	.044		.000
	N	676	669	681	680
Financial situation	Pearson Correlation	.232**	.423**	.211**	1
	Sig. (2-tailed)	.000	.000	.000	
	N	681	674	680	686

Source: Own processing.

4 Discussion and conclusion

The results of our research show that 43,4% of respondents are average at managing their income. When choosing between the polar opposites of "managing well" and "managing poorly", the respondents tend to choose the positive trait – 22.1% of respondents evaluate that they manage their income well; 14.7% evaluate their income management skill as poor. These results are similar to the results presented by the Czech Statistical Office for the entire Czech Republic that state that 17.6% of Czech residents struggle with managing their income [6].

The results were similar (average) for the question of evaluating the subjective economic status, for which respondents tended to choose average values – "5 points" (on the scale from 1 to 10). Poverty is one of the fundamental global issues [15] that additionally influence economic decision-making [16].

The results of our research show that the ability to manage income statistically does not correlate with any of the researched variables (age, education level, sex, un/employment) or with the net income of the household. So far, there is no Republic-wide study for this issue. Only partial outputs are available, such as the study on the differences between urban areas and the countryside [17; 18]. The results of similar studies show however that the income or socio-economic status do not impact the subjective evaluation of wealth and affluence [19]. Macroeconomic indicators such as GDP or economic crises do however impact such evaluation [20].

We agree with the statement from Czarenski [21] that it is necessary to use an interdisciplinary approach to the researched issue and to focus on the cooperation betweem economic and other humanities in practice.

References

1. P. Beno, P, Stukovsky, L. Cintulova, M. Samohyl, Analysis of the povery risk European countries. *Acta Missiologica* [online]. **14**(1), pp 92-101 Available from : https://apps-webofknowledge-com.ezproxy.techlib.cz/full_record.do?product=WOS&search_mode=GeneralSearch&qid=5&SID=D27kQRmY6S7GhnzfnpF&page=1&doc=3&cacheurlFromRightClick=no (2020)

2. J, Benard, Community Satisfaction in Czech Rural Communities: A Multilevel Model. *Sociologie Ruralis* [online] **55**(2) pp 205-226. Available from : https://apps-webofknowledge-com.ezproxy.techlib.cz/CitedFullRecord.do?product=WOS&colName=WOS&SID=D27kQRmY6S7GhnzfnpF&search_mode=CitedFullRecord&isickref=WOS:000352587100006&cacheurlFromRightClick=no (2015)

3. B, J, L, Berry, A, Okulicz-Kozaryn, An Urban-Rural Happiness Gradient. *Urban Geography* [online] **32**(6) pp 871-883. Available from : https://apps-webofknowledge-com.ezproxy.techlib.cz/CitedFullRecord.do?product=WOS&colName=WOS&SID=D27kQRmY6S7GhnzfnpF&search_mode=CitedFullRecord&isickref=WOS:000295125800007&cacheurlFromRightClick=no (2011)

4. R. Wilkinson, R., M. Marmot, *Social determinants of health: the solid facts*. 2nd ed. Copenhage: World Health Organization (2003)

5. A. Šplíchalová, H. Šlachtová, P. Fejtková, H. Tomášková, P. Polaufová, P. *Ukazatele zdravotního stavu a indikátory socioekonomické deprivace v epidemiologických studiích* [online]. Available from: https://www.zuova.cz/Content/files/ukoncene-projekty/sesindexy007clanek.pdf (2020)

6. ČSÚ. *Příjmová chudoba ohrožuje necelou desetinu obyvatel* [online]. Available from: https://www.czso.cz/csu/czso/prijmova-chudoba-ohrozuje-necelou-desetinu-obyvatel (2019)

7. ČSÚ. *Příjmy a životní podmínky domácností* [online]. Available from: https://www.czso.cz/csu/czso/prijmy-a-zivotni-podminky-domacnosti-2019 (2020)

8. M. Brázdilová, Domácnosti vycházejí se svými příjmy lépe. *Statistika&My*. Magazín Českého statistického úřadu [online]. **17**(6) Available from: https://www.statistikaamy.cz/2017/06/26/domacnosti-vychazeji-se-svymi-prijmy-lepe/ (2017)

9. CVVM. *Příznivé hodnocení ekonomické situace zůstalo na maximu* [online]. Available from: https://www.investicniweb.cz/news-cvvm-priznive-hodnoceni-ekonomicke-situace-zustalo-na-maximu/ (2019)

10. CVVM. *Veřejnost o svých obavách z různých hrozeb a o očekávání do budoucna – listopad 2019* [online]. Available from: https://cvvm.soc.cas.cz/media/com_form2content/documents/c2/a5079/f9/po200106.pdf (2020)

11. STEM. *Subjektivní hodnocení finanční situace vlastní domácnosti se v české populaci zlepšuje* [online]. Available from: https://www.stem.cz/subjektivni-hodnoceni-financni-situace-vlastni-domacnosti-se-v-ceske-populaci-zlepsuje/ (2016)

12. ČNB. *Systém datových řad ARAD* [online]. Available from: https://www.cnb.cz/cnb/STAT.ARADY_PKG.VYSTUP?p_period=1&p_sort=2&p_des=50&p_sestuid=1538&p_uka=1%2C2%2C3&p_strid=AAD&p_od=200512&p_do=201912&p_lang=CS&p_format=0&p_decsep=%2C (2020)

13. ČSÚ. *Životní podmínky českých domácností* [online]. Available from: https://www.czso.cz/csu/czso/zivotni_podminky_ceskych_domacnosti (2011)

14. Disman. *Jak se vyrábí sociologická znalost: Příručka pro uživatele.* Praha: Karolinum (2002)

15. P. Novacek, The Sustainable development phenomeon : A way out of the global problems of the first half of the 21st century ? *Acta Missiologica* [online]. **9** (2) pp 33-50. Available from : https://apps-webofknowledge-com.ezproxy.techlib.cz/CitedFullRecord.do?product=WOS&colName=WOS&SID=D27kQRmY6S7GhnzfnpF&search_mode=CitedFullRecord&isickref=WOS:000421910300004&cacheurlFromRightClick=no (2015).

16. M. Adamkovic, M. Martoncik, A Review of Consequences of Poverty on Economic Decision-Making: A Hypothesized Model of a Cognitive Mechanism. *Frontiers in Psychology* [online]. **8**(1) a.n. 1784. Available from : https://apps-webofknowledge-com.ezproxy.techlib.cz/CitedFullRecord.do?product=WOS&colName=WOS&SID=D27kQRmY6S7GhnzfnpF&search_mode=CitedFullRecord&isickref=WOS:000412710500001&cacheurlFromRightClick=no (2017)

17. Hubatkova, B, Older adults' well-being by the type of residential setting: The case of small municipalities in the Czech Republic. *Journal of Rural studies* [online] **72**(1) pp 196-204. Available from : https://apps-webofknowledge-com.ezproxy.techlib.cz/full_record.do?product=WOS&search_mode=GeneralSearch&qid=5&SID=D27kQRmY6S7GhnzfnpF&page=1&doc=6&cacheurlFromRightClick=no 2019)

18. J, Bernard, European Countryside. *View Web of Science ResearcherID and ORCID* [online] **10**(2) pp 2018-0012. Available from : https://apps-webofknowledge-

com.ezproxy.techlib.cz/CitedFullRecord.do?product=WOS&colName=WOS&SID=D27kQRmY6S7GhnzfnpF&search_mode=CitedFullRecord&isickref=WOS:000438868500002&cacheurlFromRightClick=no (2018)

19. J. D. Fang, B. R. Chang, Socioeconomic Status Effects on Subjective Well-being of College Students: The Mediating Effect of Negotiable Fate. *Proceedings of The Second Summit Forum of China's Cultural Psychology*. pp. 181-186 (2016)

20. G. Gonza, A. Burger, Subjective Well-Being During the 2008 Economic Crisis: Identification of Mediating and Moderating Factors. *Journal of Happiness Studies*. **18**(6), pp. 1763-1797 (2017)

21. P, Czarnecki, Social pedagogy, social policy and the system of care and social assistance- How to keep human dignity at the center of the discussion ? *Acta Missiologica* [online]. **12**(1) pp 54-62. Available from : https://apps-webofknowledge-com.ezproxy.techlib.cz/CitedFullRecord.do?product=WOS&colName=WOS&SID=D27kQRmY6S7GhnzfnpF&search_mode=CitedFullRecord&isickref=WOS:000437065300006&cacheurlFromRightClick=no (2018)

Business valuation in times of crisis

Tomas Krulicky[1], Jakub Horak[1,*], and *Kristyna* Skulcova[1]

[1]Institute of Technology and Business, School of Expertness and Valuation, Okruzni 517/10, 370 01 Ceske Budejovice

Abstract. Business valuation is very important and complex process, mainly in the current global situation. To determine the company value, we will base our calculations on an analysis of the internal environment of the company, which is affected by macroeconomic and microeconomic influences. The aim of this paper is to determine the value of a selected company using FCFF in an environment of negative and unplanned interventions in the economy. The main source for determining the value of the company is data obtained from the financial statements and annual reports of the limited liability company Pepco, s.r.o. from the period of 2013-2018. The method applied to the selected company is the discounted cash flow method, which is one of the most popular and most frequently used earnings methods. In the first years before the company became known on the Czech market, the value of the FCFF was more often than not in negative numbers. This also affected the value of the business. During next years, with the stable growth of the company, the value of the company also grew. We can expect that due to the global problem of COVID-19, the value of the company will decrease in 2020, but it will not have a very significant effect on its overall value.
Keywords: company value, cash flow, business valuation, gross domestic product, financial indicators, prediction

1 Introduction

Business valuation, or determining the value of a company, is a complex process in which various experts from various fields may participate. But why is the value of a company determined in the first place and for what purposes can it be used?

A business should know how it will continue to grow and develop. It can lead to a decision whether to offer the company up for sale, liquidation or a merger. If the owner of a company decides for merger, they should be well aware of its value, so that they know whether the merger is advantageous for then or not. To determine this value, we will base our calculations on an analysis of the internal environment of the company, which is affected by macroeconomic and microeconomic influences [1-4].

We can demonstrate this on the example of the current global situation, which is deeply affecting the global economic market. This current predicament is the disease COVID-19 (coronavirus), which has spread throughout the world. Due to the disease, many countries

* Corresponding author: horak@mail.vstecb.cz

have taken certain measures that have prevented some businesses from operating. This is a very difficult situation for self-employed individuals and businesses, as their main source of income is still limited. If they do not have sufficient financial reserves, they cannot keep the company running long-term (payroll, rent, etc.). This sudden problem will force businessmen to think and decide in which direction they will lead their company [5-6].

Buus [7] argues that the simulation of company value and capital structure in the presence of taxes, risk and growth also shows that there are unique optimal levers for each combination of the aforementioned factors. For this reason, it is important to understand the influences that affect the company and determine the value of the company during the decision-making period.

At present, we have many methods available. Choosing the right method to determine the value of a business can lead to a competitive advantage. One of these methods is the discounted cash flow method, which is one of the most widely used.

The aim of this paper is to determine the value of a selected company using FCFF in an environment of negative and unplanned interventions in the economy.

2 Literature research

The purpose of a company's financial valuation is a way of expressing the company's value using a monetary amount. It is a complex process that can involve many experts from different fields of profession. The company as such may be valuated as a whole, or only some of its parts can be valuated [8-10].

This value can be determined in various ways. These are interesting especially for investors, owners and creditors. The subject of the valuation is therefore the market price, which is referred to as the market value [11]. According to Kislingerová and Krabec [12], the valuation of a company takes place primarily using methods based on revenue analysis, methods based on the analysis of the current market situation and methods based on the valuation of company assets.

Companies should not just drift along with external influences and events, so planning is a relatively important activity. Vochozka and Machova [13] state that if companies plan ahead, they can be better prepared for various situations and thus more easily respond to them. Vrbka, Rowland and Suler [14] also leans towards this, stating that it is necessary to take into account the macroeconomic factors that directly affect the company.

With global economic progress, planning is a prerequisite for survival, reducing the risk of future threats. Vrbka, Suler a Horak [15] also states that a company's financial plan contains all financial statements. These are, by default, the balance sheet, profit and loss statement and cash flow statement.

If the company is affected by external influences to such an extent that it reaches the so-called bottom, it is appropriate to consider the liquidation of the company. In this regard, it is possible to use liquidation methods. It tells us what proceeds can be obtained from the sale of all assets, the monetary amount of the company's liabilities and what the costs of liquidation will be. The liquidation value is determined in cases where the company can no longer continue its business activities in the future [16-18].

Behera [19] gives two basic directions today, which are among the most common methods of valuation. The first method is based on economic value added (EVA), which is linked to stock prices. However, its weakness is in future constant yields. The second of Behera's [19] methods is the discounted cash flow (DCF) method. Nyvltova and Marinic [20] also mention that these DCF models are the most used models in practice, and this is also confirmed by Aljifri and Ahomad [21].

Marik and Marikova [22] suggest that if we aim to obtain the market value of the company, then it is most appropriate to choose the DCF method. A variant of the DCF method is, for example, the FCFF model, which is cash flow available to suppliers after taking into account all operating costs, investments into inventory and equipment of the company. According to Mařík and Maříková [22], this is cash flow from operating activities exempt from the cost of capital, ie free cash flow for owners and creditors.

3 Data and methods

The main source for determining the value of the company is data obtained from the financial statements of the limited liability company Pepco, s.r.o. from the period of 2013-2018. An equally important source is the company's annual reports from the same time frame. The information obtained in this way is publicly available and this was decisive in the choice of the method used.

We distinguish three basic techniques of DCF calculation:

1. Entity method – society as a whole,
2. Equity method – the size of equity,
3. APV method – adjusted present value.

We will be dealing with the calculation in the form of Entity.

To calculate earnings before interest and tax (EBIT), we add together earnings before tax (EBT) and interest (I).

$$EBIT = EBT + I \tag{1}$$

After the subsequent operation, the final FCFF value will be calculated according to the following formula:

$$FCFF = EBIT * (1 - t) + A - INV \tag{2}$$

In order to calculate the total value of a company, it is necessary to find out the average cost of capital (WACC) in advance. The formula below will be used to calculate these costs.

$$WACC = r_f + r_{LA} + r_{POD} + r_{FS} \tag{3}$$

r_f – risk-free yield rate (can be determined on the basis of the yield on long-term government bonds in the Czech Republic, which are available at www.cnb.cz); r_{LA} – surcharge for the size of the company determined on the basis of the amount of borrowed capital:

- if the value of C> 3 bil. CZK, then $r_{LA} = 0$,
- if the value of C< 100 mil. CZK, then $r_{LA} = 5\%$,
- if the value of C is inbetween these two values, then $r_{LA} = \frac{(3-C)^2}{168.2}$.

r_{POD} – business risk surcharge (determined on the basis of EBIT/assets)

- if the value of EBIT/A < 0, then $r_{POD} = 10\%$,
- if the value of EBIT/A > r_d * C/A, then $r_{POD} = 0\%$,
- if the value of EBIT/A is greater than 0, but lower than r_d * C/A, then $r_{POD} = \left(\frac{r_d*C-EBIT}{r_d*C-10*A}\right)^2$.

r_{FS} – financial stability surcharge (determined on the basis of current liquidity; L1 = current assets/current liabilities) :

- if the value of L1<1, then $r_{FS} = 10\%$
- if the value of L1> industry average[7](XL) and at the same time >1,25, then r_{FS} = 0%,

- if the value of L1>1, but lower than XL, then $r_{FS} = (\frac{XL - \frac{OA}{KZ}}{XL-1})^2 * \frac{1}{10}$.

The total value of the company, also known as the market value of the company, is obtained by discounting the cash flows from operationally necessary assets.

$$Company\ value_1 = \sum_{t=1}^{n} \frac{FCFF_t}{(1+WACC)^t} \tag{4}$$

WACC – weighted average cost of capital,
FCFF – free cash flows to firm in year *t*,
n – number of years of expected operation.
The continuing value of the analyzed company can then be determined according to the following formula:

$$PH = \frac{FCFF_{n+1}}{WACC - g} \tag{5}$$

When planning, we assume that the company will exist indefinitely. However, it is not possible to plan cash flows for this assumption. In practice, this phase deals with a two-phase method. The future period will be divided into two phases. The first phase includes the period in which we are able to determine the free cash flow prediction for each year. The second phase contains the period from the end of the first phase to infinity. We determine the total value of the company using the following formula:

$$Company\ value = \sum_{t=1}^{n} \frac{FCFF_t}{(1+WACC)^t} + \frac{FCFF_{n+1}}{wacc - g} \tag{6}$$

g – expected growth rate of FCF throughout the second phase (determined on the basis of a constant growth rate based on GDP and inflation)

4 Results

The method applied to the selected company is the discounted cash flow method, which is one of the most popular and most frequently used earnings methods. This was also the reason for choosing the appropriate method for determining the value of the analyzed company.

4.1 Company value in method phase one

The first table shows the development of earnings before tax achieved over 6 years by the company Pepco s.r.o. Data from publicly available financial statements were taken to calculate EBT. All necessary data was taken from the entire period of operation of this business in the Czech Republic and are given in thous. CZK.

Table 1. EBT calculation for years 2013-2018

	2013	2014	2015	2016	2017	2018
Operating profit	-13,047	5,936	12,6915	507,972	592,490	668,762
Financial results	-3,347	-6,679	7,164	-13,464	-11,423	-8,017

Extraordinary profit or loss	2,862	-760	0	0	0	0
EBT	**-13,532**	**-1,503**	**134,079**	**494,508**	**581,067**	**660,745**

Source: Own processing.

The following table already quantifies the company's earnings before paying interest and taxes. Fluctuations over the years were influenced by many factors, especially operating costs in the first years and the increase in sales of products.

Table 2. EBIT calculation for years 2013-2018

	2013	2014	2015	2016	2017	2018
EBT	-13,532	-1,503	134,079	494,508	581,067	660,745
+ I	10	3,345	1,329	1	0	0
EBIT	**-13,522**	**1,842**	**135,408**	**494,509**	**581,067**	**660,745**

Source: Own processing.

Table 3 shows the evolution of free Cash Flow, which changes over the years as the company grows.

Table 3. FCFF calculation for years 2013-2018

	2013	2014	2015	2016	2017	2018
EBIT	-13,522	1,842	135,408	494,509	581,067	660,745
- taxes	8	29	1,155	1,982	1,587	8,911
+ depreciation	1,936	11,089	28,362	52,220	61,123	76,727
- change in working capital	n/a	-57,212	30,428	368,319	396,458	469,578
- investments	n/a	54,264	73,512	34,111	70,449	64,665
FCFF	**n/a**	**-98,574**	**58,675**	**142,317**	**173,696**	**194,318**

Source: Own processing.

Table 4 reflects the statistical and financial indicators recommended by the Ministry of Industry and Trade of the Czech Republic. Here is a quantification of the average cost of capital of Pepco s.r.o.

Table 4. Calculation of average cost of capital for years 2013-2018

	2013	2014	2015	2016	2017	2018
Risk-free interest rate	2.20%	0.67%	0.49%	0.53%	1.50%	2.01%
Surcharge for the size of the company	5%	5%	5%	3.7%	2.5%	1.4%
Surcharge for business risk	0%	0%	0%	0%	0%	10%
Financial stability surcharge	10%	10%	10%	0%	0%	0%
WACC	**15.02%**	**15.01%**	**15%**	**3.75%**	**2.49%**	**11.37%**

Source: Own processing.

The value of the company in the first phase is shown in the table below. When using the discount rate, the weighted average cost of capital (WACC) is derived, free cash flows (FCFF) are discounted to present value and then summarized.

Table 5. Calculation of the total value of the company in phase one

	2013	2014	2015	2016	2017	2018
FCFF	n/a	-98,574	58,675	142,317	173,696	194,318
WACC	15.02%	15.01%	15%	3.75%	2.49%	11.37%
Company value	**n/a**	**-85,709**	**51,021**	**137,173**	**169,476**	**174,479**
Total company value	**446,440**					

Source: Own processing.

4.2 Method phase two

The resulting value of free cash flows for the following period and the growth rate is determined from the expected year-on-year GDP growth and inflation. The GDP growth (2.4%) and inflation (2.6%) rates were determined using statistical data. In the final sum, the value is 5.2%. The growth rate of cash flows for the following period is therefore:

$$FCFF_{2019} = 194{,}318 * (1+0.052) = 204{,}422 \quad (7)$$

The continuing value of the company in the next year allows us to calculate the formula described in the methodological part. It is :

$$PH_{2019} = 204{,}422 / 6.17\% = 3{,}313{,}160 \quad (8)$$

4.3 Total company value

The total value of the company is given by the sum of the two previous results in individual steps.

$$446{,}440 + 3{,}313{,}160 = \mathbf{3{,}759{,}600\ thous.\ CZK} \quad (9)$$

5 Discussion

The free cash flow discount method was used to value the company. With the help of data about the company, the amount of free Cash Flow was determined during the company's operation in the Czech Republic, which was crucial for the company's valuation. In the first years before the company became known on the Czech market, the value of the FCFF was more often than not in negative numbers. This also affected the value of the business in that year. What is especially obvious and what significantly affected the application of the method is the number of periods included in the calculation phase. During these years, with the stable growth of the company, the value of the company also grew, rising to positive numbers several times. In the second phase of the method, the value of free cash flows for the following year was predicted, assuming the future values of GDP and inflation. It is obvious that the first years of calculation of the value of the company have a negative impact on the final value, because they point to the development stage of the company. Over the last 4 years, year-on-year growth has been relatively constant and the total value of the company is still increasing. The resulting value of Pepco, s.r.o. amounted to 3,759,600,000 CZK.

The FCFF method is one of the most widely used. However, its correct results depend on the appropriate setting of many indicators that are needed for the calculation. If we work with a longer time horizon and the prediction of future development, it is necessary to have the knowledge of predicting the appropriate values. This is not only cash flows and discount factors, but also the company's future financial plan. The longer the time horizon, the higher the probability that the predicted values will differ from real ones. Furthermore, we must not forget the influence of the microeconomic and macroeconomic environment on the company, which also affects the accuracy and calculation of the current value of the company [23-24].

For the selected company, the data was sufficient to calculate the total value. However, now, in the midst of an unexpected situation, it is difficult to choose the final value. The accuracy of the result depends mainly on the amount of data available to the evaluator for the selected company and on the possibilities of predictable, stable development of a market economy.

6 Conclusion

The aim of the paper was achieved. The method of determining the value of a company using discounted free cash flows can be used. This was demonstrated through a case study on calculating the value of a randomly selected company.

Estimating the future development of society and the economy is a very difficult task, where the deviation between real and predicted values also increases in a situation of negative and unplanned interventions in the economy. The DCF model is therefore limited by the risk of not reaching the predictions on which the valuation is based. This year, from the possible effects of coronavirus, preventive measures may have an effect on the stagnation of the Czech economy. It is now difficult to quantify damage estimates, as we still do not know how quickly the coronavirus will recede and whether it will not be back on the scene when we abandon preventive measures. In the forecast for 2020, which was set at the end of last year, the value of Czech GDP growth was estimated at 2%. Gradual supplementation of data and gradual deterioration of the situation now reduces the GDP growth estimate to 1.3%. Therefore, we can expect that due to the global problem of COVID-19, the value of the company will decrease in 2020, but it will not have a very significant effect on its overall value.

Findings regarding the applicability of the method for determining the value of companies can be helpful in the decision-making of owners and potential shareholders. If the owners want to know the value of their company in this difficult situation and have up-to-date and accurate information, I recommend valuating the business once again.

References

1. Z. Rowland, V. Machova, J. Horak, J. Hejda, Determining the market value of the enterprise using the modified method of capitalized net incomes and Metfessel allocation of input data. *Ad Alta: Journal of Interdisciplinary Research*, **9**(2), 305-310 (2019)
2. M. Vochozka, V. Stehel, Z. Rowland, Determining development of business value over time with the identification of factors. *Ad Alta: Journal of Interdisciplinary Research*, **9**(2), 358-363 (2019)
3. S. Haskova, P. Suler, V. Machova, T. Krulicky, Determining the price of the business share of a business in a group. *Ad Alta: Journal of Interdisciplinary Research*, **9**(2), 60-70 (2019)
4. M. Vochozka, Z. Rowland, P. Suler, The specifics of valuating a business with a limited lifespan. *Ad Alta: Journal of Interdisciplinary Research*, **9**(2), 339-345 (2019)

5. J. Horak, J. Vrbka, P. Suler, Support vector machine methods and artificial neural networks used for the development of bankruptcy prediction models and their comparison. *Journal of Risk and Financial Management*, 13(3) (2020)
6. V. Stehel, J. Horak, M. Vochozka, Prediction of institutional sector development and analysis of enterprises active in agriculture. *E+M Ekonomie a Management*, 22(4), 103-118 (2019)
7. T. Buus, A general free cash flow theory of capital structure. *Journal of Business Economics and Management*, **16**, 675-695 (2015)
8. T. Kliestik, K. Valaskova, G. Lazaroiu, M. Kovacova, J. Vrbka, Remaining financially healthy and competitive: The role of financial predictors. *Journal of Competitiveness*, **12**(1), 74-92 (2020)
9. K. Valaskova, T. Kliestik, M. Kovacova, Management of financial risks in Slovak enterprises using regression analysis. *Oeconomia Copernicana*, **9**(1), 105-121 (2018)
10. Kovacova, M., T. Kliestik, K. Valaskova, P. Durana, Z. Juhaszova, Systematic review of variables applied in bankruptcy prediction models of Visegrad group countries. *Oeconomia Copernicana*, **10**(4), 743-772 (2019)
11. V. Machova, J. Vrbka, Value generators for businesses in agriculture. *The 12th International Days of Statistics and Economics Conference Proceedings*, pp. 1123-1132 (2018)
12. E. Kislingerova, T. Krabec, Valuation methods for the advanced. A deeper look at selected issues. *Politická Ekonomie*, **61**(3), 439-444 (2013)
13. M. Vochozka, V. Machova, Enterprise value generators in the building industry. *SHS Web of Conferences - Innovative Economic Symposium 2017: Strategic Partnership in International Trade* (2017)
14. J. Vrbka, Z. Rowland, P. Suler, Comparison of neural networks and regression time series in estimating the development of the EU and the PRC trade balance. *SHS Web of Conferences: Innovative Economic Symposium 2018 - Milestones and Trends of World Economy (IES2018)* (2019)
15. J. Vrbka, P. Suler, J. Horak, Analysis of companies operating in the construction industry in the Czech Republic based on Kohonen networks - identification of leaders in the field. *Proceedings of the 7th International Conference Innovation Management, Entrepreneurship and Sustainability (IMES 2019)*, pp. 1017-1028 (2019)
16. N. Borah, L. Pan, J. C. Park, N. Shao, Does corporate diversification reduce value in high technology firms? *Review of Quantitative Finance and Accounting,* **51,** 683-718 (2018)
17. T. Dogru, M. Kizildag, O. Ozdemir, A. Erdogan, Acquisitions and shareholder´s returns firms: The effects of cash flow, growth opportunities, and franchising. *International Journal of Hospitality Management*, **84** (2020)
18. L. Raifur, A.F. De Sousa, Financial ratios and market value of company. *Revista Contabilidade e Controladoria-RCC,* **7**(1) 8-24 (2015)
19. S. Behera, Does the EVA valuation model explain the market value of equity better under changing required return than constant required return? *Financial Innovation*, **9** (2020)
20. R. Nyvltova, P. Marinic, *Finanční řízení podniku: moderní metody a trendy [Financial management of a company: Modern methods and trends]*. Prague: Grada Publishing (2010)

21. K. Aljifri, H. I. Ahmad, Choosing valuation models in the UAE. *Lecture Notes in Information Systems and Organisation*, **30**, 191-203 (2019)
22. M. Marik, P. Marikova, Capital structure - the problem of income business valuation. *Journal of Economics,* **63**(3), 259-277 (2015)
23. Y. Dong, Enterprise value evaluation based on FCFF model - A case analysis of Beixinyuan company. *Proceedings of the 2nd International Conference on Economics and Management, Education, Humanities and Social Sciences (EMEHSS 2018),* pp. 615-620 (2018)
24. P. Mielcarz, F. Mlinaric, The superiority of FCFF over EVA and FCFE in capital budgeting. *Economic Research*, **27**(1), 559-572 (2014)

Determining business value according to FCFE

Zuzana Rowland[1,*] and *Tomas* Stanek[1]

[1]Institute of Technology and Business, School of Expertness and Valuation, Okruzni 517/10, 37001 Ceske Budejovice, Czech Republic

Abstract. The large number of joint stock companies causes that investors strive for better awareness and information in the stock area. The objective of the contribution is to apply the method of discounted FCFE in the valuation of a specific company and to determine the parameters for the model application in practice. The data for the analysis are obtained from the database of Kofola ČeskoSlovensko a.s. The data are taken from the published annual reports for the years 2018 and 2019. First, a financial plan for a period of stability was created. Subsequently, the discount value of the company was calculated using the FCFE model. Finally, the calculation of the company value in the continuation phase was calculated. To achieve the research goal, the individual FCFE need to be discounted by the value of equity using the CAMP model. In conclusion, it can be assumed that the discount rate will remain stable and the prediction of the growth rate will not change. If there are no fundamental changes in the company e.g. in terms of dividend payment, investment activities, or changes in the working capital, it can be assumed that the value of the company should be stabilized until 2021.
Keywords: stock companies, discounted FCFE, financial plan, discount rate, dividend policy

1 Introduction

The current period of economic welfare, financial staility, and availability of financial markets encourages new investments [1]. One of the many possibilities are equity funds. Many joint stock companies issue a large volume of stock. The large number of joint stock companies causes that investors strive for better awareness and information in this area so that their investments are proven and safe if possible [2]. Therefore, the question is how to find out the most accurate information about the fair value of a company. A clear and reliable method from the area of business valuation is only the FCFE (Free Cash Flow to Equity) model, that is, free cash flow for owners [3-4].

It shall be said in the introduction that this model is used especially in companies which do not achieve high monetary value only due to their high indebtedness. The FCFE model is based mainly on the DDM (dividend discount models).

In some cases, the company shall not be evaluated on the basis of actual dividends paid only. A more realistic view is offered by a model that deals with a potential dividend paid,

* Corresponding author: rowland@mail.vstecb.cz

which includes retained earnings the company retains for the purposes of future investments. In the following chapters, it will thus be examined whether the company pays too high or too low dividends, how this could be measured, and subsequently, how much money can be paid to shareholders after the reinvestment activities are carried out.

The objective of the paper is to apply the method of discounted FCFE in the valuation of a specific company and to determine the parameters for the model application in practice.

2 Literature review

The valuation methods are currently a common method in everyday business activities. Companies, regardless of their size, are regularly valued in order to review their progress. Progress is an integral part of performing the future plans and determining the company´s actual competitive position, as Patelis and Grose [6] say. A special approach to property valuation is in the case of the companies, whose assets are divided into a certain number of shares. Joint stock companies represent a unique group of companies that decide for the so-called "corner solution" in the case of their capital structure. Each company has a different motivation for setting all shares related to free cash flows and growth opportunities [7].

As already mentioned in the introduction, the criterion for assessing the value of a project using a method of discounted cash flow is FCFE. This was confirmed by Chu, Wang, and Feng [8]. In terms of addressing the modelling method, it can be said that economic modelling is not suitable for using FCFE; on the other hand, FCFE is the best method to explain the financial aspect [9].

The nature of long-term management and investment in the company´s future leads to the distribution of funds to shareholders by means of repurchase of shares. The question to be asked is whether the repurchase of shares could be considered equity undervaluation. This is not only investment in the repurchase of shares but also a confirmation that regular payment of dividends to shareholders creates more funds for challenges the company receives [10]. Incorrect interpretation of capital budget may cause considerable problems in the redistribution of capital and structure of its components. The fact that there are at least ten methods to estimate the free cash flow and discount rate can be confusing and inaccurate in specific cases. When using these models, application risks may arise, without calculating the overall capital costs of the company. This problem leads to value-reducing investment decisions. FCFE or free cash flow to equity combined with the EVA model is the most efficient method of analysis [11].

Burns, McTiera, and Midnicka [12] argue that the effects of compensation and protection of shareholders in the policy of dividend payment, limited stock and dividend compensation are negative. If the motivation compensation is protected against dividends, the actual payment of dividends incresases. In countries with low investor protection, higher dividends are paid, in an effort to maintain the reputation of distrivution of excessive free cash flow. This reduced the dividends (since they increase the purchase), with regard to increased compensation of incentive stock.

Within this issue, several difficulties may appear. This is confirmed by Jackowicz, Meilcarz, and Wnuczak [13], who claim that the determination of the correct method of capital cash flow includes many doubts. These doubts may result in serious mistakes. The method of potential dividends leads to incorrect valuation of investments. The problems with free cash flow result from the mandatory payment of capital. However, if they repurchase the shares within five years, they show stagnant management [14].

The opinion of Jiménéz and Paucual [15] is very inspiring. They claim that the dividend yields are used for linking the stock market valuation to basic cash flows. Dividend yields thus must be stationary. Beschwitz [16] mentions the fact that the fied and fair value of a company may be affected e.g. by the sale of shares without the capital duty, as in the case of Germany after the introduction of the tax reform.

Naturally, all these situations are closely interlinked with the accounting sector. Various incentives within setting of shares, which are interlinked with free cash flows and growth opportunities, influence and affect the quality of accountings. Significantly different quality of accountings occurs in extreme situations of free cash flow compared to others. Fair capital strategy should not be generalized for all companies [17].

The state of differentiation of diversified companies is addressed by Nguyen, Cai, McCalgan [18], who believe that diversified companies retain much less cash that specialized companies. Companies adjust their cash flows to reach their target cash balance. Diversified companies show a higher free cash flow due to the same cash flows in the operation. Diversified companies save less money also due to lower debt and equity and the distribution of higher dividends.

Caban [19] emphasizes that the equity structure is affected by two factors: free cash flows and the so-called growth opportunities. The company shows higher costs compared to its competitors, if it has a high level of FCFE and low level of growth opportunities. The financial management usually does not value all assets at the same time, including the financial assets, which are not used for production. The value of equity, however, should be the same for boh methods of calculating FCF valuation [20].

Finally, it can be said that the company must be aware of its financial situation as well as its future plans and goals. In this respect, it is necessary to carry out regular valuation, which can take many forms.

The methodology for the FCFE model analyses the basic phenomena, theoretical and practical information leading to the best possible investigation of this issue and findings. Even in this FCFE model, there are clearly defined procedures, which will be addressed from the whole to their parts. The analysis of the creation of this model must comply with its system, which enables the separation of the significant and the insignificant, and identification of the properties of phenomena and processes. Specifically, in this issue, it is necessary to use the mathematical formulation of theoretical hypotheses which are subject to mathematical operations. This enables a precise formulation of phenomena and their mutual relations. It is necessary to follow the theoretical guidelines and carefully determine the examined company, which has a financial leverage for this model, since otherwise the model could not be applied. It is necessary to rely on a fixed discount rate, which is linked to the financial flow risk, to have sufficient information for all procedures. Companies which are tradable on capital market, pay dividends, o have relatively low financial leverage should be strictly avoided, as the FCFE model would not have informative value as e.g. dividend model. Finally, it is recommendable to use synthesis and to combine the findings from the aforementioned methods into a whole.

3 Data and methods

The data for the analysis were obtained from the database of Kofola ČeskoSlovensko a.s. [22]. The data are taken from the published annual reports for the years 2018 and 2019. The annual report is released after the end of accounting year in every calendar year, and it is a 180-page report containing the information provided by the management of the company, consolidated financial statement, and separate financial statement.

The data from the annual report will be used for the calculation of the company value using the discounted model FCFE. The table below show the individual items. The annual reports are from the years 2018 and 2019; it includes all products manufactured under the brand Kofola ČeskoSlovensko a.s. The data will be used for the calculation of FCFE (free cash flow for equity). This cash flow is riskier than the FCFF cash flow.

Table 1. Values for calculation

Items for calculating FCFE	Data from 2018 (thousand CZK)	Data from 2019 (thousand CZK)	Data from 2020 (thousand CZK)	Data from 2021 (thousand CZK)
Net profit	166,888,-	156,921,-	164,690,-	171,653,-
Investment	143,430,-	146,264,-	149,306,-	152,439,-
Change in working capital	92,021,-	-259,-	-278,-	-287,-
Repayment	1,426,-	1,425,-	1423,-	1422,-
Depreciation	159,181,-	155,789,-	153,722,-	152,685,-
Payment of preferred dividends	68,186,-	66,258,-	67,603,-	71,294,-

Source: Own processing according to Kofola ČeskoSlovensko a.s. [22].

From the annual reports of Kofola ČeskoSlovensko a.s. [22], there were selected data for the years 2018 and 2019 necessary for the calculation of business value using FCFE.

The calculation will be carried out using a mathematical method, that is, entering the relevant data in a numerical form in the formula. Subsequently, the calculation will be carried out and thus the required result (calculated business value) will be achieved.
The following formula explains in detail the calculation.

$$FCFE = \text{net profit} + \text{depreciation} - \text{investments} - \text{change in working capital} - \text{repayment} - \text{payment of preferred dividends} \quad (1)$$

The data used for the calculation are presented in Table 1 above. In the calculation the data from all four years (2018, 2019, 2020, 2021) will be considered. The data for the years 2020 and 2021 were obtained from the financial survey of the company carried out by Komerční banka. First, a financial plan for a period of stability will be created (the years 2018 and 2019). Subsequently, the discount value of the company will be calculated using the FCFE model. Finally, the calculation of the company value in the continuation phase will be calculated.

This calculation enabled the determination of the value of the company. The calculation based on the data from one year would be distorting; therefore, the values of all four years will be averaged for greater accuracy of results. However, the objective was the calculation of the discounted value of the company using the FCFE model. Since in this method, only equity is considered, the discount rate equals the costs of equity. It thus eliminates any financial and operating risk. This indicates the current indebtedness of the company.

To estimate the cost of equity, the CAPM model can be used. It was originally used for securities tradable on the capital market. For its calculation, the formula given below is used.

The data for the individual variables are obtained from the financial survey conducted by KB in 2018 and 2019 [23]. The discount rate will be calculated using the formula:

$$R_e = R_f + \beta_i * [E(R_m) - R_f] \quad (2)$$

A sum of risk-free rate and systematic risk of the i-th security multiplied by expected yield of market portfolio reduced by risk-free rate. The individual variables are R_e = discount rate, R_f = risk-free rate, β_i = systematic risk of the i-th security, and $E(R_m)$= expected yield of market portfolio.

This formula will be calculated for discounting FCFE and subsequently for the results in individual years. The data for the calculation were taken from Damodaran [21] and the financial survey of KB [23]. After the basic FCFE model for the valuation of the company is calculated, the discounted FCFE is calculated. The value of the FCFE in the first phase will be obtained by the sum of discounted FCFE for individual years in the period of stability. Furthermore, the terminal value will be calculated using the following formula: $PH=FCFE_t / (r_e\text{-}g)$. Another step is entering the variables in the formula for calculating the value of the company in the continuing phase: $PH / (1+R_e)^{T}$. After that, the value of the company from the first phase will be added up with the value of the company in the continuing phase. This way it will be possible to obtain the resulting value of the company.

In this case, the calculation of the following years includes the prediction of the growth rate of 1.4% up to the year 2023. The data for all years were provided by KB on the basis of the financial survey of Kofola ČeskoSlovensko [23]. The results will be then presented in a summary table with a necessary explanation. Subsequently, the value of the company will be assessed to gether with the comparison of year-on-year growth or decrease of value and final conclusion.

4 Results

The data analysis will be performed in accordance with the aforementioned methodology. The data from Table 1 will be used for the calculation using the following formula for the calculation of FCFE already mentioned in the methodology:

The result for 2018 is calculated as follows:

$$FCFE_{2018} = 166{,}888+159{,}181-143{,}430-92{,}021-1{,}426-68{,}186 = 21{,}006 \quad (3)$$

The result for 2019 is calculated as follows:

$$FCFE_{2019} = 156{,}921+155{,}789-146{,}264-(-278) -1{,}425-66{,}258= 99{,}041 \quad (4)$$

The result for 2020 is calculated as follows:

$$FCFE_{2020} = 164{,}690+153{,}722-149{,}306-(-278) -1{,}423-67{,}603= 100{,}358 \quad (5)$$

The result for 2021 is calculated as follows:

$$FCFE_{2021} = 171{,}653+152{,}685-152{,}439-(-287) -1{,}422-71{,}294= 99{,}467 \quad (6)$$

To achieve the set goal, it is necessary to discount the value of the FCFE calculated of the individual years. Discounting will be based on the calculated value of R_e, using the following formula (2).

Following the calculation of the FCFE value and R_e, the discounted FCFE can be calculated using this formula:

$$Equity\ value = FCFE_t / (r_e\text{-}g) \quad (7)$$

Below, you can see the calculation of *Re*. For the year 2018, R_e will be calculated as follows:

$$R_{e\,(2018)} = 7.12+0.79*(1.4-7.12) = 2.6\% \quad (8)$$

For the year 2019, R_e will be calculated as follows:

$$R_{e\,(2019)} = 7.12+0.79*(2.1-7.12) = 3.1542\% \quad (9)$$

For the year 2020, R_e will be calculated as follows:

$$R_{e\,(2020)} = 7.12+0.79*(2.1-7.12) = 3.1542\% \quad (10)$$

For the year 2021, R_e will be calculated as follows:

$$R_{e\,(2021)} = 7.12+0.79*(2.1-7.12) = 3.1542\% \quad (11)$$

For discounting, the values of FCFE and R_e must be known. As the values were calculated within the previous steps, the discounts can be calculated:

Table 2. Calculation of discount

	2018	2019	2020	2021
FCFE	21,006	99,041	100,358	99,467
Discount rate	2.6 %	3.1542 %	3.1542 %	3.1542 %
Discounted FCFE	20,474	93,077	91,431	87,848
Overall value	292,830			

Source: Own processing.

The terminal value *(TV)* calculated as:

$$FCFE_{2021}/\,(r_e\text{-}g) = 87{,}848/\,(3.1542\text{-}1.4) = 50{,}079 \tag{12}$$

The value of the company in the continuing phase is calculated as:

$$PH\,/\,(1+R_e)^{\,T} = 50{,}079\,/\,(1+\,0.031542)^3 = 45{,}624 \tag{13}$$

The resulting value of the company for the year 2021 = 292,830 + 45,624 = 338,454 thousand CZK.

5 Discussion

As already mentioned in the previous paragraphs, the object of the study was the company Kofola ČeskoSlovensko a.s. The objective of the paper was to determine the value of the company using the model of discounted FCFE and to determine the parameters for the application of this model in practice. The initial values for the calculation for the years 2018, 2019, 2020, and 2021 are presented in Table 1. The values in the table show a considerable change in the working capital of the company between 2018 and 2019, which affected the free cash flow for shareholders. Using a simple mathematical operation and entering the relevant values to the formula for calculating FCFE for individual years (the sum of net profit and depreciation reduced by costs of investments, change in working capital, repayment, and payment of dividends). To achieve the research goal, the individual FCFE need to be discounted by the value of equity using the CAMP model. Using the formula for calculating discount rate, where the risk-free rate (R_f) and beta (β_i) multiplied by the risk premium ($E(R_m)$-R_f). this model enabled to determine the values presented in Table 2, which show that the discount rate increased by 0.5542% between 2018 and 2019. This indicates that the value of discounted FCFE for the years 2019, 2020, and 2021 reduced and affected the value of the sum of the discounted FCFE in the period of stability. In the following step, the constant growth model and the growth rate were used to determine the terminal value of the company, which was necessary to calculate the value of the company in the continuing phase. The resulting value of Kofola ČeskoSlovensko a.s. for the year 2021 was determined on the basis of the sum of the discounted FCFE for the individual years and the value of the company in the continuing phase at CZK 338,454.

The FCFE model is applicable in the analysis of a company in practice provided that it is a company with a stable financing structure, so that given values can be reliably determined. There can be doubts whether to use the dividend discount models or FCFE. The decision is based on available information and data on the company [24]. FCFE has a better informative value and is applicable e.g. for the companies that cannot be traded on the capital market. These can be companies where there is a big difference between the dividends and FCFE, or public limited companies that do not pay dividends. The FCFE model can be used even in the situation where there is a significant number of majority shareholders who are allocated a large share of cash flows. These shareholders will most likely what that the cash flows are invested in profitable projects; for this reason, the dividends paid will be lower. The FCFE model can be used even in the case of low profit or loss [25].

6 Conclusion

The objective of the paper was to determine the value of a selected company using the model of discounted free cash flow to equity (FCFE) and to determine the parameters for its application in practice.

First, it was necessary to gather data from annual reports of the company Kofola ČeskoSlovensko a.s. and from publicly available financial surveys described in the methodology part of the paper. After obtaining all the necessary information, a basic formula for calculating FCFE and the value of equity was used to determine the value of discounted FCFE for all considered years (2018-2021). In another step, the formula for the calculation of the terminal value was used to determine the overall value of the company, which is estimated to CZK 338,454 for the year 2021. All the collected data are described, summarize, and presented in the Results of the paper. The performed analysis enabled the determination of the individual parameters that could be used by companies and valuers in practice. The basic parameters necessary for the application of the FCFE model are the net profit, depreciations, repayments, payment of dividends, and the degree of changes in the working capital. After obtaining the result on the basis of the formula used, the value of the company is significantly influenced by the discount rate and the expected growth rate.

The research has shown that to determine the value of a business, it is necessary to use a suitable model of analysis of free cash flows with regard to the situation of the company or the information available. The contribution of the work could be the specification of all parameters that can affect the behaviour of shareholders and managers when using the FCFE model in practice. Moreover, it can also help potential investors in decision making. A positive result of the FCFE model can be a signal for a company that decides to start its dividend policy or is going to invest the free cash flows in alternative projects where they can be valued.

In the future, it would be interesting to see how the global crisis caused by COVID 19 will affect the shares of the company and what influence their change will have on the value of the whole company.

In conclusion, it can be assumed that the discount rate will remain stable and the prediction of the growth rate will not change. If there are no fundamental changes in the company e.g. in terms of dividend payment, investment activities, or changes in the working capital, it can be assumed that the value of the company should be stabilized until 2021. The objective of the paper was thus achieved.

References

1. K. Valaskova, T. Kliestik, L. Svabova, P. Adamko, Financial risk measurement and prediction modelling for sustainable development of business entities using regression analysis. *Sustainability*, **10**(7) (2018)
2. K. Valaskova, T. Kliestik, M. Kovacova, Management of financial risks in Slovak enterprises using regression analysis. *Oeconomia Copernicana*, **9**(1), 105-121 (2018)
3. M. Vochozka, J. Horak, T. Krulicky, Innovations in management forecast: Time development of stock prices with neural networks. *Marketing and Management of Innovations*, **2020**(2), 324-339 (2020)

4. B. Groda, J. Vrbka, Prediction of stock price developments using the Box-Jenkins method. *SHS Web of Conferences - Innovative Economic Symposium 2017: Strategic Partnership in International Trade* (2017)
5. J. Vrbka, Z. Rowland, Stock price development forecasting using neural networks. *SHS Web of Conferences - Innovative Economic Symposium 2017: Strategic Partnership in International Trade* (2017)
6. G. Patelis N., CH. Grose, Employing valuation tools for public and private companies. The food sector in Greece. *Procedia Economics and Finance*, **33**, 491-505 (2015)
7. B. Ayash, The origin, ownership and use of cash flows in leveraged buyouts. *Quarterly Review of Economics and Finance,* **77**, 286-295 (2019)
8. X.L. Chu, S.G. Wang, K. Feng, Evaluating demand guarantee for PPP projects by real-option pricing. *ICCREM: Prefabricated buildings, Industrialized Construction, and Public-Private Partnership*, 155-162 (2017)
9. M. Lotti, G. Bonazzi, Profitability and financial sustainability analysis in Italian aquaculture firms by application of economic and financial margins. *American Journal of Agricultural and Biological Science*, **10**, 18-34 (2015)
10. Y.N. Bhagwat, M. DeBruine, Do firms shave capital expenditures when repurchasing shares? *Academy of Accounting and Financial Studies Journal*, **22**(6), (2018)
11. P. Mielcarz, F. Mlinaric, The superiority of FCFF over EVA and FCFE in capital budgeting. *Economic Research-Ekonomska Istra*, **27**, 559-572 (2014)
12. N. Burns, B. McTier, K. Minnick, Equity-incentive compensation and payout policy in Europe. *Journal of Corporate Finance,* **30**, 85-97 (2015)
13. K. Jackowicz, P. Mielcarz, P. Wnuczak, Fair value, equity cash flow and project finance valuation: ambiguities and a solution. *Managerial Finance,* **43**, 914-927 (2017)
14. W. Bessler, W. Drobetz, M. Seim, Equity issues and stock repurchases of initial public offerings. *European Financial Management,* **22**, 31-62 (2016)
15. L.G. Jiménez, L.B. Pascual, Enterprise valuation with track-record ratios and rates of change. *European Journal of Finance*, **16**, 57-78 (2010)
16. B. von Beschwitz, Cash windfalls and acquisitions. *Journal of Financial Economics*, **128**, 287-319 (2018)
17. D. Cabán, R. College, Managerial motivation behind an all-equity structure: The role of free cash flows and growth opportunities on accounting quality. *Academy of Accounting and Financial Studies Journal*, **24**(1), (2020)
18. T. Nguyen, C. Cai, P.Mc. Colgan, How firms manage their cash flows: An examination of diversification's effect. *Review of Quantitative finance and accounting* **48**, 701-724 (2017)
19. D. Caban, The motivational heterogeneity of an all-equity capital structure. *Review of Accounting and Finance*, **17**, 215-237 (2018)
20. P.D. Easton, G.A. Sommers, Two different ways of treating corporate cash in FCF valuations—and the importance of getting the cost of capital right. *Journal of Applied Corporate Finance*, **29**, 71-79 (2017)
21. Aswath Damodaran. *Damodaran* [online], Available at: http://people.stern.nyu.edu/adamodar/New_Home_Page/home.html (2020)
22. Kofola ČeskoSlovensko a.s. *ČR: Ministry of Justice of the Czech Republic* [online], Available at: https://or.justice.cz/ias/ui/vypis-sl-firma?subjektId=715045 (2018)

23. Ekonomický a strategický výzkum [Economic and strategic research]. *ČR: Komerční banka* [online], Available at: https://www.investicniweb.cz/Files/kofola-cs-doporuceni-kb-cervenec-2019/ (2019)
24. S. Haskova, P. Suler, V. Machova, T. Krulicky, Determining the price of the business share of a business in a group. *Ad Alta: Journal of Interdisciplinary Research*, **9**(2), 60-70 (2019)
25. M. Vochozka, V. Stehel, Z. Rowland, Determining development of business value over time with the identification of factors. *Ad Alta: Journal of Interdisciplinary Research*, **9**(2), 358-363 (2019)

A Basic Technical Analysis of Shares on the Example of a Specific Company

Jakub Horak[1,*] and *Jiri* Sulek

[1]Institute of Technology and Business in České Budejovice, Okružní 517/10, 370 01 České Budějovice, Czech Republic

Abstract. This paper deals with the technical analysis of The Coca-Cola Company (KO) shares. The first signal for the stock price movement was created on the basis of the intersections of moving averages, namely after intersecting the shorter average through the longer average. This signal signifies a drop in the share price. The same signal was generated after comparing the trend of the share price and the trend of the trading volumes. It is these particular trends that have a different direction. Consequently, this is an indicator for a decline in the share price. The last signal created a graphical pattern called symmetric bear triangle which also predicts a drop in the stock price. Only one indicator did not generate a signal for price increase or decrease, namely the RSI with a 14-day period. The examined data collection contains data from 2.1.2015 to 29.4.2020. At the same time, all presented results and predictions are based on the date of April 29, 2020. Finally, the use of the given indicators for different titles, assets or for different types of trade with different lengths is evaluated and proposed. The paper provides suggestions of measures for investors and speculators which should be applied before the start of the analysis.
Keywords: stock market, time series, prediction, technical indicators, volatility

1 Introduction

Considering the global predictions of the impending next global crisis, the subject of stock market analysis is very topical. Should a greater or smaller financial crisis occur, it is almost certain that the stock values of most companies will fall by tens of percent. And this is exactly the right time for both large and small investors or small speculators who can increase the value of their capital in this period of economic depression [1-2]. The valuation can be solved by both short positions and long positions directly on the share or by the respective option on the share. Another possibility is a speculation on a commodity, a currency pair, a precious metal etc. It important to be familiar with the mechanisms of the stock market not only during a crisis but also at other times.

At this time, a lot of information can be found for free without much delay. That is the reason why people's interest in this type of business is growing. There are many ways to

* Corresponding author: horak@mail.vstecb.cz

make money on stocks. They can be kept for a long time, which can generate passive income for the investor or increase the value of the invested capital. Quality stocks protect against inflation, and even trend more than inflation. Other investors may use the equity portfolio as a way to saving for themselves or for future generations. Another possibility is a short-term speculation on the movement of the stock price or stock index, in the time frame reaching from a few minutes to several months [3-4].

The analysis of the share time series is performed by a technical analysis, which represents the basic methodology for finding the optimal entry into the trade. It uses many indicators which include supports and resistance, trend periodical, the number, strength and type of open trades of individual positions and many others.

The advantage of stock trading is the unlimited potential of choosing a strategy, as well as a theoretically unlimited profit. On the other hand, it is possible to lose the whole invested capital in a few minutes. Another advantage is the fact that the speculator's result is not influenced by any direct colleague affecting the result. Everything is up to the speculator's or investor's ability to analyse, predict, experience, etc.

This paper will be dealing with the shares of The Coca-Cola Company (ticker KO). The reason for choosing this particular company is the strength of this brand on its respective market where it is the number one player and its strength on the global stock market as well. The quality and strength of the company is easy to present as the company ranks among the dividend kings. This means that for more than 50 years, the company has been increasing the number of dividends paid per share year-on-year. In the case of Coca-Cola, the company managed to keep this state for 57 years. The Coca-Cola shares are traded on the New York Stock Exchange (NYSE).

The aim of the paper is to find the optimal position to enter the trade, either long position or short position for swing trade, depending on the analysis of the historical trending of the particular stock and the trending of other dividend kings, indices and competitors in the same industry.

2 Literature Research

The first indicator that is examined by a stock trader without the need for employing mathematical procedures is a trend. It is either rising (bull) or falling (bear) [5]. To decide whether a trader will speculate on the increase (bull) or rather on a decrease (bear) in the share price. Trend recognition is a basic pattern of a technical analysis. Trends are very often repeated, either cyclically or seasonally [6]. The cyclical component of the trend is repeated after more than a year, the seasonal component, on the other hand, usually within one year. It is important to be aware of the fact that the price trend appearing on the price chart, be it rising or falling, is never endless. Every trend has its beginning and end. The end of the trend can be observed based on the combination of Japanese candles on the price chart [7]. The Japanese candle is one of the most popular price level representations in a chart. The candle itself contains four pieces of information: the price at the time of opening, the price at the time of closing, the minimum price in the time interval and the maximum price in the time interval. The Japanese candle appears in two colour variants depending on the rising or falling tendency of the price [5]. The trends are further divided into short-term and long-term ones [8-9]. It may occur that the limited trend is bearish but the long-term trend is bullish and the other way around.

One of the most used technical indicators is the moving average (MA). The moving average can be defined according to the required conditions. It is an average that calculates the past data serving as a foundation for the calculation of the average. The 50-day and 200-day simple moving average (SMA) are the two most used types. Other moving averages are

WMA - weighted moving average, KAMA - Kaufman's adaptive moving average, EMA - exponential moving average, LWMA - linear weighted moving average, etc. These averages have a modified calculation [10]. The moving average is used for a clearer detection of a trend which may seem confusing on a classic price chart, for example due to high price volatility [11]. Combining two moving averages with different periods creates a signal for opening a long or a short position (the same types of MA). For the long position, the signal is created when the MA with a shorter period intersects the MA with the longer period from the bottom. For a short position, the signal is generated when an MA with a shorter period intersects an MA with a larger period from above [9].

The RSI, or relative force index, is one of the very popular oscillators that ranks among the technical indicators. This indicator graphically represents the internal strength of the underlying asset. On this basis, it sends signals about the overbought or oversold state of a particular asset [12]. RSI works with past data and is one of the short-term to medium-term indicators [13]. RSI can detect deviations in market behaviour. At the same time, it operates with fixed indication levels and it is only up to the trader whether they use the signal at the first hint or whether they wait for a repeated signal [14]. The RSI values range from 0 to 100. The signal generation levels are set at level 30 (growth signal) and 70 (decrease signal) by default. The 30/70 levels are the most suitable to use for a 14-day RSI [14]. For 20-day RSI, the levels are adjusted to 40/60 [15].

Another factor frequently used for trading on the stock market is volatility which indirectly indicates how risky a specific underlying asset is. Volatility is a fragmentation of the market and it is directly linked to the amount of open positions. The bigger and more numerous the trades, the higher the volatility and, consequently, the potential risk [16-18]. The higher the volatility of an asset, the riskier the trade but also more profitable at the same time [19]. Volatility can be influenced by large investors, especially investment funds, states, hedge funds, etc. disposing of large amounts of money triggering price movements of a given asset [20]. This, on the other hand, can lure small retail investors and speculators away. Small speculators avoid great volatility as the significant movement of an asset can drain their capital in a few hours or days [21]. High volatility on the market occurs especially during a financial crisis and a recession where there is an increased risk of bankruptcy [20]. Another factor that increases uncertainty and market volatility is political uncertainty in both local and global areas, mainly for assets that are directly dependent on the behaviour of the state apparatus [22].

Currently, the opportunity to analyse the time series of exchange-traded assets using neural networks has widely spread. Market analysis based on neural networks is especially suitable for intraday trading or scalping. Neural networks can extract specific functions from a large amount of source data, without knowledge of past predictors and success indicators.

From the available sources, it is obvious that it is possible to perform a time series analysis using basic technical indicators that are suitable for beginners [10] but also for portfolio managers in various companies, option speculators but also for small experienced investors [8].

Three indicators will be used for the basic technical analysis of the time series in this paper, namely moving averages, RSI and trade volumes. For their simplicity and efficiency, these indicators are mentioned in the literature dealing with this issue the most often. Based on the use of these indicators, it will be possible to determine whether the trend in the development of the share price will continue, go sideways or reverse. Thanks to this, it is possible to determine the optimal share price for entering the trade, the type of trade and the length of the trade.

3 Data and Methods

The price movement in one trading day or one monitored interval is determined by four values that are most often graphically displayed in form of the Japanese candle (Figure 1). These values are: minimum price, maximum price, price at the time of opening and price at the time of closing.

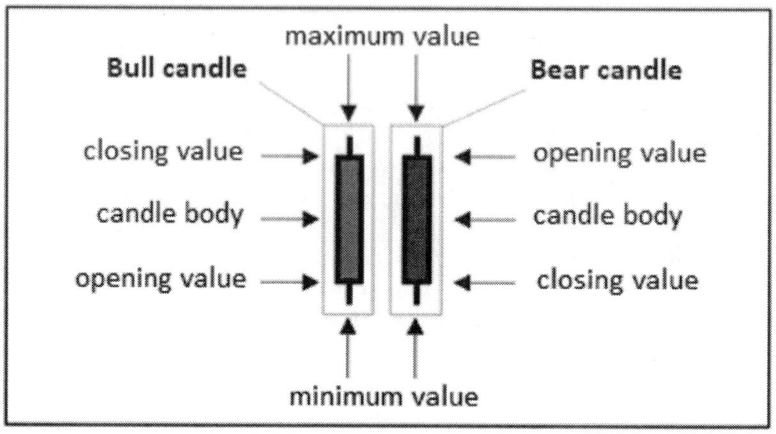

Fig. 1. The Japanese Candle
Source: Authors.

It is exactly the combination and size of these candles that determines the trend on the market. The data file was obtained from Yahoo Finance [23]. It contains data for the last 5 years, more precisely 1,339 stock exchange days. The start date is 2015/01/02 and the end date is 2020/04/29. The data file contains information about the sums that can be graphically represented by the Japanese candle. The graphic representation will show the development of the stock price from 2020/07/01. In addition to that, the data file contains information about the amount of open trading positions. All price levels are given in USD / 1 stock format. An example of the data structure is shown in the Table 1, with the data in the example stemming from one-month long intervals. All calculations will be performed from the daily data.

Table 1. Structure of the data file classification

Date	Open	High	Low	Close	Volume
29.04.2019	48.17	48.45	48.11	48.42	9,731,500
28.05.2019	49.60	49.67	49.03	49.10	18,614,000
27.06.2019	51.29	51.40	50.97	51.08	12,374,900
26.07.2019	53.34	54.31	53.19	54.17	11,451,900
27.08.2019	54.70	54.95	54.41	54.72	13,537,000
27.09.2019	54.56	54.64	54.00	54.31	7,939,900
28.10.2019	53.74	53.95	53.37	53.57	8,866,100
27.11.2019	53.93	54.10	53.60	53.95	9,595,200
27.12.2019	55.01	55.45	55.00	55.35	6,895,500
27.01.2020	57.21	57.85	57.13	57.48	14,359,900
27.02.2020	57.15	57.89	54.88	54.93	23,754,000
27.03.2020	43.66	44.03	42.40	42.81	23,297,900
27.04.2020	46.00	46.81	45.71	46.62	15,175,000

Source: Authors.

In order to perform a technical analysis, it is necessary to use a basic indicator which is the moving average. The 50-day, 100-day and 200-day simple moving averages are used to analyze The Coca-Cola Company stock by applying the following formula:

$$SMAn_p = \frac{1}{n} \sum_{t=p-n+1}^{n} c_i \tag{1}$$

Where:
SMA – simple moving average,
n – length of the moving average calculation period,
p – current position of the observed period,
c_i – the closing price of the share on the stock exchange day.

An exponential moving average (EMA) is very often used as well. The difference between EMA and SMA is the following: EMA puts more emphasis on newer data and "forgets" older data. EMA is, therefore, more sensitive to current market developments. EMA is calculated by the following formula:

$$EMAn_p = \left(\frac{2}{n+1}\right) * c_i - 1\left(1 - \left(\frac{2}{n+1}\right)\right) * EMAn_{p-1} \tag{2}$$

Where:
EMA – exponential moving average,
n – length of the moving average calculation period,
p – current position of the observed period,
c_i – closing price of the stock in the stock exchange,
$EMAn_{p-1}$ – moving average from the previous day.

Further information about a given asset can be read from the values of the relative strength index (RSI) which belongs to the group of leading indicators. This means that such indicator provides a signal even before the events on the market start to develop in a certain direction. At the same time, this indicator is measurable and belongs to the group of oscillators. Mathematically and graphically, RSI represents the oversold or overbought state of a given asset. It can also be used to confirm a particular trend. Another feature of the RSI indicator is the fact that under standard conditions, its development on the chart has a similar trend as the price chart of the examined asset. To calculate the RSI value, the following formula applies:

$$RSI = 100 - \left(\frac{100}{1 + RS}\right) \tag{3}$$

Where:
RSI – relative force index,
RS – the ratio of the average change in the asset price.

To calculate RSI, it is necessary to know the RS ratio calculated by the following relation:

$$RS = \frac{AVG\ gain}{AVG\ loss} \tag{4}$$

Where:
AVG gain – average increase in the asset price for the observed period,
AVG loss – average decrease in the asset price for the observed period.

As far as RSI is concerned, the following applies:

a) RSI <30 .... the market is oversold; signal for trend growth.

b) 30 <RSI <70 ...... the market is neutral; the trend continues.

c) 70 <RSI .... the market is overbought; signal for a downward trend.

For the purposes of this paper, the calculation is made with a 14-day period, which is also the most used period in general.

The movement of the stock price trend can be predicted from the volume of trades of a given stock. As trading volumes increase, so does the volatility of the asset price, and so does the risk of the trade. Dow Theory argues that volumes must confirm the trend. There are four trend relationships between trading volumes and asset price developments. Specifically, it describes four situations that are recognized and used even today [24]. These rules are:

Table 2. Relationship between trading volumes and asset price development

Actual Trend	Trading Volumes	Future Trend Prediction
Rising (bull) trend	Rising	Rising (bull) trend
Rising (bull) trend	Falling	Falling (bear) trend
Falling (bear) trend	Rising	Falling (bear) trend
Falling (bear) trend	Falling	Rising (bull) trend

Source: Authors.

From this model, it can be concluded that the increasing trade volumes confirm the trend while declining trade volumes stop and reverse the trend.

The trend has its imaginary boundaries which determine the already mentioned sideways direction, bull trend or bear trend. These boundaries are represented by support and resistance. Supports limit the trend from below and do not drop the price below this level. Resistance limits the trend from above and does not drop the price above this level. These two levels can go parallel, in opposite directions or towards each other. Based on the direction of these boundaries, a pattern of trend movement can be recognized and it can be predicted how and when the trend will change. In addition to that, these patterns help set the stop loss. On the price chart, the presence of any of the patterns will be examined, such as head-shoulders, bear or bull rectangle, rising or falling, or symmetrical triangle.

After the application of the above-mentioned methods, the results will be presented in a graphic form where moving averages and trading volumes will be illustrated. Furthermore, a separate graph for the RSI values will be created. If a pattern based on supports and resistances is indicated, it will be graphically highlighted in the price chart. Subsequently, the development of the stock price time series will be graphically indicated. Finally, an overview representing all the used indicators and predictions based on their value will be created.

4 Results

Technical indicators were calculated from the data set using the above-mentioned mathematical apparatus. The calculations were performed using Excel software from Microsoft. First, the moving average for its 50-day, 100-day and 200-day variant was calculated. For an easier presentation, all three averages were projected into a graph of the asset price. At the same time, all three averages are calculated for each day of the observed period and they change dynamically. It can be confirmed from the chart that the average with a shorter period reacts faster to a change in the price of an asset. Currently, the value of the averages with the shortest period is the lowest. With the increasing period, the value increases as well.

Next, the RSI indicator was calculated with a 14-day period, which is generally the most used one. The resulting model values are calculated for each day of the observed period in order to graphically display the trend of the RSI development. The limit for the overbought signal is set at 70, for the oversold signal then at 30. The resulting value amounts to approximately one month at a neutral value around 50.

Other values reflected in the chart are the volumes of trades executed in one day. These values were included in the basic data file and are marked as volume in the graph. The resulting display shows a clear declining trend in volumes.

The next indicator is not calculated mathematically, the values of the asset price and the development of trading volumes are represented only in a graphical way. The chart shows the opposite trend between the price of the asset and the development of volumes in the short term (2020/03/2020 – now).

Finally, the trend limits, one resistance and one support were projected into the graph. These boundaries converge over time forcing the price of the asset to trend sideways in an ever-smaller range. Such a push is classically followed by a burst of support or resistance, and the price moves quickly in one direction.

Fig. 2. The development of the KO stock price and the representation of technical indicators

Source: Authors.

5 Discussion

In the preceding chapter, five signals were evaluated which indicated the total of four signals for the short position. One result was at a neutral value, which indicates price movement to sideways or a slight trending which is not suitable for scalping, intraday or swing trading.

The first two signals are caused by the movement of the moving averages, the always shorter simple moving average intersecting the average with the higher period. This signal was confirmed in the case of moving the SMA200 and SMA100, but also in the case of moving the SMA100 and SMA50. Both SMA-based signals apply for the short position. Another indicator did not generate any trading signal, as the value of the RSI with a 14-day period stayed for 1 month around the value 50. According to this indicator, it is not appropriate to open any trading position now, as there is no indication of future movements in asset prices. Another signal was indicated on the basis of the movement development of the asset price and the movement development of the trading volumes, at the same time. With the help of a graphical representation of these waveforms, the opposite trend is noticeable in the short term. According to the Dow Theory [24], given the different course of these two trends, it is expected that the asset price trend will stop and reverse, which, in the case of KO, should turn down, thus supporting the short position. The last signal was indicated after the projection of support and resistance for the last two months. A symmetrical triangle was created on the price chart. Support and resistance force the price to trend sideways in an ever-smaller range. Such compression is followed by the rupture of one of these boundaries. The price of the asset began to form this triangle around 2020/03/01, with a sharp decline. The direction in which the asset enters the triangle is the one that tends to break through the resistance line and move sharply in that direction. In this case, there is a signal that the price will break through the support and will fall, i.e. a signal to open a short position. Over time, there is the possibility of developing resistance with a simple 50-day moving average. In the case of KO shares, four short signals out of five possible were created, so there is a great assumption of a decline in the share price. An overview of indicators and signals is clearly written in the table below.

Table 3. Indicators and signals of the KO share price development – an overview of the Coca-Cola company (KO) technical analysis

Indicator	Development	Signal
Simple moving average SMA200 – SMA100	SMA100 intersected SMA200 from above and falls slightly	Short
Simple moving average SMA100 – SMA50	SMA50 intersected SMA100 from above and falls steeply	Short
RSI index RSI14	One month on a neutral level at about 50	Do not trade
Price trend vs. Volume trend	Asset price rises but trade volumes fall	Short
Graphical pattern	Symmetrical bear triangle	Short

Source: Authors.

Although 80% of the signals indicate a fall in the price of an asset, it is important to keep in mind that the value of a stock is also affected by fundamental factors which may indicate the same as well as the opposite signals. It is always better to perform both a technical and a fundamental analysis. Especially in the time of a coronavirus pandemics, it is important to take into account fundamental factors. At the same time, it is appropriate to wait for the official publication of economic results for the first quarter of 2020 and to observe its effect on the share price. The results were published on April 21, 2020. Indicators that are more suitable for short-term speculation were used in the technical analysis of KO. In the case of an analysis with the intention of holding shares in the long run, other indicators would be more suitable.

Before performing a technical analysis, a speculator or investor must consider what type of trade he wants to open. If they want to open a position for scalping or for intraday trading, they use the above-mentioned indicators with a shorter period, for example RSI with period 9. Furthermore, for such a short trade, it is not appropriate to examine the ratio between moving averages. For such a short trade, it is more suitable to focus on the graphic patterns that create the price of the asset. On the other hand, for longer-term trades, the speculator will make more use of long periods for indicators and will put more emphasis on the fundamental analysis. It is also important to consider what level of risk the speculator wants to take and whether they will look for more volatile assets that generate higher profits but also losses. A more conservative speculator will use assets with lower volatility and will trade with lower leverage. For such type of trade, stock indices are a good option, for example in the form of ETF or option spread strategies.

6 Conclusion

The aim of this paper was to analyze the time series of the stock and then to predict its future development. In this paper, the shares of The Coca-Cola Company (KO) were analyzed, in which case five technical indicators were further examined. Their use is minutely described in the chapter concerned with data collection and methodology. Subsequently, the results of the analysis were presented and evaluated. Out of the five examined indicators, a signal for a decrease in price appeared four times creating a prediction of a decrease and fulfilling the goal of the paper. The suitability and effectiveness of using the indicators for technical analysis is confirmed by their high popularity with large speculators and investors or investment and hedge funds for decades. Another factor that adds credibility to the indicators is based on the large number of professional, scientific or conference papers that are available on professionally recognized databases, which is confirmed in the chapter presenting the literary research. The technical analysis based on the same indicators is also suitable for other stocks or for other types of assets, as all used indicators work only with the price of the asset. There is no limiting time frame for technical analysis determining when the analysis is to be performed. The values change dynamically over time so the asset can be analyzed at any time. In the future, it would be interesting to perform a technical analysis using the same indicators on different types of assets, repeatedly. Other assets include cryptocurrencies, currency pairs, commodities or options, and finally compare the effectiveness of applying the same analysis to different assets.

In conclusion, it can be stated that, on the basis of the performed analysis, the KO share expects a price decrease, which is predicted by four out of the five indicators that were used in the paper. The goal of the paper has been met.

Acknowledgements

This article was funded within the project SVV202007 "Creation of a research report for scientific research activities of students at the Institute of Technology and Business in České Budějovice".

References

1. M. Vochozka, J. Horak, T. Krulicky, Innovations in management forecast: Time development of stock prices with neural networks. *Marketing and Management of Innovations*, **2020**(2), 324-339 (2020)
2. P. Suler, J. Horak, T. Krulicky, Validation of the prediction of ČEZ stock prices. *Littera Scripta*, **13**(1), 194-210 (2020)

3. J. Vrbka, Z. Rowland, Stock price development forecasting using neural networks. *SHS Web of Conferences - Innovative Economic Symposium 2017: Strategic Partnership in International Trade* (2017)
4. B. Groda, J. Vrbka, Prediction of stock price developments using the Box-Jenkins method. *SHS Web of Conferences - Innovative Economic Symposium 2017: Strategic Partnership in International Trade* (2017)
5. H. Zhou, Y. He, Y.X. Jin, Modelling for forecasting of pattern recognition - based on comparison and analysis be-tween U.S. stock market and Chinese stock markets. *Proceedings of the 2016 International Conference on Applied Mathematics, Simulation and Modelling*, **41**, pp. 1-5 (2016)
6. D. Stramirovic, D. Sarvan, V. Miljkovic, S. Blesic, Analysis of cyclical behavior in time series of stock market returns. *Communications in Nonlinear Science and Numerical Simulation*, **54**, 21-33 (2018)
7. L.J. Chmielewski, M. Janowicz, A. Orłowski, *Advances in intelligent systems and computing*. Springer Verlag, **403**, pp. 641-647 (2016)
8. A. Kolkova, Indicators of technical analysis on the basis of moving averages as prognostic methods in the food industry. *Journal of Competitiveness*, **10**, 102-119 (2018)
9. M. Frömmel, K. Lampaert, Does frequency matter for intraday technical trading? *Finance Research Letters*, **18**, 177-183 (2016)
10. T. Shalini, S. Pranav, S. Utkarsh, Picking buy-sell signals: A practitioner's perspective on key technical indicators for selected Indian firms. *Studies in Business and Economics*, **14**, 205-219 (2019)
11. V.A. Sobreiro, T.R.C.C. da Costa, R.T.F. Nazario, J.L.E. Silva, E.A. Moreira, M.C. Lima, H. Kimura, J.C.A. Zambrano, The profitability of moving average trading rules in BRICS and emerging stock markets. *North American Journal of Economics and Finance*, **38**, 86-101 (2016)
12. P. Marek a B. Sediva, Optimization and testing of RSI. In: *Financial management of firms and financial institutions*. Ostrava: VŠB-Technical University of Ostrava, Czech Republic, pp. 530-537 (2017)
13. F. García, F. Guijarro, J. Oliver, R. Tamošiūnienė, Hybrid fuzzy neural network to predict price direction in the German DAX-30 index. *Technological and Economic Development of Economy*, **24**, 2161-2178 (2018)
14. B.W. Crowell, Y. Bock, Z. Liu, Single-station automated detection of transient deformation in GPS time series with the relative strength index: A case study of Cascadian slow slip. *Journal of Geophysical Research: Solid Earth*, **121**, 9077-9094 (2016)
15. P. Rihova, M. Svoboda, Profitability of selected technical analysis indicators. *European Financial Systems 2018: Proceedings of the 15th International Scientific Conference*, pp. 591-598 (2018)
16. I. Podhorska, K. Valaskova, V. Stehel, T. Kliestik, Possibility of company goodwill valuation: Verification in Slovak and Czech Republic. *Management & Marketing-Challenges for the Knowledge Society*, **14**(3), 338-356 (2019)
17. K. Valaskova, T. Kliestik, L. Svabova, P. Adamko, Financial risk measurement and prediction modelling for sustainable development of business entities using regression analysis. *Sustainability*, **10**(7) (2018)
18. A. Moreira, T. Muir, Volatility-managed portfolios. *Journal of Finance*, **72**, 1611-1644 (2017)
19. S. Basak, A. Pavlova, A model of financialization of commodities. *Journal of Finance*, **71**, 1511-1556 (2016)

20. S.R. Baker, N. Bloom, S.J. Davis, Measuring economic policy uncertainty. *Quarterly Journal of Economics*, **131**, 1593-1636 (2016)
21. A. R. Arévalo Murillo, Short-term forecasting of financial time series with deep neural networks. Thesis: *Ingenieria de Sistemas* (2016)
22. E. Chong, C. Han, F.C. Park, Deep learning networks for stock market analysis and prediction: Methodology, data representations, and case studies. *Expert Systems with Applications*, **83**, 187-205 (2017)
23. The Coca-Cola Company (KO). *Yahoo Finance* [online]. Available at: https://finance.yahoo.com/quote/KO/history?p=KO (2020)
24. A. Hayes. *Investopedia* [online]. New York. Available at: https://www.investopedia.com/terms/d/dowtheory.asp (2019)

The applicability of FCFF method evaluating an enterprise of Real Estate segment

Jaromír Vrbka[1,*] and *Pavla* Vitková[1]

[1]Institute of Technology and Business, School of Expertness and Valuation, Okružní 517/10, 37001 České Budějovice, Czech Republic

Abstract. The article aims at evaluating a specific enterprise of the Real Estate segment using FCFF (Free Cash Flow to Firm) method. This technique determines the company's value through free cash flows. Enterprise valuation presents a distinct discipline requiring appraiser's deep understanding not only of the evaluated enterprise but also other external decisive influences. The theoretical part focuses on calculation procedures using The CAPM (Capital Asset Pricing Model) model quantifying separate variables that determine discount rates. The suggested technique deals with specific financial data of the company and is applicable in evaluating small and medium-sized enterprises.
Key words: enterprise valuation, FCFF, CAPM, cash flow, real estate, risk rate

1 Introduction

Enterprise valuation currently presents an extensively discussed issue in terms of economy and finance. Adopting a good method facilitates a thorough valuation of the enterprise, which is a very important factor regarding finance as companies certainly need equity evaluated for purposes such as loans, mergers and seeking or making public tenders. Developed economies use the technique of property valuation upon business mergers or acquisitions. On the other hand, developing countries apply this method to foreign investments in a specific country. Growing globalization, market liberalization and cut-throat competition prevent academic workers, company owners, and managers of advisory companies from precisely setting the value of equity of the enterprise [1-2].

Successful determination of the value of an enterprise is subject to many aspects. Thriving companies have a high market value guaranteed only by long-term and complete stabilization, which is achieved by permanent financial balance and continuously positive economic results [3-4]. Financial soundness has a priority as yield methods apply only to the future potential of the company [5]. Upon conducting analyses, prognosis generators, e.g. revenue estimates, operating profit margin, working capital and investments, will deliver the valuation. These generators carry out a preliminary valuation of the company which is, however, informative and its value indicates only provisional estimate [6].

* Corresponding author: vrbka@mail.vstecb.cz

The property valuation method presents a distinct and complex discipline, considering the enterprise value from different aspects. The first depends on the purpose of the valuation and the appraising subject; the second being the reason [7].

The article aims at applying the method of valuation by discount free cash flows conducted on a specific enterprise. The result suggests strong sides and shortcomings of the technique to recommend the use of this method of valuation.

2 Literature Research

Dong Yangig [8] regards the company valuation as a part of an economic discipline that grows in importance within the economic transformation. Free cash flow method contains information that may objectively and comprehensively reflect the ability of enterprises to generate the value, but also help companies clearly understand their market value. The market appraisal depends on the profit rate from previous periods and corporations' current free cash flow [9]. The company value should thereby be of vital interest to future investors and managers. Measuring performance plays an essential role in the companies' value generation [10]. The most challenging task for the management is to adopt an effective method of assessing accomplishments, which will precisely measure the performance of the organization within a specific period. The traditional accounting measures were subject to criticism for failing to consider overall capital costs [11-12].

Dong [8] suggest methods of calculating the free cash flow in a text focused on analysing and appraising enterprises. The analysis of the balance-sheet report and valuation, in general, begins with business operations, i.e. units of accounts dealing with enterprise's primary activities to generate incomes and subsequent addition of the value of its monies and other financial assets. The author claims that the valuation of a company's property ownership should be the same for both methods of calculating the free cash flow. The analyst should, however, improve the technique to calculate capital costs (WACC), which will allow discounting of estimated free cash flows to the current value of the company.

If free cash flow (FCF) value or the value of the cash flow is relevant, and managers submit prognoses to investors, Hoque and Rakow [13] constructed a model demonstrating the influence of these prognoses on the value of the company. On the other hand, corporations with positive free cash flows and high performance pressure their management to reduce profits to avoid the interest of state institutions and tend to lower or cover their performance by profit management.

Shaikh [14] observe the diversion from traditional performance-related assessments based on accounting to performance-related values. Economic value added (EVA) and FCF are reliable indicators showing performance that considers values derived from a long-term goal of accumulating wealth, as contrasted to a short-term approach to maximize the profit [15-18]. Performance indicators built on values (including without limitation EVA) are superior to traditional ratios such as earnings per share (EPS), return on equity (ROE), return on assets (ROA) etc. These do not express value creation and, as they closely relate to accounting, can be effectively processed by senior workers and therefore are not very useful for shareholders as reliable performance ratios.

It is crucial to understand the relationship between different performance values and empirically verify the conceptual equivalence of the FCF and EVA [19-20]. In his analysis, Shaikh [14] proves a strong correlation between FCF and EVA. EVA and FCF are favourably close, which means they yield similar results to measure the performance of the enterprise. Discounting of carefully defined FCF is conceptually equivalent to the discount of economic profits (EVA) for the output and decision-making process.

Although there are lots of approaches to determine the value of the company, experts entertain conflicting opinions as to the efficient method of valuation. The methodological part thereby shows a financial model suitable for calculating capital costs and value of the company, considering only components essential to the operation and capacity of the enterprise; i.e. those which reflect the financial soundness of the enterprise.

3 Data and methods

The case study analysed a medium-sized enterprise, "Malá Šárka plc.", with publicly available financial data from 2015-2018.

Determining the value of the enterprise employs the method of discounted free cash flow FCFF. The technique involves all cash flows of owners and creditors generated by the enterprise from operational assets, providing the company's value as a whole. What follows is the subtraction of the foreign capital to achieve the value of equity. Calculations proceed in two steps. The first part deals with discount FCFF followed by subtracting the value from the foreign capital to evaluate the equity of the enterprise.

The first step involves determining the earnings of the enterprise before paying taxes and interests.

$$EBIT = EBT + I \tag{1}$$

EBIT: Earnings before Interest and Taxes,
EBT: Earning before Taxes,
I: Interest.

To determine FCFF we modify EBIT values. The first phase monitors periods reduced by the tax rate for a specific year, amortization and other costs. The second phase considers investments in operational working capital and investments in fixed assets.

The final value of the formula for calculating FCFF is determined according to aforesaid operations.

$$FCFF = EBIT * (1 - t) + A - INV \tag{2}$$

t: Tax rate,
A: Amortization,
INV: Long-Term Investments + Invests in Working Capital.

We further determine WACC by CAPM, which fixes r_e – costs of equity calculated by risk-free profit rate on the market, risk rate and equity risk premium. The risk-free rate on the market is based on the interest rate of long-term and medium-term government bonds [21]. Specific values arise from issuing calendars of 2015, 2016, 2017, 2018 available on the internet portal of the Ministry of Finance of the CZ. Beta Parameter is determined on the value accessible on websites of Professor A. Damodaran. Our case explores Real estate operations & services. The equity risk premium also originates from websites of the respective professor as a value of the entire Czech Republic valid for 2015, 2016, 2017, 2018.

$$r_e = r_f + \beta * (r_m - r_f) \tag{3}$$

r_f: Risk-free Rate,
β: Volatility of Company´s Stock Price in the Capital Market,
$r_m - r_f$: Equity Risk Premium.

To determine weighted costs of the total market value of the company, we apply a formula considering both – the equity and foreign capital.

$$\text{WACC} = \frac{E}{C} * r_e + \frac{D}{C} * r_d * (1 - t) \quad (4)$$

r_e: Cost of Equity,
r_d: Cost of Debt,
C: Total Market Value of the company,
D: Market Value of the company's debt,
E: Market Value of the company's equity,
t: Corporate tax rate.

The first step encompasses determining FCFF value for individual years and discount WACC value to calculate the enterprise value.

$$EV1 = \sum_{t=1}^{t=n} \frac{FCFF_t}{(1+WACC)^t} \quad (5)$$

EV1: Enterprise Value,
t: the Specified Year,
n: the Number of Years.

The second step consists in fixing a going value of the company. This calculation uses Gordon Model including growth rate g, assessing gross national product (GNP) growth and inflation. The essential requirement is also a steady growth of cash flows.

$$FCFF_{n+1} = FCFF_t * (1 + g) \quad (6)$$

g: Growth Rate of FCFF.

The specific discount rate is derived from future costs of the capital WACC, requiring the enterprise's stability in cash flow growth.

The applied formula is as follows:

$$GV = \frac{FCFF_{n+1}}{WACC - g} \quad (7)$$

GV: Going Value

The total value of the enterprise involves the sum of partial results of both steps.

$$EV = \sum_{t=1}^{t=n} \frac{FCFF_t}{(1+WACC)^t} + \frac{FCFF_{n+1}}{WACC - g} \quad (8)$$

EV: Enterprise Value,
t: the Specified Year last year of first step,
n: the Number of Years in first step,
g: Perpetual Grow.

4 Results

"Malá Šárka plc." was the selected enterprise for the application of accounting data and validation of the devised method. The company's core business is to provide services and is located in Prague.

4.1 Enterprise value in the first step

Publicly available balance-sheet statements reported the profit rate before interest and EBIT tax deductions.

Table 1. Calculating earnings before interest and tax payments (EBIT) within 2015-2018 (in thousands CZK)

Value	2015	2016	2017	2018
Economic result before taxation (+)	87	196	233	194
Interest costs (+)	0	0	0	0
EBIT	87	196	233	194

Sources: Own processing.

Table 1 depicts the perpetual growth of earnings of the enterprise – the economic result before taxation and, subsequently, before paying interests and taxes (EBIT). The year of 2017 saw an increase in its earnings caused by revenue growth for provided services.

Table 2. Values of free cash flows (FCFF) within 2015-2018 (in thousands CZK)

Value	2015	2016	2017	2018
EBIT	87	196	233	194
Tax – tax rate t=19% (-)	17	37	44	37
Amortization (+)	5 612	5 983	6 026	6 264
Working capital change (-)	n/a	313	524	477
Investment changes in fixed assets (-)	n/a	5 759	5 622	5 879
FCFF	n/a	70	69	65

Sources: Own processing.

Table 2 suggests the free Cash Flow rate. The tax rate 19% and company's amortization is overall equal, with the only fluctuation in 2017 caused by revenue growth and subsequently influenced by the change in the working capital.

Table 3 shows statistical values within separate periods as published in annual balance-sheet reports within the business sphere of the Ministry of Industry and Trade. Beta Coefficient and risk premiums are available on websites of Professor A. Damodaran. Calculations determined the rate of average capital costs paid by the company to ensure its business activities (WACC). Furthermore, costs of capital and equity fixed WACC for individual periods.

Table 3. Weighted average capital costs (WACC) within 2015-2018 (in thousands CZK)

Value	2015	2016	2017	2018
Surcharge for the financial stability; rf	0.54%	0.46%	0.79%	1.90%
Beta	0.89	0.99	0.62	0.80
Premium for market risk in CZ; rm-rf	5.60%	6.30%	6.20%	5.90%
Equity yield; r	5.52%	6.70%	4.63%	6.62%

Cost of debt; rd	0.00%	0.00%	0.00%	0.00%
Market value of the company's debt; D	0	0	0	0
Market value of the company's equity; E	34 287	34 376	34 496	34 588
Total market value of the company; C	34 287	34 376	34 496	34 588
WACC	5,52%	6,70%	4,63%	6,62%

Sources: Own processing.

Table 4. Enterprise value EV in the first step (in thousands CZK)

1st step	2015	2016	2017	2018
FCFF		70	69	65
WACC	5.52%	6.70%	4.63%	6.62%
EV 1st step		65	63	54
EV 1st step	**182**			

Sources: Own processing.

Table 4 demonstrates enterprise values in the first step of the calculation. Discount rates applied to calculate weighted average capital costs. The resulted value discounted and subsequently added up free cash flows FFCF.

4.2 Calculating values in the second step

The resulting value of free cash flows for the following period ($FCFF_{n+1}$) and the growth rate is based on the estimated year-to-year GNP growth and inflation. GNP rate 2.4% and inflation 2.6% was built on statistical data with final sum 5.2%.

$$FCFF2019 = 65 * (1 + 0.052) = 69 \text{ } thousands \text{ } CZK \quad (9)$$

The going value for the consecutive year equals to: GV = 4,826 thousands CZK.

4.3 Total market value of the company

The total market value of the company is given by summarizing partial results of both steps:

$$EV = 182 + 3982 = 4{,}164 \text{ } thousands \text{ } CZK \quad (10)$$

5 Discussion

To valuate an enterprise using Discount Free Cash Flows to Firm Method, we randomly chose an organization within Real Estate segment.

The procedure considered all aspects that may influence the precision of the calculation, i.e. a large scale of variability of calculating the enterprise value and essential variables to determine the discount rate. Honque and Rakow [13] argue that we also must take into account the ability of the management to achieve high performance and free cash flows.

If the discount rate arose only from WACC calculation from equity costs from CAPM, the foreign capital and capital structure, there could be a risk of an inadequate assessment of the discount rate value. What might potentially cause this fault would be measuring the risk by CAPM, which indicated a slight deficiency, as the risk rate is an inevitable part of all investments. The risk rate, therefore, requires quantification.

A different decisive factor is the number of periods involved in the first step of assessing the enterprise value. Their lower number negatively influences the final price, as they do not inform on the perpetual growth of the company. The positive effect would be discernible in the event of recalculating the whole development of the monitored periods since the company's foundation.

The evaluated enterprise met all the decisive parameters. The method's applicability strongly depends on a precise calculation. Valuation of enterprises presents a comprehensive discipline established on appraiser's understanding not only the evaluated organization but also other external influences involved.

The validity of the result depends on the data available to the appraiser. His ability to predict the stability (or instability) of the development of the market economy is vital.

6 Conclusion

The goal was achieved. The method for evaluating an enterprise by free cash flows applied to a randomly selected corporation (small or medium-sized) within the designated segment.

The findings showed that valuation standards comply with value standards; the enterprise price agrees with the qualified estimate. Throughout the experiment of setting a discount rate and risk premium, we found out common contact points relating to enterprise capital value regarding the risk for its current shareholders and potential investors.

The selected enterprise met all required parameters for setting the enterprise value. The method used in this study has qualifications regarding the precision of the calculation. However, we can expect its application upon necessary modifications. On the other hand, some computations may give grounds for further research. The risk rate presents an issue for in-depth analysis and could become a central topic of another scholarly article.

Findings relating to the applicability of the technique in questions can serve as appraisers' decision-making criteria or clues for data collection to evaluate the enterprise. We may predict its higher potential for owners who want to know the up-to-date value of their companies and dispose of fresh and accurate information and data.

The research revealed that evaluating assets of enterprises mostly involves yield methods to fix the discount rate of the company. Disregarding the risk rate when pricing the enterprise and ignoring the value of risk premium of the discount rate may result in a simplification, as the amount of risk premium of the discount rate provides useful information for determining a discount rate within the project management's investment decision-making.

References

1. T. Kliestik, J. Vrbka, Z. Rowland, Bankruptcy prediction in Visegrad group countries using multiple discriminant analysis. *Equilibrium-Quarterly Journal of Economics and Economic Policy*, **13**(3), 569-593 (2018)
2. S. Haskova, P. Suler, V. Machova, T. Krulicky, Determining the price of the business share of a business in a group. *Ad Alta: Journal of Interdisciplinary Research*, **9**(2), 60-70 (2019)

3. M. Vochozka, V. Stehel, Z. Rowland, Determining development of business value over time with the identification of factors. *Ad Alta: Journal of Interdisciplinary Research*, **9**(2), 358-363 (2019)
4. Z. Rowland, V. Machova, J. Horak, J. Hejda, Determining the market value of the enterprise using the modified methodof capitalized net incomes and Metfessel allocation of input data. *Ad Alta: Journal of Interdisciplinary Research*, **9**(2), 305-310 (2019)
5. M. Vochozka, Z. Rowland, P. Suler, The specifics of valuating a business with a limited lifespan. *Ad Alta: Journal of Interdisciplinary Research*, **9**(2), 339-345 (2019)
6. M. Vochozka, V. Machova, Determination of value drivers for transport companies in the Czech Republic. *Nase More*, **65**(4), 197-201 (2018)
7. V. Machova, J. Horak, Value generators in forestry and logging. *SHS Web of Conferences: Innovative Economic Symposium 2019 - Potential of Eurasian Economic Union* (2020)
8. Y. Dong, Enterprise value evaluation based on FCFF model - A case analysis of Beixinyuan company. *Proceedings of the 2nd International Conference on Economics and Management, Education, Humanities and Social Sciences (EMEHSS 2018)*, **151**, pp. 615-620 (2018)
9. N. Zhang, The effects of anticipated future investments on firm value: evidence from mergers and acquisitions. *Review of Accounting Studies*, **21**(2), 516-558 (2016)
10. K. Valaskova, T. Kliestik, M. Kovacova, Management of financial risks in Slovak enterprises using regression analysis. *Oeconomia Copernicana*, **9**(1), 105-121 (2018)
11. T. Kliestik, K. Valaskova, G. Lazaroiu, M. Kovacova, J. Vrbka, Remaining financially healthy and competitive: The role of financial predictors. *Journal of Competitiveness*, **12**(1), 74-92 (2020)
12. Kovacova, M., T. Kliestik, K. Valaskova, P. Durana, Z. Juhaszova, Systematic review of variables applied in bankruptcy prediction models of Visegrad group countries. *Oeconomia Copernicana*, **10**(4), 743-772 (2019)
13. M. Hoque, K. Rakow, Do voluntary cash flow disclosures and forecasts matter to value of the firms? *Managerial Finance*, **42**(1), 3-12 (2016)
14. I. Shaik, On the examination of value-based performance measures: Evidence from Indian firms. *Springer Proceedings in Business and Economics*, pp. 527-535 (2017)
15. N. R. Abramishvili, N. A. Lvova, N. S. Voronova, A.V. Kazansky, Free cash flow potential in assessing financial stability of nonfinancial businesses (the case of Russia). *Proceedings of the 30th International Business Information Management Association Conference (IBIMA)*, **1-9**, pp. 1376-1385 (2017)
16. P. Mielcarz, F. Mlinaric, The superiority of FCFF over EVA and FCFE in capital budgeting, *Economic research*, **27**(1), 559-572 (2014)
17. D. Aharon, Y. Kroll, S. Riff, Degree of free cash flow leverage. *Review of Accounting and Finance*, **18**(3), 346-365 (2019)
18. U. Yaari, A. Nikiforov, E. Kahya, Y. Shachmurove, Finance methodology of free cash flow. *Global Finance Journal*, **29**, 1-11 (2016)
19. S. Kim, S. Park, J. Suh, A J-shaped cross-sectional relation between dividends and firm value. *Journal of Corporte Finance*, **48**, 857-877 (2018)
20. E. Hendrawaty, Value of excess cash holdings of financially unconstrained companies. *Contemporary Economics*, **13**(3), 293-303 (2019)

21. C. Mari, M. Marra, Valuing firm's financial flexibility under default risk and bankruptcy costs: A WACC based approach, *International Journal of Managerial Finance*, **27**(1), 688-699 (2019)

SHS Web of Conferences **91**, 01043 (2021) https://doi.org/10.1051/shsconf/20219101043
IES2020

Application of the classical Payout Ratio Model and determination of the limitations of the Lintner Model demonstrated on the example of Komerční banka, PLC

Veronika Machova[1,*] and *Tereza* Geierova[1]

[1]Institute of Technology and Business, School of Expertness and Valuation, Okruzni 517/10, 37001 Ceske Budejovice, Czech Republic

Abstract. Initial determination of the payout ratio should be an integral part of each company strategy; it is a cornerstone of a long-term operation of any company. The aim of this contribution is to analyse the possibilities of determining the payout ratio of the company using dividend models and to design the most optimal application of the dividend model, and to define the limitations for the application. The data required for the application of the dividend models are obtained from the annual reports of Komerční banka, PLC. for the years of 2006-2018. The obtained data are used to calculate the classical payout ratio model, and subsequently to determine the payout ratio according to the Lintner model. Using the method of comparison, the obtained results for the application of the pay ratio are contrasted. The classic payout ratio model coincided directly with the results of the payout ratios of Komerční banka, PLC. and therefore turned out to be optimal for the application. The Lintner model is considered in literature to be very well functioning. However, this research confirmed that the resulting values of payout ratios after the application of the Lintner model are unacceptable in the long run.
Keywords: dividend politics, payout ratio, company strategy, Lintner model, sustainability, comparison

1 Introduction

In the business sphere, current turbulent times demand in particular determination of a quality company strategy. Initial determination of the payout ratio therefore should be an integral part of each company strategy; it is a cornerstone of a long-term operation of any company [1-2]. There is no resource in the world that defines the ideal payout ratio. This value depends on the sector of business in which the particular company is located, and thus the results vary. It is therefore possible to determine decidedly, based on the payout ratio, the percentage of the profit that can be paid to the shareholders in dividends [3-4]. However, this apparently simple model is accompanied by a rather complex dividend policy, combined with a number of arguments and theories. Every company should aim to

* Corresponding author: machova@mail.vstecb.cz

maintain a stable pay ratio in the long term, and thus manage the competitiveness of the external environment [5-6].

The payout ratio is also very important as an indicator that creates values for the use of other indicators in the area of dividend policy. The payout ratio also generally enters other dividend discount models. Thus, the pay ratio affects the Gordon's Growth Model, the Two-Stage Dividend Discount Model, or the H Model for growth assessment. The determination of the payout ratio also applies the so-called Lintner model, which contains primarily the aspect of time and thus creates values for the future.

The aim of this paper is to analyse the possibilities of determining the payout ratio of the company using dividend models and to design the most optimal application of the dividend model, and to define the limitations for the application.

2 Literature review

Over the last five years, the issue of the pay ratio has been addressed by a number of global studies, bringing many findings. In particular, a number of entities that influence the determination of the pay ratio were identified. There are several hypotheses that explain the various factors affecting the pay ratio. ElBannan and Farooq [7] examined the influence of determinants of cross-sectional differences in stock price synchronization and dividend payments in the MENA region. He came to the results showing that both of these variables interact negatively. The higher asymmetry of information associated with companies that show high synchronicity leads to lower pay ratios. The lower asymmetry of information in companies with lower synchronicity leads to higher payout ratios. Mohd and Zaharudin [8], on contrary, examined the effect of dividend payments on the future growth of the company's revenues using panel data. They found a significant positive relationship between the payout ratio and the future growth of the company's revenues. In order to achieve a higher level of business growth, companies should pay fewer dividends to their shareholders. Bond et al. [9] came to different conclusions compared to this study. They examined whether dividend payout options are affected by the presence of a dividend reinvestment plan (DRIP), and concluded that companies paying dividends with DRIP tend to pay high dividends and maintain a stable payout policy. Sumail [10], who divided the effects into negative and positive, also addressed the relationship between the company at the level of corporate governance and the payout ratio. Among the positive influences on the determination of the pay ratio, he included the independence of the board of directors, the size of the board of directors, institutional ownership, the size of the company, and the profit before tax, while the duality of the CEO, ownership management, ownership concentration, and leverage (i.e. gearing) are related negatively to dividend payments. According to Susanti et al. [11], financial leverage and market risk, using the descriptive method, have a significant effect on the pay ratio.

The payout ratio also enters other dividend discount models, which have been the subject of a number of studies. Ahroum and Achchabse [12] intended to create a new model. The model integrates one of the key factors in the expansion of joint ventures, which is the dividend payment ratio, the impact of dividend payments on the price structure, duration and convexity of the market. They found that the higher the payout ratio, the lower the price, duration and convexity of the market. Dzidic and Orsag [13] applied Lintner's model to a sample of customary law and civil law countries. Research has shown that dividend smoothing is spread worldwide, but common law countries have a significantly higher percentage of companies that smooth dividends at a significance level of 1%. This issue was extended by Fernau and Hirsch [14], who created a guide to future dividend smoothing research using the meta-regression analysis of the Lintner model. Hanousek and Tresl [15] deepened the existing research on the findings of overall characteristics of payments in private companies. They examined the payout behaviour, and the way in which private company shareholders' payouts were smoothed in four Central European countries, concluding that the Lintner model indicated a shift over time to a higher target payout ratio and the absence of payout payments for private companies. After the financial crisis, higher payout ratios are set, and there is also a shift away from the smoothing of payments for investors. Kilincarslan [16] also shows the negative impact of the global financial crisis and the tendency to adjust dividends immediately in response to changes in earnings in the first years of the period concerned. He says that despite the credit crunch and earnings volatility, British companies set high payout ratios. Andres et al. [17] analysed Lintner's model for both dividends and total payouts. The results prove a contradiction with the hypothesis that dividends and share repurchases are perfect substitutes. In contrast, Fliers [18] used Lintner's model to examine the relation between financial flexibility and smoothing of dividends. He concluded that the financial flexibility, measured by the unused debt capacity of the company and the costs of adjusting the capital structure, is an important aspect for companies in determining the policy of dividend smoothing. The financial flexibility of a company is at least an important factor in understanding the differences in dividend smoothing behaviour. Significant findings were provided by Ranajee, Pathak and Saxena [19], who identified the determinants of setting the payout ratio, and examined the consistency of dividends over time (stable, crisis, and recovery) in a multivariate regression framework. They supplemented the literature with changing trends and factors contributing to the payment of dividends by companies. Badruzaman and Kusmayadia [20] added to the current research the effect of the dividend payout ratio on the share price of LQ45 joint stock companies on the Indonesian stock exchange Efurs Efek with the use of descriptive analysis. They concluded that the share price is significantly affected by the dividend payment ratio. Ernayani and Robiyanto [21] obtained empirical results regarding the effects of investment return, cash ratio, and credit risk indicator on the payout ratio. The results showed that the investment return has a significant and partially positive impact on the payout ratio; the credit risk indicator has a significant and negative impact on the payout ratio, and the cash ratio has no significant impact on the payout ratio.

The research of the last five years focused mainly on the determinants influencing the pay ratio, and the determination of the pay ratio using the Lintner model or its innovations. It would therefore be appropriate to supplement the current literature with other models for determining the payout ratio. To determine the payout ratio, a large number of studies is inclined to determine the Payout Ratio based on the Lintner model. This is a very suitable model and it can be applied to a wide range of companies.

3 Data and Methods

The data required for the application of the dividend models were obtained from the annual reports of Komerční banka, PLC. for the years of 2006-2018 [22]. The annual reports are published on the bank's official website. Subsequently, selected statistical indicators were calculated using the MS Excel program within the period of 2006-2018.

Table 1. Basic descriptive statistics

Variable	Median	Standard deviation	Mean	Dispersion
Dividends per share in individual years	180.00	88.63	179.85	8510.8
Net earnings per share (EPS) in individual years	295.30	107.32	260.8	12477.36
Net profit for the year (in millions of CZK)	12,424	2,035.94	12,219.92	4,490,449
Dividends paid per year (in millions of CZK)	12,954	1,906.15	12,414.15	3,936,180

Source: Komerční banka [22].

The first part of the specification of the payout ratio determination analysis is based on the classical model, which according to Rejnuš looks as follows [23]:

$$Payout\ ratio = Dividends\ per\ share\ /\ Net\ earnings\ per\ share\ (EPS) \tag{1}$$

The second part of the specification of the payout ratio analysis is based on the Lintner model, which according to Strnad provides the following formula [24]:

$$D_{it} = r_i * E_{it;} \tag{2}$$

where D_{it} is the size of the dividend paid in year t,
$E_{it;}$ is the company's profit in year t,
r_i is the targeted pay ratio.

The data for the application of both models are obtained from the annual reports of Komerční banka, PLC. for the years of 2006-2018 [22]. The obtained data are used to calculate the classical payout ratio model, and subsequently to determine the payout ratio according to the Lintner model. Using the method of comparison, the obtained results for the application of the pay ratio are contrasted. The differences are displayed graphically. The obtained results are further compared with the procedure of Komerční banka, PLC. Afterwards, the optimal application of the dividend model is determined. The conclusion defines the limitations for the application of the models.

4 Results and Discussion

With the use of the classical model for determining the pay ratio, according to Rejnuš [17], Table 2 shows the values of dividends per share and net earnings per share, as well as the resulting values of the determined pay ratio in individual years.

Table 2. Determination of the payout ratio using the classical model

Year	Variable		
	Dividends per share	**Net earnings per share**	**Payout ratio (%)**
2018	51	78.61	64.88
2017	47	79.05	59.46
2016	40	72.48	55.19
2015	310	337.8	91.77
2014	310	343.8	90.17
2013	230	331.68	69.34
2012	230	369.44	62.26
2011	160	249.97	64.01
2010	270	351.2	76.88
2009	180	289.99	62.07
2008	180	348.7	51.62
2007	180	295.3	60.95
2006	150	242.5	61.86

Source: Own processing.

Table No. 3 shows the values of variables and the values of the resulting payout ratio using the Lintner model according to Strnad [24].

Table 3. Determination of the payout ratio using the Lintner model

Year	Variable		
	Net earnings (in millions CZK)	**Dividends paid per year (in millions CZK)**	**Payout ratio (%)**
2018	15,238	14,846	97.43
2017	14,914	14,930	100.11
2016	14,119	13,688	96.95
2015	12,424	12,758	102.69
2014	12,768	12,954	101.46
2013	13,123	12 ,528	95.47
2012	12,249	13 ,954	113.92
2011	7,951	9,475	119.17
2010	12,035	13,330	110.76
2009	10,369	11,007	106.15
2008	13,233	13,161	99.46
2007	11,225	9,605	85.57
2006	9,211	9,148	99.32

Source: Own processing.

The Figure 1 shows significantly different results of determining the payout ratio using these two different models.

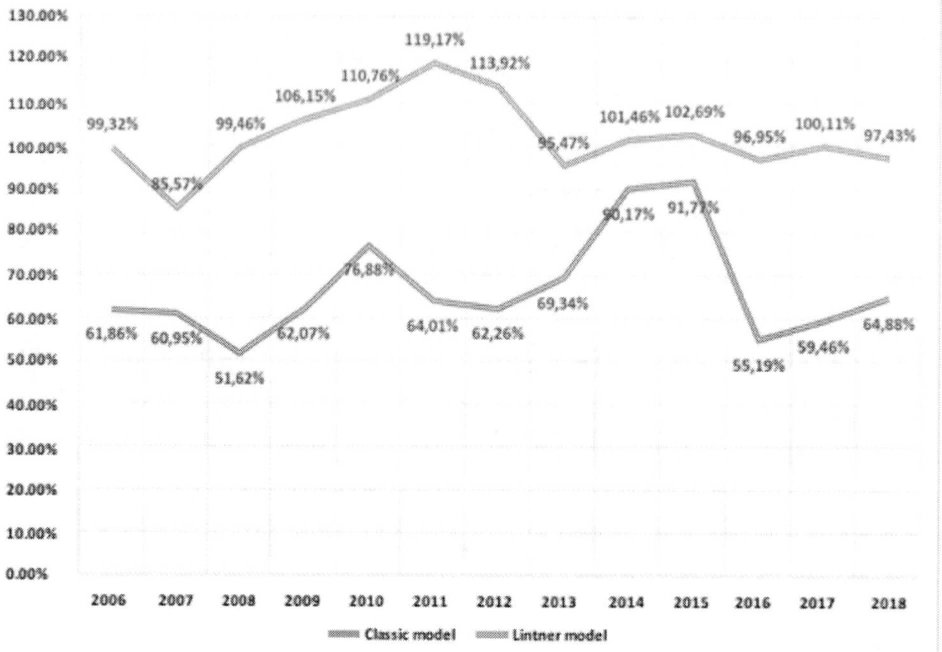

Fig. 1. Comparison of the results of the payout ratio determination

Source: Own processing.

While using two different models for determining the payout ratio of Komerční banka, PLC., it turned out that the Lintner model leads to the determination of much higher payout ratios compared to the classical model. In 2011 and 2012, this is a difference of up to 50%. After applying the Linter model, some of the resulting pay ratios are higher than 100%. This would mean that Komerční banka, PLC. pays more in the dividends than it earns. These results are therefore undesirable. The biggest correspondence in the values of the payout ratio using the different models is recorded in 2014 and 2015. The values of the payout ratio differ in these years by about 10%. In comparison with the procedure of Komerční banka, PLC., the results are completely identical to the application of the classical model for determining the payout ratio according to Rejnuš [23]. Komerční banka, PLC. thus shows its willingness to share its revenues with shareholders in individual years. From the obtained results follows the optimal application of the payout ratio according to the classical model by Rejnuš [23]. Apart from the years 2014 and 2015, Komerční banka, PLC. reported generally stable values of payout ratios.

Thus, when choosing a payout strategy, all companies should take into account that the Lintner model leads to the determination of higher target payout ratios in the long term, as confirmed by this research and other authors. It is not possible to determine exactly the ideal value of the pay ratio, as there are certain limitations in determining the payout ratio. Businesses should pay particular attention to the pay ratio values that are close to, or above, 100%. These fixed pay ratios are unsustainable in the long run, as companies would pay more in dividends than they earn. As a rule, companies should return part of the earned money to the company. The payout ratio shows a reflection of the company's dividend policy, which should be a strategic decision of the company's management. The determination of this indicator is also necessary for the entry into other subsequent indicators. A general recommendation for all companies is the application of more than one model to determine the target payout ratio, and to compare the results. In this way, the determination of undesirable payout ratios can be reduced. The literature states that the Lintner model is a very well-functioning model. However, this research refuted this claim, as the model showed long-term unacceptable results. It is therefore desirable to apply several models to determine the pay ratio, and to make a comparison based on the results. Subsequently, it is necessary to define the undesirable results and to obtain the optimal model.

5 Conclusion

This research expands the existing knowledge on setting the target payout ratio. The aim of this research was to analyse the possibilities of determining the payout ratio of the company using dividend models, to design the most optimal application of the dividend model, and to define the limitations for the application. The goal was met. Achieving the goal consisted mainly in the application of the Lintner model and the classical model of determining the payout ratio on the example of Komerční banka, PLC. The data for the application of these models were obtained from the annual reports of Komerční banka, PLC. for the period of 2006-2018. Different payout ratios were found for individual years. These results were compared, with different values of target payout ratios. The classic payout ratio model coincided directly with the results of the payout ratios of Komerční banka, PLC. and therefore turned out to be optimal for the application.

However, certain limitations for the application are also related to the determination of the target payout ratio. Lintner's model yielded undesirable results, which indicated certain final values of the payout ratio higher than 100%. These results would thus be unsustainable in the long run, as the company would pay more in dividends than it earns. Hanousek and Tresl [9] also concluded that Lintner's model indicates a shift over time to a higher target payout ratio and the absence of payment smoothing. When choosing the application, companies should choose several models for determining the payout ratio and compare them based on the results obtained. Thus, companies can avoid undesirable payout targets.

It is the Lintner model that is considered in literature to be very well functioning. However, this research confirmed that the resulting values of payout ratios after the application of the Lintner model are unacceptable in the long run. In general, it is therefore appropriate to first apply several models and compare the results. Based on this, the most optimal model for determining the payout ratio can be obtained.

The research of the last five years has mostly focused on determinants influencing the payout ratio and the determination of the payout ratio using the Lintner model or its innovations. Further scientific interest should therefore be directed primarily at deepening the application of other models for determining the payout ratio. Furthermore, it would be appropriate to examine in detail the limitations of the Lintner model on a wider sample of companies.

References

1. V. Stehel, J. Horak, M. Vochozka, Prediction of institutional sector development and analysis of enterprises active in agriculture. *E+M Ekonomie a Management*, **22**(4), 103-118 (2019)
2. T. Kliestik, M. Misankova, K. Valaskova, L. Svabova, Bankruptcy prevention: New effort to reflect on legal and social changes. *Science and Engineering Ethics*, **24**(2), 791-803 (2018)

3. M. Vochozka, J. Horak, T. Krulicky, Innovations in management forecast: Time development of stock prices with neural networks. *Marketing and Management of Innovations*, **2020**(2), 324-339 (2020)
4. J. Horak, T. Krulicky, Comparison of exponential time series alignment and time series alignment using artificial neural networks by example of prediction of future development of stock prices of a specific company. *SHS Web of Conferences: Innovative Economic Symposium 2018 - Milestones and Trends of World Economy* (2019)
5. B. Groda, J. Vrbka, Prediction of stock price developments using the Box-Jenkins method. *SHS Web of Conferences - Innovative Economic Symposium 2017: Strategic Partnership in International Trade* (2017)
6. J. Vrbka, Z. Rowland, Stock price development forecasting using neural networks. *SHS Web of Conferences - Innovative Economic Symposium 2017: Strategic Partnership in International Trade* (2017)
7. M.A. ElBannan, O.M. Farooq, Simultaneous determination of stock price synchronicity and dividend payout ratios: evidence from the MENA region. *Journal of Applied Business Research,* **32**(4), 1025-1032 (2016)
8. K.N.T. Mohd, K.Z. Zaharudin, Future earnings growth and dividend payout: Evidence from Malaysia. *Management Science Letters,* **9**, 347–356 (2019)
9. S.A. Bond, Y.J. Pai, P. Wang, S. Zheng, The impact of dividend reinvestment plans on firm payout choices—Evidence from real estate investment Trusts. *Real Estate Economics,* **47**(1), 178-213 (2019)
10. L.O. Sumail, Corporate governance and dividend payout ratio in non-financial firms listed in Indonesian Stock Exchange. *Business and Economic Horizons (BEH),* **14**(4), 851-861 (2018)
11. N. Susanti, R. Rahmayanti, R.R. Padmakusumah, R.A.D.A. Sujai, Influence investment opportunity set, leverage and market risk of dividend payout ratio. *International Journal of Psychosocial Rehabilitation,* **24**(2), (2020)
12. R. Ahroum, B. Achchab, Pricing of Sukuk Musharakah with joint venture as underlying, beyond the use of PLS ratio. *Journal of Islamic Accounting and business research,* **8**(4), 406-419 (2017)
13. A. Dzidic, S. Orsag, Dividend smoothing and investor protection. *Zagreb International Rewiew of Economics & Business*, **22**(2), 55-70 (2019)
14. E. Fernau, S. Hirsch, What drives dividend smoothing? A meta regression analysis of the Lintner model. *International Review of Financial Analysis*, **61**, 255-273 (2019)
15. J. Hanousek, J. Tresl, Firms' total payout characteristics in Visegrad countries. *Politická Ekonomie*, **67**(5), 458-475 (2019)
16. E. Kilincarslan, Smoothed or not smoothed: The impact of the 2008 global financial crisis on dividend stability in the UK. *Finance Research Letters,* 101423 (2019)
17. C. Andres, M. Doumet, E. Fernau, E. Theissen, The Lintner model revisited: Dividends versus total payouts. *Journal of Banking & Finance,* **55**(C), 56-69 (2015)
18. P.T. Fliers, What is the relation between financial flexibility and dividend smoothing? *Journal of International Money and Finance*, **92**, 87-111 (2019)
19. R. Ranajee, R. Pathak, A. Saxena, To pay or not to pay: what matters the most for dividend payments? *International Journal of Managerial finance,* **14**(2), (2018)
20. J. Badruzaman, D. Kusmayadia, Dividend payout ratio influence towards stock price. *Advanced Science Letters*, **23**(9), 8793-8795 (2017)

21. R. Ernayani, O. Sari, The effect of return on investment, cash ratio, and debt to total assets towards dividend payout ratio (a study towards manufacturing companies listed in Indonesia stock exchange). *Advanced Science Letters,* **23**(8), 7196-7199 (2017)
22. Komerční banka. *Výroční a pololetní zprávy [Annual and semi-annual reports]* [online]. In: Výkazy a výsledky [Reports and results]. Available at: https://www.kb.cz/cs/about-the-bank/for-investors/vykazy-a-vysledky/vyrocni-a-pololetni-zpravy (2020)
23. O. Rejnuš, *Finanční trhy [Financial markets]*. 4th ed. Prague: Grada (2014)
24. J. Strnad, *Signální efekty dividend [Signaling effect of dividends]*. Bachelor thesis, Prague: Charles University (2013)

Analysis of government bonds and prediction of their development after the pandemic caused by COVID-19

Petr Suler[1,*] and *Vaclav* Polan[1]

[1]Institute of Technology and Business, School of Expertness and Valuation, Okruzni 517/10, 37001 Ceske Budejovice, Czech Republic

Abstract. This paper describes the current state of the government bond market and predicts the future development of government bond yields using the yield curve to bond maturity, spot yield curve, credit rating and simple prediction. The ongoing economic crisis caused by the COVID-19 pandemic is changing the lives of many people. In order for each individual country to help its households, prevent mass job lay-offs and high mortality, their fiscal budget deficits are growing to unexpected heights. The aim of this paper is to analyse government bonds as one of the tools that can help both the state and individual households at this time. Government bond yields are analysed and compared with other countries based on the development of government bonds using credit ratings, yield curves to maturity, spot yield curves and simple historical development of government bonds from the previous economic crisis in 2008. Based on the results, we conclude that countries severely affected by the COVID-19 pandemic and subsequent mortality, such as Italy, have a relatively stable yield. In contrast, for countries such as the Czech Republic and South Korea, yields to maturity at both ends are relatively declining.
Keywords: government bond market, prediction, economic crisis, households, yield curve, credit rating

1 Introduction

The world is in crisis, businesses are closing, employees are being laid off. Neighbouring countries are facing record unemployment, and Austria in particular had an unemployment rate of 12.2% in the first quarter of 2020. These numbers are frightening, and after years of economic abundance comes a great wake-up call. This is the highest unemployment rate in Austria since the war. The state of emergency affects restaurants, shops, hotels, accommodation facilities, mass events and more. The economy has slowed down, the number of construction orders is declining, engineering is running out of material and there are no customers for transport and supply industries. Statistics already indicate this, and the steps of individual companies confirm that the time is probably approaching when other countries will encounter similarly high record unemployment. What will be the

* Corresponding author: petr.suler@cez.cz

government's measures? Is it possible to predict the development of the economy for when this situation ceases? How will this situation affect the country, businesses, and the citizens themselves? Can individual subjects mitigate the effects of this crisis and deal with these conditions?

The first to react should be the national government, and each one chose their own approach. The Czech Republic predicts a deficit approaching 300 billion CZK. This is the largest deficit in history. And it must be added that this may not be the final number. What will be the next steps and how will the individual countries differ from each other? Is it possible to predict how much will the support of companies and self-employed individuals decrease or increase? We can find financial instruments that the government will use to respond to this situation.

Prediction of economic development in times of crisis is important. Many companies have expected a crisis and are now activating their crisis plans. However, it can be assumed that few expected such a large-scale crisis. At the time of writing, financial institutions are suspending loans and models estimating the bankruptcy of individual entities and the institutions themselves are sounding the alarm. The importance of predicting the crisis and the development of the economy after its decline became apparent in 2008, when we encountered an economic crisis. The consequences of these failures are far-reaching and affect not only the company itself but also affect all of its stakeholders and participants [1-2].

Bonds are a type of financial derivative that can help mitigate these effects. In a way, it is a win-win type of tool, where citizens and successful businesses themselves can take part in economic recovery [3]. In view of the development of the situation after 2008, the development of government bonds in recent years and the individual financial steps that governments are taking in the fight against coronavirus, it will be important to predict their development after the COVID-19 crisis has subsided.

Can the Czech Republic and other countries be expected to issue new government bonds? Is it possible to predict that the interest rate and yield of these issued government bonds will increase? Will countries that face fewer financial issues due to COVID-19 than others offer higher interest rates than countries such as Italy or Spain? The aim of this paper will be to analyse the future development of the price and yield of government bonds in areas with significantly differing mortality caused by COVID-19 and thus predict the development of the market after COVID-19.

2 Literary research

The recent global financial crisis caused by the COVID-19 pandemic is weakening households, businesses, and the government at the same time. There are only few who can profit from this crisis. The largest companies in the world have lost around 30% of their value and now they have a long way to go. Berkshire Hathaway Inc. fell from $340,000 per share to $240,000, agricultural and food company Seaboard Corporation from $4,000 per share to $2,600 per share and the third most valuable company on the stock exchange, NVR, Inc., even fell from $4,000 per share to $2,000 per share. Similar figures can be found after the crisis in 2008. Stock markets fell by 50% because of non-standard mortgages and a slight recession was observed all over the world. These changes, which one does not always expect, cause disintegration and bankruptcy of many households. Every crisis is an opportunity for education, and households must address how to maintain finances and, if necessary, invest them. In contrast, the government must address how to reduce the newly created deficit and where to obtain funding for the necessary investments. Harvey et al. [4] deals with this topic and addresses what opportunities the government has and how they can affect individual households.

According to Akgun et al. [5], the government has the following options. First of all, the government can increase the tax burden and thus get its investment back. Therefore, if the government supports businesses, their employees or individual entrepreneurs and gives them an injection in the form of forgiveness of certain expenses, or pays a part of their expenses itself, one way to get these funds back will be to increase the tax burden. What businesses and people gain by this help will later be lost. The second option are reforms. According to Madsen, Archard and Thorvaldsen [6], the world is developing mainly technologically, and governments are not reacting quickly enough. Thanks to sometimes too lengthy procedures or rough solutions, companies are forced to slow down their growth. The third option offered to the government is the restructuring of their ministries and state-owned businesses. Especially at a time when unemployment is low, the government is keeping the people the private sector would need for themselves. The government has the opportunity to take these steps, and each one of these steps would help the state treasury. The fourth and, according to Engel [7], most preferred option, is the issuance of government bonds. Robyianto [8] sees government bonds as a certain guarantee for the state and a guarantee for households. Nevertheless, according to Pilar and Helena [9], markets must be regulated to avoid undesirable fluctuations.

According to He, Krishnamurthy and Milbradt [10], bonds have shown us their stability and are a security in the financial market. In general, according to Huong [11], it can be assumed that the size of the economy is positively related to the size of the government bond market. Countries must therefore focus on strengthening the economy itself before strengthening the government bond market. Strong economies predict stable bonds. Nevertheless, a bond may not always be the most appropriate solution and it also poses risks that investors are not aware of. Some companies go bankrupt, others fight for survival, and of course the government bond market is affected by this [12-15]. On the other hand, there is the opinion of Ammer et al. [16], who prefers bonds even at a time when the market is going up. His view is based on the simple statement that bonds in every household's portfolio are important simply because when the market goes down, they are a lever that reduces portfolio loss. Yes, the opposite is also true, as the market goes up, they slow down portfolio growth. Arribas et al. [17] considers knowledge of diversification to be the basis of every successful investor. On the S&P 500, we can see that when stocks lose value, bonds tend to gain it. We can even find that a portfolio composed of a certain amount of bonds and stocks can outperform the stock index. Bonds lose much less than stocks and can make a profit for the investor even with lower volatility. According to Ammer [16], investors are looking for ways to find foreign bonds at lower yields, as confirmed, for example, by research on USA bonds. At the same time, Blommestein [18] adds that it is necessary to consider all risks and monitor, for example, the monetary policy of a country that significantly affects bond yields. Thus, for households or businesses that deal with future financial stability after each financial crisis, bonds may seem like a suitable type of investment. But can we somehow predict the value of bonds in individual countries in the future, what kind of yield the investor will have, or whether there is a risk of losing finances?

Bonds and their development after several financial crises will serve as a suitable example. Harvey et al. [4] defined these following crises as "Black Monday" (1987), the Gulf War (1998), the Asian Crisis (1998), the Technology Bubble (2000-2002), the World Global Crisis (2007-2009), the first European crisis (2010) and the second European crisis (2011). Based on data from these crises, we can predict the price of government bonds after the COVID-19 crisis.

Vukovic, Hanic and Hanic [19] deals with the question of whether and what effect the crisis has on the financial market. It is important to note that financial markets are experiencing great volatility in times of crisis, and it remains there for some time even after the crisis is gradually receding. Investors should take this information into account and hold or sell their bonds accordingly. However, it is not possible to generalize and firmly determine whether it is more appropriate to hold long-term or short-term bonds in a crisis. The value of bonds is determined by many factors, the most important of which are the economic and political stability of the country. According to McKay and Peters [20], various methods can be used to arrive at the answer to how the yields of government bonds of various compared countries will differ after COVID-19. Specifically, for example, by calculating the yield curve from yields to maturity, by calculating the coupon yield curve, the spot yield curve, the rating, the annuity yield curve and many others. Choudhry [21] adds that the markets are not always completely perfect and therefore each method can offer us slightly different results. And even though in times of crisis it is a bit more difficult to predict the development of bonds, according to Martin and Ross [22], the development of bonds with longer maturities can aid us with the prediction.

3 Data and methods

First of all, we will compare the development of bonds in three areas. We will look at the development of bonds where COVID-19 hit the most, i.e. Italy, then a moderately affected country, i.e. the Czech Republic, and one of the least affected countries, South Korea. We selected the aforementioned countries according to mortality per 100,000 inhabitants.

To get the most accurate results, we will approach this in 5 different ways. We first predict the development and yield of government bonds according to the yield curve from yields to maturity, then according to the spot yield curve and finally, we use less scientific methods, namely prediction according to credit rating and simple prediction according to the development of government bonds after the 2008 crisis. We will compare the results of these methods with each other and determine the development of government bonds.

We will obtain the data we need from the indices of the rating agencies S&P Global Ratings, Moody's, and Fitch Ratings, and according to current data on world stock exchanges. We will also base our calculations on national markets. We will need to find out the country's rating, the market price of the bond, the annual coupon payment, the nominal value of the bond, the yield to maturity and the number of years to maturity and the spot interest rate. According to this data, we will calculate everything determined above.

We will analyse bonds with maturities of 1 year, 2 years, 3 years, 4 years, 5 years, 6 years, 7 years, 8 years, 9 years, and 10 years. In the case of South Korea, we will only use a period of 1-5 years until maturity. To obtain the desired results, we will proceed as follows.

3.1 Prediction of the yield curve from yields to maturity

Yield to maturity is the rate at which the present value of any future income from a bond equals its current price. Therefore, the following applies to coupon bonds:

$$P_{TR} = \frac{C}{1 + r_{DS}} + \cdots + \frac{C}{(1 + r_{DS})^n} + \frac{NH}{(1 + r_{DS})^n} \tag{1}$$

P_{TR} – market price of bond,
C – annual interest payment,
r_{DS} – yield to maturity,
n – number of years,
NH – nominal value of bond.

3.2 Prediction according to the spot yield curve

It is important to proceed year by year to calculate the spot yield curve. First, we calculate the spot interest rate for the first year. We then insert it into the next formula and calculate the spot interest rate for the second year. Afterwards, we use this interest rate again and keep repeating the same procedure.

$$P_1 = \frac{C_1 + NH}{1 + r_{0.1}};\ P_2 = \frac{C_2}{1 + r_{0.1}} + \frac{C_2 + NH}{(1 + r_{0.2})^2};\ \ldots \tag{2}$$

P_n – bond price for the given year,
C_n – annual interest payment,
$r_{x,y}$ – yield to maturity for period,
NH – nominal value of bond.

3.3 Simple prediction of bond development according to rating

We base our prediction from the simple knowledge that the better the rating, the lower the expected yield.

3.4 Simple prediction of bond development according to the values of bonds during the crisis of 2008

We base our prediction on the known yield rates of two-year government bonds for the period from 8/2008 to 7/2009, we create a yield curve and compare them with the selected countries.

4 Results

As already described, in this paper we will focus on three different areas. We will look at the development of bonds where COVID-19 hit the hardest, i.e. Italy, then we will continue with the Czech Republic, which is ranked somewhere in the middle regarding the mortality caused by COVID-19, and finally move on to the country that can be considered an example of how to cope with this pandemic, South Korea. We will analyse each country separately and apply several different methods to the obtained data. Finally, we will use simple methods of comparison with the development of two-year government bonds during the economic crisis in 2008. The prediction of government bond yields according to yields to maturity and according to the spot yield curve will be among the most appropriate methods.

4.1 Prediction of government bond yields in the Czech Republic

When monitoring yield calculations from the world's leading agencies, we usually get a yield curve using yield to maturity. But in order to have our results as accurate as possible, I also chose to calculate the spot yield curve. This curve is more accurate, and we obtain it by calculating the spot interest rate for the previous year from the selected formula each time and gradually adding it to the calculation of the following years.

4.1.1. Prediction of government bond yields according to yields to maturity and according to spot interest rate

Table 1 presents the base data for determining the yield and spot curve as of May 5th, 2020. We also see these results in the following charts and compare them with the previous period. We can see from the results for the Czech Republic that the yield decreased significantly compared to the previous year. The results suggest that the Czech Republic's bonds are by no means attractive to investors and are declining. We can observe a smaller difference in multi-year bonds, but even here the yield decreased.

Table 1. Base data for determining the yield and spot curve as of May 5th, 2020

Sale price	Yield to maturity in %	Spot interest rate in%	Coupon	Maturity in years
104.11	-0.35%	-0.35%	3.75	1
107.22	0.23%	0.24%	3.85	2
113.16	0.29%	0.30%	4.7	3
100.20	0.40%	0.40%	0.45	4
125.73	0.48%	0.50%	5.7	5
109.29	0.81%	0.84%	2.4	6
101.02	0.85%	0.86%	1	7
95.04	0.90%	0.90%	0.25	8
113.10	0.97%	1.01%	2.5	9
116.00	1.06%	1.22%	2.75	10

Source: Own processing.

Table 2 presents the base data for determining the yield and spot curve as of May 7, 2019.

Table 2. Base data for determining the yield and spot curve as of May 7, 2019

Sale price	Yield to maturity in%	Spot interest rate in%	Coupon	Maturity in years
102.01	1.70%	1.70%	3.75	1
104.05	1.77%	1.77%	3.85	2
109.00	1.60%	1.60%	4.7	3
95.39	1.65%	1.66%	0.45	4
118.81	1.74%	1.75%	5.7	5
104.44	1.62%	1.61%	2.4	6
94.67	1.82%	1.82%	1	7
88.20	1.85%	1.85%	0.25	8
105.34	1.85%	1.86%	2.5	9
107.99	1.87%	2.09%	2.75	10

Source: Own processing.

The yield curve according to yields to maturity is shown in Figure 1 for comparison between the dates of May 7, 2020 and May 7, 2019.

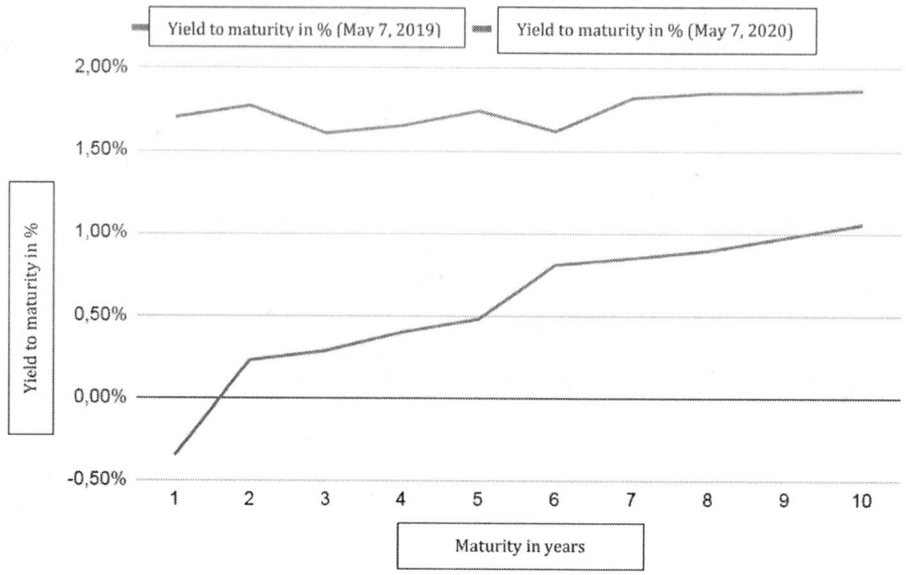

Fig. 1. Yield curve according to yields to maturity

Source: Own processing.

The spot yield curve is shown in Figure 2 with a comparison between May 7, 2020 and May 7, 2019.

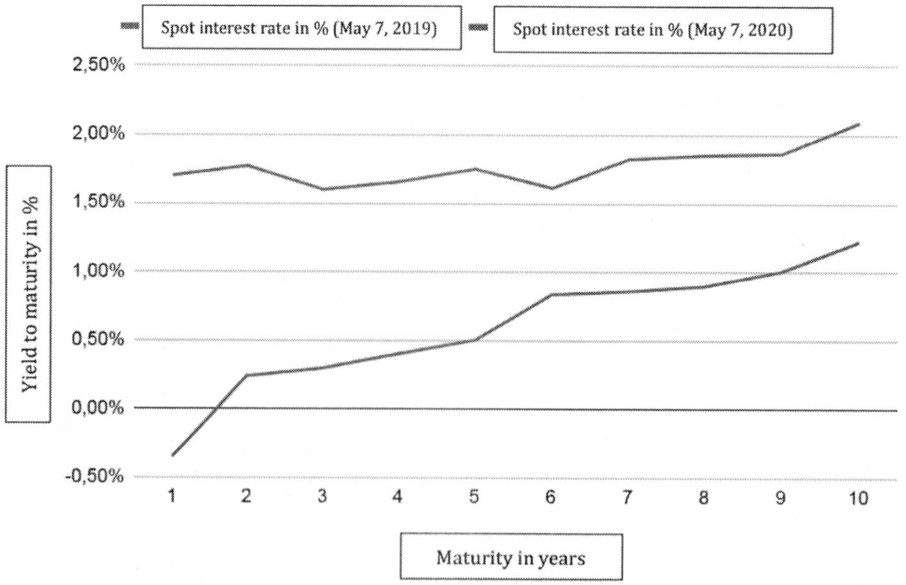

Fig. 2. Spot yield curve

Source: Own processing.

4.1.2 Simple prediction of bond development according to rating

According to well-known statements, we know that a country's rating depends on the rate of inflation, the political and economic situation in the country and many other conditions. The rating of the Czech Republic according to selected rating agencies is shown in Table 3. The rating agencies view the Czech Republic as trustworthy; it is a stabilized environment, and a potential investor can thus perceive the bonds as more or less risk-free. On the other hand, no high returns can be expected, and if we base our investment only on the rating, we would assume a stable, but still quite low return.

Table 3. Rating assessment of the Czech Republic according to selected credit rating agencies

Rating agency	Rating
Standard & Poor's	AA-
Moody's Investors Service	Aa3
Fitch Ratings	AA-

Source: Own processing, according to www.worldgovernmentbonds.com.

4.2 Prediction of government bond yields in Italy

4.2.1. Prediction of government bond yields according to yields to maturity and according to the spot interest rate

According to the results below, we can see that the short-term maturity yield is growing compared to the previous year. On the other hand, with increasing maturity, the yield decreases compared to 2019. These results suggest a minimal effect caused by COVID-19. While the Czech Republic has the status of a stable country for the time being and yields are falling proportionally from shorter to longer maturities, we see slightly different results in Italy. In Italy, we see that the country is currently more attractive to investors, both in the short and long term. In particular, short-term yield with a maturity of 1 to 3 years is more interesting than in the previous year. Given the management of the current pandemic, it can be assumed that we will see a similar yield in Italy in the near future.

Table 4 presents the base data for determining the yield and spot curve as of May 7, 2020.

Table 4. Base data for determining the yield and spot curve as of May 7, 2020

Sale price	Yield to maturity in %	Spot interest rate in %	Coupon	Maturity in years
99.79	0.21%	0.21%	0	1
100.42	0.79%	0.79%	1	2
98.94	0.96%	0.96%	0.6	3
101.83	1.28%	1.30%	1.75	4
101.73	1.49%	1.51%	1.85	5
102.78	1.61%	1.63%	2.1	6
94.75	1.65%	1.66%	0.85	7
106.96	1.86%	1.91%	2.8	8
108.80	1.93%	1.98%	3	9

94.74	1.93%	2.18%	1.35	10

Source: Own processing.

Table 5 presents the base data for determining the yield and spot curves as of May 7, 2019.

Table 5. Base data for determining the yield and spot curve as of May 7, 2019

Sale price	Yield to maturity in %	Spot interest rate in %	Coupon	Maturity in years
99.95	0.06%	0.06%	0	1
100.92	0.54%	0.54%	1	2
98.81	1.01%	1.01%	0.6	3
101.64	1.33%	1.36%	1.75	4
100.88	1.67%	1.69%	1.85	5
101.05	1.91%	1.95%	2.1	6
92.32	2.04%	2.06%	0.85	7
104.08	2.24%	2.31%	2.8	8
105.66	2.30%	2.37%	3	9
89.03	2.61%	2.96%	1.35	10

Source: Own processing.

The yield curve according to yields to maturity is shown in Figure 3 with a comparison between May 7, 2020 and May 7, 2019.

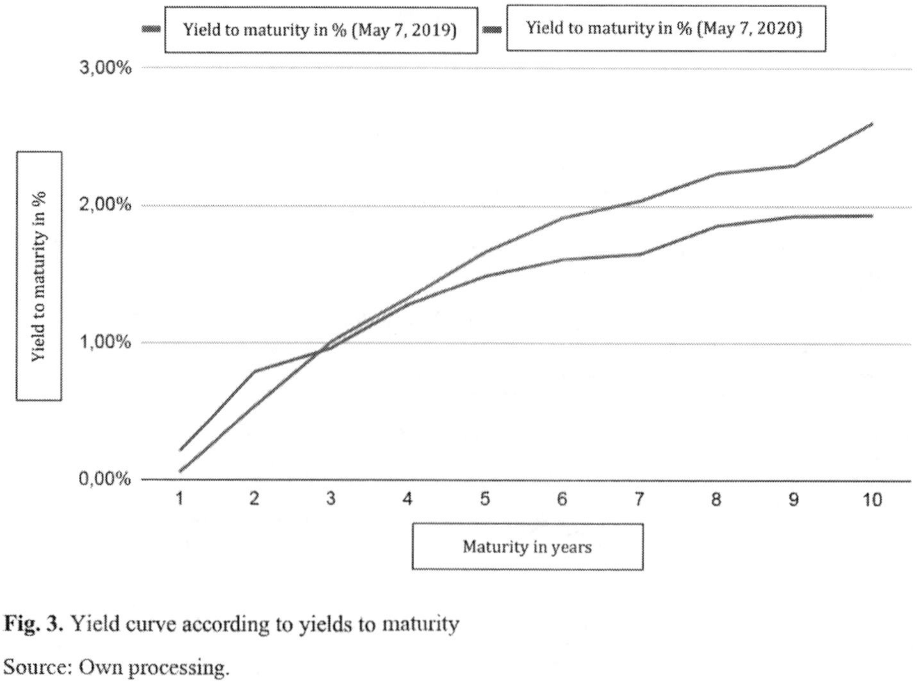

Fig. 3. Yield curve according to yields to maturity

Source: Own processing.

The spot yield curve is shown in Figure 4 with a comparison between May 7, 2020 and May 7, 2019.

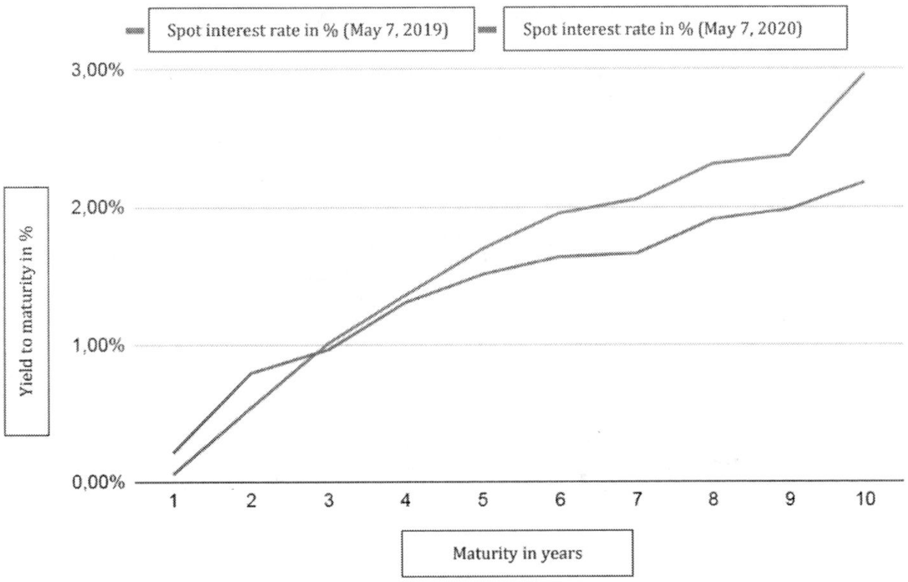

Fig. 4. Spot yield curve

Source: Own processing.

4.2.2 Simple prediction of bond development according to rating

Italy's rating confirms that this country is one of the most affected by the current pandemic. The rating of Italy according to selected rating agencies is shown in Table 6. Credit rating agencies still value this country as a safe opportunity for investment, therefore not speculative, but at the last possible level. Nevertheless, if we look at this assessment, it is still interesting for the investor. It is not yet a speculative investment, but relatively high and safe yields can still be expected.

Table 6. Rating of Italy according to selected credit rating agencies

Rating agency	Rating
Standard & Poor's	BBB
Moody's Investors Service	Baa3
Fitch Ratings	BBB-

Source: Own processing, according to www.worldgovernmentbonds.com.

4.3 Prediction of government bond yields in South Korea

4.3.1. Prediction of government bond yields according to yields to maturity and according to the spot interest rate

South Korea is often considered the model country that managed the current pandemic in the best way possible. These views are also reflected in the yields of their bonds, which fell compared to the previous year, so we can expect early stability of the country. On the other hand, South Korea still offers interesting returns starting from the first year to maturity and, for example, for the first year to maturity, it is still more attractive to investors than Italy and the Czech Republic, despite the aforementioned decline. South Korea offers a stable yield that deviates by a maximum of 0.5 percentage points over 5 years and there is no indication that this yield should change in the future.

Table 7 presents the base data for determining the yield and spot curves as of May 7, 2020.

Table 7. Base data for determining the yield and spot curves as of May 7, 2020

Sale price	Yield to maturity	Spot interest rate	Coupon	Maturity in years
101.46	0.77%	0.77%	2.25	1
101.46	0.89%	0.89%	1.625	2
100.88	0.95%	0.95%	1.25	3
102.94	1.12%	1.14%	1.875	4
101.38	1.21%	1.22%	1.5	5

Source: Own processing.

Table 8 presents the base data for determining the yield and spot curves as of May 7, 2019.

Table 8. Base data for determining the yield and spot curves as of May 7, 2019.

Sale price	Yield to maturity	Spot interest rate	Coupon	Maturity in years
100.53	1.72%	1.72%	2.25	1
99.79	1.73%	1.73%	1.625	2
98.65	1.72%	1.71%	1.25	3
101.57	1.47%	1.48%	1.875	4
98.80	1.75%	1.76%	1.5	5

Source: Own processing.

The yield curve according to yields to maturity is shown in Figure 5 with a comparison between May 7, 2020 and May 7, 2019.

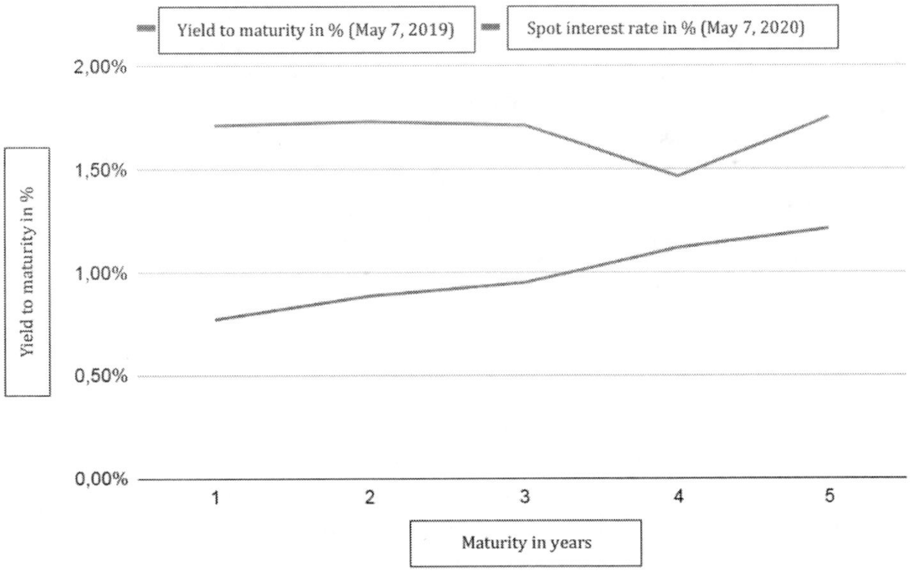

Fig. 5. Yield curve according to yields to maturity

Source: Own processing.

The spot yield curve is shown in Figure 6 with a comparison between May 7, 2020 and May 7, 2019.

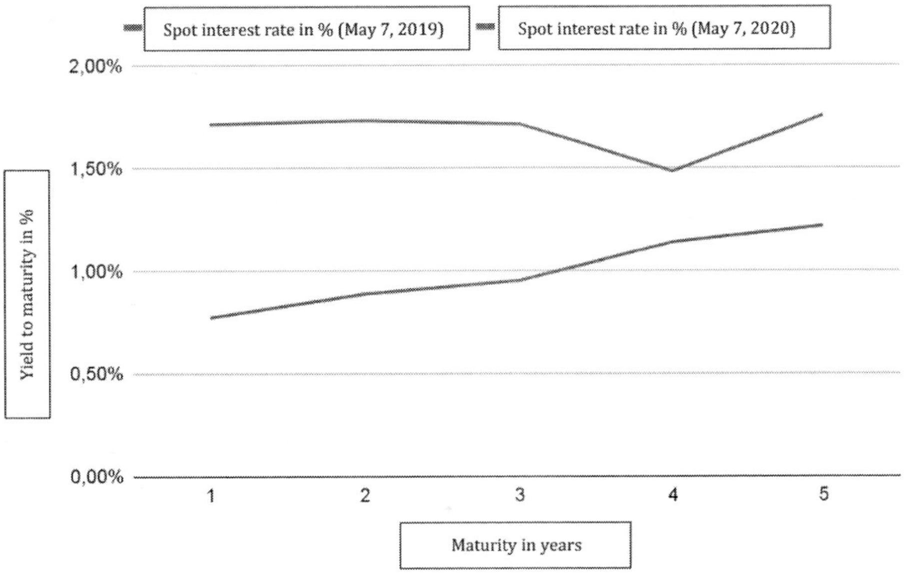

Fig. 6. Spot yield curve

Source: Own processing.

4.3.2 Simple prediction of bond development according to rating

The stability of the country and the successful fight against COVID-19 are also confirmed by rating agencies, which place this country in the category of bonds with an investment opportunity and a relatively strong position. South Korea's credit rating according to selected rating agencies is shown in Table 9. Based on this rating, investors cannot expect high returns, on the other hand, according to the rating it is a politically and economically stable country and investors' finances are relatively safe here.

Table 9. Rating assessment of South Korea according to selected rating agencies

Rating agency	Rating
Standard & Poor's	AA
Moody's Investors Service	Aa2
Fitch Ratings	AA-

Source: Own processing, according to www.worldgovernmentbonds.com.

4.4 Simple prediction of government bonds and international comparison according to the development of bonds after the economic crisis in 2008

In order to predict the development of government bonds after the COVID-19 pandemic subsides, we compared their development in the period of August 2008 to July 2009, i.e. the beginning and subsequent gradual easing of the economic crisis. An international comparison of the development of 2-year government bonds for the period of 08/2008 to 07/2009 is shown in Figure 7. If we assume that the lower the yield, the stronger the rating and the stability of the country, we see from the chart below that Italy's bonds performed the best and did not record yield growth even once. So, the results slightly contradict the previous ones, where we see that Italy is managing the current crisis the worst out of the selected countries. Nevertheless, if this data on two-year government bonds during the economic crisis at the turn of 2008 and 2009 should lend us help in predicting their current development, the chart shows that after surviving the initial boom, all 3 of these countries stabilized relatively quickly and the same can be expected now. Spot yield curves or forward yield curves, for example, will be a more suitable form of predicting the future value of government bonds.

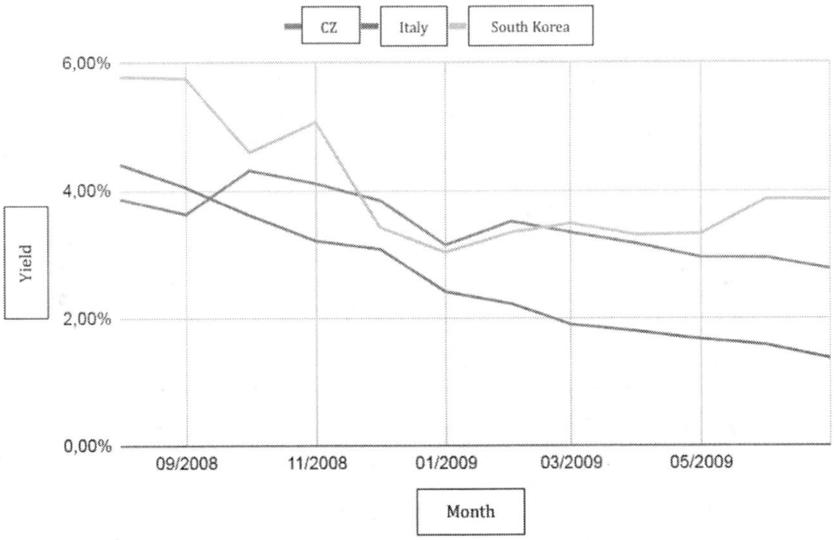

Fig. 7. International comparison of the development of 2-year government bonds for the period of 08/2008-07/2009

Source: Own processing.

5 Conclusion

This paper describes the current state of the government bond market and predicts the future development of government bond yields using the yield curve to maturity, spot yield curve, rating and simple prediction. The yield curve itself is a unique tool that can predict the development of the bond market for the following period. If we consider the normal rising shape of the yield curve, its slope is important. A steeper yield curve suggests that short-term yields are lower and money market investments are less profitable than long-term bond investments. In conclusion, we can certainly refute the hypothesis that the countries hit hardest by the COVID-19 pandemic and with the highest mortality per 100,000 inhabitants are also the most affected ones when it comes to government bonds in terms of rising yields. Based on the results, we find that the hardest hit countries, such as Italy, have a relatively stable yield. In contrast, in countries such as the Czech Republic and South Korea, yields to maturity at both ends are declining relatively.

Bond securities are referred to as a conservative type of investment and are popular mainly because they provide investors with a reasonable return on the investment in comparison with their low riskiness. Yield curves to maturity and spot yield curves show us that even though a year ago Italy was the strongest of the countries observed and thus had the strongest yields, the current situation has changed everything and Italy is the one that offers the highest yields of the countries analysed.

The results above are also confirmed by the credit rating of the selected countries. The better the rating, the lower the yields can be expected. Looking back at the results on the current rating, it is clear that the least stable country is Italy, which should therefore also offer the highest yields. Nevertheless, these are still the most attractive and, in a way, still relatively safe bonds.

References

1. A.P. Balcerzak, T. Kliestik, D. Streimikiene, L. Smrcka, Non-parametric approach to measuring the efficiency of banking sectors in European Union countries. *Acta Polytechnica Hungarica*, **14**(7), 51-70 (2017)
2. Z. Virglerova, L. Homolka, L. Smrcka, K. Lazanyi, T. Kliestik, Key determinants of the quality of business environment of SMEs in the Czech Republic. *E & M Ekonomie a Management*, **20**(2), 87-101 (2017)
3. A. Kasych, S. Kizhemiakina, M. Vochozka, O. Romanenko, V. Glukkhova, A world model of social entrepreneurship in a crisis. *Journal of Entrepreneurship Education*, 22(Special Issue), 1-6 (2019)
4. C.R. Harvey, E. Hoyle, S. Rattray, M.Sargaison, D. Taylor, O. Van Hemert, The best of strategies for the worst of times: Can portfolios be crisis proofed? *Journal of Portfolio Management*, **45**(5), 7-28 (2019)
5. O. Akgun, D. Bartolini, B. Cournede, The capacity of governments to raise taxes. *OECD Economics Department Working Papers*, No. 1407, OECD Publishing [online], Available at: https://www.oecd-ilibrary.org/economics/the-capacity-of-governments-to-raise-taxes_6bee2df9-en
6. S. S. Madsen, S. Archard, S. Thorvaldsen, How different national strategies of implementing digital technology can affect teacher educators. *Nordic Journal of Digital Literacy*, **13**(04), 7-23 (2019)
7. C. Engel, Safe U.S. assets and U.S. capital flows. *Journal of International Money and Finance*, **102** (2020)
8. R. Robiyanto, Gold VS bond: What is the safe haven for the Indonesian and Malaysian capital market? *Gadjah Mada International Journal of Business*, **20**(3), 277-302 (2018)
9. A. Pilar, C. Helena, European Government bond market contagion in turbulent times. *Czech Journal of Economics and Finance (Finance a uver)*, **66**(3), 263-276 (2016)
10. Z. He, A. Krishnamurthy, K. Milbradt. What makes US government bonds safe assets? *American Economic Review*, **106**(5), 519-523 (2016)
11. T.T.T. Huong, The determinants of government bond market development in ASEAN+3. *International Journal of Science and Research*, **9**(3), 1284-1289 (2020)
12. V. Stehel, J. Horak, M. Vochozka, Prediction of institutional sector development and analysis of enterprises active in agriculture. *E+M Ekonomie a Management*, **22**(4), 103-118 (2019)
13. J. Horak, J. Vrbka, P. Suler, Support vector machine methods and artificial neural networks used for the development of bankruptcy prediction models and their comparison. *Journal of Risk and Financial Management*, **13**(3), (2020)
14. T. Kliеštik, K. Valaskova, G. Lazaroiu, M. Kovacova, J. Vrbka, Remaining financially healthy and competitive: The role of financial predictors. *Journal of Competitiveness*, **12**(1), 74-92 (2020)
15. T. Kliestik, J. Vrbka, Z. Rowland, Bankruptcy prediction in Visegrad group countries using multiple discriminant analysis. *Equilibrium-Quarterly Journal of Economics and Economic Policy*, **13**(3), 569-593 (2018)
16. J. Ammer, S. Claessens, A. Tabova, C. Wroblewski, Home country interest rates and international investment in U.S. bonds. *Journal of International Money and Finance*, **95**, 212-227 (2019)
17. I. Arribas, M.D. Espinós-Vañó, F. García, R. Tamošiūnienė, Negative screening and sustainable portfolio diversification. *Entrepreneurship and Sustainability Issues*, **6**(4), 1566-1586 (2019)

18. H. Blommestein, Impact of regulatory changes on government bond market liquidity. *Journal of Financial Regulation and Compliance*, **25**(3), 307-317 (2017)
19. D.B. Vukovic, E. Hanic, H. Hanic, Financial integration in the European Union - The impact of the crisis on the bond market. *Equilibrium*, **12**(2), 195-210 (2017)
20. D.R. Mckay, D.A. Peters, Bond basics and the yield curve. Plastic Surgery, **27**(1), 83-85 (2018)
21. M. Choudhry, The yield curve. *The Moorad Choudhry Anthology*, 272-343 (2018)
22. I.W.R. Martin, S.A. Ross, Notes on the yield curve. *Journal of Financial Economics*, **134**(3), 689-702 (2019)

Methodology of calculating risk premiums in the environment of the Czech Republic and its comparison with Damodaran

Jiri Kucera[1,*] and *Lenka* Maskova[2]

[1]University of Zilina, The Faculty of Operation and Economics of Transport and Communications, Department of Economics, Univerzitna 8215/1, 01026 Zilina, Slovakia
[2]Institute of Technology and Business, School of Expertness and Valuation, Okruzni 517/10, 37001 Ceske Budejovice, Czech Republic

Abstract. Investors' decisions are largely influenced by the riskiness of the country. Several different approaches are available to calculate this risk, but even so, the values set by Damodaran are usually used, even for non-US states. The aim of the paper is to propose a methodology for creating a risk premium in the environment of the Czech Republic and then compare it with Damodaran [1]. Methods applicable in the Czech Republic and Damodaran methods are used, then these methods are compared. For Czech as well as foreign investors, the easiest way to obtain a risk premium is to use the company's investment rating. In the case of determining the risk premium of the Czech Republic, the easiest method is the CRP (country risk premiums) model. If the country's market does not have a long history or does not have such a developed capital market, it is recommended to apply data from the US capital market. However, there are significant differences in the economy between Europe and the USA, so the data of an European country such as Germany, which has historical risk premium calculations, should be used.
Keywords: risk premium, risk countries, Damodaran´s methods, rating, CRP

1 Introduction

Recently, foreign investment has been growing. Investors make their decisions based on the reward they receive for taking the risk of making an investment. The efficiency of the investment is related directly to the risk, also called the risk premium, which is included in the discount rate [2].

According to Damodaran [3], a financial theorist, risk is characterized in terms of the difference between actual and expected returns. The investment is risk-free if the actual returns are always the same as the expected returns. The risk premium is what investors require on average as a premium for risk-free investment rate with an average risk for each factor.

* Corresponding author: kuceraj@mail.vstecb.cz

Kruschwitz et al. [4] mention the Country Risk Premiums (CRP) calculation model proposed by Damodaran. This model also influences the investor's decision on the appropriate placement of capital. There are different ways to assess country risk, but the simplest method would be to use the ratings of the relevant agencies. Although Damodaran's model is applied in many countries, it is criticized by several theorists. Marik and Marikova [5] state that this model requires a database containing, if possible, longer time series. However, the problem is that the Czech capital market does not provide the necessary statistically sufficient database. Both Czech and foreign investors who would like to invest in the Czech Republic have this problem.

The aim of this paper is to propose a methodology for creating a risk premium in the environment of the Czech Republic and then compare it with Damodaran [1].

2 Literature research

2.1 Risk premiums

Damodaran [1] characterized the risk premium as an essential and critical part of portfolio management, business finance and valuation. Given its importance, it is surprising that estimation issues are not given much attention in practice. The Equity Risk Premium, ERP, is specified by Damodaran [6] as the price of risk on the stock markets and is also a key input in estimating the cost of capital and capital in business finance and valuation. It is also a central part of the risk and reward model in finance. According to Chen [7], all financial funds are based on the claim that investors do not like risk and therefore demand higher returns as compensation for risk-bearing. The equity risk premium can be considered as an additional rate of return that investors consider to be returns in exchange for a burden bearing volatility and the associated risk of loss. Economists are still looking for a simple model that justifies the premium given the much lower volatility of aggregated economic data. According to Siegel [8], the search for the right equity risk premium model has yielded insights that can help professionals structure their clients' portfolios. First, habit formation means that the investor's short-term and long-term attitudes to risk may differ. Second, the investor's earnings can cause significant differences in asset allocation. Third, the idea that risk aversion becomes extremely high at low levels of consumption has prompted "investing in liabilities." Investors, especially the older generation, finance what they consider to be the absolute minimum expenditure on risk-free assets and allocate the remainder according to the usual risk.

According to Damodaran [9], risk can be both a threat to a company's financial health and an opportunity to have an advantage over competition. Most analysts focus on the threat when managing risks and emphasize protection against that threat. The risk associated with an investment is generally reflected in the discount rate used in conventional discounted cash flow models. Kliestik, Kocisova and Misankova [10] therefore also focused on the analysis of methods for predicting the failure of society on the basis of publicly available data. Failing companies thus become high-risk for investors. Kliestik [11] further extended this topic to predict the financial health of a company based on mathematical methods.

Furthermore, Damodaran [1] argued that the premiums that are estimated can vary widely in different approaches. Three approaches are used to estimate equity risk premiums. The first is the exploratory approach, where investors or managers are asked to provide estimates of the equity risk premium for the future. Another is a historical return approach, where the bonus is based on how well the stock has performed in the past. The last is the implied approach, where future cash flows are used to estimate the current equity risk premium. Corelli [12] also adds that the estimate of the equity risk premium comes only after a careful analysis of the individual factors that determine it.

2.2 Country risk premium

The term rating is used in connection with country risk assessment. The Ministry of Finance of the Czech Republic [13] describes the country's rating as an assessment of its creditworthiness and economic ability to meet its own obligations, but also as a synthetic expression of the quality of the Czech state as a debtor. Potential investors rely not only on their own judgment, but especially on the country's rating.

An Aaa-rated country is considered almost risk-free, while a C-rated country is very risky [14]. For the most part, rating companies agree in their ratings, but there may be significant differences between ratings of the countries. These differences may result from different assessments of political and economic risk. The Czech National Bank [15] compiled a clear table 1, which is given in the appendix, with scores ranging from the best, AAA to the worst, CC. The country ratings from three rating agencies, referred to as the "Big Three" - Moody's, Standard & Poor's and Fitch, are recognized and used around the world. These agencies work with the central bank and ministries of a particular country. Current information on the rating of individual countries dates from January 2020. According to the table of rating agencies, the Czech Republic belongs to the group with the rating score Aa3, or also AA-. It is therefore the only country in the former Eastern bloc to achieve a high rating.

In 2019, the international rating agency Standard & Poor's confirmed the excellent rating of the Czech Republic for long-term liabilities. The agency positively assessed mainly the low indebtedness of the public sector, but also healthy economic growth. According to them, the Czech Republic is currently one of the European Union countries with the lowest share of government debt to GDP [13].

In the same year, Fitch also confirmed the country's high creditworthiness. Confidence in the Czech Republic is based on strong public finances, a stable banking sector and low indebtedness. The country is considered a net foreign creditor. Based on three scales of the rating agencies, Chart 1 shows the historical development of the Czech Republic's rating from 1992 to 2020. Historically, the country received the first rating in 1992. The chart shows that, according to S&P, the rating has been stable since 2011.

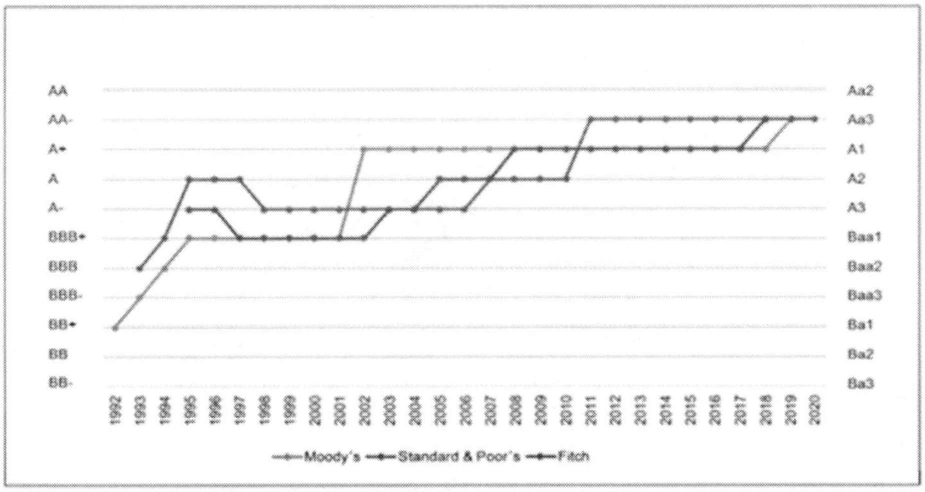

Fig. 1. Rating history of the Czech Republic

Source: Authors, according to Czech National Bank [16].

The credit insurance company Euler Hermes [17] published individual risks, which are assessed in a country's risk. The overall risk assessment is based on two elements - an ongoing assessment of the country and a short-term assessment of the business environment. This data is valuable information when making decisions to potential investors.

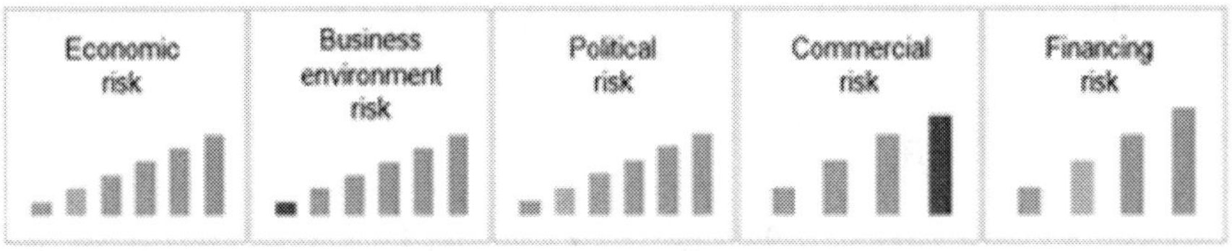

Fig. 2. Individual risks

Source: [17].

When investing in emerging markets, country risk is important and is a key part of estimating the capital cost of investment. Investors in developed countries can diversify country risk fluctuations, and therefore risk does not require an additional premium [18].

In Figure 1 below, the current country risk assessment map is divided into four levels on a scale from low to very high. These levels are based on four components, which are analyzed globally for all economic sectors [17]. These are demand, profitability, liquidity, and the business environment. The map also shows the Czech Republic, which has a low risk score. It is therefore on the same level as, for example, the United States of America, Australia or Western and Northern Europe.

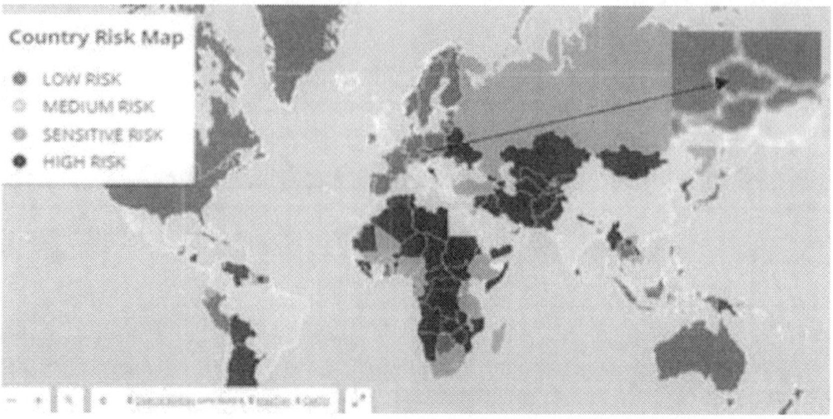

Fig. 3. Country risk assessment map

Source: [17].

Palic, Posedel and Vizek [19] point out that the "overheating" of the economy, together with unexpected increases in public debt, inflation and interest rates, all increase the volatility of the country's risk premium. A sudden increase in volatility oppresses the economy, exerts deflationary pressures on consumer prices, and then a strong and sustained rise in public debt follows [20-22].

The premium for equity risk, but also the premium risk of the country, is an important factor for investors' decisions. The most reliable risk assessment can be found through three world-renowned rating agencies, working with the country's central bank. As the Czech Republic does not have a sufficient database, it is necessary to adjust the calculation methods to determine the risk premium, which will then be compared with Damodaran's methods.

3 Data and methods

The methodology is divided into two parts – the methods used in the Czech Republic and Damodaran's methods, which are clearly described and for which formulas with explanations are given. These applied methods will then be compared with each other.

As is well known, Damodaran [23] deals with risk premiums, and therefore the data needed for the calculation comes directly from his website, where it is regularly updated. Statistical data is secondary and is presented in a table, but also in a line chart, which shows the historical development of the country's risk premium values. The data contained in the example for the Czech Republic, which relates to inflation, is obtained from the Czech National Bank [24]. These are time series with annual observation. The data is publicly available and updated for the year 2020.

3.1 Risk premium methods in the Czech Republic

Fotr and Soucek [25] stated that determining a company's risk premium is a bit more difficult in terms of the difficulty of determining the extent of business risk. The basic approach is based on the so-called CAPM (Capital Asset Pricing Model), i.e. the valuation of capital assets. The risk premium shall be determined based on the following formula:

$$RP = \beta * (R_m - R_d) \qquad (1)$$

R_m – average annual return on the equity market portfolio,
R_d – average annual return on government bonds.

In practice, the so-called modular method is widely used, determining the risk premium as a sum of several components. There are multiple versions of this method. The first is the so-called comprehensive modular method, which determines the company's risk premium based on the evaluation of a set of at least twenty risk factors. These are divided into two groups of financial and business risk factors. Another version of the modular method is the INFA method, which determines the risk premium as the sum of four individual premiums, defined by financial indicators [25].

The CRP model can be used to determine the country risk premium. Petrik [26] points out that in the case of less developed and emerging markets, it is not easy to obtain the necessary volatility parameters, which are expressed by standard deviations of returns analyzed from a long-term perspective. This situation also applies to the Czech Republic. A universal starting point is proposed for these cases, which is based on the term in the formula below being replaced by a flat rate of 1.5.

$$CRP = CDS_{nt} * (vsm_{nt}/vbm_{nt}) \tag{2}$$

$$CRP = CDS_{nt} * 1{,}5 \tag{3}$$

CDS_{nt} – Credit Default Spread risk,
Vsm_{nt} – Volatility of the stock market,
Vbm_{nt} – Volatility of the bond market.

3.2 Damodaran's methods

In this part, three approaches according to Damodaran [14] will be discussed. In the first approach, Damodaran assumed that the default margin for a country was a good measure of the additional risk faced when investing in equity and in a selected country. The simplest and most widespread method is the standard spread that investors charge for the purchase of bonds issued by countries. This range can be estimated in one of three ways. The first is the current standard spread in the government bond market or in the CDS (Country Default Spread) market. The second method is the average (normalized) bond spread and the last method is the synthetic spread.

The second approach assumes a linear relationship between equity risk premiums and stock market standard deviations. It is therefore assumed that if the US risk premium can be calculated (e.g. through historical data), then the country risk premium and the equity risk premium for country X are as follows:

$$RSD_{country\,X} = standard\ deviation_{country\,X}/standard\ deviation_{USA} \tag{4}$$

$$ERP_{country\,X} = risk\ premium_{USA} * RSD_{country\,X} \tag{5}$$

$$CRP_{country\,X} = ERP_{country\,X} - risk\ premium_{USA} \tag{6}$$

The third approach combines the first two approaches and uses information from the country default spread and stock market volatility. This examines the relationship between stock market volatility and the volatility of the bond market used to estimate the spread. In this way, an estimate is obtained for the country equity risk premium. This relationship has already been expressed in the methods used also in the Czech Republic.

4 Results

4.1 Risk premium methods in the Czech Republic

Marikova and Marik [27] found that in the Czech Republic it is necessary to examine the return on government bonds with a maturity of more than 10 years. It is more efficient in the calculation to use the average of several bonds with a similar maturity, as they provide more reliable data than when using only one bond.

Marikova and Marik [27] outline the calculation procedure typical for the Czech Republic. This procedure consists of four steps. For each step, current data is provided, which leads to the final result of the risk premium.

1. Find out the rating of the Czech Republic, which is published on the website of the Czech National Bank or on the individual websites of rating agencies. Currently, the rating is Aa3 or also AA-.
2. Project the country's rating into the value of the risk premium or the country's default risk. The difference between the return on bonds with the same rating as the Czech Republic and the US government bonds with the highest (Aaa) rating is used. Damodaran [23] states that the value of the country's default risk is 0.51%.
3. Adjust the country default risk by the difference between stock market volatility and bond market volatility according to formula (2). For easier calculation, the fraction is replaced by an approximate coefficient of 1.5 according to formula (3). After reaching the numerical values, the country risk premium formula looks like this:

$$\text{CRP} = 0.51\% * 1.5 = \mathbf{0.765\%} \tag{7}$$

4. Adjust the country risk for differences in long-term projected inflation between the Czech Republic and the USA. The rate of inflation in the Czech Republic as of May 2020 is 2.6% and 2% in the USA. The difference in inflation is therefore 0.6%. The country risk premium adjusted by the difference in inflation is:

$$\text{CRP (adjusted by the inflation difference)} = 0.765\% + 0.6\% = \mathbf{1.365\%} \tag{8}$$

5. The last part of the calculation is the sum of the average value of returns on US bonds, which is set at 4.88%, and the country's risk premium.

$$\text{Total risk premium of the Czech Republic} = 4.88\% + 1.365\% = \mathbf{6.245\%} \tag{9}$$

4.2 Damodaran's methods

On his blog, Damodaran [23] provides data on individual countries, where the values of the equity risk premium (ERP), the country risk premium (CRP) and the country default spreads (CDS) are already calculated directly. Table 1 compares the data of the Czech Republic from 2018 and 2020, where there was a change in the rating. It's possible to say that the better the rating, the lower the risk premium.

Table 1. Data regarding the Czech Republic

	Year 2018	**Year 2020**
Moody´s rating	A1	Aa3
S & P´s rating	AA-	AA-
CDS	0.72%	0.51%

CRP	0.81%	0.60%
ERP	5.89%	5.80%

Source: Own processing, according to Damodaran [23].

Table 2 presents the standard deviations of stocks and bonds.

Table 2. Standard deviations of stocks and bonds

	Value
σ_{equity}	13.82%
Σ_{bond}	4.93%
$\sigma_{equity} / \sigma_{bond}$	2.80
σ(CDS)	0.3475%
$\sigma_{equity} / \sigma_{CDS}$	0.32

Source: Own processing, according to Damodaran [23].

Chart 2 below shows the values from the archive of Damodaran [23] in the period of years 2001 to 2019. The output of the graph is the historical development of the country's risk premium value, which after 2010 is around 1.

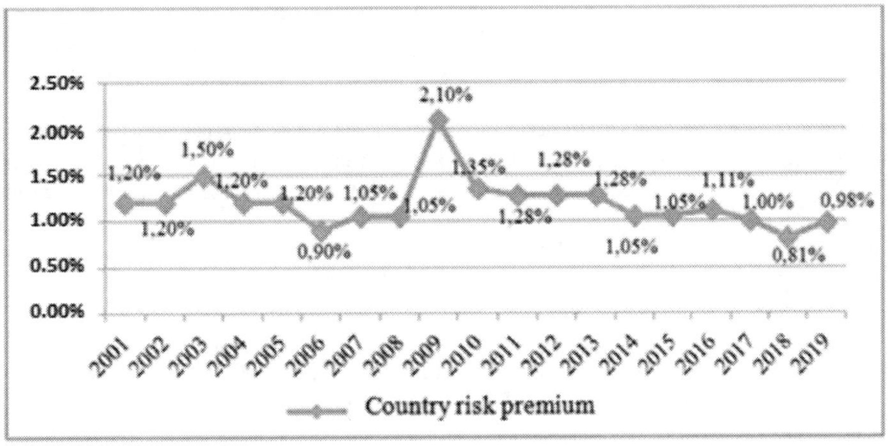

Fig. 4. Country risk premium development

Source: Own processing, according to Damodaran [23].

4.3 Comparison between methods used in the Czech Republic and Damodaran's methods

The CAPM model is used in the Czech Republic. It is used to value assets on capital markets and at the same time to inform the investor about the risk of assets. In the case of the country risk model, the value of summed volatilities is replaced by a coefficient of 1.5. This method of calculation is easier but less accurate, as the coefficient is only indicative. Damodaran's methods mainly feature data from the US capital market, as it has a long history.

5 Discussion

For Czech as well as foreign investors, the easiest way to obtain a risk premium is to use the company's investment rating. Most large companies are rated by world-renowned rating agencies.

In the case of determining the risk premium of the Czech Republic, the easiest method is the CRP model. Most of the statistical data is stored together on Damodaran's website.

If the country's market does not have a long history or does not have such a developed capital market, it is recommended to apply data from the US capital market. However, I am inclined to the opinion of Marikova and Marik [27] that there are significant differences in the economy between Europe and the USA, so the data of an European country such as Germany, which has historical risk premium calculations, should be used.

According to Haskova and Vochozka [28], the economy of European countries is greatly influenced by the economic development of Germany. Austria, Belgium, France, Luxembourg, the Netherlands and Finland also have similar economic performance, with a similar impact on neighboring European countries.

6 Conclusion

Based on the achieved results of the risk premium in the environment of the Czech Republic, there is a noticeable difference in the results of this value when using individual methodological calculation procedures. One of the reasons for this inequality may be the coefficient of 1.5 used in the calculation of the CRP for the Czech Republic. Another problem may be the sum of US bond yield values and a country-specific risk premium. This conclusion therefore contributes to the criticism of the use of the values set by Damodaran on the basis of the economic situation of large US companies in setting the risk premium of other countries.

The aim of the contribution was thus met. In the future, other theoretical economists could focus on the methodology of determining the risk premium of a particular country, or at least for the European individual continents. However, when using the Damodaran calculation methodology, in many cases the problem is the absence of individual input data for the calculation of the risk premium, which would be relevant for a particular country. Even so, in countries outside the US, this result could be considered more accurate when using more relevant data.

References

1. A. Damodaran, Equity risk premiums (ERP) – Determinants, estimation and implications. *Financial Markets, Institutions and Instruments*, **18**(5), 289-370 (2009)
2. M. Psarska, M. Vochozka, P. Sheng, Determining discount rate and investment evaluation in small and medium companies within Czech and Slovak conditions within the context of possible macroeconomic risks. *Journal of Valuation and Expertness*, 2(1), 28-47 (2017)
3. M. Marik et al., *Metody oceňování podniku: proces ocenění, základní metody a postupy [Business valuation methods: valuation process, basic methods and procedures]*. 2nd ed. Prague: Ekopress (2007)
4. L. Kruschwitz, A. Löffler, G. Mandl, Damodaran´s country risk premium: A serious critique. *Business Valuation Review*, **31**(2-3), 75-84 (2012)

5. M. Marik, P. Marikova, Přirážky k diskontní míře – teoretické a praktické problémy modelu rizika země [Discount rate surcharges - theoretical and practical problems of the country risk model]. *Odhadce a oceňování podniku [Appraiser and valuation of the company]*, **20**(1), 5-17 (2014)
6. A. Damodaran, Equity Risk Premiums (ERP): Determinants, estimation and implications – The 2019 edition. *SSRN Electronic Journal*, (2019)
7. J. M. Chen, The equity risk premium and the equity premium puzzle. *Finance and the Behavioral Prospect*, 137-179 (2016)
8. J. Siegel, Perspectives on the equity risk premium. *Financial Analysts Journal*, **61**(6), 61-73 (2005)
9. A. Damodaran, Value and risk: Beyond betas. *Financial Analysts Journal*, **61**(2), 38-43 (2005)
10. T. Kliestik, K. Kocisova, M. Misankova, Logit and probit model used for prediction of financial health of company. *Ekonomika a finance procedur*, **23**, 850-855 (2015)
11. T. Kliestik, The application of mathematical modeling to predict the financial health of bussinesses. *Advances in Economics, Business and Management Research*, **56**, 298-302 (2018)
12. A. Corelli, Company valuation. *Analytical Corporate Finance*, 257-295 (2018)
13. Ministry of Finance of the Czech Republic [online], Available at: www.mfcr.cz (2020)
14. A. Damodaran, Country risk: Determinants, measures and implications – the 2015 edition. *SSRN Electronic Journal*, (2015)
15. Czech National Bank [online], *Ratingové ohodnocení vybraných zemí [Rating of selected countries]*, Available at: www.cnb.cz (2020)
16. Czech National Bank [online], *Globální ekonomický výhled – únor 2020 [Global Economic Outlook - February 2020]*, Available at: www.cnb.cz (2020)
17. Euler Hermes [online], *Risk maps by country*, Available at: www.eulerhermes.com (2020)
18. A. Naumoski, Estimating the country risk premium in emerging markets: the case of the Republic of Macedonia. *Financial Theory and Practice*, **36**(4), 413-434 (2012)
19. P. Palic, P. Posedel, M. Vizek, The determinants of country risk premium volatility: Evidence from a panel VAR model. *Croatian Economic Survey*, **19**(1), 37-66 (2017)
20. P. Rousek, M. Vochozka, M. Psarska, Analysis of the consumption on the goods and services market in the cybernetic model of the 15 older EU member states. *Ad Alta: Journal of Interdisciplinary Research*, 9(1), 252-254 (2019)
21. P. Rousek, Analysis of Consumption on the Goods and Services Market in the Cybernetic Model of Selected Country. *SHS Web of Conferences: Innovative Economic Symposium 2018 - Milestones and Trends of World Economy (IES2018)* (2019)
22. P. Rousek, Cybernetic view of multipliers in macroeconomic theory. *12th International Days of Statistics and Economics*, pp. 1506-1515 (2018)
23. A. Damodaran [online], Available at: http://pages.stern.nyu.edu/~adamodar/ (2020)
24. Czech National Bank [online], *Prognóza ČNB z května 2020 [CNB forecast for May 2020]*, Available at: www.cnb.cz (2020)
25. J. Fotr, I. Soucek, *Podnikatelský záměr a investiční rozhodování [Business plan and investment decisions]*. Prague: Grada (2005)

26. T. Petrik, *Ekonomické a finanční řízení firmy [Economic and financial management of the company].* 2nd ed. Prague: Grada (2009)
27. P. Mariková, M. Marik, *Diskontní míra pro výnosové oceňování podniku [Discount rate for revenue valuation of a company].* 1st ed. Prague: Oeconomica (2007)
28. S. Haskova, M. Vochozka. Duality in cyclical trends in European Union confirmed. *SAGE Open*, **8**(1), 1-7 (2018)

Appendix

Moody's		Standard & Poor's		Fitch
Investment grades				
Aaa	Australia, Canada, Denmark, Germany, Luxembourg, N. Zealand, Norway, Singapore, Sweden, Switzerland, USA	AAA	Australia, Denmark, Germany, Liechtenstein, Luxembourg, Netherlands, Norway, Singapore, Sweden, Switzerland	Australia, Canada, Denmark, Germany, Luxembourg, Netherlands, Norway, Singapore, Sweden, Switzerland, USA
Aa1	Austria, Finland	AA+	Austria, Finland, Hong Kong, USA	Austria, Finland
Aa2	France, Kuwait, S. Korea, United Kingdom	AA	Belgium, France, Kuwait, N. Zealand, S. Korea, United Kingdom	Belgium, France, Hong Kong, Kuwait, N. Zealand, S. Arabia, United Kingdom
Aa3	Belgium, **CZECH REPUBLIC**, Hong Kong, Chile, Qatar, Taiwan	AA-	**CZECH REPUBLIC**, Estonia, Ireland, Israel, Qatar, Slovenia, Taiwan	**CZECH REPUBLIC**, Estonia, S. Korea
A1	Estonia, China, Israel, Japan, S. Arabia	A+	Chile, China, Japan, Slovakia	Chile, China, Ireland, Israel, Malta, Slovakia, Taiwan
A2	Ireland, Malta, Poland, Slovakia	A	Iceland, Lithuania	Iceland, Japan, Slovenia
A3	Iceland, Latvia, Lithuania, Malaysia, Mexico	A-	Latvia, Malaysia, Malta, Poland, S. Arabia, Spain	Latvia, Lithuania, Malaysia, Poland, Spain
Baa1	Slovenia, Spain	BBB+	Mexico	
Baa2	Bulgaria, India, Indonesia	BBB	Hungary, Indonesia, Italy, Portugal	Bulgaria, Hungary, Indonesia, Italy, Kazakhstan, Mexico, Portugal, Russia
Baa3	Hungary, Italy, Kazakhstan, Portugal, Romania, Russia, S. Africa	BBB-	Bulgaria, Croatia, Cyprus, India, Kazakhstan, Romania, Russia	Croatia, India, Romania
Non-investment grades				
Ba1		BB+	Serbia, S. Africa	Cyprus, Northern Macedonia, Serbia, S. Africa
Ba2	Brazil, Croatia	BB	Vietnam	Brazil, Greece, Turkey, Vietnam
Ba3	Cyprus, Serbia, Vietnam	BB-	Brazil, Greece, Northern Macedonia	
B1	Albania, Greece, Montenegro, Turkey	B+	Albania, Montenegro, Turkey	Mongolia
B2		B	Argentina, Belarus, Bosnia-Herzegovina, Egypt, Ukraine	Belarus, Egypt, Ukraine
B3	Belarus, Bosnia - Herzegovina, Egypt, Moldova, Pakistan	B-	Mongolia, Pakistan	Pakistan
Caa1	Mongolia, Ukraine	CCC+		
Caa2	Argentina, Cuba	CCC		Argentina
Caa3		CCC-		
Ca		CC		

Fig. 5. Credit rating of countries

Source: Own processing, according to Czech National Bank [15].

An Analysis of the Time Series of the PepsiCo, Inc. (PEP) Share Price and a Prediction of its Development

Eva Kalinová[1,*] and *Michal* Tlustý[1]

[1]Institute of Technology and Business, School of Expertness and Valuation, Okruzni 517/10, 37001 Ceske Budejovice, Czech Republic

Abstract. The paper deals with the topic of stock time series and their forecasts. The aim of the paper is to provide a detailed analysis of the time series of PepsiCo, Inc. (PEP) shares and subsequently, to use machine tools to predict its further development. The results of the paper in their first part identify the boundaries of RSI indicators for the oversold or overbought market and moreover, with the help of these indicators, predict further possible development of the time series. Daily share price data from PepsiCo, Inc. from the end of April 2019 to the end of April 2020 are used. Two indicators have been recorded since the beginning of 2020. The first one, recorded at the end of January, recommended the trader to sell the position, and further developments suggest that this step would be correct, as stock values then began to fall sharply. The second indicator was identified at the beginning of March, when the world was shaken by the coronavirus crisis, and this indicator recommended the trader to buy a position. Further developments indicated that this would be the right move, as stock values rose in the following period. Moreover, the moving average method is employed as well. Values are calculated for 50, 100 and 200 days. With this tool, only one recommendation is identified, in March 2020, when the fast moving average intersected both slower moving averages, and the ideal solution was to sell the position, which could be described as the right step with the further development of the time series, because stock values began to fall again after this period.
Keywords: prediction, stock price development, time series, moving average, RSI indicators, strategy

1 Introduction

The issue of stocks is currently one of the most topical matters, due to the coronavirus pandemics. Most companies are closed down or produce only at half capacity, which reduces their market value. Countries have closed their borders as a result of the pandemics, which leads to a slowdown in most economies [1]. In the current situation, these restrictions seem logical and proportionate, but in the long run, they may have a much greater and more devastating impact on national economies. It will start with the collapse of smaller

* Corresponding author: kalinova@mail.vstecb.cz

companies that do not have the capital to compensate for the absence of sales for months, despite the fact that they still have fixed and other expenses. This will be followed by the downfall of companies that could produce, but because 80% of their customers are larger companies that have also stopped their production, they will have no one to sell to and will be forced to stop [2-4]. This will lead to redundancies, the inability of people to provide for their families, foreclosures, loss of housing etc. However, every pandemics subsides over time and the aforementioned fall of shares may be an opportunity for investors to buy shares and increase their financial resources. In the current situation, however, it is very difficult to analyse and predict the future development of stocks. Despite the fact that the world already has experience with the economic crisis, the situation is different now.

The stock time series indicates how the stock value of a particular company has changed over time and helps to retrospectively evaluate contexts that may not be clear at first glance [5-6]. This data, often plotted, help people predict future developments in the data studied and prepare for these situations in advance. However, the reliability of the forecast is determined by various other factors and often may not correspond to actual developments. The purpose of predicting the time series of a stock is to create a model that, based on mathematical data, will provide an indicative outline of future developments. Time series can be divided into short-term and long-term ones [7-8]. Short-term periods are shorter than one year. For example, the time series can be analysed daily, weekly, monthly, etc. Long-term time series indicate a research period of one year or more. Another division of time series is interval and instantaneous. Interval time series are those that report data for a predetermined interval that we want to monitor. It can be, for example, the volume of the company's production in one month. Instantaneous time series are data for a specific moment, not an interval [9-11]. This variant can document, for example, the number of inhabitants of the Czech Republic at a specific point in time, e.g. 31 December 2019.

The purpose of this paper is to analyse time series o of PepsiCo, Inc. (PEP) shares. The company was founded in 1898 in the United States of America and entered the Czech market in 1993. It is the second most important producer of soft drinks in the world and its leading product is the world-famous cola flavoured lemonade, Pepsi. Among other things, the company sells beverages such as Mirinda, 7UP or, for example, the energy drink Rockstar.

The aim of the paper is to provide an analysis of selected methods of time series machine prediction on the example of the PepsiCo, Inc. share price development.

2 Literature research

The correct determination of a trend and its use is an absolutely key component for trading in shares, without which the investor can hardly succeed and value their financial resources. The value of every stock evolves in time and for the trader, it is important to determine the actual trend and, consequently, to make their decisions accordingly [12].

These trends are determined by the rising and falling elements of a stock. If the stock price rises and then turns and starts to fall, the highest point reached is called "high". Conversely, if the value of a stock falls and turns at a particular point and begins to rise, we call the lowest point "low" [13]. The first type of trend is the so-called "rising trend". It indicates that rising values of the particular stock are higher than the falling ones, which means that the value of a share is rising in the long run. In this case, it is a signal for the trader that the stock is developing appropriately and should, after evaluating other factors, be considered for a purchase [14]. In the opposite case, there is the so-called "declining trend" which means that the increasing values are inevitably shorter than the declining values and the long-term share value heads, in this case, rather downward. Consequently, businessmen thinking rationally should avoid the purchase of such stocks or evaluate other factors carefully [15]. The last type of stock trend is "sideways movement", which signals rising and falling values at roughly the same intervals, which in the long run means that the stock is moving at about the same level and its value is, therefore, not rising nor falling [16].

From the point of view of time, stock trends can be divided into long, medium and short-term. Long-term trends are sometimes referred to as key trends and have the most fundamental influence in analysing the value of a stock because they define the long-term development of the stock value. If this trend is increasing, the further developments can be expected to be similar. However, this is not a rule [17]. The medium-term trend takes place in a shorter time horizon than the long-term trend, which means that the medium-term trend may not only show the increasing values, but also the decreasing ones, even if the value of the stock increases in the long run. Out of the above-mentioned trends, it is the short-term trend that indicates the shortest period of time. This trend fluctuates the most because it is affected by "fresh" information published in hour or days. It gives the investor the opportunity to assess whether the long-term, medium-term and short-term trends are in line and their development is similar [18].

Another important element that has to be defined is the moving average (MA). This statistic figure can be calculated for any data set, however, in connection with the technical analysis, stock price data samples are taken into account the most often. The moving average is a numerical value of the arithmetic mean in the course of time [19]. This means that if it is necessary to calculate the moving average for one month, we will use thirty different stock prices from the previous thirty days and every other day the value of the moving average will change, because one day at the beginning will no longer be taken into account and will be replaced by the current day added to the data file [20].

For every investor, it is also important to be familiar with the RSI indicators. Using this tool, the relative strength of the market can be quickly and transparently calculated. The main criteria for measuring a market using indicators are either an overbought or vice versa an oversold market. The indicator values range from 0 to 100 %. Most investors prefer the limit at 80 % for an overbought market and 20 % for an oversold market [21]. In case that the value of the index exceeds 20% and more, the investor is informed that he should consider buying a share. Similarly, if the value of the index falls under 80%, the investor should consider selling the investment. RSI is an essential tool included nowadays in each analytical software, which helps businessmen read the stock time series [22-23].

Until recently, European investors were not familiar with the term candlestick graph and the related Japanese candle. It originated in Japan around the 18th century, when its founder, Munehisa Homma, was able to record a tremendous return on investment with its help. The candlestick graph is used with a specific time limit, often one day, and the candle itself has four parts that provide information about the minimum, maximum, opening and closing share prices in the observed time horizon [24]. The Japanese candle has a column structure and its colour varies depending on whether the value of the stock decreased or increased during the reviewed period. If the opening price is lower than the closing price, it means that the value of the stock increased during the day and the candle will therefore be green. In the opposite case, where the opening price is higher than the closing, the candle has the colour red and this represents a decrease in the value of the shares [25].

In general, traders deal with two types of analyses that can be used to monitor stock developments. The first type is the fundamental analysis. It deals mainly with data that comes from the macroeconomic sector of the company, the country etc. An example of data can be the gross domestic product rate of a country, the percentage of unemployment or, for example, the political situation of a particular country [26]. The second type is a technical analysis, which deals more with stock trends, RSI indicators etc. [27]. The main difference from the fundamental analysis lies in the fact that the technical analysis works with the assumption that history repeats itself. This means that a stock showing a long-term growth can be expected to grow further and respond accordingly to developments. This condition can be applied even to more complex situations, for example, if there was a situation A in the past followed by situation B, it can be assumed that now, when the market is in situation A, situation B will follow again. This piece of information can be of a key importance for a potential investor [28].

The stock time series can be evaluated with a wide range of tools. Experts agree that the basic practice is to monitor trend graphs, whether short-term, medium-term or long-term. Furthermore, in the scientific literature, it is also stated that it is important to monitor the RSI indicators, which can make it easier for the investor to decide whether to buy or sell a share. There is also a mention of the famous candle graphs, which facilitate the clarity of the most common daily results. Furthermore, the difference between fundamental and technical analysis is often described. Experts present the opinion that the technical analysis is more accurate and reliable. Therefore, this paper will deal with the technical analysis of the stock time series and its machine prediction.

3 Data and methods

For each analysis and prediction of the stocks time series, it is necessary to use specific data. All data necessary for this paper were obtained from the website yahoo! finance [29]. The data selection will include data in the annual time period with a daily frequency. The calculations will be performed using the closing price "close". The first value of the time share will be from 30th April 2019 and the last value will be the formula from 29th April 2020. An example of the data, in this case in a period of one year, can be seen in the following table. For illustrative purposes, the values of the monthly intervals are shown. The final calculations will be performed with daily data.

Table 1. Example of the data

Date	Open	High	Low	Close	Adj Close	Volume
05/01/2019	127.85	130.90	124.85	128.00	124.48	88149400
06/01/2019	127.86	135.24	127.49	131.13	127.52	88686800
07/01/2019	131.99	135.24	126.24	127.81	125.20	96373300
08/01/2019	128.96	137.11	123.88	136.73	133.94	95177800
09/01/2019	136.35	139.18	134.10	137.10	134.31	78024000
10/01/2019	136.93	140.45	133.49	137.17	135.30	80469000
11/01/2019	137.55	138.34	131.59	135.83	133.98	64684200
12/01/2019	136.65	138.25	134.32	136.67	134.81	82358000
01/01/2020	136.87	144.65	133.95	142.02	141.07	89890800
02/01/2020	142.77	147.20	128.35	132.03	131.14	94688300
03/01/2020	132.00	142.71	101.42	120.10	119.29	219268700
04/01/2020	117.27	138.63	115.67	136.32	136.32	101828900

Source: Yahoo ! finance [online]. New York City, New York, USA [quoted: 2020-05-05]. Available from: https://finance.yahoo.com/

One way to evaluate and analyse a time series of stocks is by using simple moving averages (SMAs). The difference between this tool and the usual arithmetic mean lies in the fact that the moving average changes with the passing time. Moving averages can be calculated for thirty, fifty or even a hundred days. The purpose of this quantity is the competence of the moving average to change with time. With the shift of time by one unit (most often a day), the oldest value is removed from the calculation and the formula is supplemented by the most current value (therefore "moving"). The values for this calculation are therefore irregular and changing, with only the number of time units for which the quantity is calculated. The tool can be used for any data set, which is to change over time, but in the field of technical analysis is most often used for price values. It is recommended to use the moving average especially when analysing a longer time series for which the use of the trend function would be insufficient. Experts recommend always using an odd number of moving values. The relationship that applies to the calculation of simple moving averages is as follows:

$$SMA = \frac{(P_1 + P_2 + \dots + P_n)}{n} \quad (1)$$

Where:
SMA = moving average value,
P = share price in a specific time unit (most often the closing price of the day),
n = total number of values (most often days).

Due to the fact that moving averages work with past data, these are delayed characteristics. For this reason, in addition to simple moving averages, a modified version of the calculation was created. These are exponential moving averages (EMA). While for simple moving averages all values have the same weight, in the calculations of exponential moving averages the weight of values is significant. In practice, this means that the old values have a lower weight for the calculation, and the weight then increases linearly the more current the data is. The latest data then have the most significant weight. However, this does not mean that exponential moving averages are better or more valuable indicators. It always depends on the current situation and the preference of the investor, which of these two types will suit them better.

The moving average characteristics can be compared with the current value of the stock. In the case where the stock price moves above the value of the moving average, it is a so-called bull trend (i.e. growth). The opposite is the bear trend which occurs when the stock price is below the moving average.

In practice, it is still possible to encounter strategies of two or three moving averages. In the case of two moving averages, one of them is "slow" and the other one is "fast". Slow means that the moving average describes more values that are in a shorter time interval, while the fast moving average generally describes the same time period, but the interval of its values is shorter. For example, if the values of the slow moving average are represented in daily intervals, the values of the fast moving average are stated in hours. In practice, this means that a fast MA outlines the trend curve more. The outputs of this strategy are as follows. If a faster MA crosses a slow one upwards, it is an indicator for the trader to purchase. In the opposite case, the faster MA crosses the slow MA downwards, it is an indicator for sales positions. The principle of the three moving averages strategy is very similar, with the difference that in addition to the slow and fast ones, there is also a medium moving average.

Analysing stock time series using RSI indicators is also a very common tool used in technical analysis. There is an inexhaustible number of indicators, but only a few of them are proven. This is a universal tool that can be used on any market for any length of the time period. The calculation of the RSI indicator is usually based on the closing price of the share and for its calculation, it is necessary to know or determine the appropriate period. The period provides information about how many specifically long time periods the calculation will contain. For example, if the period is twelve and the RSI indicator is required to be calculated in hourly graphs, the last twelve hours will be calculated. The relationship for calculating the RSI indicator is as follows:

$$RSI = 100 - \frac{100}{(1+RS)} \qquad (2)$$

Where:
RSI = RSI indicator value,
RS = average closing price gain/average closing price loss.

To calculate the value, it is therefore necessary to first calculate the value of RS. In this particular case, mentioned above, it would be necessary to calculate how many closing prices out of a total of twelve were higher than their previous closing price and then divide this number by twelve. Furthermore, it needs to be found out how many closing prices were lower than their previous closing price and divide the value again by twelve. Finally, the result from the first point should be divided by the value from point two. In the case that there are more closing prices higher than their previous price than prices lower than their previous value, the RS result value will reach more than 1.

The resulting value of the RSI indicator will range in the interval 0-1 , which means 0-100%. The basic meaning of using RSI indicators can be found in deciding whether to buy a position or, conversely, give it up and sell it. The approach recommended by experts is that if the resulting value of RSI indicators exceeds the limit of 20% upwards, the price has gotten out of the oversold area and it is advantageous to make a purchase position. In the opposite case, when the value exceeds the limit of 80% downwards, the price reached the overbought area and is generally recommended to give up the position on the market. If the resulting value is between 20 and 80% and none of them exceeds, it is recommended to hold the position and wait for further changes in the share price on the market. However, experts warn that it is not optimal to make a purchase or sale immediately after the limit of 20 and 80% is exceeded. If one of these values is exceeded, it is recommended to wait until the stock price returns from overbought or oversold area and to then make a purchase or sale.

4 Results

Data containing values of PepsiCo, Inc. (PEP) shares from April 30, 2019 to April 29, 2020 were downloaded from the above-mentioned internet source. The data was subjected to mathematical calculations which are explained in the methodological part of the paper. All calculations, as well as the displayed graphs were created using Excel.

First, the RSI values were calculated using a fortnight period. The 20% limit was chosen for the overbought market and the 80% limit was chosen for the oversold market.

Subsequently, the moving averages were calculated. The time period chosen for the calculation was set at 50, 100 and 200 days. All types of moving averages were based on the closing share price. For a simpler presentation, the moving averages have been calculated for a period of two years, so that the values of the moving averages were displayed from the very first day in the inserted graph presenting the values of the last half year.

Last but not least, the graph shows also the trading volumes of shares with a period of one day in the lower part.

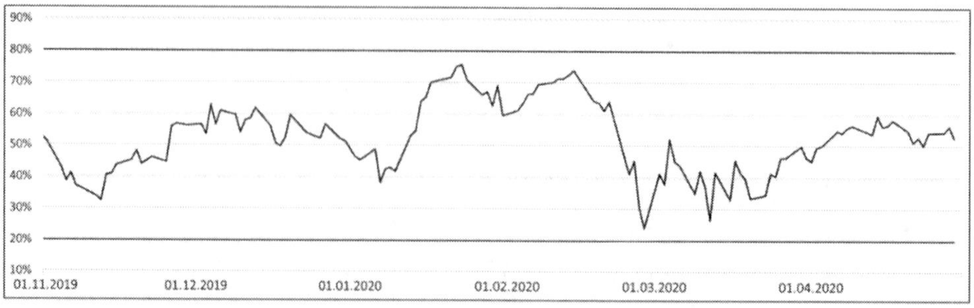

Fig. 1. RSI Indicators

Source: Own processing.

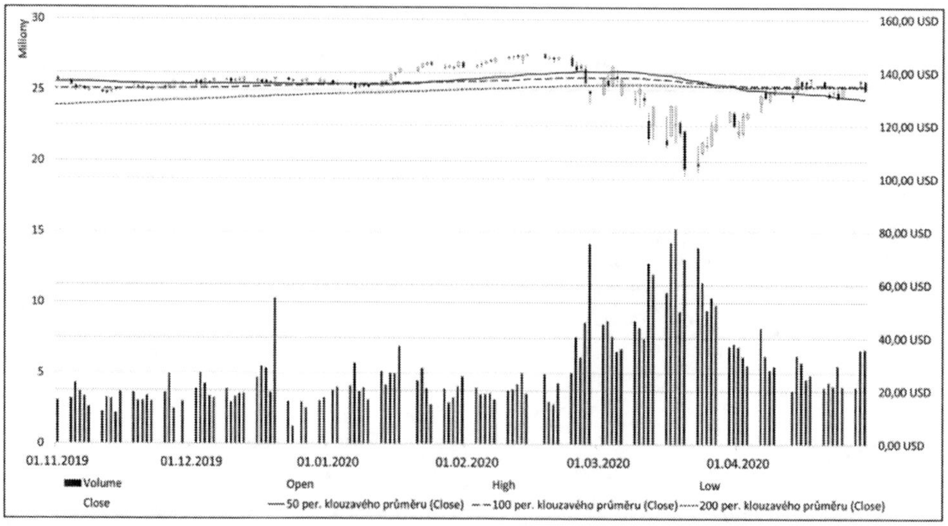

Fig. 2. Trend development, moving averages and values of trade volumes

Source: Own processing.

5 Discussion

Generally speaking, the values of most shares have fallen sharply since the beginning of 2020 due to the coronavirus crisis, and some stock market experts are talking about the worst values in the last seven years. The surveyed company PepsiCo, Inc. is no exception to this. This company was also affected by the coronavirus crisis, as it is evident in the values of shares and its trading volumes. This slump in the world stock market could have caused major problems, sometimes even bankruptcies for smaller businesses despite economic help from the government. However, huge companies such as Pepsi undoubtedly can get out of these problems without much difficulty and may even profit from this situation.

In the first part of the results, RSI indicators were examined. Limits of 80% were set for the overbought market and 20% for the oversold market, and the indicators were calculated with a period of fourteen days. It is clear from the documented graph that the development curve does not cross any of these boundaries at any time. This means that the value of the shares is neither in the oversold area nor in the overbought area. Therefore, the investor should not make any purchases if they are considering making an investment, and if they already own a certain volume of shares, they should not give up their positions and should wait for further development. In scientific literature, it is stated that investors often use different criteria for the boundaries of oversold or overbought market as well, namely 30% and 70%. Taking into consideration these limits, it can be argued that from the end of January to early February 2020, the boundary for an overbought market was crossed, which would be an indicator for the trader to sell their position. Looking at further developments, it is clear that this consideration would be correct, as stock values began to decline in the future. They reached their lowest levels at the beginning of March 2020, when the whole world was shocked by the above-mentioned coronavirus crisis and the shares were in the oversold area. In the following weeks, the world began to recover and the curve crossed the border of the oversold area, and at that moment it was an indicator for the investor to purchase. It is clear from further developments that the share values have grown even further, so this statement would be correct and profitable in this case as well. Currently, the values are around 50-60% and whether the limits are set at 20 and 80% or 30 and 70%, there are currently no clear indicators for buying or selling positions, so the recommendation is to wait and monitor further developments in stock values.

The second part of the results was devoted to examining the development of the trends, moving averages and trading volumes. Moving averages for 50, 100 and 200 days were calculated. From the above-mentioned graph, it can be concluded that the course of all examined trends did not oscillate much from the beginning of November 2019 to the beginning of March 2020 and more or less copied the stock prices. Only during January, it can it be argued that the fast average was close to the medium-fast, but did not intersect it and rose again. This situation, however, changed abruptly in the course of March 2020 and culminated in April 2020 when a 50-day (fast) moving average crossed both the slower moving averages (the 200 and 100 days). According to the recommendation outlined above, in a situation where the fast moving average crosses the slower one downwards, it is recommended to give up the position held and expect prices to fall. It is clear from further developments that the aforementioned fall in stock prices has indeed occurred so it can be concluded that the development would have been predicted correctly. For the next period, it seems to be the best recommendation to wait for a situation where the faster moving average, which is now below the medium-fast and below the slow, intersects one of these and indicates the right time for buying a position.

Due to the fact that both RSI indicators and moving averages for different periods and trading volumes do not currently produce any signal that would recommend an investor to buy or sell their positions, it is difficult to predict the further development of the time series. If the fundamental analysis was taken into account, it could be argued from the development so far that the stock values were at higher levels by the beginning of March 2020, then fell sharply due to the coronavirus pandemics and currently, they are recovering from the crisis and rising steadily. Unless another wave of the coronavirus crisis hits the world, further growth back to normal levels can be expected. At the moment, careful market monitoring and expectations of a signal to buy or sell positions seem to be the ideal solution. Both RSI indicators and moving averages for different periods and volumes of trades have proven in practice as tools that can be used to successfully analyse time series and investors can base their purchases and sales on them. Primarily the RSI indicators seem to be the most beneficial, as they estimated the further course correctly in all cases and thus would save money of the potential investor, or vice versa, earn them a fortune. Moving averages for 50, 100 and 200 days provided only one recommendation signal, however, even in this case the estimate was successful. Both of these methods appear to be functional at any time, however, in exceptional cases, such as the current coronavirus crisis, it is recommended to supplement them with a fundamental analysis that these tools do not take into account. Therefore, their prediction could be incorrect without supplementing the fundamental analysis.

6 Conclusion

The aim of this paper was to use selected tools to evaluate and predict the time series of PepsiCo, Inc. shares.

The methodology of the paper described various opinions of other authors and presented basic information that is used for analysing and forecasting time series of shares. In the methodological part, the most suitable tools were selected and both their essence and other more extensive strategies were outlined. Specifically, two main methods of stock time series analysis were chosen, namely moving averages and RSI indicators, or trading volumes of the company. In the results part, the previously outlined tools were elaborated in detail and the output of the calculations was presented, namely two graphs showing both the RSI indicators and the trading volumes, Japanese candles and three types of moving averages (50-day, 100-day, 200-day). In the next part, a discussion of the results was created from these outputs where the results were presented in a clear form. At the same time, this examination of past data provided a proof whether individual strategies, whether in the form of RSI indicators or moving averages, would provide a correct evaluation. Furthermore, it was assessed whether would be successful in investing provided that he acted accordingly. In the end, these strategies proved to be successful, and the further development of the share price corresponded with the prediction.

From the above-mentioned results, it can be argued that both of these tools, i.e. RSI indicators and moving averages for different periods, would work in practice and appear to be optimal helpers for time series analysis. In the case that it would be necessary to provide only one most suitable instrument, the RSI indicators would be the most favourable choice, for the reason that this method gave three specific indicators to buy or sell positions and in all these cases, the prediction was successful.

To conclude with, it can be argued that the stated goal of the paper was met using the above information and both RSI indicators and moving averages appear to be beneficial in the technical analysis of time series.

Acknowledgements

This article was funded within the project SVV202007 "Creation of a research report for scientific research activities of students at the Institute of Technology and Business in České Budějovice".

References

1. E. Stoye, 'No one is allowed to go out': your stories from the coronavirus outbreak. *Nature*, **578**, 499-499 (2020)
2. T. Kliestik, K. Valskova, G. Lazaroiu, M. Kovacova, J. Vrbka, Remaining financially healthy and competitive: The role of financial predictors. *Journal of Competitiveness*, **12**(1), 74-92 (2020)
3. M. Kovacova, T. Kliestik, K. Valaskova, P. Durana, Z. Juhaszova, Systematic review of variables applied in bankruptcy prediction models of Visegrad group countries. *Oeconomia Copernicana*, **10**(4), 743-772 (2019)

4. T. Kliestik, J. Vrbka, Z. Rowland, Bankruptcy prediction in Visegrad group countries using multiple discriminant analysis. *Equilibrium-Quarterly Journal of Economics and Economic Policy*, **13**(3), 569-593 (2018)
5. M. Vochozka, J. Horak, T. Krulicky, Innovations in management forecast: Time development of stock prices with neural networks. *Marketing and Management of Innovations*, **2020**(2), 324-339 (2020)
6. P. Suler, J. Horak, T. Krulicky, Validation of the prediction of ČEZ stock prices. *Littera Scripta*, **13**(1), 194-210 (2020)
7. J. Vrbka, Z. Rowland, Stock price development forecasting using neural networks. *SHS Web of Conferences - Innovative Economic Symposium 2017: Strategic Partnership in International Trade* (2017)
8. B. Groda, J. Vrbka, Prediction of stock price developments using the Box-Jenkins method. *SHS Web of Conferences - Innovative Economic Symposium 2017: Strategic Partnership in International Trade* (2017)
9. M. Vochozka, J. Horak, P. Suler, Equalizing seasonal time series using artificial neural networks in predicting the Euro-Yuan exchange rate. *Journal of Risk and Financial Management*, **12**(2), (2019)
10. J. Horak, T. Krulicky, Comparison of exponential time series alignment and time series alignment using artificial neural networks by example of prediction of future development of stock prices of a specific company. *SHS Web of Conferences: Innovative Economic Symposium 2018 - Milestones and Trends of World Economy (IES2018)* (2019)
11. J. Vrbka, Z. Rowland, P. Suler, Comparison of neural networks and regression time series in estimating the development of the EU and the PRC trade balance. *SHS Web of Conferences: Innovative Economic Symposium 2018 - Milestones and Trends of World Economy (IES2018)* (2019)
12. A. Picasso, S. Merello, Y.K. Ma, L. Oneto, E. Cambria, Technical analysis and sentiment embeddings for market trend prediction. *Expert Systems with Applications*, **135**, 60-70 (2019)
13. A.P. Ratto, S. Merello, L. Oneto, Y.K. Ma, L. Malandri, E. Cambria, Ensemble of technical analysis and machine learning for market trend prediction. *2018 IEEE Symposium Series on Computational Intelligence (SSCI)*, **3**, 2090–2096 (2018)
14. D.P. Gandhmal, K. Kumar, Systematic analysis and review of stock market prediction techniques. *Computer Science Review*, **34** (2019)
15. D. Shah, H. Isah, F. Zulkernine, Stock market analysis: A review and taxonomy of prediction techniques. *International Journal of Financial Studies*, **7** (2019)
16. B. Weng, L. Lu, X. Wang, F.M. Megahed, W. Martinez, Predicting short-term stock prices using ensemble methods and online data sources. *Expert Systems with Applications*, **112**, 258-273 (2018)
17. G. Kumar, S. Jain, U.P. Singh, Stock market forecasting using computational intelligence: A survey. *Archives of Computational Methods in Engineering*, (2020)
18. M.P. Naeini, H. Taremian, H.B. Hashemi, Stock market value prediction using neural networks. *2010 International Conference on Computer Information Systems and Industrial Management Applications (CISIM)*, 132-136 (2010)
19. W. Wang, W. Li, N. Zhang, K. Liu, Portfolio formation with preselection using deep learning from long-term financial data. *Expert Systems with Applications*, **143** (2020)

20. V. Zakamulin, J. Giner, Trend Following with Momentum Versus Moving Average: A Tale of Differences. *SSRN Electronic Journal* (2018)
21. P. Marek, B. Sediva, Optimization and testing of RSI. *11th International Scientific Conference on Financial Management of Firms and Financial Institutions*, (2017)
22. M. Culik, Financial management of firms and financial institutions. *11th International Scientific Conference on Financial Management of Firms and Financial Institutions,* **319** (2017)
23. Y. Wan, Y.W. Si, A formal approach to chart patterns classification in financial time series. *Information Sciences*, **411**, 151-175 (2017)
24. S. Thammakesorn, O. Sornil, Generating trading strategies based on candlestick chart pattern characteristics. *Journal of Physics: Conference Series, 2018 11th International Conference on Computer and Electrical Engineering*, **1195** (2018)
25. R. Naranjo, M. Santos, *Advances in intelligent systems and computing*. Springer Verlag, **527**, 323-333 (2017)
26. K. Li, P. Mohanram, Fundamental analysis: Combining the search for quality with the search for value. *Contemporary Accounting Research*, **36**, 1263-1298 (2019)
27. K.S.K. Lam, L. Dong, B. Yu, Value premium and technical analysis: Evidence from the China stock market. *Economies*, **7** (2019)
28. J.N. Wang, H.C. Liu, J. Du, Y.T. Hsu, Economic benefits of technical analysis in portfolio management: Evidence from global stock markets. *International Journal of Finance and Economics*, **24**, 890-902 (2019)
29. Yahoo! finance [online], Available at: https://finance.yahoo.com/ (2020)

Creating efficiency model of higher education institutions

Jana Janikova[1,*]

[1]Institute of Technology and Business, School of Valuation and Expertness, Okruzni 517/10, 37001 Ceske Budejovice, Czech Republic

Abstract. The paper deals with assessing the efficiency of the faculties of universities of applied sciences. A model based on Data Envelopment Analysis was created, using specifically the CCR alternative. It is an input-oriented method, assuming the constant returns to scale. The model was subsequently validated on an example of a specific university. The results clearly show that the model was able to determine how and to which extent a given faculty is effective. It also provides the assessor with a tool to manage the efficiency of the faculty, as it is able to identify the causes of possible ineffectiveness. Further research focuses on the model´s applicability in the environment of all universities of applied sciences ideally all over the world.
Keywords: Data Envelopment Analysis, measuring efficiency, efficiency, university of applied sciences

1 Introduction

Measuring of management efficiency in the environment of universities of applied sciences (hereinafter referred to as UAS) is carried out in a specific way different from commonly used tools, and may even give room for their creative interpretation [1]. In the absence of market mechanisms for price educational outputs, Avkirant [2] states that UAS have to find alternative methods for measuring their management efficiency. A different form of financing UAS brings various views of assessing the efficiency of using public resources.

There are a number of methods to be used, e.g. historical and logical method, systemic approach, documentary analysis, observations, and interviews. Mora et al. [3] use these methods to analyse the distribution of funds to universities in Ecuador and find that UAS in Ecuador have a suitable tool for distributing these funds. Tran [4] deals with the analysis of the management of Japanese UAS and considers the mere distribution of state funds provided to UAS to be inefficient due to their different characteristics. In this context, in the CR, there was created "White Book", which served as a basis for the reform of the Czech tertiary education system. According to the Ministry of Education, Youth, and Sports [5], the goal was the stabilization and competitiveness of UAS on a global scale. There were created individual segments of UAS, thus improving the efficiency of UAS management in the CR.

* Corresponding author: janikova@mail.vstecb.cz

Fund from the state budget are allocated to individual UAS based on the share they have on the overall performance of all UAS.

Kowang et al. [6] deal with the massive reduction of the operating budget in Malaysia after 2016 and seeks new approaches to sustain stable growth in education and research. Sustainability is in the continuous creation of values in line with the Lean Six Sigma concept (hereinafter referred to as LSS). According to Kowang et al. [6], the goal of the LSS concept is to categorize UAS expenditures with added value, without added value, and operational added value.

The efficiency of public tertiary education across the EU countries, Japan, and the USA is analysed by St. Aubyn et al. [7] using semi-parametric methods and stochastic boundary analysis. The authors also compare resources with outcomes and results of education and research.

Zhang et al. [8] compare inputs in the form of research expenditures with outputs on the example of three research universities in Atlanta and Georgia in the period of 2002-2015, which covers the period before and after the economic crisis. Zhang et al. [8] argue that despite more intensive research funding after the economic crisis, it has not been possible to add value to the research outputs; according to the authors, there are many relevant factors to be considered which limit research activities of the academic and scientific workers.

In their study on the efficiency of UAS in Romania, Munteanu and Andrei-Coman [9] present an original overview of modern methods for assessing the efficiency of UAS. Mikušová [10] deals with the efficiency of resource processing and the question to what extent inefficient UAS in the CR should change the outputs to be efficient, for which she uses Data Envelope Analysis (hereinafter referred to as DEA). In DEA, Mikušová [10] uses the efficiency of education as an outcome. DEA was also used by Avkiran [2], Abbot and Doucouliagos [11] to measure the efficiency of individual Australian universities. The efficiency of Thai UAS using DEA is examined by Kantabutra and Tang [12]. According to Mikušová [10], DEA assesses the technical efficiency of homogeneous units such as schools, hospitals, banks, etc.; however, there are aspects to be considered. Pietrzak [13], Mikušová [10] recommend the division of the units into sub-groups by their specialization or faculties/departments. This ensures better homogeneity, thus providing better informative capability of the analysis results. On the other hand, Mikušová [10] or Klieštik [14, 15] believes that the availability of data of faculties or departments may play a decisive role. Flegg et al. [16] define the technical efficiency of homogeneous units as a ratio of a weighted sum of outputs to a weighted sum of inputs. The set of permissible options in DEA models is determined by the efficient frontier. According to Jablonský and Dlouhý [17], efficient units are those at this efficient frontier. Due to empirical orientation and the lack of necessity to include a wide range of a priori assumptions, Cooper et al. [18] defines DEA as a multilateral instrument to assess problems in various contexts. Jablonský and Dlouhý [17] argue the following prerequisites need to be met in order to use the DEA method:

- homogeneous production units,
- preferably lower inputs and higher outputs,
- appropriate inputs and outputs,
- semi-positive data (each production unit has at least one positive input and one positive output).

DEA meets all criteria for a proper determination of UAS efficiency, including the assessment of the individual UAS segments. The method is suitable also due to well-prepared methodology; however, as already mentioned, the data for certain organizational segments of UAS are not easily available.

The objective of the paper is to create a model of assessing efficiency of UAS based on the DEA method.

2 Data and methods

The outputs will be set according to their positive influence on the efficiency of management of the assessed faculties at the Institute of Technology and Business (hereinafter referred to as ITB). The analysis inputs are the funds used for personnel costs, costs of material, services, and fixed assets. The outputs will be the funds used for pedagogical, and contract activities of the individual faculties. For their processing, there will be used ITB data obtained from the iFIS (Integrated Financial Information System – an accounting module).

The efficiency of individual faculties will be measured and compared for the period of 2017 - 2019. For assessment of efficiency, an input-oriented CCR model with assumed constant return on scale will be used.

Efficiency is defined as a ratio of one input and one output, calculated using the following formula:

$$\text{efficiency} = \frac{\text{output}}{\text{input}}.$$

To assess the efficiency rate in practice, it is necessary to consider more than one input or output; the following formula is used:

$$\text{efficiency rate} = \frac{\text{weighted sum of outputs}}{\text{weighted sum of inputs}}.$$

Effective units with the highest value of efficiency determine the efficiency frontier. The units at this frontier are referred to as efficient, and will be compared with the others. Efficient unit is a unit that cannot use less inputs for the production of the same outputs, or a unit that cannot achieve higher outputs using the same inputs.

In the paper, there is used an input-oriented CCR model with assumed constant return on scale, that is, the number of outputs is directly changed with the change in the number of inputs. Technical efficiency is calculated as a ratio of a weighted sum of outputs to a weighted sum of inputs. The coefficients sought are those at the interval of <0, 1>, where the units with the coefficient of 1 are efficient, while those with lower coefficients are considered inefficient. The model determines the necessary reduction of inputs to ensure the efficiency of the monitored units.

The efficiency is assessed for four inputs and three outputs, with the inputs being personnel costs, costs of material, services, and fixed assets (tangible and intangible). The outputs are funds obtained for creative, pedagogical, and contract activities of faculties. There are three ITB faculties, whose efficiency is being examined: Faculty of Corporate Strategy (hereinafter referred to as FCS), Faculty of Technology (FT), and School of Expertness and Valuation (SEV). Their efficiency is examined for the period of 2017-2019.

In the CCR model, the unknown variables are assigned input weights I and output weights j with a unit k. For each unit (faculty), a separate model is created. The model for a set of units is based on the following objective function:

$$e_H = \frac{\sum_{j=1}^{n} u_{jH} y_{jH}}{\sum_{i=1}^{m} v_{iH} x_{iH}} \rightarrow \max \tag{1}$$

where the set contains the p number of units, while H is one of the p units. The function maximizes the ratio of weighted inputs and weighted outputs.

The function has the following limits:

$$\frac{\sum_{j=1}^{n} u_{jH} y_{jH}}{\sum_{i=1}^{m} v_{iH} x_{iH}} \leq 1, \forall k = 1, 2, \ldots, p. \tag{2}$$

The limits ensure that the ratio of inputs to outputs with the weights for the k-th unit was lower than or equal to 1.

Another limit is a non-negative condition, which ensures non-negativity of the unknown variables sought, specifically the weights:

$$u_{jH} \geq 0, \forall j = 1, 2, \ldots, n, \quad (3)$$

$$v_{iH} \geq 0, \forall i = 1, 2, \ldots, m. \quad (4)$$

The model can be modified to linear by maximizing the numerator of the objective function if the denominator equals 1.

$$e_H = \sum_{j=1}^{n} u_{jH} y_{jH} \rightarrow \max \quad (5)$$

$$\sum_{i=1}^{m} v_{iH} x_{iH} = 1 \quad (6)$$

$$- \sum_{i=1}^{m} v_{iH} x_{ik} + \sum_{j=1}^{n} u_{jH} y_{jk} \leq 0, \forall k = 1, 2, \ldots, p, \quad (7)$$

$$u_{jH} \geq 0, \forall j = 1, 2, \ldots, n, \quad (8)$$

$$v_{iH} \geq 0, \forall i = 1, 2, \ldots, m. \quad (9)$$

The assumption is that the unit H is inefficient. By creating a dual model for the primary model, it is possible to identify the units that form a set of model units of the inefficient unit, thus obtaining the coefficients λkH of the combination of model units that create a virtual efficient unit to the unit H.

The objective function of the dual model is as follows:

$$z_H \rightarrow \min. \quad (10)$$

The limits for this function are as follows:

$$x_{iH} z_H - \sum_{k=1}^{p} \lambda_{kH} x_{ik} \geq 0, \forall i = 1, 2, \ldots, m, \quad (11)$$

$$\sum_{k=1}^{p} \lambda_{kH} y_{jk} \geq y_{jH}, \forall j = 1, 2, \ldots, n, \quad (12)$$

$$\lambda_{kH} \geq 0, \forall k = 1, 2, \ldots, p. \quad (13)$$

The variable zH takes arbitrary values.

The value of inputs and outputs of the unit is calculated as a combination of inputs and outputs of model units.

For inputs:

$$x'_{iH} = \sum_{k=1}^{p} \lambda_{kH} x_{ik}; i = 1, 2, \ldots, m. \quad (14)$$

For outputs:

$$y'_{jH} = \sum_{k=1}^{p} \lambda_{kH} y_{jk}; j = 1, 2, \ldots, n. \quad (15)$$

The tasks will be solved in the add-in program MS Excel Solver. After the creation of a table from the inputs and outputs, variables – individual weights will be designated in MS Excel. The column "Conditions" will contain the individual functions, which are based on the selected model.

3 Results

The model was applied to the example of the Institute of Technology and Business in České Budějovice for the period of 2017-2019.

3.1 Assessment of faculty efficiency for the year 2017

The inputs for the assessment of ITB faculties for the year 2017 are the data given in Table 1.

Table 1. Inputs for specific example of ITB faculties for the year 2017

Faculty	Material consumption (in CZK million)	Personnel costs (in CZK million)	Fixed assets (in CZK million)	Services consumption (in CZK million)
Identifier of input	v_1	v_2	v_3	v_4
FCS	0.293	24.402	0.000	2.990
FT	0.167	24.155	0.000	2.954
SEV	0.097	4.575	0.000	0.571

Source: iFIS ITB, own processing.

The outputs of specific assessment of ITB faculties for the year 2017 are the data presented in Table 2.

Table 2. Outputs for specific case of ITB faculties for the year 2017

Faculty	Funds for creative activities (in CZK million)	Funds for pedagogical activities (in CZK million)	Funds obtained from contract activities (in CZK million)
Identifier of input	u_1	u_2	u_3
FCS	9.605	6.714	0.010
FT	8.965	12.453	0.598
SEV	2.787	0.735	1.019

Source: iFIS ITB, own processing.

As three faculties are being assessed, three models need to be created. Faculty efficiency is marked e and an order number.

The first faculty assessed in 2017 is FCS. In the case of the primary model, the result of the efficiency assessment is given in Table 3.

Table 3. Results of assessing primary model for FCS

Variable	Value
e_1 – faculty efficiency	0.866
u_1 –funds for creative activities	0.090
u_2 – funds for pedagogical activities	0.000

u_3 – funds obtained from contract activities	0.000
v_1 – material consumption	1.468
v_2 – personnel costs	0.000
v_3 – fixed assets	0.000
v_4 – services consumption	0.191

Source: Own processing.

The table above indicates that FCS is not efficient, as the value of e_1 is lower than 1. For this reason, a dual model is created. Its creation enables obtaining a suitable combination to achieve the efficiency of FCS. The result of the dual model for assessing FCS efficiency for the year 2017 is presented in Table 4.

Table 4. Result of assessing dual model for FCS

Variable	Value
z_1	0.866
λ_1	0.000
λ_2	0.556
λ_3	1.658

Source: Own processing.

The results indicate that the model using for FCS is FT and SEV, as the value of their variables is not zero. To obtain the optimal value of inputs for FCS, the coefficients are multiplied by the inputs of FT and SEV (efficient units). The optimal values of inputs are shown in Table 5.

Table 5. Optimal values of inputs for FCS

Input	Optimal value
Material consumption (in CZK million)	0.254
Personnel costs (in CZK million)	21.013
Fixed assets (in CZK million)	0.000
Services consumption (in CZK million)	2.589

Source: Own processing.

To be efficient in comparison with FT and SEV, FCS should reduce the consumption of material from the initial CZK 0.293 million to CZK 0.254 million, i.e. by CZK 39 thousand. In the case of personnel costs, there should be a decrease by CZK 3.389 million (from CZK 24.402 million to CZK 21.013 million). No reduction is necessary for fixed assets, as there was no fixed asset purchased in 2017. In the case of services consumption, the faculty should reduce the initial CZK 2.990 million to CZK 2.589 million, i.e. by 0.401 million.

The second faculty assessed in 2017 is FT. According to the results obtained, FT is efficient, as the value of e_2 equals 1. The important input for FT is v_2, personnel costs (with a value of 0.041). The important source for the FT model is the output u_2, funds for pedagogical activities (with a value of 0.080).

The third faculty assessed in 2017 is SEV. The results show that SEV is efficient, since the value of e_3 is 1. The important input for the model of SEV is v_2 – personnel costs (with a value of 0.219), while the important source is the output u_1 – funds for creative activities (with a value of 0.359).

3.2 Assessment faculty efficiency for the year 2018

The inputs for specific assessment of ITB faculties for the year 2018 are the data given in Table 6.

Table 6. Inputs for ITB faculties for the year 2018

Faculty	**Material consumption (in CZK million)**	**Personnel costs (in CZK million)**	**Fixed assets (in CZK million)**	**Services consumption (in CZK million)**
Identifier of input	v_1	v_2	v_3	v_4
FCS	0.144	25.150	0.000	2.871
FT	0.492	35.973	15.608	2.544
SEV	0.040	6.205	0.000	1.037

Source: iFIS ITB, own processing.

The inputs for the specific assessment of ITB faculties for the year 2018 are the data given in Table 7.

Table 7. Inputs for ITB faculties for the year 2018

Faculty	**Funds for creative activities (in CZK million)**	**Funds for pedagogical activities (in CZK million)**	**Funds obtained from contract activities (in CZK million)**
Identifier of input	u_1	u_2	u_3
FCS	9.795	12.137	0.000
FT	9.742	15.771	1.913
SEV	3.458	0.643	1.202

Source: iFIS ITB, own processing.

As three faculties are assessed, three models need to be created. Faculty efficiency is marked e and an order number.

The first faculty assessed in 2018 is FCS. The results of the primary model show that FCS is efficient, since the value of e_1 is 1. The important input for the model of FCS is v_2 – personnel costs (with a value of 0.040). The important source for the model of FCS is the output u_2 – funds for pedagogical activities (with a value of 0.082).

The second faculty assessed in 2018 is FT. The results indicate that FT is efficient, since the value of e_2 is 1. The important input for the model of FT is v_2 – personnel costs (with a value of 0.028). The important source for the model of FT is the output u_2 – funds for pedagogical activities (with a value of 0.058) and u_3 – funds obtained from contract activities (with a value of 0.048).

The third faculty assessed in 2018 is SEV. The results show that SEV is efficient, since the value of e_3 is 1. The important input for the model of SEV is v_2 – personnel costs (with a value of 0.161). The important source for the model of SEV is the output u_1 – funds for creative activities (with a value of 0.289).

3.3 Assessment of faculty efficiency for the year 2019

The inputs for the specific assessment of ITB faculties for the year 2019 are the data presented in Table 8.

Table 8. Inputs for ITB faculties for the year 2019

Faculty	Material consumption (in CZK million)	Personnel costs (in CZK million)	Fixed assets (in CZK million)	Services consumption (in CZK million)
Identifier of input	v_1	v_2	v_3	v_4
FCS	0.167	22.015	0.212	3.134
FT	0.948	43.052	5.788	4.439
SEV	0.052	7.794	0.000	0.991

Source: iFIS ITB, own processing.

The outputs for the specific assessment of ITB faculties for the year 2019 are the data presented in Table 9.

Table 9. Outputs for ITB faculties for the year 2019

Faculty	Funds for creative activities (in CZK million)	Funds for pedagogical activities (in CZK million)	Funds obtained from contract activities (in CZK million)
Identifier of input	u_1	u_2	u_3
FCS	5.312	10.624	0.011
FT	12.061	16.525	2.334
SEV	5.290	1.417	1.933

Source: iFIS ITB, own processing.

As three faculties are assessed, three models need to be created. Faculty efficiency is marked e and an order number.

The first faculty assessed in 2019 is FCS. The results show that FCS I efficient, since the value of e_1 is 1. The important input for the model of FCS is v_2 – personnel costs (with a value of 0.045). The important source for the model of FCS is the output u_2 – funds for pedagogical activities (with a value of 0.094).

The second faculty assessed in 2019 is FT. The results show that FT is efficient, since the value of e_2 is 1. The important inputs for the model of FT are v_2 – personnel costs (with a value of 0.015) and v_4 – services consumption (with a value of 0.079). The important sources are the outputs u_1 – funds for creative activities (with a value of 0.026) and u_2 – funds for pedagogical activities (with a value of 0.042).

The third faculty assessed in 2019 is SEV; the results show that SEV is efficient, since the value of e_3 is 1. The important input for the model of SEV is v_2 – personnel costs (with a value of 0.128), while the important source is the output u_1 – funds for creative activities (with a value of 0.189).

4 Discussion

To assess the efficiency of the ITB management, the data envelope analysis was carried out, which is used for assessing the management of sources of financing. The input-oriented CCR model was used since constant return on scale was assumed. There were identified efficient units, which were compared with the others. By analysis, the efficiency of individual ITB faculties (FCS, FT, SEV) in the period of 2017-2019 was assessed on the basis of four inputs and three outputs. The inputs included personnel costs, costs of material, services, and fixed assets (tangible and intangible). The outputs included funds for creative, pedagogical, and contract activities of the individual faculties. Appropriate selection of inputs and outputs plays an important role in terms of the informative capability of the analysis performed. To the analysis outputs, there should be always assigned the inputs that have links to the selected outputs. The monitored units are thus assessed on the basis of the resources used to achieve the outputs – inputs necessary to achieve the required results (outputs).

The analysis shows that in 2017, not all faculties were efficient. This applies to FCS, since the value of e_1 was lower than 1. For this reason, a dual model was created, which enabled obtaining a suitable combination to achieve the efficiency of FCS and identification of the optimal value of inputs in comparison with FT and SEV (efficient faculties). The analysis also determined a required reduction of material consumption from the initial CZK 0.293 million to CZK 0.254 million, i.e. a reduction by CZK 39 thousand. In the case of personnel costs, the required reduction was by CZK 3.389 million (from CZK 24.402 million to CZK 21.013 million). No change was required in the case of fixed assets, as there was no purchase in that year. In the case of services consumption, the faculty was recommended to reduce the value from the initial CZK 2.990 million to CZK 2.589 million (by CZK 0.401 million). Both FT and SEV were efficient in that year, since the values of e_2 and e_3 was equal to 1.

The analysis for the years 2018 and 2019 confirmed the efficiency of all ITB faculties concerned, since the values of e_1, e_2, and e_3 were equal to 1 in all cases.

The analysis shows that since 2018, the management of all ITB faculties has become more efficient.

5 Conclusion

The objective of the paper was to create a model of assessing the efficiency of UAS faculties based on the DEA method.

The basic model was created on the basis of the literary research; for this reason, the DEA method is mentioned in the definition of the objective of the paper. The specific model was described in the methodology part of the paper. The methodology also includes the procedure for the subsequent validation of the model on an example of a specific university and its faculties. The results clearly show that the model provides interesting findings easily applicable in practice. It could be stated that the faculties concerned work very efficiently.

Nevertheless, there was a case when the situation was different. In such a case, the model enables us to identify the cause of inefficiency of a given organizational unit and the direction on which the attention of the faculty management should be focused. The objective of the paper was thus achieved.

There was thus created a model that allows universities to manage the individual organizational units on the basis of valid input data, providing outputs in the form of specific measures to be taken in order to increase the efficiency of their activities.

So far, the model has been validated on the example of one institution, the Institute of Technology and Business in České Budějovice. Although the results are considered very positive, it will be necessary to validate them on a larger sample of universities and thus modify the model to a form and state that is applicable and useful for all universities in the world. This is a topic of further research.

References

1. S. Ashmarina, E. Kandrashina, A. Zotova, M. Vochozka, Z. Rowland, J. Vrbka, P. Suler, *Research on development trends and tendencies of the higher education system in Russia and in the world.* 1st ed. České Budějovice, Czech Republic: Institute of Technology and Business in České Budějovice, Samara State University of Economics (2018)
2. N. K. Avkiran, Investigating technical and scale efficiencies of Australian universities through data envelopment analysis. *Socio-Economic Planning Sciences*, **35**(1), 57-80 (2001)
3. M. M. Mora, M. R. Espinosa, M. R. Delgado, Approach of processes for the distribution of economic resources in public university of Ecuador. *International Research Journal of Management, IT and Social Sciences*, **5**(1), 25-35 (2018)
4. C.-D. T. T. Tran, Cost efficiency – one size fits all? A university-level analysis of economies of scale and scope in Vietnamese higher education. *Asia Pacific Journal of Education*, 1-20 (2020)
5. Ministry of Education Youth and Sports. *Bílá kniha terciárního vzdělávání [White book on tertiary education]* [online], Available at: https://www.msmt.cz/reforma-terciarniho-vzdelavani/bila-kniha (2009)
6. T. O. Kowang, S. A. Samsudin, L. K. Yew, O. C. Hee, G. C. Fei, C. S. Long, Factors affecting car purchase intention among undergraduates in Malaysia. International *Journal of Academic Research in Business and Social Sciences*, **8**(8), 80-88 (2018)
7. M. St. Aubyn, A. Pina, F. Garcia, J. Pais, Study on the efficiency and effectiveness of public spending on tertiary education, *Economic Papers No. 390*, European Comission (2009)
8. Q. Zhang, P. A. Clayton, S. M. Breznitz, The hope fulfilled? Measuring research performance of universities in the economic crisis. *International Regional Science Review* (2019)
9. V. Munteanu, N. Andrei-Coman, Modern methods for underlying the revenue and expenses budget in higher education institutions from Romania based on the criteria of cost-efficiency. *Procedia - Social and Behavioral Sciences,* **15**, 2445-2448 (2011)
10. P. Mikušová, Measuring the efficiency of the Czech Public higher education institutions: An application of DEA. *Journal on Efficiency and Responsibility in Education and Science*, **10**(2), 58-63 (2017)
11. M. Abbott, C. Doucouliagos, The efficiency of Australian universities: A data envelopment analysis. *Economics of Education Review*, **22**(1), 89-97 (2003)

12. S. Kantabutra, J. C. S. Tang, Efficiency analysis of public universities in Thailand. *Tertiary Education and Management*, **16**(1), 15-33 (2010)
13. M. Pietrzak, P. Pietrzak, J. Baran, Efficiency assessment of public higher education with the application of Data Envelopment Analysis: The evidence from Poland. *Online Journal of Applied Knowledge Management*, **4**(2), 59-73 (2016)
14. T. Kliestik, Quantification effectiveness activities traffic company by the rules of data envelopment analysis. *E & M Ekonomie a Management*, **12**(1), 133-145 (2009)
15. T. Kliestik, The application of mathematical modelling to predict the financial health of businesses. *Proceedings of the Third International Conference on Economic and Business Management*, **56**, pp. 298-302 (2018)
16. A. T. Flegg, D. O. Allen, K. Field, T. W. Thurlow, Measuring the efficiency and productivity of British universities: An application of DEA and the Malmquist approach. *University of the West of England, Department of Economics, series Discussion Papers*, No. 304. (2003)
17. J. Jablonský, M. Dlouhý, Models for efficiency evaluation of decision making units. *Politická Ekonomie*, **52**(2) (2004)
18. W. W. Cooper, L. M. Seidford, K. Tone, *Data Envelopment Analysis – A comprehensive text with models, applications, references and DEA-solver software*. New York City, United States: Springer (2007)

Conflict between the Accounting Act and Czech Accounting Standards in land

Kristina Kabourkova[1,*]

[1]Institute of Technology and Business, School of Expertness and Valuation, Okruzni 517/10, 37001 Ceske Budejovice, Czech Republic

Abstract. The paper deals with the analysis of existing Czech accounting procedures in the case of land accounting. The method of process analysis is used, where the currently valid methodological procedures are examined. The strengths and weaknesses of this methodology are defined. It is found that the accounting and valuation of land according to Czech Accounting Standards is in conflict with the Accounting Act, according to which accounting data should provide a true picture of the status and value of the entity's assets. The purchase price of the land is entered into bookkeeping and any decrease or increase in value is never accounted for again. In the case of forest land, the value of the land and the value of the plants grown on it are not accounted for separately. The accounting therefore does not provide users with true and current information. The good thing is that these accounting policies do not burden the accounting entity in any way. Furthermore, a proposal for new methodological procedures is presented. According to them, it would be possible to update the accounting data depending on the condition of the land and the development of market prices. Based on an expert opinion, a significant increase or decrease in the current value of the land would be accounted for. In the case of a forest, it would be possible to use analytical accounts to account separately for the condition of the land and the condition of the forest foliage. These new procedures would make it possible to reflect in bookkeeping the actual state of the entity's assets. However, the disadvantage is the higher financial and time demand. Subsequently, the advantages and disadvantages of the currently effective and proposed methodology are compared.

Keywords: accounting procedures, valuation of land, market price, enterprise´s assets, tangible fixed assets

1 Introduction

The issue of accounting for land that is not a commodity is not addressed very often in literature. This is probably, among other things, because economic theory considers land to be a fixed factor of production that is not consumed. Nevertheless, valuation practice shows that the price of land changes over time. However, land accounting is a relatively important issue, as the price of land can in some cases be an important item in deciding how to deal with it.

* Corresponding author: kabourkova@mail.vstecb.cz

In the Czech Accounting Standards, which currently regulate accounting in the Czech Republic, land is considered as tangible fixed assets that do not depreciate. It is based on the opinion that the value of land does not change over the years or, on the contrary, increases. According to Czech accounting regulations, the increase in the price of land cannot be accounted for. The author believes that this rule in some cases makes it impossible to show the actual value of land. This is, for example, a plot of land on which there is a forest. It is not accounted for separately, as the forest is considered to be a part of the land. However, as the forest grows, so does the amount of wood and other forest functions also improve. The value of this land is objectively increasing. However, this fact will not be reflected in Czech accounting.

The opposite example can be a decrease in the value of land. If the land is contaminated, its value will be drastically reduced. In the case of successful reclamation of such land, allowances can be accounted for, which are used in the event of a temporary decrease in the value of the property and when the value is expected to increase back to the original level. After successful reclamation, allowances will be cancelled.

The situation is more complicated if reclamation does not take place at all or if a complete repair is not possible. In such a case, there is a permanent decrease in the value of the land, which is not reflected in the accounting at all. The land is still registered at its original value. The consequence is again the fact that the book value of the land has been distorted and the accounting does not show this fact. This creates a conflict with the basic principles of accounting, which then do not fulfil one of its important tasks.

It can be assumed that the value of land will increase in the future. For this reason, it is necessary to pay attention to the valuation of land and the accounting presentation of changes that have resulted in a change in their true value.

The aim of the article is to analyse those land accounting procedures that currently apply according to Czech Accounting Standards. In the next part, such a solution will be proposed, the result of which will be an accurate representation of reality.

The following research questions are asked in this paper:

1. Is the current land accounting procedure according to Czech Accounting Standards in accordance with the Accounting Act?
2. Is there a method of accounting for land that corresponds to the principle of actual representation of reality?

2 Literature review

Fixed assets are an essential part of an entity's assets. Hence the need to deal with them efficiently and to account for it properly.

Kouřilová [1] states that accounting is currently in the dimension of data manipulation. This can only be agreed with, as the overestimation and the method of said overestimation of certain components of assets, especially fixed assets, can be done on purpose.

The efficiency of the organization's activities is determined by the level of securing fixed tangible assets and their rational use in economic processes [2]. This is mainly because tangible fixed assets are effective in the basic activities of the company and are of special importance [3].

Land is an important part of fixed assets. It is an asset still undervalued in the Czech Republic, mainly due to the functions that the land performs and because the area of land is limited and in our conditions it is not possible to increase the area of the country (e.g. by building islands in the sea).

At present, we see the consequences of poor land management, when it was and still is being occupied due to construction, although it is possible to rebuild old unused areas instead. It is always cheaper for the investor to build and thus destroy a meadow or field forever, but it is extremely expensive for society as a whole. Unfortunately, politicians who decide on territorial plans are not aware of this and are preparing big problems for future generations with their short-sighted decisions. Already today, many smaller municipalities have major problems with the lack of water, which due to the massive development cannot be absorbed into the soil, but completely unnecessarily drained by streams and rivers out of our country. Other problems, such as agricultural drought, etc., are also associated with poor soil management.

The land is multifunctional because it can serve simultaneously or consecutively for different purposes [4]. It can therefore be stated that land is a special type of property and with regard to its irreplaceability, it is necessary to handle it with the care of a proper manager.

If the land owned by a company is intended for long-term use, for capital appreciation or for obtaining revenue, we classify it as tangible fixed assets. It is included in the balance sheet in assets, used for more than one year and the valuation amount is determined by the accounting entity. It gradually wears out, its value transforms into costs in the form of amortization [5].

Land values are often close to or even higher than building values [6]. It can be expected that the value of land in the world will increase significantly in the future, as its area is limited and with the growing population of the Earth and with increasing needs, there will be an increasing hunger for land. There is already a large trend in investing in land [7].

Investments in real estate development are riskier than investments in stabilized real estate assets. The Development Assets Value Index (DAVI) reveals the behavior of developers and landowners using the options contained in land ownership [8].

Land ownership and accounting is regulated in the Czech Republic by a number of legal norms. They are characterized by legal data (e.g. ownership rights, land decisions), technical parameters (e.g. availability, capacity of utilities), other parameters (e.g. buildability, size) [9].

According to the Cadastral Act No. 256/2013, land is understood as a part of the earth's surface separated from neighbouring parts by the boundary of a territorial unit or the boundary of a cadastral territory, the boundary of ownership, the boundary determined by a regulatory or territorial plan, etc.

Accounting based on Decree 500/2002 Section 7 considers land as fixed assets.
The indicative chart of accounts classifies land to account 031 land. This account records land that is a tangible fixed asset from the point of view of the accountant, always regardless of the amount of valuation, as long as it is not a business commodity. The valuation of the land includes forest vegetation or planted trees and shrubs unless they are considered cultivational units of permanent crops. Land does not depreciate. This account does not include parts of the land that are depreciated and reported as assets or parts thereof in the accounts of buildings, etc.

Land (like all other assets and liabilities) is valued at historical prices in our country, which means that it is valued at the price that was spent on its acquisition. This price refers to the time of acquisition [10].

According to the Czech Accounting Standard No. 013, technical improvement is not accounted for in the case of cultivational units of permanent crops, adult and labour animals and assets in account group 03 - tangible fixed assets not depreciated. However, this is a shortcoming, as in many cases the value of the relevant assets may increase.

Technical improvements can be made to the land. This can occur, for example, as a result of landscaping, when the appearance of the environment or water drainage conditions change significantly. Accessories that are not legally considered a building by the Building Act, such as ground tiles and stones that are not an integral part of the land, may also be added to the land. However, the installation of these utilities cannot increase the value of the land, because it is a separate tangible fixed asset [11].

The current legal regulation of land accounting shows that in many cases the book value of land in the Czech Republic is underestimated, as its actual value increases over time. These are mainly forest lands, which are valued according to the predominant sets of forest types. In this case, the accounting does not take into account the increase in timber over time at all. The value of mature trees is reflected in the accounting only at the time of sale.

The Property Valuation Act further states that building plots of land are valued by a multiple of the land area and the price per m2 according to the price map of the municipality. Agricultural land is valued on the basis of rated soil ecological units. Reservoirs, streams and other land is valued in accordance with the relevant decree.

Land is also defined in International Accounting Standards in IAS 16. It is an asset that an enterprise holds because of the need to use that asset as part of a production process. The lifespan of its utility is longer than one accounting period and the entity acquires it for this purpose. It is expected that in the future, these assets will bring economic benefits to the company, including all risks and added value. It must be possible to reliably measure the acquisition costs and thus determine the acquisition price for which the land will be entered in bookkeeping. It is valued at acquisition costs, which also include acquisition-related costs (attributable). After initial recognition as an asset, the asset is accounted for at an overestimated amount, which is usually determined by reference to market price based on a qualified appraisal.

Revaluations must be performed at regular intervals. However, there is a risk that revaluation will be performed for the sake of revaluation. It would probably be more appropriate to stipulate that revaluation is necessary in significant cases and when significant values are reached. This would be specified in more detail by the entity's internal guidelines. IAS 16 states that tangible assets are plots of land, buildings and equipment held by an entity for the purpose of producing and supplying goods or services, leasing to other persons or for administrative purposes. They are expected to be used for more than one accounting period [12].

In 2018, the International Accounting Standards Board published its revised Conceptual Framework for Financial Reporting. It describes two valuation bases, namely historical costs, and present value. Historical cost reflects the price of the transaction that led to the creation of the asset. Present value then includes actual value, value from use and current costs. The conceptual framework does not favour one valuation basis over another.

In Czech accounting, land is reported and valued separately and is not depreciated in accordance with Section 28 of the Accounting Act [13]. This is a serious shortcoming, as in fact the value of the land is sometimes reduced in a permanent or temporary form (e.g. contamination due to an accident). Here, it would be appropriate to account for either allowances or accounting amortization. Due to the fact that in these cases we do not account for the decrease in the value of the land, the accounting does not show a true picture of the assets of the entity. However, if the value of the land is not actually reduced, its amortization is incorrect.

The tax regime of costs incurred to the taxpayer in connection with the acquisition, holding, appreciation and sale of land is given primarily by the fact that the land is not defined in terms of income tax as depreciated tangible assets. The tax-effective cost is generally the acquisition price of the land, at the time of sale, and in certain cases only up to the amount of income from the sale. (In the case of land acquired by a deposit, the eligible cost is determined differently, in most cases as the acquisition price registered with the depositor) [14].

Amortization is the systematic allocation of the depreciable amount of an asset over its lifespan of utility. This corrects the valuation of assets in relation to their use and the balance sheet expresses the situation of the entity more realistically [15].

Amortization of tangible fixed assets is an important tool when looking at a company's profit or loss. They are a part of the costs incurred [16].

An entity may choose one of the following options for amortization: even, accelerated or decelerated accounting. The chosen variant should be closest to the most faithful way of expressing the actual physical and moral wear and tear of the property [17].

Amortization puts the value of assets into costs. The total amount applied in individual years may not exceed the value expressed in the entry price, or with regard to technical evaluation [18].

In practice, there may be so-called negative technical appreciation, which will reduce the usability of assets [11].

An inventory check is necessary for the entity to determine the actual condition of the land. This is exceptional in this case in that it must take place in physical and documentary form, on the same date. The physical inventory check should find out the actual condition of the land in the field and compare it with the records in the books [19]. The revaluation of the land should take place as a result of the findings.

An important component of land is forest land. According to our legislation, in the case of a forest, the value of the land and the vegetation on it is recorded by one aggregate item and the value of these assets does not change in the accounting over the years.

Bartůňková [20] states that the current methodology of accounting of forest vegetation does not provide a true picture of the company's management of the land. The shortcoming of the current Czech regulation is also the inability to record the current value of the growing stock on forest land into bookkeeping. The interested party can obtain more detailed information only in the notes to the financial statements. Conversely, International Accounting Standards separately report and measure forest land and forest vegetation, with forest vegetation being considered a biological asset until harvested. The forest vegetation must be revaluated to its current value at the time the financial statements are prepared.

Higher value of forest roads can cause positive and negative side effects [21].

The aim of the article will be achieved by a process analysis of those procedures for accounting for land, which currently apply according to Czech accounting standards. Furthermore, the results will be compared with the Accounting Act. It is obvious that Act No. 563/1991, On Accounting, contradicts itself in Section 7 paragraphs 1 and 2. It is stated here that the representation is true if the content of the items in the financial statements corresponds to the actual situation, while the representation is in accordance with the accounting methods, the use of which is imposed on the entity by this Act. In the next part, such a solution will be proposed, the result of which would be an actual depiction of reality.

3 Materials and methods

First, a process analysis of the current methodology of accounting for land and their valuation will be performed in accordance with the Accounting Act and Czech Accounting Standards. It will analyse the currently valid accounting procedures for accounting and valuation of land, namely Act No. 563/1991, On Accounting, Section 7, Czech Accounting Standards: No. 005 allowances, 013 tangible and intangible fixed assets, 019 costs and revenues. The method of process analysis will be used. The principles of using specific accounts and valuations in accounting for typical selected economic operations will be defined. The output will be currently valid pre-accountments.

Next, the entry price at the time of purchase will be compared with the price at the time of sale and the price development over time will be monitored. It will be determined whether there is a change in the valuation of the land due to time or temporary or permanent impairment of the land. The comparison process will take place using bookkeeping pre-accountments according to the valid methodology. The output will be the determination of any differences.

The next step will be the analysis of the strengths and weaknesses of accounting according to standards.

The advantages and disadvantages of the current methodological procedure for individual bookkeeping pre-accountments will be defined. The output will be an assessment of whether the current system complies with the Accounting Act and whether it is beneficial for executive decisions.

A proposal for a new methodological procedure will follow. It will include new accounting procedures, including valuations. The whole process of land revaluation will be defined with the necessary accounting documents. The output will be pre-accountments, including valuations, a list of necessary accounting documents and a proposal to change the methodology in Czech Accounting Standards. The next step will be to analyse the strengths and weaknesses of the new approach. The advantages and disadvantages of the proposed procedure will be defined here. The output will be an assessment of whether the proposed procedure is in accordance with the Accounting Act.

This will be followed by a comparison of the strengths and weaknesses of both approaches. The strengths and weaknesses of the current and proposed accounting procedures will be compared. The complexity of the application of the existing methodology in comparison with the proposed methodology will be evaluated. The output will be a comparison of the benefits of both procedures in decision-making processes.

4 Results

Accounting for land is the subject of Act No. 563/1991, On Accounting, and the Czech Accounting Standard for Entrepreneurs No. 013. According to these legal norms, land is a tangible fixed asset which is not depreciated. Decree No. 500/2002, which implements the Accounting Act, states that land is a tangible fixed asset, regardless of the valuation, if it is not a business commodity. It is recorded on the balance sheet asset account 031 land. Act No. 563/1991, On Accounting, states that these assets in the case of acquisition (except for own-account acquisition) are valued at acquisition cost, which is the price at which it was acquired, including acquisition-related costs.

Another option is a valuation at reproduction cost, which is the price at which it would be acquired at the time the asset is accounted for. The reproduction purchase price may also be determined by a partnership agreement or a deed of incorporation. The reproduction acquisition price is used, for example, in the case of an investment of assets in an accounting unit or if the assets are newly identified and not yet recorded in the accounts. In subsequent periods, if the land is not sold, neither the account nor the value on it changes. Only when the land is sold will the acquisition price be subtracted from account 031 and added to account 541, sold intangible and tangible fixed assets. At the same time, the revenue from this economic operation is recorded in the revenue account 641, revenue from intangible and tangible fixed assets sold and the receivable on the account 311 customers is increased. The difference between the values on accounts 541 and 641 then determines the profitability or loss.

According to currently valid standards, land is not depreciated. This follows from Section 7 of Decree No. 500/2002, Which states that accounting item land does not contain parts of the land that are depreciated, those that do are reported in construction, units of permanent crops or other tangible fixed assets. The possibility of creating accounting adjustments is neither ordered nor prohibited. However, the Czech Accounting Standard for Entrepreneurs No. 5 states that adjustments may be made to depreciated fixed assets, the value in use of which is normally reduced by wear and tear. It is therefore not entirely clear whether the possible creation of a provision for land would not be criticized.

It follows from the above that account 031 according to the methodology currently in force does not reflect any changes that may occur on the land itself or as a result of changes in the real estate market over time. The land is basically forever valued at the historical purchase price. This is in conflict with the Accounting Act Section 7 paragraph 2, which stipulates that the items of the financial statements should correspond to the actual situation. In reality, however, the value of land has increased over the years. There may also be, for example, contamination of the land, when, on the contrary, its value decreases temporarily or permanently, which should again be reflected in the valuation on account 031.

The Accounting Act contradicts itself in Section 7 para. 2, where it is claimed that the presentation is true if the content of the items of the financial statements corresponds to the actual state, which is presented in accordance with accounting methods, the use of which is imposed on the entity on the basis of this of the law. The display cannot be faithful in all circumstances unless it is possible to account for an increase or decrease in the value of the land. At the same time, however, the regulations based on this Act do not allow an entity to account for this change in value.

The impossibility of accounting for an increase or decrease in the value of land is significantly reflected in the accounting for forest land. Here, the total price including the forest vegetation is accounted for on account 031 with the value of the land at the time of purchase, if this land was purchased for the purpose of further forest cultivation. In the following years, the amount of timber increases, and thus so does the actual value of the forest, but this is not reflected in the accounting at all. The price of the relevant land does not change even when the valuation regulations change. The forest vegetation will only be accounted for at the moment of harvesting, when the harvested timber is valued as a stock in account class 1 as calculated own costs and the amount is marked as an asset in account group 58 change in stock of own activities and assets. The sale of this timber is then an income for the entity in account class 6 and at the same time means an increase in receivables from customers. Account 031 land remains valued at the original historical price at all times.

It is therefore clear that the methodology of accounting and valuation of land is not comprehensively resolved in the Czech Republic and that in practice a number of situations may arise that no existing legal norm addresses. However, this is a serious shortcoming, because in real life there is a real change in the price of land almost constantly. Land owned by an entity may represent a significant value item in its balance sheet. However, shareholders or other stakeholders will not find out, as they only have a historical price at a value that is far from reflecting the current price of the land. The only possible solution at present is to state the current market price of the land in the notes to the financial statements, but the question is whether the authors of the notes do so.

Pre-accountment for selected accounting transactions is shown in the table 1.

Table 1. Accounting for land according to regulations valid in the Czech Republic

Economic operation	Pre-accountment - valuation
Purchase of land from a supplier	031/321 - purchase price – acquisition price
Deposit of land into company ownership by an individual entrepreneur	031/491 - price determined by an expert - acquisition price
Sale of land - decommissioning from bookkeeping	541/031 - acquisition price
Accountment of revenue for sold land	311/641 - sale price – market price

Activation of wood mass in the forest after its extraction	1xx/58x - calculated own costs
Issuance of an invoice for the sale of timber	311/601 - market price

Source: Own processing.

From the above overview, it is clear that account 031 land is constantly working with the acquisition price, which never changes and does not respond to changes in the quality of land or market developments.

The advantages of the current land accounting system are low labour and time costs and financial simplicity. It is not necessary to seek expert valuation.

The disadvantage of the current method of accounting for land is the fact that accounting does not provide true information about the current state of these assets, which does not fulfil one of the basic functions of accounting. Distorted information can be misleading to shareholders and the market.

The current situation could be changed by an amendment to the accounting rules in the area of land. Land revaluation would not have to take place on an annual basis for each entity, but the obligation to do so could be required by the entity's internal guidelines. The revaluation should take place no more than once a year as at the date of the annual accounts, based on the findings of taking inventory. The inventory committee should assess the condition of the relevant land and propose a revaluation if a fundamental change compared to the previous accounting period is found.

In the event that a decrease in the value of the land is detected during the accounting period, e.g. by contamination in the event of an ecological accident, an expert valuation should be prepared. It should determine whether it is possible to restore the land to its original condition or not. Furthermore, the current value of the land should be determined in it.

If it were stated in the expert opinion that the land can be restored to its original condition after the relevant measures have been taken, it would be possible to account for provisions in the account in group 55 creation and settlement of adjustments in the operating area in relation to the account in group 09 adjustment items of fixed assets. After the land has been restored to its original condition, the adjustment would be cancelled in its entirety.

If for some reason it was not possible to restore the land to its original condition, then its value would have to be reduced to the appropriate extent by depreciation, using account 549 deficit, damages and extraordinary operating costs and reducing the value on account 031 land. However, some value would always remain on account 031. In an extreme situation, for example, if an entity owns land on a riverbank and during a flood this part is completely washed away, then account 031 land would be cancelled completely.

However, these accounting operations would always have to be performed on the basis of an expert opinion or a document from the cadastral office. These documents would also be considered accounting documents.

The opposite is the increase in the value of the land. This may be due to price movements in the real estate market, landscaping, etc.

In terms of market influence, an increase in the accounting value of the land would be recognized only for the annual balance sheet date if the increase in the market price would be significant. This issue would be more precisely defined by the company's internal guidelines. The increase would concern account 031 land and the corresponding account change in balance. This is currently posted in account group 58, which is not ideal. The former method of accounting for changes in revenue was more appropriate, but the current rules must be respected.

In the event that the land value increases for another reason, e.g. due to landscaping, then the balance of account 031 land would increase again and the account 321 suppliers would be used in the case of the purchase of this activity. If the entity carried out this activity itself, then it would use the account in group 58. The basis for accounting would be an expert opinion, which would be an accounting document.

In my opinion, accounting for forest vegetation is completely incorrect. A distinction should be made between the value of land itself and the value of forest vegetation, e.g. by analytics to account 031 land. When buying land, the expert should already distinguish these two components in his opinion.

Wood mass grows every year and on the basis of an expert opinion, its current value should be determined at specified intervals. An increase would then be charged to the relevant analytical account 031 land in relation to the account in group 58 change in balance. In the case of logging or destruction by storms, bark beetles, etc., the relevant analytical account 031 of land would be reduced in relation to the cost account in group 58 change of state or 549 deficit, damages and extraordinary events. When selling wood, the account 311 customers would be used in conjunction with the revenue account 601 sales of own products. The condition of the forest after a storm or bark beetle infestation would again have to be assessed by an expert in his opinion.

In my opinion, these proposed procedures would respect the Accounting Act, which mandates a true and fair view of the real situation. Executives, shareholders and the market would obtain accurate and up-to-date information about the entity and its components of assets. On the other hand, it would be much more time and money consuming to obtain accounting documents.

The table 2 contains a proposal for accounting pre-accountment of selected economic operations.

Table 2. Proposed pre-accountments

Economic operation	Pre-accountment - valuation
Creation of an adjustment for land	55x/09x - value determined by an expert
Decrease in the value of the land due to irreparable damage	549/031 - value determined by an expert, or the acquisition price
Revaluation of land due to a significant change in market price	031/58x - value determined by an expert
Land value increase due to landscaping - by supplier	031/321 - invoiced price
Land value increase due to landscaping - in-house	031/58x - value determined by an expert
Purchase of forest - the division of the price into land and timber	031/1/2 AE/ 321 - invoiced price including expert opinion
Growth of wood in the forest	031AE/58x - value determined by an expert
Logging - reducing the value of forest property	58x/031 AE - calculated price
Destruction of forest vegetation due to natural disaster	549/031 AE - value determined by an expert
Sale of wood - issued invoice	311/601- invoiced price

Source: Own processing.

When applying the proposed accounting procedures, it will be necessary to often use the services of experts, where their opinion would also fulfil the function of an accounting document. It would have to contain all the requisites arising from Section 11 of the Accounting Act and it would also have to be kept properly.

The newly proposed accounting procedures are in accordance with the Accounting Act, but in conflict with Czech Accounting Standards. However, as mentioned above, this follows from the fact that these standards themselves are in conflict with the Accounting Act in this area.

The advantage of the newly proposed methodological procedures is a realistic view of the entity's assets, where managers, owners and the market obtain true information about the current value of land owned by the entity.

The disadvantage is higher labour, the need for more frequent monitoring of the development of land prices on the market and the condition of the land. A significant negative can be the need to use the services of experts often, the cost of preparing an opinion and the simple fact that there are a limited number of such experts.

If the current and newly proposed procedure of accounting and valuation of land is compared, we will find that the procedures still in force are not laborious and do not require the provision of additional documents over time. However, the information we obtain from accounting is becoming less and less relevant over time.

The proposed procedure is significantly more time-consuming and labour-intensive than the application of the current methodology and the entity will undoubtedly incur costs in seeking up-to-date expert opinions, and it may be a problem to obtain an expert. It will also be necessary to ensure that any revaluation of land does not lead to malpractice.

The advantage of the proposed methodology is to achieve a state where the accounting will actually show the current situation of the accounting entity, the assets will be recorded at fair value and thus the managers of the entity, owners and the market will obtain true information.

It should also be noted that the proposed change in the accounting and valuation of land is not systemic in relation to the accounting for other tangible fixed assets. However, this is justified by the fact that these are special, non-depreciable assets. This is more or less the same problem as in the case of accounting for works of art and collections (Table 3).

Table 3. Comparison of the current and proposed states

Current methodology	**Proposed methodology**
Not in accordance with the Accounting Act, it is in accordance with Czech Accounting Standards	In accordance with the Accounting Act, but conflict with Czech Accounting Standards
Over time, the unrealistic value of the land can negatively affect decision-making	Over time, the value of the land is updated according to its actual condition and market
Low workload, it is not necessary to obtain new expert opinions and spend funds on their processing	Significantly labourious, the need to provide an expert who prepares the report, higher costs
The data obtained from the accounts over time are not in line with reality	The data obtained from the accounts are in line with reality

Source: Own processing.

5 Discussion

The first research question was whether the current procedure of accounting for land according to Czech Accounting Standards is in accordance with the Accounting Act.

It is found that Czech accounting standards in this area are in conflict with Act No. 563/1991, On Accounting. In Section 7 paragraphs (1) and (2), it requires entities to keep accounts in such a way that the financial statements prepared on its basis give a true and fair view of the subject matter and financial position of the entity. According to this Act, the display is true if the content of the items in the financial statements corresponds to the actual situation.

Czech Accounting Standard No. 13 for Entrepreneurs states in Part 4 that tangible fixed assets are depreciated in accordance with Section 56 of Decree No. 500/2002. However, according to this standard, land is not depreciated, so that in accounting practice it is not possible to document a decrease in the value of land in any way.

Nor is it possible to account for the increase in the value of the land, which undoubtedly actually happens very often. Section 27 of the Accounting Act defines the assets that can be measured at fair value, but land is not listed here. Section 25 of this Act orders the valuation of acquired tangible fixed assets at acquisition cost. It follows that the purchased land will be valued once and for all at the acquisition price, regardless of any devaluation of the land in the future or the growth of its value due to market influence. In the case of land with mature forest vegetation, the price recorded in the accounts and the market price are completely different and the condition of the land itself and the condition of the forest vegetation is not reflected at all. The information about the forest is, in fact, untrue and misleading from the very beginning, as the value of the land itself and the value of the vegetation growing on it are not separated at all.

It must therefore be stated that the rules of law governing land accounting are contradictory and, as a result, accounting in this area does not give a true and fair view of that component of assets.

The answer to the first question must therefore be that Czech accounting standards in the area of land accounting are in complete conflict with the Accounting Act.

The second research question asked whether there is a method of accounting for land that corresponds to the principle of actual representation of reality.

The answer to the second question is that according to the valid Czech legislation, it is currently not possible to account for land so that the accounting gives a true and fair view of the facts.

Possible accounting procedures, which would result in a realistic presentation of the facts, are suggested in the previous part of the article. However, a prerequisite for their application in practice is a significant amendment to Czech Accounting Standards and Decree No. 500/2002.

Legal norms in this area are from earlier times when land prices were not very high. However, land prices are rising, as citizens and businesses consider buying land as a certain investment. It is also necessary to take into account the fact that the amount of land is limited.

6 Conclusions

The aim of the paper is to analyse the accounting procedures for accounting for land according to applicable Czech legislation and to propose a new methodology.

It was found that Czech accounting standards in the area of land accounting are in conflict with the Accounting Act. This Act is then in conflict with Decree No. 500/2002.

Land is recorded in the accounting at acquisition price and does not reflect any decrease or increase in the value of the land. In the case of a forest, the accounting procedures are erroneous and insufficient from the very beginning, as the condition of the land itself and the condition of the vegetation growing on it are not taken into account at all. The accounting does not show an increase in the value of the forest due to the increase in timber.

The current accounting procedures for accounting and valuation of land are in all respects undemanding, the entity does not have to obtain new expert opinions on an ongoing basis or spend additional funds to pay for these expert opinions. However, the information provided in the accounts is becoming less and less up-to-date over time.

Managers, shareholders, and the market currently obtain completely outdated or untrue information from accounting data in this area, which can be the cause of fatal decisions.

The article proposes new procedures for land accounting. According to them, it would be possible to account for a permanent and temporary decrease in the value of land as well as for an increase in the value of land, which occurs mainly due to rising land prices. A procedure for accounting for forest land was also proposed, which would use analytical accounts to separate the value of the land itself and the value of the vegetation. Then it would be possible to account for an increase in the value of the forest itself due to the increase in wood mass.

The newly proposed procedures would be much more laborious and costly for the entity, as it would be necessary to regularly assess the current price of the land and provide expert services.

The application of the proposed procedures would comply with the requirements of the Accounting Act and would provide true and up-to-date information on the value of these assets, which would be beneficial for users of the information. The aim of the paper was therefore achieved.

The newly proposed procedures deviate from the currently valid system of accounting for tangible fixed assets, which is forever (with the exception of technical improvement) recorded at acquisition price, while its residual accounting value is updated through accounting amortization. However, land is such a specific type of property that this exception can be justified.

According to the Czech methodology, it is not possible to account for land in any other way than stated above. Undoubtedly, it would be interesting to compare these accounting practices with International Accounting Standards. This comparison will be the content of the next article.

References

1. J. Kouřilová, K vývoji a pojetí účetnictví, vnímání předmětu [The development and concepts of accounting, the perception of the subject]. *Ekonomické dopady legislativních změn v oblasti financí, účetnictví a daní (2. díl) [Economic Impacts of Legislative Changes in Finance, Accounting and Taxes (Part 2)]*, (2015)
2. V. Mykolaiotiene, G. Vecerskiene, K. Jankauskine, L. Valanciene, Peculiarities of tangible fixed assets accounting. *Inzinerine Ekonomika – Engineering Economics,* **21**(2), 142-150 (2010)
3. B. Sabuncu, Measurement and accounting of tangible fixed assets in terms of the finantial reporting standard for large and medium sized enterprises. *Journal of Economics and Administrative Sciences*, **14**(3), 713-727 (2019)
4. M. Pekárek, J. Dudová, J. Hanák, I. Průchová, J. Tkáčiková, *Pozemkové právo [Land Law]*. Brno: Masaryk University Brno (2003)
5. J. J. Weygandt, P. D. Kimmel, D. E. Kriso, *Acounting principles: International student version*. 11th ed. Hoboken: Wiley (2014)
6. U. Ozdilek, Land and building separation based on Shapley values. *Palgrame Communications*, **6**(1), Art. No.: 68 (2020)

7. R. Zdenek, J. Lososová, Investments of Czech farms located in less favoured areas after EU accesion. *Agricultural Economics – Zemědělská ekonomika*, **66**(2), 55-64 (2020)
8. D. Geltner, A. Kumar, A. M. Van de Minne, Riskiness of real esstate development: A perspektive from urban economics and opticon value theory. *Real Estate Economics*. **48**(2), 406-445 (2020)
9. P. Ort, *Oceňování nemovitostí - moderní metody a přístupy [Real estate valuation - modern methods and approaches]*. Prague: Leges. 32-37 (2013)
10. J. Sedláček, *Základy finančního účetnictví [Basics of financial accounting]*. Prague: Ekopress (2005)
11. M. Skála, *Technické zhodnocení a opravy [Technical evaluation and repairs]*. Ostrava: Sagit (2008)
12. M. I. Wiecenk, M. N. Joung, *IFRS Primer International Gaap Basics*. The United States of America: George Hoffinan (2009)
13. P. Ryneš, *Podvojné účetnictví a účetní závěrka 2020 [Double entry accounting and financial statements 2020]*. Olomouc: Anag (2020)
14. J. Nigrin, *Pozemky v daňových nákladech [Land in tax costs]*. Verlag Dashofer (2016)
15. J. Strouhal, *Lexikon účetních pojmů [Lexicon of accounting terms]*. Prague: Wolters Kluwer (2015)
16. A. Suhanyiova, L. Suhanyi, Analysis of depreciation of intangible and tangible fixed assets and the impact of depreciation on the profit or loss and the tax base of the enterprise. *18th International Scientific Conference on International Relations. Current Issues of World Economy and Politics*, 939-950 (2017)
17. P. Valouch, *Účetní a daňové odpisy [Accounting and tax depreciation]*. Prague: Grada Publishing (2012)
18. P. Prudký, M. Lošťák, *Hmotný a nehmotný majetek v praxi: komentář, příklady, výklad změn [Tangible and intangible assets in practice: commentary, examples, interpretation of changes]*. Olomouc: Anag (1997)
19. J. Svobodová, *Inventarizace - praktický průvodce [Inventory - a practical guide]*. Olomouc: Anag (2013)
20. L. Bartůňková, *Účetní systém vybraného lesnického podniku [Accounting system of chosen forestry company]*. Brno: Mendel University in Brno, diploma thesis (2010)
21. P. Toscani, W. Sekot, F. Holzleitner, Forest road from the perspective of managerial accounting – Empirical evidence from Austria. *Forests*, **11**(4), Art. No.: 378 (2020)

AUTHOR INDEX

Aleshkova D. V. 281
Alexandre Paulo 271
Ashmarina S. I. 291
Bartosova Viera 46, 256
Belayeva I. Yu. 1
Bílek Stanislav 364
Bogomolov Sergey 95
Caha Zdenìk 364
Chernikova Lyudmila 113
Chernyshev Valentyn 340
Chikileva L. S. 144
Chutka Jan 136
Danilova O. V. 1
Dias Rui 271
Dlask Petr 328
Dušek Radim 30
Èerná Marie 318
Eryomin Ivan 113
Farinha Ana 271
Gajanova Lubica 104, 228, 238
Geierova Tereza 421
Gogolova Martina 197
Gusev Andrey 174
Hašková Simona 37
Heliodoro Paula 271
Hinke Jana 307
Horak Jakub 383, 401
Hraskova Dagmar 217
Hromada Eduard 264, 328
Ilyina Larisa 95
Invdeeva Alina 174
Janèíková Kateøina 126
Janikova Jana 469
Janoskova Katarina 206
Jurickova Veronika 78
Kabourkova Kristina 480
Kajanova Alena 375
Kajanová Alena 354
Kalinová Eva 458
Kapitanov V. A. 144
Karpac Dusan 46
Kiselakova Dana 65
Kmecová Iveta 354
Komorousová Veronika 307
Korobov Yury 95
Kosyakova I. V. 21
Kovalyov Anatoliy 340
Krizanova Anna 78, 85
Krulicky Tomas 383
Krulický Tomáš 328
Kubjatkova Anna 78
Kucera Jiri 446
Latyshev Kostyantyn 154
Likhacheva Olga 113
Lvov A. S. 12
Machova Veronika 421
Majerova Jana 104, 228
Mantulenko V. V. 281
Maskova Lenka 446
Matskul Valerii 340
Milichovský František 126
Mirzabalaeva F. I. 186
Motuzenko Polina 174
Murzagalina G. M. 291
Nadanyiova Margareta 104, 228
Okara Diana 340
Olexova Veronika 197
Onuferova Erika 65
Osipova O. S. 144
Plotnikova Marina 95
Pochtovyuk Andriy 154
Polan Vaclav 430
Poradova Monika 206
Rebetak Filip 256
Rosnerova Zuzana 217
Rowland Zuzana 392
Sankova L. V. 186
Santos Hortense 271
Schneiderová Heralová Renáta 328
Shalneva M. S. 12
Simonova M. V. 186
Skulcova Kristyna 383
Slepneva T. A. 165
Sofrankova Beata 65
Stanek Tomas 392
Strakova Lenka 52
Streltsov A. V. 21
Sulek Jiri 401
Suler Petr 430
Svabova Lucia 52
Svishcheva N. V. 165
Tlustý Michal 458
Tretiakova V. V. 12
Urban David 375
Urban Jan 247
Uskov K. V. 1
Vagner Ladislav 136
Vallišová Lucie 318
Velková Vendula 364
Vitková Pavla 412
Vrbka Jaromír 412
Vrtana David 85
Yakovlev G. I. 21
Zaika Kateryna 154